Ecological Studies

Analysis and Synthesis

Volume 53

Hypersaline Ecosystems

The Gavish Sabkha

Edited by
G. M. Friedman and W. E. Krumbein

In Cooperation with
M. R. Buyce and G. Gerdes

With 246 Figures

Springer-Verlag
Berlin Heidelberg New York Tokyo

Prof. Gerald M. Friedman
Department of Geology
Brooklyn College of the City University of New York
Brooklyn, N.Y. 11210, USA
and
Rensselaer Center of Applied Geology Affiliated with
Brooklyn College of the City University of New York
15 Third Street, Box 746
Troy, N.Y.12180-0746, USA

Prof. Wolfgang E. Krumbein
Universität Oldenburg
– Geomikrobiologie –
Carl-von-Ossietzky-Str. 9–11
2900 Oldenburg, FRG

ISBN-13:978-3-642-70292-1 e-ISBN-13:978-3-642-70290-7
DOI: 10.1007/978-3-642-70290-7

2131/3130-543210

Contents

A. Introduction

B. General Framework of Hypersaline Environments with Special Reference to the Red Sea

C. The Gavish Sabkha – A Case Study

D. Applied Aspects and Paleoecology

Contributors

BIVIN, DON	Cardiovascular Research Institute, Department of Biochemistry and Biophysics, University of California, San Francisco, CA 94143, USA
BOON, J. J.	FOM-Instituut voor Atoom-en Molecuulfysica, Kruislaan 407, 1098 SJ Amsterdam, Netherlands
COHEN, YEHUDA	Marine Biological Laboratory, Hebrew University, P.O. Box 469, Eilat, Israel
DIMENTMAN, CHANAN	Institute of Life Sciences, 91904 Department of Zoology, Hebrew University Jerusalem, P.O. Box, Jerusalem, Israel
DOR, INKA	Human Environmental Science Laboratory, Hebrew University Jerusalem, P.O. Box, Jerusalem, Israel
EHRLICH, ALINE	Geological Survey of Israel, Malchei Israel Street, Jerusalem, Israel
EVENARI, M.	Department of Botany, Hebrew University Jerusalem, P.O. Box, Jerusalem, Israel
FRIEDMAN, G. M.	Department of Geology, Brooklyn College of the City University of New York, Brooklyn, NY 11210, and Rensselaer Center of Applied Geology Affiliated with Brooklyn College of the City University of New York, 15 Third Street, Box 746, Troy, NY 12180-0746, USA
GAUDETTE, H. E.	College of Engineering and Physical Sciences, Department of Earth Sciences, James Hall, University of New Hampshire, Durham N.H., 03824, USA
GERDES, GISELA	Geomicrobiology Division, University of Oldenburg, Postfach 2503, D-2900 Oldenburg, FRG
GUTTERMAN, Y.	I. Blaustein Institute of Desert Research, Sde Boqer Campus, Israel 84980
HALEVY, JACOB	Department of Geophysics and Planetary Science, University of Tel Aviv, Ramat Aviv, Tel Aviv 69978, Israel

Holtkamp, E.	Geomicrobiology Division, University of Oldenburg, Presently: I.O.L.R., P.O. Box, Elat, Israel
Kessel, Martin	Hadassah Medical School, Department of Electron Microscopy, P.O. Box 1172, Jerusalem, Israel
Klok, J.	Delft University of Technology, Department of Chemistry and Chemical Engineering, De Vries van Heystplantsoen 2, 2628RZ Delft, Netherlands
Knoll, A.	Paleobiology Department, Harvard University, Divinity Ave, Cambridge, MA, USA
Krumbein, W. E.	Geomicrobiology Division, University of Oldenburg, Postfach 2503, D-2900 Oldenburg, FRG
de Leeuw, J. W.	Delft University of Technology, Department of Chemistry and Chemical Engineering, De Vries van Heystplantsoen 2, 2628RZ Delft, Netherlands
Lyons, W. B.	College of Engineering and Physical Sciences, Department of Earth Sciences, James Hall, University of New Hampshire, Durham N.H., 03824, USA
Matzigkeit, Udo	Max-Planck-Institut für Chemie (Otto-Hahn-Institut), D-6500 Mainz, FRG
McGinnis, Kathleen	Cardiovascular Research Institute, and Department of Biochemistry and Biophysics, University of California, San Francisco, CA 94143, USA
Mook, W. G.	Laboratorium v. Algemene Natuurkunde, Rijks Universiteit, Groningen, Netherlands
Owen, Roy W.	L.S. Steel Corporation, Pittsburg, PA, USA
Por, F. D.	Institute of Life Sciences, 91904 Department of Zoology, Hebrew University Jerusalem, P.O. Box, Jerusalem, Israel
Purser, B. H.	Laboratoire de Petrologie sédimentaire et Paléontologie, Université de Paris Sud, Centre d'Orsay Batiment 504, Cedex F-91405 Orsay, Paris, France
Schenck, P. A.	Delft University of Technology, Department of Chemistry and Chemical Engineering, De Vries van Heystplantsoen 2, 2628RZ Delft, Netherlands
Schidlowski, M.	Max-Planck-Institut für Chemie (Otto-Hahn-Institut), D-6500 Mainz, FRG
Sinninghe Damsté, J. S.	Delft University of Technology, Department of Chemistry and Chemical Engineering, De Vries van Heystplantsoen 2, 2628RZ Delft, Netherlands

SNEH, A.	Geological Survey of Israel, Malchai Israel Street, Jerusalem, Israel
SPIRA, JONATHAN	Institute of Life Sciences, 91904 Department of Zoology, Hebrew University Jerusalem, P.O. Box, Jerusalem, Israel
STOECKENIUS, W.	Cardiovascular Research Institute, Department of Biochemistry and Biophysics, University of California, San Francisco, CA 94143, USA
WALSBY, A. E.	Department of Botany, University of Bristol, P.O. Box, Bristol, England

A. Introduction

Obituary

Gerald M. Friedman

Eliezer Gavish
(1937–1981)

Eliezer Gavish, known to his friends and colleagues as Eli, was torn from our midst at the age of only 44. During the past 11 years Eli studied in depth sedimentary processes and products of sea-marginal sabkhas of the northern gulfs of the Red Sea. These pioneering studies have drawn the attention of the international sedimentological community to Eli's specific model along the Gulf of Elat, which other scientists have cited as the "Gavish Sabkha." A Symposium in his honor was held in 1981 at the Oil Institute in Tel Aviv. This memorial volume on the ecology, biology, geology, geochemistry and mineralogy of sabkhas, using the example of the Gavish Sabkha, is another tribute to an outstanding scientist.

Eli, a national of Israel, appeared in my office in the spring of 1965 to join the Rensselaer Sedimentology program. He had obtained a B. Sc. from Tulane University and a degree of M. A. from Dartmouth University. With high recommendations from his previous academic sponsors he embarked on a Ph. D. program. When he completed it in 1968 and published it in the Journal of Sedimentary Petrology in 1969 it drew wide acclaim and was a candidate for an Outstanding Paper Award. Because of its scientific impact the SEPM republished the paper *(Progressive Diagenesis in Quaternary to Late Tertiary Carbonate Sediments: Sequence and Time-scale)* in its Reprint Volume in 1976. According to figures in

Science Citation Index this paper is among the most widely quoted publications in the field of carbonate diagenesis.

Eli served as president of the James Hall Society of Rensselaer Polytechnic Institute, as abstractor for *Mineralogical Abstracts*, as an active coordinator of the Geological Society of Israel, on Committees of the Israel Academy of Science, and received the Grader Award for best Contribution to Geologic Studies of the Geological Society of Israel, and an Alexander von Humboldt Stipendium for Research in Germany. In addition, he was honored as a Fulbright Scholar in the United States, and as a visiting lecturer of many universities in the United States and Europe.

Between 1975 and 1978 many of his working hours were devoted to the International Association of Sedimentologists. He served as national correspondent (1971–1975), Vice-President (1975–1978), Secretary of the Organizing Comittee for the Tenth International Congress on Sedimentology (1975–1978), and Field Trip leader of Excursion Y9 at the same Congress (1978).

He inspired several generations of students as Lecturer at the Hebrew University of Jerusalem (1968–1973), as Senior Lecturer at Tel Aviv University (1973–1981), and as visiting Associate Professor at the University of Texas (Dallas) (1979–1980). Eli accomplished more in his brief scientific career of 13 years than most professional geologists in a much longer life-span.

Eli as an individual was a delightful person. His death at so young an age shocked many in our profession and beyond. As his friend and teacher I feel a special loss. His wife Ayala and two daughters survive him.

Purpose of this Book

Wolfgang E. Krumbein

Geology and Microbiology in the widest sense, as they are interconnected through new disciplines, such as Geomicrobiology, Biogeochemistry, Organic Geochemistry, Paleocology and Environmental Biogeochemistry, are increasingly creating awareness of the complex history of the "Green Planet Earth." Almost daily we receive new information on biological controls on the atmosphere, the hydrosphere and – astonishingly – also about the biological influence on such important processes as the genesis of ore, petroleum and salt deposits. This can only be achieved through the cooperation of geoscientists, microbiologists, and chemists. The traces of bygone biospheres are studied by paleoecologists and microbial ecologists with the goal of understanding the microbial ecology and biogeochemistry of weathering and new formation of mineral deposits, and of the functional cooperative relationships between geosphere and biosphere.

The main purpose of this book is to aim at an understanding of the functions of a unique ecosystem, characterized by special geological setting, sedimentology and geochemistry, as well as by very special life conditions. Such individual separate ecosystems may serve the same purpose in paleoecological research and in the understanding of the evolution of the planet and its living coating through time as the famous "living fossils" like Archaebacteria, Cyanobacteria, Gingko trees and Limulus, to give a very brief selection. We may even expand this view of a living fossil to the term sabkha itself, when applying the term "living stromatolites". Sabkha-type stromatolitic environments have recently been described as paleoecosystems reaching back in Earth History as far as the 3.5 billion-years-old North Pole Series in Warrawoona, West Australia and the Bulawayan in South Africa. The comparison of today sabkha microbial mat systems with those ancient analogs may help us to understand the tremendous stability and persistance of specific ecosystems.

The second purpose of this book is to honor the memory of the sedimentologist/geochemist Eliezer Gavish. Eli Gavish did most of his scientific work on coastal systems in Israel and especially on the shores of the Northern Red Sea gulfs in the years between 1967 and 1981. In order to explain why I am writing the preface to this memorial volume for my late friend Eli Gavish, I have to go a little further back in time.

Shortly before the war of 1967, I was invited by Michael Evenari to do my post-doctoral work at the Hebrew University in order to study the biogeochemistry of petroglyphs and rock varnishes in the Negev desert. I accepted this invitation and during this stay I experienced the Sinai and the coasts of the Gulf of Elat for the first time.

Although I was not an active participant in the first scientific excursions after the war in August and September 1967, I was already in Israel when the first scientific data were reported. These first excursions were wonderful examples of interdisciplinary scientific work. Geologists, botanists, soil scientists, zoologists, archeologists and ethnologists were jointly traveling the Sinai and the beautiful coasts of the Red Sea in the footsteps of their forefathers for the first time in more than 2,000 years.

The outcome of one of those joint efforts was never published. Michael Evenari and the young sedimentologist Eliezer Gavish, who had just graduated from Rensselaer Polytechnic under the guidance of G. M. Friedman, cooperated in the study of the root systems of halophytes and the soil chemistry of the groundwater of the Sabkha plains in which the plants were growing.

After more than 15 years these data are now published as a joint Chapter of M. Evenari (now eighty) and Eliezer Gavish, who was taken from us in his early forties. When I asked Professor Evenari whether he was willing to write such an article, he suggested at first a younger person who had also participated in the Sinai trip of 1967. It was Prof. Abinoam Danin, who has been my guide in the Negev. But later Michael Evenari called me and said that he had found all the data on the sabkhas in his files together with hundreds of pore water analyses carried out by the young Eli Gavish and that he would suggest publishing Eli's data rather than a general survey of the halophytes of sabkhas.

In 1969, I started to plan a new period of scientific work in Israel. My contacts with Professor Steinitz were promising. Professor Bentor this time suggested cooperation with a young sedimentologist. Professor Shilo suggested that I should work with Yehuda Cohen, a young M. Sc. student on the microbial ecology of a heliothermal pond.

At this time Eli Gavish, Yehuda Cohen and my group started to work on the microbial ecology, the sedimentology and biogeochemistry of several supratidal and often hypersaline ecosystems, all of which were characterized by evaporative stress, special sedimentological and paleocological problems and exciting geomicrobiological aspects. These were (1) Beach Rocks, (2) the Solar Lake and (3) two Sabkha Systems in the southern area of the Gulf of Elat. One of them was the Ras Muhammed Pool, the other a sabkha south of the Bedouin village of Nabq, later called the Gavish Sabkha.

With the years, mainly through the Biogeochemical Cycle Group of SCOPE (Scientific Committee on problems of the Environment) and IGCP (International Geological Correlation Project), 157 "Early Organic Evolution and Mineral and Energy Resources", several other groups including the past and present Chairman of the IGCP project got interested in our work on the sabkhas and joined with their individual projects. Volkswagen Foundation, Deutsche Forschungsgemeinschaft, and several Israeli and American science funding agencies have helped us to widen the scope and goals of our work.

Thus the 13 short years the scientific career of Eli Gavish lasted were by coincidence the 13 years of most intensive study of the Gulf of Elat ever to be done since the times of Solomon and Herodotus (see Chap. 1). International scientists and congress excursions were constantly traveling to the Sinai area and a constant flow of scientific publications was the outcome of all these visits. These were the years of fruitful scientific and more or less peaceful work of Israeli and international scientists in the Sinai. I know from many discussions with Eli and other scientists from the Elat Laboratory, who worked in the Gulf area, that all of us wish nothing more fervently than that this international joint scientific exploration of the biology, sedimentology and biogeochemistry of the Sinai and of the shores of the Red Sea adjacent to the desert may be continued and that the scientists from Cairo, Alexandria and Amman may join this venture.

When I planned another visit to the Gavish Sabkha early in 1981, I was glad to hear that Eli was back from Dallas, where he had spent a sabbatical. A year before I had spent my sabbatical at the Hebrew University working on "his" sabkha. Both of us intended to study the influence of the desert floods in that area which stormed down in the winter of 1979/80 and 1980/81. These rainfloods are very rare in this small area south of Nabq. The hundreds of bedouins and thousands of animals coming together in the springs of 1980 and 1981 in this area were witnessing the importance of the desert rainfloods.

Eli Gavish by then must have already known that he might never come back to the Sabkha, but he was talking to nobody about his fate at that time.

One morning in spring, after coming back from the Gavish Sabkha, I called his home in Ra'anana because I could not get in touch with Eli at his office at Tel Aviv University. I wanted urgently to talk with him about our findings and get him to come down for a joint field trip. That morning Ayala, his wife, informed me that Eli had died the day before. Two days later I met with all geologists of Israel on the cemetery in Ra'anana.

It was then that I decided to honor Eli with an integrated view of the coastal sabkha system he first described and liked so much, as everybody knows who knew him, or who had the chance to see him explain the system to the participants of one of the excursions of the International Congress of Sedimentology in 1979 or as a teacher to Nature Reserve guides or students of Tel Aviv University. Eli Gavish was above all a wunderful teacher to young students. For this reason many people not mentioned in this preface will keep him in their memories as their guide into science and the understanding of nature.

The number of people who have cooperated in the study of the Gulf of Elat in these 13 years counts in the hundreds, if not thousands. It is literally impossible to acknowledge all their input or to name them. An important research effort in the study of the Gulf of Elat was that of the Geological Survey of Israel involving many geologists, among them G. Gvirtzman, A. Sneh, B. Buchbinder, E. Zohar, and as a Visiting Scientist G.M. Friedman. The editors wish to express their gratitude to all those involved in this research. Without them we would not have been able to compile the data presented in this book. In this book we try to reflect some of the coastal research on a broader background of sabkha systems and their importance in general. The main scientific lines of the book will be given in the Synopsis.

This book is also dedicated to the memory of Eliezer Gavish. It is not an official memorial volume inasmuch as it is focusing on the Gavish Sabkha and the study of hypersaline environments along the shores of the Gulf of Elat. It deals with several different scientific aspects of one single environment. It shall be mentioned that we have preferred to stay close to this topic, which was only one of the topics of Eli's work. Therefore, we could not invite some of his best friends to contribute to the volume. Many colleagues have cooperated in the final book plan and have dealt with the various aspects of sabkhas in general, and the Gavish Sabkha in particular. Above all and lastly we wish to express our feeling of friendship and memory to a friend, who agreed to be a friend: Eli Gavish.

Synopsis

Wolfgang E. Krumbein

In this compilation of papers on the ecology of sea-marginal sabkha environments we are trying to summarize 15 years of intensive geological, geochemical and ecological work done on the shores of the two northern gulfs of the Red Sea. This compilation could be regarded somehow as the shoreline counterpart of the study of Hottinger and Reiss (1984, this series).

Although many different ecosystems exist along the shores of the Gulf of Elat (Aqaba) and the Gulf of Suez, all of them have in common that they are adjacent to a sea which is slightly more saline throughout the year than the ocean (approximately 44‰ instead 35‰). They also have in common that they occur along desert coasts. Most of the contributions of this book deal with the Gavish Sabkha, which is south of the bedouin village of Nabq and a few kilometers north of the Straits of Tiran. Other contributions deal with several other hypersaline environments, like the saline pool of the Ras Muhammed Peninsula (the southernmost tip of the Sinai) or the Solar Lake, which lies close to the northernmost end of the Gulf of Elat. The main purpose was to give a comprehensive survey of an important sedimentary system, its ecological balance, its accumulative force and aspects and its potential for mineral resource formation. Also the antiquity of biocoenoses shall be stressed, which have practically not changed their aspects through billions of years!

A general section (Sect. B) deals with the Red Sea area and its exploration through historical time and its oceanographic, geological and climatological setting. It gives an account of the term sabkha and its meaning, as well as of its geological and biological importance. Some of the most important sabkha systems of the world are quoted and compared to the coastal systems of the Gulfs of Elat and Suez. The special geological setting of the two Gulfs is described and related to the sabkha environments. General botanical and zoological aspects of hypersaline systems are described and illustrated by examples from the Red Sea area.

In the special case study (Sect. C) of our work, which was stimulated by the International Geological Correlation Program 157 on fossil mineral resources, 12 chapters deal with the special microbial, biogeochemical and mineralogical aspects of the Gavish Sabkha. Although some of the studies seem to be farfetched and very specialized at first sight, we hope that it becomes evident that only such detailed analyses will enable us to understand the past and future of life on Earth and its impact on the biogeochemical cycles and on bygone and future biospheres. Membrane studies, proton pumps, heavy metal enrichment and its catalyzation by microbial activities presently become topics of molecular biology, genetic engineering and future biotechnologies. The study of this nonpolluted but highly accumulative hypersaline microbial ecosystem enables also the estimation of environmental hazards created by man.

Biotracers, analyzed by highly sophisticated pyrolysis mass spectroscopy and multivariate analyses of the individual compounds, may reveal the fingerprints of ancient hydrocarbon-producing environments and the amount of organic substances in sediments of a preindustrial world. Isotope studies and their implications may enable us to have further insight into the onset of life on this planet and its earliest documents.

In Section D an attempt is made to connect our findings on the Gavish Sabkha with similar environments of past periods of Earth History, especially of the Precambrian. On the other hand, we try to stress the economic importance of the study of such isolated and small spots as the Gavish Sabkha. It is presently well agreed that many if not all of the salt, oil and heavy metal deposits, on which our technical civilization relies so heavily, have been produced under conditions similar to those recognizable in the Gavish Sabkha and described in this book. Finally, the study of this environment may enable us to trust more in the regularity of nature and its potential to self-regulate deviations from a coevolutionary path beginning in the earliest Precambrian through geological time up to our small section of time. If this cooperative model of biosphere–geosphere balance has been in successful equilibrium for 3.5 billion years, we may hope that our living planet will also survive the disequilibration of the cooperative system between Earth's geodynamics and biological dynamics seemingly produced by our species.

B. General Framework of Hypersaline Environments with Special Reference to the Red Sea

1. Introduction and Definitions

WOLFGANG E. KRUMBEIN

The limits of life on this planet have been the subject of many studies. One of the interesting results of these studies was the phenomenon that water availability and water activity under conditions extremely deviating from the average surface conditions on the Earth are key factors for the establishment and success of very special ecosystems. We understand today that extremely cold, extremely hot and extremely saline environments are similar in this respect. The mere fact that glycerol is used in the preservation of bacterial strains in deep-frozen state and at the same time is one of the many compounds produced for adaptation to hypersaline conditions demonstrates this relationship.

In this book we deal with the geology, sedimentology, biology, and biogeochemistry of some of the most saline natural environments registered so far. The general framework section deals with general aspects and a comprehensive approach to hypersaline environments along the coasts of the Northern Red Sea.

During summer the Gavish Sabkha and many other hypersaline desert depressions come to almost complete evaporation of standing water. Water is, however, available under different conditions. Firstly, it is well known that many of the alkali elements form extremely hygroscopic salts and thus stay wet at normal sun irradiation and at relative humidities as low as 10 and less. Secondly, many of the microorganisms settling and thriving in such hypersaline environments have evolved extracellular compounds capable of holding large amounts of water. Long before the camel they have so to speak developed adaptational mechanisms in order to survive under desert conditions.

The study of such microbial ecosystems had a first peak during the times when salting of food was much more important than modern food technologies as, e.g., vacuum freezing or deep-freezing. Later the interest of microbiologists turned back to these biotopes, when it became clear that the food resources of the Earth were getting scarce to nourish the ever increasing population and when first calculations were made about exhaustion of freshwater as one of the most important compounds necessary in agriculture. At a certain time the State of Israel used more freshwater than was supplied by the natural climatic and hydrological conditions. Hereby, more and more interest was attracted to salt-water agriculture and the study of plant types not only resisting high salinities but even thriving better under elevated salinities. Studies of hypersaline environments, such as mangroves, the Solar Lake, and sabkhas indicated soon that astonishingly high primary productivities are supported by such systems.

Not only biologists were increasingly interested in the study of hypersaline ecosystems. Geologists and geochemists became aware of the fact that many of the important deposits of phosphorus, uranium, gold and several other heavy metals were associated with evaporative cycles and even more so with biocatalyzers for the accumulation of mineral resources.

Finally, the main oil-producing countries produce hydrocarbons and gas from oilfields close to ancient and recent hypersaline environments as, e.g., the Persian Gulf, Permian hypersaline systems and other evaporative periods in Earth History. Even the oilfields of Abu Rhodeis, not far from the environments described and studied in this book, have probably formed in a hypersaline shallow-water environment.

Thus the study of sabkhas and other hypersaline coastal systems is important for several fields and attractive also for economic reasons.

Sabkha is an old Semitic-Arabic term. Its spelling in international transcription is complex. Kinsman (1969 a) suggested to use the spelling sabkha rather than sebkha, sabakha, sabach, sabath, sabka, sabkhah, sebcha, sebkra, sebja, sebjet, sebchet, sebchat, sabkhat, sebkaha, sabkhet, shott, schott, chott. All these words refer to different dialects of the Arabic language. According to the grammar of the Arabic and Hebrew languages, it would be appropriate to refer to the Gavish Sabkha as the "Sabkhat Gavish," with a connecting "t" inserted between the two words. Unfortunately, none of the terms is phonetically a valid transliteration of the foreign term. The correct form should be sabHa. The H symbol represents a guttural consonant. The sound is produced by a slight friction of air in the gullet and it differs from the symbol KH which represents the sound formed by friction of air against the palate. The H sound actually differs very little from the h, when spelled, i.e., the sounding h like in who. In this book, we use the word sabkha as a normal English word, because it has thus the widest distribution among all other spellings.

The term sabkha refers to intertidal coastal salt marshes, salt flats and evaporative pumping systems (see Chaps. 3 and 9). The term usually describes salt environments, dry lakes, lagoons and depressions which seasonally, periodically or sporadically get wetted and flooded and present great problems for traffic and water in these desert areas of the Near and Far East. But they are also places were various salts are mined for human use. Besides sand dunes, hamadas and sserirs (i.e., rock-covered slopes and plains with specific soil structures underneath) sabkhas are among the most typical structures in desert areas. They have been studied in detail by geologists and sedimentologists. Biologists and biogeochemists, however, rarely focused on these peculiar biotopes. It is the aim of this book to combine geological and biological features of sabkhas into one picture.

Scientifically, the term sabkha defines (1) smooth, flat, usually saline plains, sometimes occupied after a rain by a marsh or a temporary shallow lake; a playa, or the dry salt-encrusted bed of a lake; (2) a salt marsh; (3) a salt flat or low salt-encrusted coastal area of a normally tropical sea. The etymology of the word refers to salt infiltration (*Glossary of Geology* 1972). G. M. Friedman and Sanders (1978) distinguish between continental and coastal sabkhas (pp. 213 and 335, respectively). They define continental sabkhas as "equilibrium deflation sedimentation surfaces or ‚deflation-sedimentation windows‘ through to the local water table. The capillary fringe above the water table marks the base level of wind deflation. Sediment above this capillary fringe is removed by the wind, hence a flat surface is formed that is related to the groundwater table."

This description is true only as long as the water table is of uniform quality. When different water tables of different salinities are interspersed, no flat surface will be created. The forces generated by wind, plant roots and capillary water action, and cementation within the range of the water tables may also modify the picture.

If not covered with treacherous white salt efflorescences, sabkhas may often be detected by brown-colored spots on the surface. This indicates that the underground is wet from salt water, which does not evaporate completely. Hereby the sabkha in contrast to the surrounding desert sand remains wet. At times, a hard bitumen-like cover forms the surface, when many cars have passed over a sabkha plain. This effect is frequently used in the maintenance of dust roads in arid countries. Often wadis interfere with the flow of groundwater from the sea. In this case, clear white surface areas will cut through the areas wetted by inward-flowing sea water. The freshwater coming from the mountains and from desert floods rapidly, dries up, and usually eventually crystallizing salts are washed into the sea during heavy rains. Thus there is no salt left to keep the sand wet. This promotes wind erosion and creates a wavy surface pattern. The sparse desert flora also modifies the situation both above and below the soil surface (see Chap. 7). In contrast to the above described coastal sabkhas, inland sabkhas may be generated also from freshwater sources alone. This is explained by the fact that the small amounts of salt are accumulating over the years in mountain plains or playas, because the drainage system does not reach the sea. Hereby salt accumulates and whenever freshwater is supplied during rainy seasons it is immediately dissolving the salt crusts and forming a salty brine with the same effects as coastal sabkhas.

Some of the aerial photographs in Chapters 9 and 11 illustrate this phenomenon. Plant growth is also correlated to the sabkha and wadi systems (see Chap. 7). It is very astonishing that Ehrenberg (1830, 1834a, b and later publications), and Hemprich and Ehrenberg (1828), never mentioned the biology of sabkhas. Ehrenberg was well aware of planktonic cyanobacteria in the Red Sea and even named one after the Red Sea (see Chap. 2). Yet he never mentioned the reddish waters of the sabkha pools of Ras Muhammed or the Gulf of Suez, even though he must have seen these places on his collecting trips from Tur. We found the first real descriptions of sabkhas or salinas, including references to the anoxygenic phototrophic purple and green sulfur bacteria, the red-to-green oxygenic cyanobacteria, and *Dunaliella,* in Darwin's descriptions of the Rio Negro Salinas (1939) and in Russegger's reports on Egypt and the Sinai (1835–1841).

Oerstedt (1841) was the first to study and name these micrnorganisms, which were then still called Confervae or Animalculae (after Leeuwenhoek). Thiopedia, the platelet forming phototroph, has been first described by Oerstedt (1841). He also mentions *that Microcoleus* and *Oscillatoria* and either *Aphanocapsa* or *Synechococcus* occur in saline sediments (see Chaps. 8 and 11).

Carless (1837) first mentions the sabkha south of Nabq and noted that the Bedouins used it as a natural salt pan. Russegger, although he was a mining engineer, was apparently the first person to suggest clearly that microorganisms participate in the development of sabkha sediments. He found organic pigments, red colors and organic residues when he chemically analyzed the sediments from the sabkhas of Wadi Natrun. He also mentioned that the open water of the Natrun lakes was purple-red.

Darwin and Russegger imply the biogenic formation and transformation of carbonates (including dolomite) and gypsum. The first precise experiments concerning the microbial transformation of carbonates into sulfides and vice versa were made by Nadson (1903) (see Chaps. 5, 9, and 22). The isotopic ratios of biogenic carbonates and their implications for fossil sabkhas or stromatolites have only recently been studied (see Chaps. 19 and 21).

Hoye (1907), Woronichin (1926), and Hof (1935) were the first biologists to study seriously the microbiology of sabkha-type microbial mats. They stated that anoxygenic photosynthesis of cyanobacteria is an important pattern of ecological adaptation and described the tremendous pleomorphy of many microorganisms, especially cyanobacteria, from sabkha pools and salt-covered microbial mats. They did not connect these findings to Kalkowsky's (1908) descriptions of stromatolites and oolites. Walter (1976) and also Krumbein (1983a) have connected the geological and biological studies which were done simultaneously in the early years of the twentieth century and were only recently treated in integrated views (Krumbein and Cohen 1974, Cohen et al. 1983, Krumbein 1984b, Stal et al. 1984).

The term microbial mat is in use since 1979 (Krumbein et al. 1979a). Microbial mats – sometimes regarded as potential or recent stromatolites – are single- to multi-layered microbial systems growing at the surface of or within a sediment or rock. They are intimately connected to and with the systems above and below. By photosynthesis, chemosynthesis or both, by degradative activities and mechanically they rule the relationships between the underlying systems and those above the microbial mat. They represent biological, physical and chemical traps, filters, and catalysators and can be regarded as multilayered biofilms. They never consist of one single species. Often several individual but interrelated microbial systems build complex multilayered laminated mats. Photosynthetic (e.g. cyanobacterial) mats presently seem to be more frequent than chemoorganotrophic or chemolithotrophic prokaryote (e.g., Beggiatoa) or eukaryotic (e.g., fungal) mats. Microbial mats often are the sediment agglutinating, organic matter and mineral producing precursors of laminated rock systems called stromatolites.

An interesting feature of ancient and present salt production in marine salinas whiter salt (Fürer 1900, Chaps. 9 and 11). The Bedouins use the salt for conserving fish, for tanning, and for medically treating men and animals. It is also well known since Venetian times that a relatively high benthic productivity of sabkha-type microbial mats in highly saline environment reduces the number of plankton organisms in the salt pans and thus guarantees large salt crystals, which are easier to exploit and sell.

Although in the wrong context, Russegger (1841–1848) mentioned interesting cross-connections between sabkhas, wadis, primary productivity and the formation of petroleum and heavy metal deposits (Krumbein 1983 a, Dexter-Dyer et al. 1984, Chaps. 16 and 19). Russeger noted that the reefs and the saline pools are nutrient traps that produce carbohydrates and mercaptanes (see also Chaps. 18 and 19). Later, it was found that the petroleum of the Abu Rhodeis oilfields was formed in older sediments. These, however, are also of coastal origin. Thus, this book involves research on recent coastal phenomena which have been recognized as the precursor environments of petroleum deposits, ore deposits, and salt and phosphate deposits of great economical value.

Geologists and travelers usually believe these sabkhas are sterile. In fact, they are extremely complex ecosystems with a very interesting faunal food web in which the terrestrial and marine faunal kingdoms interact in a unique way (see Chaps. 6 and 15).

The geological importance of sabkhas and their evaporative function as well as a differentiation in different coastal systems are described in Chapters 4 and 5. The special coastal geological and sedimentological setting is described in Chapters 3 and 5. General botanical and zoological aspects with special sections on the coastal sabkhas of the Gulf of Elat are given in Chapters 6 and 7. These chapters are connecting the general framework of sabkha environments with the special cases analyzed in the next section.

Altogether this general framework section represents the background and basis for the understanding of the more specialized case study contributions.

In Chapter 2 an attempt is made to sum up and recollect some of the spirit of this area of the world, which already many times before our technological period has challenged scientists and travelers to considerations and explanations of the unique situation of desert coasts. To use a term of Bernhard of Chartres, often said to have been coined by Sir Isaac Newton (Krumbein 1984 a, Merton 1965): "If I have seen further, it is by standing upon the shoulders of Giants."

2. The Northern Red Sea, a Historical Sketch

Wolfgang E. Krumbein

2.1 The Origin of the Name Red Sea

The origin of the name Red Sea always has been a reason for speculations. These go back to the very beginnings of scientific literature. Although the shores of the Red Sea have been always sparsely populated throughout human history, the Red Sea has played a significant role in it. Even the people living on the banks of the Euphrates, the Tigris, the Nile, and the Danube, and on the shores of the Mediterranean Sea, the Black Sea and the Caspian Sea, have referred to the Red Sea in legendary and fabulous terms.

Traders and adventurers traveling overland from China and India through Persia, the Euphrates valley, Colchis and Northeast Turkey carried and distributed both goods and knowledge to Egypt and the "Fertile Crescent", which surrounds the Red Sea. Goods from Ceylon (also called Sri Lanka or Taprobane) or from the Danakil and Ethiopian areas usually came by sea through the Red Sea to Egypt and the countries of the Fertile Crescent.

Two other situations which have persisted until today have added to the mystical and magical nature of the Red Sea. Its shores are infertile and have been settled mainly by nomads and other peoples uncivilized in comparison to those which inhabited the Nile Valley, Mesopotamia and the coasts of the Mediterranean. Also, sea travel on the Red Sea was difficult because wood to build the ships had to be imported from Mediterranean regions. This was possible only if there was a high degree of cooperation between the nations dwelling between the Mediterranean and the Red Sea, or if this whole region was politically united under one ruler for a considerable span of time, or if a navigable canal was dug from the Red Sea through the Bitter Lakes to the Mediterranean Sea.

The Red Sea has been scientifically studied only in times when peace and liberal ways of thinking prevailed in the area, since scientific investigation presupposed that people were open-minded enough to challenge earlier views of the world in which all lands surrounded the Mediterranean Sea and were themselves surrounded by a continuous river. The main periods of exploration in the Red Sea area were the Mesolithic (Petroglyphic), the times of Sesostris, Solomon, the great Nabatean kings, the Greek Christian missionaries and monks, Saladin, the period of Arab expansion into Spain, the times of Mehmed Ali under Turkish rule (approx. 1820–1850) and the years since the 1967 war. Hopefully, the current period of Red Sea exploration will continue because of the Israeli–Egyptian peace agreement.

Thus the origin of the name Red Sea may go back as far as men were travelling along its shores, perhaps even before people were able to build ships. Unfortunately we have few reports remaining from that time. The ancient cosmogonies and Herodotus do not really discuss the name. Herodotus was a famous historian, who can also be regarded as a natural scientist. Much of his knowledge was derived from earlier scientists, as we will see later. Herodotus originated from Halicarnassos, and lived probably from 484 to 425 B.C. One of his contemporaries was Socrates. Of the famous Greek scientists only Thales and Pythagoras lived before him. He travelled through practically the whole known world and collected enormous amounts of information from books and libraries. I have used the translation of Braun (1958).

Strabo was apparently the first scientist to inquire into the origin of the name Red Sea. Strabo was a Greek geographer, born about 60 B.C. in Amasia in Cappadokia. After extended scientific research trips through most parts of the known world he came to Rome about 26 B.C. and wrote his *Geography in 17 Books.* I have used the translation of Jones, H. L. in Warmington, E. H. (ed). The Loeb Classical Library, Harvard, University Press (1959) there also exists an excellent translation by Groskurd, Berlin, 1831–1833, 4 volumes.

The two most important passages in Strabo about the origin of the name RED SEA are: *For example, Eratosthenes says that some writers call the sea Erythra from the color it presents as the result of reflection, whether from the rays of the sun when it is in the zenith, or from the mountains, which have been reddened by the scorching heat, for, he continues, conjecture runs both ways about the cause; but Ctesias, the Cnidian, reports a spring, consisting of red- and ochre-colored water, as emptying into the sea; and Agatharcides, a fellow citizen of Ctesias, reports from a certain Boxus, of Persian descent,... that a certain Persian, Erythras by name, built a raft to recover horses from an island. He drove the herd back to Persis, sent forth colonists to that island and to the others and to the coast, and caused the sea to be named after himself. Other writers claim that Erythras was the son of Perseus and that he ruled over his region.* At another place Strabo stresses the latter origin of the name Red Sea and does not mention the other theories: *Both Nearchos and Orthagoras state that the island Ogyris* (most probably Bahrein) *lies in the high seas at a distance of 2000 stadia from Carmania and that on it is to be seen the grave of Erythras, a large mound planted with wild palm trees; and that Erythras reigned as king over that region and left the sea named after himself.*

Thus Strabo already firmly stated that the name of the Red Sea is related to mythological rather than to scientific causation. This is strongly supported by the fact that in ancient science mythology and science were intimately mixed – as is the case in modern science if we look at "taoistic physics", "the paradigm of evolution", the question of the "origin of life" and such important questions as "continental drift", "plate tectonics" and "global ecology" or "man-made pollution".

It is not astonishing therefore, that geological and geographical knowledge, even biological descriptions of authors before Strabo have to be excerpted from mythologically orientated books like the Odyssey of Homer, the Indian Vedas or Upanishads and other mythologies. Homer, who probably lived around 800 B.C. knew about the Red Sea and it is stated in his text that the Earth is round and the equator runs through Africa. Some modern text analysts believe that a scientist, a historian and a poet may have cooperated over the years in order to accomplish the two books of Homer.

Problems with the name Red Sea occurred also for the reason that the Greek usually circumscribed colors rather than using specific adjectives for them as is the case in the Romanic and Germanic languages. Herodotus and his successors also used the term Mare Erythraea for the whole Indian Ocean. The present-day Red Sea is only one of five or six branches of the ancient Mare Erythraea.

Movers (1849) and Ritter (1820, 1861) have later derived from the Greek and Roman authors that a joint origin of the Erythras/Dionysos legends may well go back to ancient Indian and especially Ceylonese myths.

Later several other suggestions were made about the origin of the name Red Sea which were mostly related to the Biblical text about Yam Suph (which was translated as algae, sargassum or reed) or to observations made in this area. Also, many authors have made connecting theories between the Red Sea or Yam Suph and the red waters of the seven plagues: *"And all waters, which were in the river were changed into blood and the fish died and the river stank and the Egyptians were not able to drink the water of the river."* (Exodus, 7, 20–21).

According to all our knowledge the red waters were probably stinking of hydrogen sulfide and mercaptanes in which at certain periods fish and frogs die by the poisonous substances produced in an increasingly hypersaline evaporative, i.e., sabkha environment. In arid regions rain and desert sheetfloods, as well as river floods coming down from far, often fill many rivers and arms of rivers as, e.g., the Nile and its side arms, with freshwater. A few weeks or months after the rain these waters evaporate into saline waters with all the ecological consequences, which will be described in the subsequent chapters.

Astonishingly, despite all my search I did not yet find any serious description of such environments in ancient Greek or Roman reports. Strabo mentions salt marshes, salt lakes and salt works but never mentions red water blooms observed in them. He knows about estuaries and describes them as tidal flats with tidal canals within them, and also knows about red rock salt colored by iron oxides. Still, he does not mention the salt pans and sulfureta related to them. On the other hand, when referring to Herodotus' lentil-shaped fossils of the Red Sea area (Nummulites) he also mentions ooidal and oncoidal tufa occurrences as typical evaporation products.

Coming to present-day scientific studies along the shores of the Gulf of Elat one may also mention that Strabo describes very clearly "Aquaculture": *But some, who have shell fish, fatten them by throwing them down into gullies and pools of sea water, and then, throwing in minnows as food for them, use them for food when there is a scarcity of fish. They also have all kinds of places for hatching and feeding fish, from which they parcel them out.*

The finest descriptions, however, Strabo gives of the scientific aspects of Red Sea research are not about the differences in water level, not about the fossils, and also not about aquaculture. They directly relate to some of the most astonishing facts for those people living on the shores of the Mediterranean or the Black Sea, i.e., the "civilized" people of that period. These relate to coral reefs, large seagrass areas, marine algae and mangroves. He describes all of them in detail in different sections of his narrative and certainly does not mix them, as two quotations may indicate: *Along the whole of the coast of the Red Sea, down in the deep grow trees* like *the laurel and the olive, which at the ebb tide are wholly visible above the water but at the full tides are wholly covered; and while this is the case, the land that lies above the sea has no trees, and therefore the peculiarity is all the greater... and the* surface *is covered grass-like with sea weeds, and rockweeds are still visible below the surface, a thing still more visible at the strait, where among the plants even trees grow down below the water*. Strabo makes very cautious stratements about the seaweeds dwelling at the surface (like grass) and the coral trees (like laurel and olive trees).

Later scientists, copying very much from their forerunners, were less careful. Pliny the Elder (Gaius Plinius Secundus) was a Roman scientist, researcher and encyclopedist, who was born in the year 23 and died during scientific research at the Mount Vesuvius outbreak that destroyed Pompeii. His *Natural History* filled 37 volumes and was based on his own observations and more than 2,000 books. He was supervising 500 writers, who copied everything rather without reflection and produced many copying mistakes. Thus Pliny writes in one place: *"In the Red Sea laurel trees and green brushes are growing..."* We do not know whether Pliny and his co-workers were miscopying from Strabo or whether they copied from the much older but wrong observations and reports of Theophrastes (König and Winkler 1927–1977).

Theophrastes was a Greek botanist born on Lesbos in 372 B.C., a very close friend of Aristotle. He inherited the library of Aristotle and focused mainly on botany. He died in Athens about 287 B.C.

Theophrastes wrote: *"In the so-called Red Sea, in Arabia, grow plants, which are named laurel and olive."* On the other hand, he clearly distinguishes at least five different and clearly discernible seaweeds, seagrasses and mangroves sitting at low tide on large roots like the polyps. He describes the airroots as well as the leaves and flowers and fruits of *Avicennia* very clearly. When one reads the explanations of Sprengel (1822) on Theophrastus, marine plants and follows back, how Strabo was misinterpreted by changing from *like* laurel trees to the firm statement of laurel and olive trees in Pliny it may not be astonishing that younger authors than came back to Strabo and Pliny and Theophrastus and tried to "correct" their apparent misunderstandings.

Strabo clearly describes the stony coral "trees" on various occasions and even mentions the umbrella or hat-shaped isolated reef patches as well as the black gorgonians. Zanardini (1858!) interprets them as misunderstood *Sargassum, Fucus* or *Laminaria* and derives the name of the Red Sea from these algae. He does not realize that on another occasion Strabo describes these algae as being mixed with sea grasses and coral trees. Zanardini wrote his notes on the flora and fauna of the Red Sea in Latin. He refers also to Montagne who ascribed the red color of the Red Sea to *Trichodesmium erythraeum* Ehrenberg. He did not, however, read Ehrenberg's report on blood-red phenomena in the Red Sea (Ehrenberg 1830) nor did he refer to Ehrenberg's original description of the phenomenon. C. G. Ehrenberg (1830) writes after describing the whole phenomenon of red tides in the Red Sea for the first time (Ehrenberg 1830): *More specialities are described in the Symbolis physicis of mine and Dr. Hemprich's travels*. These however were never published for many reasons, some of which are described by Stresemann (1959), who also published the letters of Hemprich and Ehrenberg.

Hemprich and Ehrenberg (1828) and Ehrenberg (1832) briefly mention the red cyanobacterium and the circumstances of identification in these papers but never with the same precision as in Poggendorf's Annals of Physics (1830). In addition, a long debate was started about *Trichodesmium* and *Oscillatoria*, which can be followed up in the taxonomical literature (e.g., Geitler 1932 or Desikachary 1959). Interestingly, this *Oscillatoria*, which contains a lot of phycoerythrin, the red phycobiliprotein so characteristic for many cyanobacteria, is also recently well studied for its capacity to fix atmospheric nitrogen without the aid of heterocysts.

Later authors frequently refer to this red bloom and to Ehrenberg's collection of possibilities of red rains, red waters, and red soil and snow.

Another line of research related to the Red Sea and its color is always the study of the miraculous coral reefs. Ehrenberg (1834a, b) quotes a lot of the early reports on this and mentions Strachan, Monconny, Forskal, Forster, Chamisso, Shaw, Sloane, and others. A very interesting question concerning the corals is already laid down in Strabo, inasmuch as he mentions the red corals as well as the black gorgonians and the possibility of wood to petrify in lime-rich waters. Many people believed that all corals get stony only when they are exposed to air or fire.

From these "scientific" statements and the popularity of Ovid a lot of confusion was transferred to the present days. Gorgonians were hair-like black corals and derive their name from the hair of Gorgo. Red corals and the oxidizing red colors of the arid climates jointly with the many observations of tufa and oncoids led to the story that soft corals are transformed in the air or by heating into the precious red corals, which again are only one branch of the whole family. Additionally, the beautiful colors of the reefs, as viewed from above, and the scarcity of real reports from these areas led to many misunderstandings. The poem of Ovid has often been read and quoted and has therefore added to the confusion about red stony corals and their origin. In order to close this historical section I may quote the lines of Ovid relevant to this topic:

Sic et curalium, quo primum contigit auras
Tempore durescit: mollis fuit herba sub indis.

As for the red color and the origin of the name of the Red Sea, I myself believe rather in the mythological origin related to King Erythras. Several authors have talked about red tides and red waters, and especially those of the Red Sea (e.g., Currie 1955).

Through the centuries many observations were made on the possible origin of the name Red Sea related to natural phenomena. The red color was related to the following organisms:

a) Cyanobacterial blooms.
b) Blooms of the diatom *Pleurosigma*.
c) Blooms of *Noctiluca*.

d) Blooms of purple phototrophic bacteria in hypersaline lagoons and estuaries.
e) Blooms of halophilic bacteria (*Halobacterium halobium*; Kurochkin 1960).
f) Blooms of *Dunaliella*.
g) Red seaweeds of the types of corraline algae and *Porphyra*.
h) Red corals.

Furthermore, I myself have seen the northern Red Sea to turn red by
a) red dust blown into the sea (also known as red rain with finely distributed desert sand from the Sahara up to places like Hamburg and even further north in Europe,
b) totally red shallow waters of the gulf through suspended silt load after torrential rains and desert sheetfloods emptying into the light-blue waters of the Gulf of Elat,
c) mirror-like red reflections during sunset or even during the day caused by the red tint of the desert-varnished mountains,
d) Reflection and heat layers above the water producing red color even during noontime.

Many German scientists and one or the other scientist of other countries may have read the adventure stories of an author almost as famous, although much less artistic than Goethe or Shakespeare, namely Karl May, the creator of Winnetou, Old Shatterhand and Kara Ben Nemsi. It is well known that Karl May barely has seen any of the countries about which he wrote and into which he planted his imaginary heroes. Karl May, however, has made one major voyage. This voyage led him through the Red Sea. It is not astonishing therefore that he wrote the following lines on 12 September 1899, in his diary (Wollschläger and Bartsch 1971). The same observation was repeated in a letter to R. Plöhn on 26 September 1899: Some people ask themselves often in vain why the Red Sea is actually called red. Nobody can answer but myself from now... Very strange wave-structures (original German = complicates) are forming which carry white crowns of spray and if one looks against the sun the wave parts opposite to the sun rays have the appearance of consisting of liquid phosphorescent blood. This recalls some of the artistic romantic oil paintings of Feuerbach and Böcklin.

Maybe one day another author finds another „final solution“ to the origin of the red color sometimes observed in the Red Sea and to the origin of its name. Here it suffices to add that whenever one works in sabkhas one will realize that during certain seasons the color of the photosynthetic life-bearing pigments is rather red instead of the classical green of the chlorophylls. The pigments related to this phenomenon (see Chaps. 10–12) are carotenes, phycoerythrin, bacteriorhodopsin and some others. The reason why these red waters, tides, rains and snow plains have always fascinated travelers and scientists and irritated the people living in such areas or through such experiences is probably due to the fact that blood is red. If blood would be green, we might not have all these reports in literature culminating in blood rains as the forerunners of worse disasters to come. Mankind has always transferred and summarized natural phenomena into metaphorical statements. Today we are wont to refer to the subconscious, to psychology and sociology. In former times natural phenomena were transcribed into gods, godesses and mythological heroes. Thus, Mare Erythraea may be the ancient summary of all the astonishing facts related to the region of sunrise, including many blood-red phenomena carried on in the name of a god.

2.2 Traffic on the Red Sea in Ancient Times

All puzzles and inconsistencies about Sinai, the Red Sea and the fabulous land of Ophir relate at the end to practical questions, as how to become rich with new and unknown products or dangerous trading routes. The classical example is the transfer of elephants, pheasants, ivory, ebony, gold and precious stones via the Red Sea and the Arabian Peninsula as described in the Holy Script. These legendary trade routes and trading partners have evoked and still evoke interest and discussions among scientists and nonscientists. This is explained by the fact that four of the main religions presently distributed among the so-called civilized nations derive a lot of the ornamentation of their cosmogonies and semi-historical reports from the exchange of information and goods between East, West, North and South along the shores of the Red Sea. Several solid facts seem to have emerged into general accord by the compilation of about 3,000 years of recorded history. Any scientist in any of the fields, who might find some interest in this book – especially biologists and geoscientists – may, however, realize how much is truth of what is written and how much is speculation and distortion by making a simple assumption:

Let us assume for a moment that Darwin or Lyell did not write down their findings and theories straight away. Let us assume they told what they knew to friends and young students. These in turn would have translated the news into a foreign language, e.g., to German, and transferred them orally to Haeckel and J. Walther. These may have reported the facts to some students in Germany, e.g., Kayser and Heberer. About a hundred years later somebody would finally write a book about the findings and theories of Darwin and Lyell. This is the situation with which we are confronted concerning practically all ancient written reports about the Red Sea.

Given these facts, the scientific knowledge of the people who first had a firm understanding of the Sinai and the Red Sea was considerable. Therefore, they also had a chance to take advantage of the commercial possibilities concerning natural goods and trade in these regions.

In addition, one of the finest sets of information regarding the Sinai and Red Sea area was common knowledge among the Egyptians, the Euphratean people, the early Greek people (Homer, Thales), the Phoenicians (Philon of Byblos) and the Hebrews (Pentateuch). The cross-information between the biblical reports and the Phoenician reports is made by (a) excavations and (b) the writings of Sanchuniathon which have been transmitted to us by a translation and adaptation of a much younger Greek (Philon of Byblos) (Berytii 1826, Byblos 1887, Movers 1841–1856).

Therefore, it is not astonishing at all that one of the most fascinating literary fakes was made in the early period of modern geography at this crucial connecting point of histories and time lines. The scientists of the early nineteenth century (e.g., Fabricius, 1883 or Vincent 1800 quoted in Fabricius) knew already vaguely about the area and that the Periplus of an anonymous Greek author was a correct description of the Red Sea as a trade connection. This knowledge led a young teacher of the city of Bremen, who was teaching Latin and Greek, to fake a periplus for the highly fascinating time of King Salomon and the Queen of Sheba. Wagenfeld (1836, 1837a, b) claimed that he found a more complete copy of the Sanchuniathon script by Philon of Byblos than the previous copies existing. According to him, it was hidden in a small Portuguese monastery. He submitted two editions of a German translation supported with an introduction by the highly appreciated orientalist Grotefend, who was an authority on old languages and on deciphering ancient hieroglyphs.

The faked additional chapters clearly indicated that King Solomon and the Phoenicians had pilots from southern countries and that the place they reached for their trade was Ceylon (Seilan-Taprobane). This was and is one of the scientific assumptions even today. The fake was soon discovered, although Wagenfeld exhibited a detailed knowledge of Latin and Greek in the concordant publication of a Latin/Greek version of the original enriched with his fake chapters. He unfortunately incorporated copying mistakes from the copy of the fragment he had used for the fake (Farrar 1907, Cumberland 1720, Orelli 1826).

Coming back to scientific grounds I wish to mention that the knowledge of the archeological sites in Israel, Egypt, and the Euphrat area today is overwhelmingly rich. Also scientific exploration throughout the centuries has made it clear that the Pentateuch as well as Sanchuniathon are – as could be expected – mixtures of historical events, scientific descriptions and myths. The same is true for Homer's Ilia, written at about the same time. The scientific facts of the Iliad as well as their mythological ornamentation, were already discussed by Herodotus, the famous geographer. Ehrenberg has published many papers concerning Manna, the birds of the People and many other scientific facts of the Bible. Many of these facts were witnessed and documented in Petermann's Geographische Mitteilungen between 1825 and 1870 (Heuglin 1860, Steudner 1861, Raumer 1862).

These and younger analyses make it clear that the reports of the Bible about the traveling of the Israelites through Sinai are mostly true and that passages, like those about the burning bush and others, which were thought to be fairy-tales or supernatural actions, find natural explanations. Even the writings of Sanchuniathon, which have frequently been discredited as a Greek fake or crude alteration of former texts, have been shown to be basically true.

From this and many other findings the following picture may be derived:

In the thirteenth century B.C. several Israelite and Proto-Israelite tribes came into Palestine from the South (Egypt) and from the East (Engel 1979). At this time no major kingdoms were ruling the area so that nomadic tribes had a fair change to besiege and storm the then less important and less populated cities. Kenyon (1980), using much of the data of Y. Yadin, has given a very fair account of this period. W. Keller (1955) bases his general narrative mainly on the excellent archeological compilation of Parrot (1954).

About the twelfth to tenth century B.C. several things happened, which led to the first scientific exploration of this area. (1) The Hebrew people in Sinai developed a written alphabet which was refined enough in the twelfth century B.C.

to yield two or three written reports later combined into the Pentateuch. (2) The Phoenicians were creating their empire based on enormous skill in ship construction and training of pilots of the seas. (3) The camel replaced the donkey as the transport vehicle via Africa and Arabia. (4) A long peaceful period emerged after the initial fights under the rule of David and Solomon, and Hieram of the Phoenicians. (5) By this fact shipping was improved and the speed of both caravans and ships was higher by a factor of at least 5 as compared to the centuries before and many centuries after. Several calculations of Greek and Roman distances and etmals (day's run of a ship) reported in the Royal Academy, London, and Sitzungsberichte der königlichen Akademie der Wissenschaften, Berlin, between 1800 and 1850 clearly indicate that between 1,000 B.C. and approximately 300 B.C. etmals of up to 120 nautical miles were practicable. In later times (300 B.C. until 800 A.D.) the daily distances went down to about 5 miles per day and less for many reasons (see e.g., Ritter 1820, 1861, Niebuhr 1816, Ideler 1816 ff.).

The Red Sea for the first time became a really crowded shipping waterway, when also its name was firmly established as Mare Erythraea and very soon translated into Red Sea. In order to explain what must have happened at that time I shall recall the reports of Forskal (1775a, b, 1776), Niebuhr (1781, 1816) and of Hemprich and Ehrenberg (1828) as an example for changing situations through history.

All scientific travelers of the late eighteenth and early nineteenth century have reported large numbers of ships in the Red Sea and the enormously crowded harbors between Egypt, Sinai, and what is now Saudi Arabia. If we look at the population of Abu Thor of today (a few Bedouins and Arabs) and compare it with the reports of Ehrenberg, Kossmann (1877), Schubert and others we immediately realize that between 1800 and 1850 there were far more people living and traveling in this area than in the following 100 to 150 years. Such changes have frequently occurred in this area and the wave of tourism that swept through Sinai and its coasts between 1970 and 1980 is still another example of what political conditions may contribute to the knowledge and population of a certain area. The influence of Mehmed Ali, the Albanian, who had tried in so many different ways to modernize the Egyptian part of the Osman Empire is just one more example of the intimate relation between politics and development of the Sinai and the Red Sea. What Mehmed Ali tried had been tried many times before, as witnessed by the traces of ancient Egyptian and Hebrew mines, streets and harbors along the Gulf area. A well-documented review of all sources of the name of the Red Sea as well as of the different geographical compilations of Greek and Roman geographers is combined in Wissowa (1909, 1912). Anybody interested in geography and ancient geography will have read the first written report thas has been almost completely preserved, namely Herodotus of Halikarnassos (450 B.C.). This report, as mentioned before, comes from at least three different sources: (a) copy from Hekataios, (b) oral reports of Egyptian scientists to Herodotus and (c) his own narrative of places he visited. The latter is the really safe information as far as his outstanding intelligence and memory is concerned.

Herodotus starts his historical and geographical report with the conviction that the Phoenicians came to the Mediterranean from the Red Sea and that they were actually the founders of the ports of Berenice and Elat. The Gulf of Elat was formerly also called the Elanitic Bosom. This is another reason why the term Gulf of Elat used throughout this volume is more justified than the frequently used term Gulf of Aqaba. Elat actually was founded much earlier than Aqaba and this geographical name of the Gulf was mentioned earlier as well.

It is now the general conviction that already at the time of Ramses II or Sesostris India and Ceylon were known to the ancient people and that voyages which lasted longer than 3 months were actually leading to these countries rather than to the Saudi Arabian or Ethiopian Kingdoms. Imhotep (approx. 2700 B.C.) and Ahmose (about 1650 B.C.) are the most outstanding scientists and mathematicians of the Egyptian empires and much of their knowledge that was created for the annual problems in land reclamation after the Nile floods and for calendary necessities of the empire has been transferred to us via the formerly existing hermetic code. One of the largest cultural disasters was the fire in the library of Alexandria, which cut us off from many original reports.

Some time later the Phoenician Tarsis ships, which were a construction type and not related to the city of Tarsus in Spain, were very fast and efficient sailboats combined with oars for periods without wind and were able to travel more than 140 miles per day under favorable conditions.

Herodotus knew the Red Sea very well and had a fair knowledge also of the two gulfs at its ends, while he underestimated the Persian Gulf in its size, which is witnessed by his account of the enormous activity of the Nile river in land increase in the Delta, while he speaks less of Euphrates and Tigris, although he certainly knew about them as well. He clearly recognizes the fossils, which form much of the underground in Egypt, a knowledge that got lost for almost 2,000 years afterwards. When he states in the second book that in Arabia – not far from Egypt – there exists another Gulf of the Red Sea which stretches to Syria, he mentions certainly the Gulf of Elat. The later scientists however had no idea of this very narrow and very deep gulf. It comes back into the maps only in the nineteenth century, when the first expeditions had been there and when Seetzen and others had defined the latitude of Elat. Therefore it is not astonishing that later geographers have frequently misinterpreted the indications given by Herodotus. He sometimes calls the Sinai side of the Gulf of Suez the "Arabian," which increases the amount of confusion. Even Niebuhr, who was traveling with Forskal in the late eighteenth century through the Red Sea, was centering his interest on the Gulf of Suez and the southern Red Sea, i.e., the Saudi Arabian coast. They reached the Straits of Tiran, but they never entered deeper into the Gulf of Elat. No wonder, if one considers the account given by Crossland (1939):

Then we entered the Gulf of Aqaba (Elat), the most desolate sea in the world. There is hardly any maritime plain even in the south, while in the central section black and red mountains rise sheer from the water. Beside this majestic wildness the desolation of the Gulf of Suez is a mild and smiling landscape. No ships visit these waters, the only sign of there having been life on the sea is the wreck of a Turkish gunboat on a reef at the entrance of the Gulf.

In view of this report of 1939 (!!) we may imagine how Niebuhr and his companions must have felt. Niebuhr was a very stubborn farmer's son of North Frisia, who incidentally studied cartography in Göttingen because his family had a lot of trouble with the redistribution of land after the storm floods and after dike construction. Thus he renewed much of the Egyptian knowledge mentioned before. When the king of Denmark was contacted in order to finance an expedition, which, thanks to the precise sketches Niebuhr gave, ended in the deciphering of much of the ancient languages of the area, he made Niebuhr the administrator of the expedition, because he was so clear-minded, precise and accurate. Niebuhr constructed all his instruments by himself. Of the whole scientific crew of that expedition he was practically the only survivor. He therefore published also the zoological and botanical reports of Forskal who died in the Northern Red Sea area.

Already then the hiddenness and desolation of the area, but also the small Bedouin fisher settlements the population of which came down from the mountains only for the date harvesting, preparing salted fish and fishing lobsters, have been set down. The scientific exploration, however, of the "stormiest sea of the world" had been started 100 years earlier by Seetzen (1808 a–c), Rüppell (1829), Hemprich and Ehrenberg (1828), and Burckhardt (1824). It must be noted, however, that the first attempts were always made from the landside (Paultre 1804, Fraas 1867). The principle of traveling along the coast or to cross Sinai was followed by all travelers and researchers and still prevails for geological reasons and still guided the tourists between 1970 and 1980. The coast was accessible only from Elat to Nueiba, with a stretch omitted that included the Solar Lake and El Hamira bay. In this area the road parallels the sea at about 7 km distance. From Nueiba it turns into the mountains running relatively close to the Monastery of Santa Catarina, and turns back to the coast only at Dahab. The beautiful bay of Ras Abu Galum with its impressive landslide-marked steep cliff was rarely visited. From Dahab the road turns back to the mountains and comes down to the coast only near the Straits of Tiran. Thus all scientists were visiting either Ras Muhammad and the Straits of Tiran and returned to Thor (Tur) or they went up to Santa Catarina and came down to Dahab or Nueiba. Other scientists came down from the Dead Sea, reached Aqaba, but had to turn back to the north. For this reason neither the Solar Lake nor the Gavish Sabkha have been mentioned in any scientific report prior to 1968. The Ras Muhammad Sabkha pool (see Chap. 19), however, was easier for access because the road is mainly running on ancient elevated reefs instead of treachery salt and loose sand, as is the case with the Gavish Sabkha and the sabkha south of the Solar Lake.

No wonder that the reports of Pliny and some of the pilgrims of the crusader period were mostly disbelieved by later travelers. Thus, as long as Herodotus talks about the Egyptians he mentions the northernmost Gulf quite appropriately. When, in his third book, he talks about Darius the Persian and his provinces he mentions the Persian Gulf of the Red Sea but he does not realize the geographical connection and the enormous size of the deserts of Saudi Arabia in-between. The confusion is still more heightened by the fact that beginning with modern times the name Red Sea was restricted to the Sea up to the Bab el Mandeb, while the outer Red Sea which led to India, Ceylon and the Persian Gulf was already named the Indian Ocean. Herodotus, who never traveled himself along these coasts, mentions reports which seemed unbelievable to the former people and especially to later Greek, Roman and Medieval travelers. Thus, when he combines ancient Greek (Homer) and Egyptian (hermetic code) reports about circumnavigating

Africa in a ship tour of 3 years, where they had the sun to their right during their sailing along the Libyan (African) coast, even then nobody would believe him. Herodotus knew that the Red Sea somewhere along the modern Erythraean coast turns into the Southern Ocean (southern Indian Ocean excluding the Arabian Sea), and after turning around the Cape of Good Hope transforms into the Northern Ocean (Atlantic). He clearly states that all this is one huge water mass completely connected through the Straits of Hercules with the Mediterranean Sea. This interpretation of the reports of Herodotus permits as well to interpret the report of the sailing trip of Solomon and Hieram of Phoenicia as a trading expedition that touched the Arabian Peninsula (Queen of Saba), Zimbabwe (Africa) and Ceylon (Taprobane). The wind conditions in these areas make it clear that traveling on one latitude with the monsoons leads from the heights of Zimbabwe to Ceylon, and with the opposite winds in January back to the Horn of Africa. By sailing in January it was very reasonable to expect to reach Madagascar and further south, while in July one could rather be drifted or carried to Ceylon or very speedily up along the African coast from the height of Zimbabwe to the Horn of Africa. It is evident that the expedition mentioned to fail was hitting adverse winds at that time. It is also evident that the reports about making a break and cultivating food onshore are related to periods without wind about which the ancient people knew from their pilots. When Herodotus and Strabo mention the knowledge of Homer, they also talk about the sun giving shadow to the north and rain falling in summer instead of winter. Both reports can only be explained by the exact knowledge of the Equator and its astronomical importance.

At this place it should be mentioned that Ritter (1820, 1836, 1850, 1852, 1861) has compiled practically all information dating before Herodotus into his highly informative textbook about the geography of India, Arabia, the Red Sea and Sinai. In two specific books he also compiles the knowledge on the history of geography. He carefully examined all literature concerned and gives clear and well-studied comments. I do not know whether any textbook of similar precision has ever been published in English on this area. All historical terms, which later have been misinterpreted or got lost, were transferred to us namely by Strabo, Diodor of Sicily, the fragments of Eratosthenes and Pliny the Elder. We will deal here mainly with the transfer of information from Strabo to Pliny and into the Ptolemaic picture of the ancient world.

Concerning the Geography of Herodotus, Niebuhr (1816) and Ritter (1820, 1861) give more detailed accounts. Movers (1849) deals mainly with the Phoenicians and the origins of the Erythras cult, which he relates to the origin of the name Red Sea as Strabo did before.

Strabo relied very much on Eratosthenes of whom, unfortunately, very little is preserved. Eratosthenes is so important for us because he believed in the spheroidal shape of the Earth, a belief which was based on mathematical calculations rather than conviction. The authority of Ptolemy, later combined with that of the Roman Catholic church, unfortunately ruled out this knowledge for more than 1,500 years, mainly because the Christian Church adopted his view which made the Earth the center of the Universe and man the crown of Creation.

Thus Galileo and Bruno had to suffer so much because Eratosthenes was defied by Ptolemy.

Strabo (18 A.D.) tells us the following stories about the Red Sea: *There are some writers which say there are only two principal winds but they are not true as are those who say that Homer did not know about several mouths of the river Nile or the isthmus between the Arabian and Egyptian sea at the height of the Ethiopians… Homer adds the words "abiding both where Hyperion sets and where he rises"… the Ethiopians are sundered (sunned) in twain (duplicate).* This means in plain words that already Homer knew about the Equator and that the African people of the Ethiopians (including Zimbabwe) were residing along both coasts of the Arabian Gulf, the Red Sea and the Southern Ocean, and therefore he also speaks of two Ethiopian peoples. The section about the winds also indicates that ancient geographers did not consider alone the directions of the skies but also the different types of winds, such as trade winds and monsoons. Also, Strabo is absolutely clear about the fact that the Arabian Gulf of the Red Sea separates Asia from Africa, and not the Nile which is separated by an immense distance from the other coast of Libya (i.e., Africa). The problem with Homer, Herodotus and Strabo is that the ancients have known about four arms of the Indian Ocean, i.e., the Persian Gulf, the Red Sea proper, the Gulf of Suez and the Gulf of Elat. They also knew about the Straits of Gibraltar, the Straits of Tiran, the Straits of Bab El Mandeb and the Straits of Hormus. Also they knew that the isthmus between the Mediterranean and the Gulf of Suez might have been opened, as the Straits of Gibraltar (Hercules) might have been closed at certain times.

They further knew perfectly well of the possibilities and problems of a canal between the Mediterranean and the Red Sea, which was initiated by Sesostris, built by Nechos, and modified by the Persians. Strabo had a clear understanding of geodynamics, i.e., erosion and sedimentation, regression and transgression of the seas, although plate tectonics may not have been a real substantial idea for him. The Indians, however, had myths about continental drift, which may have reached the Graeco-Roman cultural world as many others did. In these myths they claim that the continents swim on the metal plates of a huge turtle, and that by moving along the plates they would make noise, i.e., earthquakes (see Humboldt's *Kosmos*). At another place Strabo claims strictly: *For even according to Eratosthenes himself the Red Sea is in contact with the Western Ocean (Atlantic).* Thus it is obvious that Strabo believed of Homer and Herodotus that both of them had been well aware of the geographical situation, positioning the Red Sea into the Arabian Gulf, and that they had named the Indian Ocean Southern Atlantic with a fair knowledge of India, East Africa and Ceylon (Taprobane). Strabo also believed that the ancients (i.e., people between 1300 and 600 B.C.) were well capable of sailing over the oceans, and not only along the coasts. Strabo, however, in some of his statements contradicts himself and claims things to be fairy-tales that actually were truth.

As for Taprobane and Libya, there is another clear statement in Strabo: *But this parallel is running on the one side to the South of Taprobane and on the other side to the most southerly regions of Libya.* Strabo in contrast to geographers of the medieval times and till the early nineteenth century (i.e., for more than 1,200 years), knew exactly about the Gulf of Elat and the Gulf of Suez, and the Rocky

Desert between them: *Then near Ascalon one comes to the Harbor of the Gazeans* (Gaza)... *Thence there is said to be an overland of 1260 stadia to Aela* (Elat or Aelanitic bosom), *a city situated near the head of the Arabian Gulf. This head consists of two recesses: One extending into the region near Arabia and Gaza which is called Aelanites after the city situated on it, and the other extending to the region near Aegypt in the neighborhood of the City of Heroes* (Gulf of Suez). In another passage Strabo mentions Moses and his people moving from Egypt to Judea and Lake Sirbonis as the origin of Asphalt. The translators believe that Strabo mixed up Lake Sirbonis with the Dead Sea, which is probably not true, since asphalt and oil were recovered since ancient times nearby at Tur on the shores of the Gulf of Suez. On the other hand, he mentions Sodom a little later. This implies that the asphalt and petroleum trade was in the hands of the Phoenicians, who had access to both places, where these materials could be recovered and traded from Gaza and Ascalon (Nissenbaum 1970, 1977, 1978 a, b, 1979, 1980).

The history of canal construction goes back to Sesostris. Many Egyptian plans and later discussions have been reported and recorded throughout history. None of the early canals was successful for long periods of time. Even the latest canal of Lesseps was blocked several times during its existence and now is relatively unimportant. Peaceful trade and exchange are absolute prerequisites for such a risky venture as the construction of a canal. Bourdon (1925) and Schleiden (1858) describe the history of all canal plans. Both of them also extensively discuss how the people of Israel crossed the Red Sea in this area. The late Prof. Heinz Steinitz, founder of the Marine Biology Laboratory in Elat, intensively studied the zoological and ecological implications of the canal (see Zeitschel 1973, Por 1978 b).

2.3 The Scientific Exploration of the Northern Red Sea with Special Reference to Coastal Hypersaline Systems

The definition of Science is rather difficult. Scientific analyses and theories are looked at very often in terms of some kind of a vague social Darwinism. In the first sections of this introductory chapter I have tried to make clear that practically all basic thoughts of scientific exploration of nature and our world were repeatedly brought forward thousands of years ago. Herodotus and his perception of the world, including his reports of former scientific statements derived from Homer's Ilias, the Egypt Hermetic Code and some of the cosmogonies of the Phoenicians and the Jewish people, which cover about 600 years before him, give a good example of the rise and fall of scientific knowledge. Although the number of words, of known places, and of plants and animals has increased considerably from Herodotus to Pliny, Strabo and Theophrastes, actually a decline can be claimed for scientific views of the world. At Herodotus' time, e.g., the world was a round planet that circled around the sun. It was evident that Africa was surrounded by water, and that ship travels were going around Africa and to Ceylon and the countries beyond. The Indians had a conception of plate tectonics and Herodotus made it clear that he knew about regressions and transgressions of the sea as well as of the eroding and sedimentating forces of the exogenic cycle. Pliny warned in serious and acrid words against environmental pollution and overexploitation of the ores of the Earth. Much of this knowledge got lost and had to be elaborated again.

In connection with the scientific history of the exploration of the Northern Red Sea I wish to quote two passages from Herodotus, which might explain some of the assumptions I have made before. Herodotus tells us (Euterpe 11): *Now in Arabia, not far from Egypt, there is a Gulf of the Red Sea stretching far inland. A ship needs 14 days to run along it, but only half a day to cross it. Daily we observe Low Tide and High Tide. Egypt formerly has been such a bosom, only that this one stretched South from the Mediterranean to Ethiopia, while the Arabian Gulf stretches North from the Red Sea until it reaches the Isthmus separating the two. If the Nile would change its course, it would easily be able to fill it within 20,000 years. Well, one could believe even in 10,000 years. I do certainly believe, what the Egyptian scientists say, namely because I see the fossil molluscs on the hills and sea water coming up, which even corrodes the Pyramids.* Herodotus knows that the Nile Delta is silt and clay and mud, Libya is made of sand and sandstones, and Sinai of magmatic and metamorphic rocks. He tells us that Sesostris was not able to go into the Gulf of Elat because the sea was too wild and reefs and shoals hindered him from entering. Thus, Sesostris went back with his ships and went across Sinai to conquer the countries. He further tells us that Nechos built the Suez Canal which was wide enough to give place to three ships side by side. He says that the Suez Canal was 1,000 stadia long, and he also tells us about the Black Sea being 11,100 stadia long, and that ships were able to travel this distance within 9 days and 8 nights. Now, if we try to transfer these scales into ours of today we have some minor problems, which will also arise when one reads the passages of Ehrenberg at the end of this article. Using Ideler's (1816) and others' equations one may find the absolute truth of these descriptions and statements about traveling speed. One stadium is 568.58 Paris feet or 184.70 m. From here we come to check the distances given by Herodotus for the Bosporus, the Dardanelles, the Black Sea, the Red Sea, and the ancient Suez Canal.

This shows that the average speed of ships at that time summed up to about 6–7 miles per hour or up to 150 miles per day. Herodotus even calculates that ships go about 10% slower during the night and at approximately one-tenth of their normal speed in the Suez Canal. Things that happen also today. Our ships go no faster than at twice the speed of the ancient ships, in average. Confusion arises only because during the time of Strabo, Pliny and Theophrastes the shipping became less fast and the precise knowledge of the world got lost due to Ptolemy. It is very interesting to note that ships at the time of Herodotus were capable of passing through the whole present-day Red Sea or the Arabian Gulf of the Red Sea of Herodotus in only 12–14 days. The reports he gives of the people shipping to and around Libya on the Red Sea make it clear that he was talking about the Indian Ocean itself. The ancient scientific explorers needed months and years for these far-ranging expeditions.

Thus the exploration of coastal areas including both, marine research and terrestrial research, was started more than 2,500 years ago. During the nineteenth and twentieth centuries we got used to separate reports on travels, expeditions and scientific expeditions, and divide these into marine and continental. Remembering the simultaneous traffic on camel backs and ships along the Red Sea we may already see that the same situation holds true for our days. Actually, all reports on the early exploration of the Red Sea, and even more so of the Gulf of Elat, combine both primitive ships and camels as vehicles of exploration. The first reports came from off-course travelers, shipwrecked or strom-deviated people. These were followed by trade exploration and trade; war exploration and war, and only finally – mostly in the engineering squadron of the war-waging nations – scientific exploration. In the brief periods between the wars scientific explorations may have been sponsored by liberal governments after successful wars. This is true for the people of Herodotus, Strabo, the Saracenes, the Turks, the Portuguese, the British and the French (e.g., Napoleon in Egypt and the scientific output of his war), and so on and so forth. It may not be necessary to note that during the period between 1948 and 1983 the same classical principles of behavior of the human species were followed. The Master's Thesis of my friend Mussik Goldberg, 1958, e.g., was a direct outcome of the British-French-Israeli Suez Campaign in 1956.

The marine exploration of the Red Sea under inclusion of the Indian Ocean was already the subject of the previous section to some extent (Rühs 1811). The history of scientific marine expeditions has recently been summarized by Wenzel (1978). Important dates are Forskal (1775a, b), Niebuhr (1781), the first Marine Station at Ghardaqa (1930) and its publication series, the foundation of the Marine Stations at Elat and Aqaba (1967 and 1974), and the expeditions of the ships *Scilla* (Southern Red Sea, 1890 and 1900), *Pola* (1895/96 and 1897/98 southern and northern Red Sea), and *Al Sayad* (Gulfs of Suez and Elat, 1928/29) under Italian, Austrian, and Egyptian-British guidance. The first Israeli Red Sea Expedition took place in 1962 and reached as far down as the Straits of Bab el Mandeb Marine biological research and Israeli geological research in the area, however, goes back to the time of the foundation of the Hebrew University of Jerusalem.

Major ship expeditions, which did not touch the Gulfs of Elat and Suez, have been covered and summarized by Wenzel (1978).

The history of scientific or science-related expeditions through Medieval times till 1850 are very carefully covered by Ritter (1850). Although I have screened most of the reports directly, I rarely found any scientifically important notes in the original, sometimes very poetic, reports of the travelers that were not mentioned by Ritter in his 1,141 pages on Sinai and its coasts along the Gulfs of Elat and Suez. I have already told about Herodotus and Strabo. The geographical notes of Diodorus of Sicily are translated in the same Loeb Library (Oldfather 1958) as also the so-called small Greek geographers. All of them have slightly modified stories and descriptions mainly for the Mediterranean. Of Ptolemy we can only note that he knew more about the Gulf of Suez than about the Eilanitic Gulf, a situation that lasted for almost another 2,000 years (Edward et al. 1975).

The early Byzantine Christian community, which settled in many ancient cities of the Nabateans, built its monasteries also on the shores of both Gulfs. Elat at that time was a bishop's residence and was well known as the place of exile and banning of Christian bishops, who had trouble with their superiors. It reminds me that some Israelis talk about the early days of Elat between 1959 and 1967.

Elat apparently was the city where people could settle instead of going to jail for some crimes they committed in the north of Israel.

Twenty kilometers south of Elat the mountains form a cliff that cuts down to the Red Sea. The route turns into a mountain wadi that winds through a fault and leads down to the beach only 15 km further south. Thus travelers traveling along the coast always report about Faroun or Coral Island but never about the Solar Lake hidden behind this mountain ridge. The same situation is true for the Gavish Sabkha. Thus, all early reports mention in most cases the sabkhas around Tur on the Gulf of Suez, the sabkha lagoon on the Ras Muhammad Peninsula, and some reports also mention the Nabq Oasis and salty springs. They do not refer, however, to the mangrove swamps or the Gavish Sabkha. The Christian and Moslem pilgrim routes went straight across the Sinai mountains and ended either directly in Elat, from where they climbed up the opposite mountains of the Araba rift valley, or they reached the Gulf of Elat close to the Coral Island and a small bay named the Fjord. Many of these pilgrim caravans and of the travelers of the nineteenth century thus must have rested less than 500 m away from the peculiar Solar Lake with its boiling hot water, without realizing it. The Bedouins of that region were probably never showing them this place. To all my knowledge it was David Friedman, director of the Elat Aquarium, who first realized the peculiarity of this sabkha type pool. F. D. Por (1968 b) then published the first paper about it, rejecting the theory of hydrothermal heat and correctly identifying the sun accumulator capacity previously described and mentioned for some Hungarian salt lakes by Kaleczinsky (1904).

The first modern scientist who finally succeeded in the nineteenth century to reach Elat after crossing the Sinai was E. Rüppell (1829). He talks about salt marshes around the ancient city of Elat. These however are at the northernmost tip of the Gulf of Elat. Until his visit many scientists, who were misled by the two cities mentioned (Elat and Aqaba), thought that the Gulf of Elat at its northernmost end was divided into two small bays. The shores of the southern parts of the bay were studied from the sea side by the expedition of the *Palinurus* under the command of Moresby (Wellsted 1840). Niebuhr (1779) made a sketch which shows it only in half of its length, which Seetzen (1812) was able to correct. But this unfortunate man never came directly to Elat, mostly due to problems with the wild Bedouins. So it was Rüppell who under the protection of Mehmed Ali 10 years later made the geographical measurements of Elat, which finally enabled the corrections necessary to establish the exact latitude of Elat. This happened between 28 April and 2 May 1819 (Rüppell 1829, 1838).

The southernmost tip of the Sinai, the Ras Muhammad peninsula, the lagoons of Sharm el Sheikh and Na'ama were visited by Rüppell, Seetzen and Ehrenberg. The most detailed investigations of the elevated Coral Reefs, the tectonic faults and seawater filled cracks were made by Ehrenberg, Rüppell and Seetzen between 1813 and 1833 and much later verified by Walther (1888 a, b). Also Burckhardt (1824) has visited the Ras Muhammad, the Straits of Tiran and Nabq. He probably got closest to the Gavish Sabkha without actually describing it. He then turned west into the mountains and climbed directly to Santa Catarina, beginning his climb in Wadi Nabq (Burckhardt 1824, pp. 840–860). He rested with a Bedouin family close to the Tiran Island. He describes the daughter of that

Bedouin family as a lovely girl, as attractive and as modest in her behavior as any European girl could be. More important, he refers to pearl-fishers of the Tiran Island and to the fossil coral reefs. He does, however, not mention the Gavish Sabkha, although also he rested 4 h further north under the palm trees of the oasis of Nabq. He relates that only a few fishermen live at Nabq all the time, while the Bedouins come down from the mountains only to harvest the dates. In the explanations at the end of his book Burckhardt refers as well to the history of the two horns of the Gulf of Elat. From Santa Catarina Burckhardt climbed down in the direction of Dahab and up again to reach Tabha less than 20 km south of Elat. Thus, he also passed along the Solar Lake without seeing it.

Carless (1837) finally has seen the sabkha near Nabq, reports about the salt brush in the wide plain between Ras Nasrani and Nabq and that the monks of the monastery as well as the Bedouins come to this place to collect salt for various purposes. He does, however, not mention the microbial flora of the sabkha or its dimension. After that no description is given of the Gavish Sabkha for another 130 years. Burckhardt saw similar salt-mining places in the sabkhas at Dahab. When he could not reach Elat and had to turn south, he reached Dahab and describes the palm gardens as well as the salt ponds of the Dahab lagoon. He explains that the Bedouins are wont to close the narrow inlets of such pools by sand and afterwards collect the salt crust that forms on top of the sediment. From our work on beach rocks and sabkhas we know that the whole inner bay of Dahab is covered with a dense carpet of cyanobacteria and other phototrophic bacteria underneath a thin layer of sand and silt, which must have kept the water for the evaporation pans, the same as they were kept intact in the Venetian salt gardens. The Venetians called this carpet the "Petula".

During 1836 and 1837 Dr. Roth and Dr. Schubert were traveling from Santa Catarina to Elat (Schubert 1839). Schubert gives a very clear and poetic report of his trip down to Nueiba and along the coast to Ras Burka. As it happened to us in the years between 1967 and 1980, he says that the Bedouins who guided them wanted to pass the night at Ras Burka. The interest to collect corals and mollusc shells on the beach, however, brought them to rest a few miles further south. Schubert got stuck in one of the salt swamps between El Hamira Bay, the Solar Lake and the so-called Fjord south of Faroun Island. But he says they were forced to go closer to the mountains in order not to sink down in the sabkha.

Thus also Schubert did not find the Solar Lake and touched the Fjord only at its western end. They reached Coral Island in the late morning. He called it correctly Jezirat Faroun, but says that Laborde called it Graia, and their Bedouin guides Abu Sanira Unda el Galga. Schubert himself had no chance to set foot on the island (people at that time were unlikely to swim well) but Wellstedt gave a very detailed report of his visit (Wellsted 1840). Schubert relates that the Arabs of Aqaba were still deeply impressed by the visit of the *Palinurus* at Faroun Island. Thus they spent 14 March 1837 resting and promenading on the shores of Elat and Aqaba. The most important business they had, however, was the change of camels, which he very vividly describes including his impression of the son of the sheikh of the Araba Bedouins, who was sent down to the Gulf of Elat with 16 camels in order to take the caravan to Hebron.

Now let us go back a few hundred miles and a few years to find Dr. U. Seetzen in April 1810 in his search for the ancient Egyptian Suez Canal. He crosses several salt plains. An eye infection he had was painfully aggravated by the clear white of the sabkha plains, to such an extent that the last 2 days of his search he had to travel with his eyes closed by a bandage. His Arab servant, however, reported to him about seven salt ponds, where the Arabs were recovering salt for trade. These seven lakes were very elongated but had all the same width, which would then correspond to the ancient canal. Thus Dr. U. Seetzen, who unfortunately never reached Elat and saw it only from a far, was practically blind when he passed the sabkhas on the place of the ancient Egypt canal. This time he recovered; but he never came back to Europe. Savigny, the French scientist, who had great experience of Egypt and Sinai, suffered through an eye infection in Egypt as well, and was blind for the rest of his life.

In June 1809 Seetzen travels to Aijn Musa and finds it less impressive than the hot springs on the shores of the Dead Sea he visited a year before. He found manna, tasted its characteristic sweetness, and verifies also many other facts of the Bible (see W. Keller 1955). Seetzen's travels were before the time of Mehmed Ali and thus he had daily trouble with his Bedouins and other Bedouins he met. In Tur (or Thor) he stays for a while, has new trouble about camels he paid for in advance, and states that the village has very few inhabitants. Fifteen years later, during the visit of Hemprich and Ehrenberg, it was already much larger.

At Tur Seetzen made the necessary measurements, which together with those he made a year before near Aqaba enabled him to state that Niebuhr had underestimated the length of the Gulf of Elat by 50%. He also visits the hill El-Nakus (hill of the bells). This is the interesting hill described later by at least a dozen of the travelers passing through this region. Darwin mentions Seetzen and Ehrenberg in his diary of the "Beagle" expedition, when he met with similar phenomena in South America. The sand moving down the slopes or being touched by the hooves of horses makes a noise much like music. It sounds very peculiar and sometimes, heard from a far, it reminds one of the ringing of bells. Actually, Nakus does not mean bell, but is a large, flat piece of wood the Sinai monks knock with a hammer when calling for the prayers. In the Moslem area only very few real church bells were allowed to be put into churches.

In the hottest time of the year (on 6 July 1810) we find Seetzen on the watershed between the Gulfs of Suez and of Elat in the middle of the Sinai desert. This was his last try to reach the ancient city of Elat, which he already saw from a far when a year before he went down the Araba valley from the Dead Sea side. On his trip through Sinai he has found the inscriptions with the ancient Hebrew alphabet going back to the time of 1300 B.C., and described them in detail. It took him exactly 4 days (4–8 July) to travel from Tur to Nueiba on the shores of the Gulf of Elat. But neither money nor pleas changed the mind of his guides. He could not convince them to enter the area of the Wassit Bedouins with whom also other travelers later had a lot of problems. He did not reach the Faroun Island which according to his guides was called El Kassar Haddid, i.e., iron fortress. With miserable feelings he turned south toward Dahab. On 12 July 1809 he reached Nabq (Nebke), describes its palm trees and the Island of Tiran, the elevated fossil coral reefs, but does not mention the sabkhas in his letters.

It shall be noted here, that Seetzen's letters were published in many issues of "von Zach's Astronomic Correspondence", in which between 1800 and 1815 many important scientific reports were published. At that time the astronomical Observatory of the King of Saxony was one of the research centers of the world. Unfortunately, many of the notes, diaries and collections of Seetzen got lost, as he himself did not return from his last voyage in Africa.

U. Seetzen originated from the Frisian city of Jever, which at that time belonged to the Russian Tsar. Some of the funds he could use besides his own modest fortune were actually donated by the Russian crown. If at that time the University of Oldenburg would have already been in existence, Seetzen had certainly been a faculty member, Oldenburg being located less than 50 km from Jever.

In the region between Sharm el Sheikh and Tur Seetzen was hunted by the Sheikh of Dahab, but people of the small trible of the Beni Washil helped him with food and water and he escaped for another time. In Tur the Christians or the Bedouins have stolen from him a precious small sextant (a gift of count Rzewusky in Vienna). From Tur he, who had already a while before embraced the Moslem religion (perhaps not only for practical reasons), starts on a successful pilgrimage to Mekka and circles the Ka'aba the prescribed seven times, runs down seven times the runway and gets his hair cut, and from then on is entitled to wear the title of a Hadji (Seetzen 1813). Later he died in that region under unknown conditions.

Thus in all reports of the early scientific exploration of the Sinai shores, the coral reefs as well as the sabkhas and the salt-mining activity of the Bedouins were described. Only one report, however, deals also with the red color of the sabkhas and with possible organic contributions to the whole system (see Introduction).

Later reports of importance are by Hume (1906) and Picard (1958), concerning geology, and by Walther (1888a, b, 1891, 1924), concerning reefs, sabkhas, denudation, rock varnish or desert varnish, and possible microbial origin of the ooids of the Gulfs of Elat and Suez. G.M. Friedman (1968) gives a detailed description of the geology of the Gulf of Elat and includes the most important references on former expeditions. Emery (1964) briefly describes the sediments of Elat, but a that time he could not travel further south. My friend Mussik Goldberg did his M. Sc. thesis (written in Hebrew) on the geology of Tiran on the occasion of the first Campaign (Great Britain, France and Israel moved to the Suez Canal and were forced to withdraw immediately by a joint action of the U.S.A. and U.S.S.R.).

Only since 1967 were scientific papers produced in ever increasing number on all aspects of the geology, biology and ecology of the coastal plains as well as the northern Red Sea, its sediments and biology. Let me close this introductory historical sketch on the scientific exploration of the Red Sea and the origin of its name with a description given by the geomicrobiologist Ehrenberg in one of his letters from Tur on the shores of the Gulf of Suez. As in many other cases, only two persons of his excursion came back to Germany. Also, much of the material collected by Hemprich and Ehrenberg got lost in transportation and through stupidity of administration back in Prussia. Much of the material had been sold long before Ehrenberg could work on it after coming back. This made him so angry that he never really finished a detailed report of all his findings during this expedition. In one of his publications he reports about working conditions close to the reef. In a storm their bark was thrown onto the reef and was hit seriously four times before by sheer luck they were thrown back into the sea. He records that the highest waves he ever has seen in his life piled up during a storm between the Straits of Tiran, Sharm el Sheikh, and Ras Muhammed. He traveled by the North Sea, the Baltic, the Atlantic, the Mediterranean, and the Caspian Sea. Ehrenberg reports also of stromatolitic sabkha type sediments among the elevated reef complexes of the Tiran Island (Stresemann 1959).

Let me finish this section about the history of scientific exploration with a section of Ehrenberg's work, which describes some of the conditions of work for these daring scientists and also deliminates our problems of analyzing the earlier reports: *If in the sampling bottle I find 800,000 millions of infusoria per cubic inch (German inch, which differs from the American inch; Author), a drop consisting of 1 cubic line will thus contain very many organisms of the size of 1 thousandth or 2 thousandth of a line. They sum up to about 500 millions of individuals just about as many as there are men on Earth... I may add that at none of the places in Egypt and West Asia where I made these investigations, the number found corresponds to the actual number of organisms. Often I could set up the microscope only very secretely, because it attracted the interest of the Arabs. Astronomical and geographical measurements are easier to do, although also here the Arabs like to touch or steal the brass instruments they take for gold; this is due to the fact that, like many other people of this area, they have a high appreciation of astronomy. The microscope and its use, however, always evokes the impression of magic and we had to be very cautious for this reason. In places where we had to fear nothing from the people with and around us, we were also heavily handicapped by the lack of housing, wind, dust, travel inconvenience, eye disease and disturbance. Also serious diseases and many other inconveniences of our expedition made the detailed study extremely difficult.*

Maybe in a future that spans as many years from us to following generations as were spanned between Herodotus and Ptolemy, or Ptolemy and Ehrenberg, new scientists will again explore the then no longer used the Suez Canal of the 20th century. Maybe they will dig a new canal and perhaps again they will not be able to find that small gulf of the Red Sea which is so difficult to reach from the Dead Sea, from the Straits of Tiran or via Sinai. But how beautiful it is!

3. Gulf of Elat (Aqaba) Geological and Sedimentological Framework

GERALD M. FRIEDMAN

The purpose of this chapter is to provide an overview of the Gulf of Elat (Aqaba) with respect to its geology and sedimentology. Scientific interest in the Red Sea and its marine embayments, the Gulfs of Elat and Suez, has been abundantly evidenced by nearly continuous studies over the years from the middle 1800's to the present day. The present high level of scientific activity suggests that more studies may be actively underway in the area today than have been at any time in the past. The subjects of these research projects, such as the many fine Gavish Sabkha studies reported in this volume, are necessarily restricted in their scopes either to a limited region or to a particular aspect of science. The geological

Fig. 3.1. View north across the Sinai Peninsula and the northern Gulfs of the Red Sea (Gulf of Suez and Gulf of Elat) taken from Gemini XI. (Photo from NASA)

and sedimentological framework of the Gulf of Elat is provided here as a partially completed jigsaw puzzle that gives the general picture of the region as it looks when most of the pieces (results of individual research projects) thus far revealed have been assembled to the best of our ability at this time. It is hoped that this overview will help the readers to see where the results of each new piece of research reported fits and how the results add to the quality of the overall mosaic.

The Gulf of Elat and its western neighbor, the Gulf of Suez, are relatively small marine embayments extending from the northern end of the Red Sea in a "V" to partially enclose the wedge-shaped Sinai Peninsula (Fig. 3.1).

The geological and sedimentological framework of the Gulf of Elat will be described in sections dealing with historical background; bathymetry and topography; tectonic framework; climate; winds and waves; currents and tides; and environments of deposition. Environments of deposition include subaerial fans; sources of terrestrial sediments; sabkhas; reefs; zonation of sedimentary deposits and organisms; carbonate sands; and the sea floor including marginal slopes and basins.

Also considered, in addition to the review of what is understood, are aspects of the sedimentology and geology of the Gulf which are most in need of further study either to fill in gaps in our knowledge or to resolve conflicts between different opinions.

3.1 Recent Scientific Studies

Studies of reefs and carbonate environments in the Red Sea date back 130 years (see Chap. 2). Recently G.M. Friedman (1968, 1972), Goldberg (1970), Nir (1971), and Sneh (1978, Ph.D. thesis) started the geological study of the modern reefs and the older reef terraces. Nesteroff (1955) studied coral reefs, carbonate sediments, and beach rock in the southern Red Sea, and summarized the recent extensive literature on these subjects. Gvirtzman et al. (1977) made a comprehensive study of the morphology of reefs of the Red Sea and the northern Gulfs.

Coral reefs along the Gulf of Suez and Elat have been, once again, extensively studied in the last several years; for the nearly 60 years between Hume's 1906 study and that of Emery in 1964 no reef studies were conducted in the Gulf of Elat. Among the recent studies are: Erez (1972), Fishelson (1973), G.M. Friedman (1965, 1968), G.M. Friedman and Brenner (1977), G.M. Friedman et al. (1973b, 1974), Gvirtzman and G.M. Friedman (1977), Gvirtzman et al. (1977), Loya (1972), Loya and Slobodkin (1971), Mergner (1971), Mergner and Schuhmacher (1974), Scheer (1971), Sneh and G.M. Friedman (1980), and Spiro (1971).

Sass et al. (1972) studied intertidal and lagoonal ooids from Ras Matarma. Beachrocks from Ras Sudr, consisting of ooids, were studied by Helsinger and G.M. Friedman (1973), and other kings of beach rocks were studied by G.M. Friedman and Gavish (1971), Gavish (1975c) and Krumbein (1979a). A brief description of sediments taken from the Ras Muhammad area is given by Gavish et al. (1969). Por (1968b) studied the hypersaline sea-marginal pool near Elat. G.M. Friedman et al. (1973a, b) studied laminated evaporites from sea-marginal pools near Elat and another at Râs Muhammad. Gavish (1974 a, 1975 a) studied the environment of the Nabq sea-marginal pool (now called Gavish Sabkha) and another small sabkha near the Belayim lagoon (Gavish 1974 b). Emery (1964) made a study of sediments along the Israeli shore; in his descriptions he included several samples of the Mahabith and Vema expeditions and two shore samples from the southwestern Gulf. Horowitz (1967) studied spores and pollen of the northern Gulf, and Perath (1966) discovered supposedly Neogene to Pleistocene continental-lacustrine sediments on its western shore. Reiss et al. (1980) studied five deep-sea cores from the Gulf of Elat and one from the northernmost Red Sea.

Studies of carbonates forming in modern environments that are also hot and dry include Persian Gulf studies by C. G. Kendall and Skipwith (1969), Kinsman (1964, 1966), Purser and Evans (1973), and Purser and Seibold (1973). Similar conditions in Shark Bay, Australia were studied by Davies (1970 a, b) and Logan and Cebulski (1970).

As early as the first half of the nineteenth century travelers and scientists made observations on meteorological and hydrographic conditions in the Gulf of Elat. Rüppell (1829, p. 185) described the wind conditions, as did Hume (1906, p. 17) and Commander Moresby of the Indian Navy (reported in Carless, 1837, p. 27). Carless (1837) also made meteorological measurements.

The bathymetry of the Gulf of Elat also received early attention; the entrance to the Gulf of Tiran Island was charted by mariners at the beginning of the nineteenth century. Ritter (1847) made a detailed compilation of this early information. In the middle of the nineteenth century the bathymetry of the Gulf of Elat was mapped by the British Admiralty (Red Sea Sheet I, Gulf of Suez, Strait of Jubal) on the scale of 1:655,000 under the direction of Captain Nares, the Technical Director of the Challenger Expedition. British surveys to 1911 on a scale of 1:645,430 provide additional bathymetric details. The U. S. Navy hydrographic map (H. O. 2812), which includes the Gulfs of Suez and Elat, is based on British surveys to 1911 with additions from other sources to 1950. The old Nares map (British Admiralty Chart No. 757) is cited as one of the sources for the up-to-date U. S. Navy H. O. 2812 map of the area.

Many expeditions have provided more recent information on the Gulf of Elat. These include the *Pola* expedition of 1895 (Luksch 1901, Natterer 1898, 1901), the *Magnaghi* expedition of 1923–1924 (Picotti 1930, Vercelli 1927, 1931), the *Mahabith* expedition of 1934–1935 (Crossland 1939, Mohamed 1940), and the *Manihime* expedition of 1948–1949 (Deacon 1952). The R/V *Vema* of Columbia University collected a core of bottom sediment from the Gulf of Elat (S. Friedman 1960). Badr and Crossland (1939) included information on the topography of the Gulf floor and Schick (1958) studied the geomorphological aspects of the Gulf. Goldberg (1958) studied the geology of Tiran Island at the entrance of the Gulf of Elat.

The bathymetry of the Gulf of Elat has now been mapped in detail (J. Hall and Ben Avraham 1979) based primarily on the data from the cruise of the R/V *Ramona*, 1976 but also data from other cruises: R/V *Vema*, 1958; R/V *Aragonese*, 1961; R/V *Atlantis* II, 1977 and detailed studies in the vicinity of the Straits of Tiran (J. Hall 1975).

Neumann (1966) and Neumann and McGill (1962) provide information on water circulation in the Gulf of Elat and the Red Sea.

In recent years, many scientists have made systematic meteorological, hydrographic, and biological studies along the northern Israeli shoreline of the Gulf of Elat. Publications on meteorological and hydrographic research include those of Rosenan (1951), O.H. Oren (1962b), and Ashbel (1951, 1963). In the biological field, many contributions are found in the publications of the Sea Fisheries Research Station, Division of Fish, Ministry of Agriculture of the State of Israel. A particularly valuable series is entitled *Contributions to the Knowledge of the Red Sea.* A comprehensive survey of the oceanography and productivity of the Gulf of Elat has been compiled by Hottinger and Reiss (1984).

Early sedimentological studies focused on the Red Sea and its gulfs simply because carbonates form there. A greater understanding of global tectonics has broadened the interest in this area where a new ocean is actually in process of forming. Studies of such phenomena as the side-by-side deposition of carbonates and terrigenous clastics are now felt to give insights into processes that must have existed during the early stages of the other oceans (e.g., the Atlantic Ocean during the Cretaceous; Epstein and G.M. Friedman 1983, Friedman 1982).

3.2 Bathymetry and Topography

The Gulf of Elat is about 180 km long and 23 to 25 km wide in the south, narrowing to 16 to 17 km in the north. A very accurate bathymetric map now exists for the Gulf of Elat (J. Hall and Ben Avraham 1978). This map was based mainly on the geophysical survey conducted aboard the R/V *Ramona* during 1976 (Avraham et al. 1978, p. 242) and is presented here as simplified by them (1979, p. 246; Fig. 3.2). Data was incorporated from other cruises: R/V *Vema* (cruise 14, 1958), *Aragonese* (cruise concrete, 1961) and *Atlantis II* (cruise 93, 1977), plus J. Hall's (1975) detailed (Straits of Tiran) surveys.

Ben Avraham et al. have described the general bathymetry as follows (1979, p. 243–246):

"The interior of the Gulf of Elat is occupied by three deep and elongated basins, striking N20–25° E, which are arranged in echelon. The basins are separated by low sills. Thus the Gulf is divided into three distinct parts. The floors of the basins are undulating, so several distinct deeps are formed (Fig. 3.2). The northern part of the Gulf has a relatively simple bathymetry, and is dominated by the flat-bottomed Elat Deep, about 50 km long and 3–8 km wide, which is the largest deep in the Gulf, but also the shallowest (only 900 m). The central basin is 35 km long and 6 km wide. It includes the deepest points in the Gulf, the Aragonese (1850 m) and Arnono (1550 m) Deeps, which are separated by a shallow

→

Fig. 3.2. Bathymetric map of the Gulf of Elat, simplified after J. Hall and Ben-Avraham (1978). The map is based on the geophysical cruises shown in Fig. 3.1 plus data in the vicinity of the Tiran Straits (J. Hall 1975). Contours are at 100 m interval. Values are in corrected meters (Ben Avraham et al. 1979)

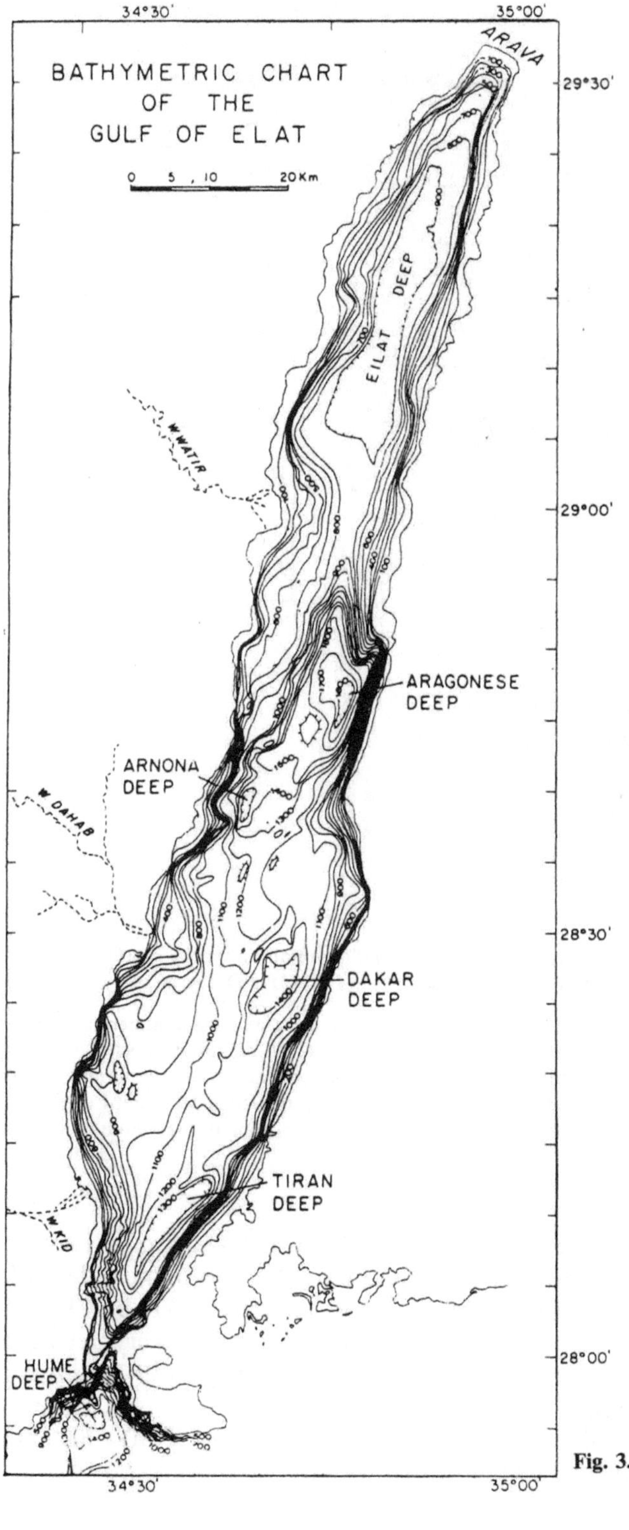
BATHYMETRIC CHART
OF THE
GULF OF ELAT
0 5 10 20Km
34°30'
35°00'
ARAVA
29°30'
EILAT DEEP
29°00'
W. WATIR
ARAGONESE DEEP
ARNONA DEEP
W. DAHAB
28°30'
DAKAR DEEP
TIRAN DEEP
W. KID
28°00'
HUME DEEP

Fig. 3.2

Fig. 3.3

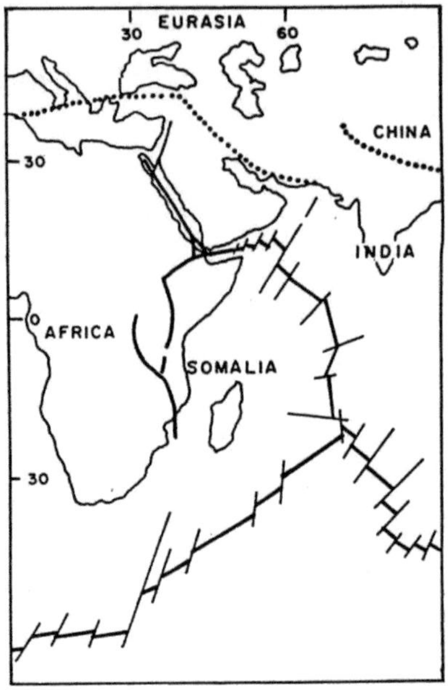

Fig. 3.4

Fig. 3.3. Coast of Elat with narrow fringing reef and mountainous rocky desert rising abruptly from the shoreline to 900 m (Sneh 1978, Ph. D. thesis)

Fig. 3.4. Mid-oceanic rift system in the Indian Ocean running into the continent as the Red Sea Rift and ending at the head of the Red Sea where it bifurcates into the Gulfs of Suez and Elat. The plate boundary extends from the Red Sea rift along the transform-fault zone extending north through the Gulf of Elat as the Gulf of Elat–Dead Sea Rift; *heavy lines:* extensional (spreading center) plate boundaries; *thin lines:* transform faults; *dotted lines:* consuming (collisional) plate boundaries. (After Le Pichon et al. 1973)

intrabasin sill. These deeps are 6–8 km by 2 km and 4–5 km by 1 km in size, respectively. The southern basin, about 55 km long and 6 km wide, includes the Dakar (1,400 m) and Tiran (1,300 m) Deeps. They are 18 km by 4 km and 18 km by 3 km in size, respectively. East of the southern basin is a very steep and narrow slope, while west of it is a terrace, 20–25 km long and up to 10–12 km wide, which slopes gently to the east.

"South of the Straits of Tiran is the Hume Deep (1,450 m) which has an oval outline. This deep links the Gulf of Elat with the axial depression of the Red Sea.

On the northern side this deep sea bottom has very steep slopes of more than 40°" (J. Hall 1975)

At the Straits of Tiran the Gulf is separated from the Hume Deep and the Red Sea by a threshold whose present maximum depth is 252 m (J. Hall 1975).

Except for an area along the southwest shore, coastal plains are all but missing, as the adjacent rocky desert lands continue the steep submarine slope upward from the deep basins, rising precipitously to heights of 1.5 km above sea level within 10 to 15 km of the shoreline (2.5 km farther inland, Fig. 3.3). Erosion has accentuated the faults bordering the graben occupied by the Gulf and its marginal blocks, but the faults show only very localized evidence of recent activity, the scarps which are so prominent are fault-line scarps.

The Gulf of Suez is very different from the Gulf of Elat in that it occupies a wide valley with gently sloping coastal plains. Water depths reach only 50 to 70 m, except near its connection with the Red Sea, and no threshold separates the two bodies of water.

3.3 Tectonic Framework

From a tectonic viewpoint, the Red Sea is well known as a place where the most extensive of the Earth's structural features, the midoceanic rift system, abuts a continental platform (Fig. 3.4). In the oceans, the rift system is a spreading center widening the ocean basins; in the Red Sea, it is breaking up the once continuous Arabian-African platform (Avraham et al. 1979). Beyond the north end of the Red Sea the rift system bifurcates, with the west branch of the V occupied by the Gulf of Suez, the eastern by the Gulf of Elat (Fig. 3.1). At its junction with the Gulf of Elat, the Red Sea spreading center is thought to end and the tectonic plate boundary continues as a transform-fault zone; a situation known to prevail at only one other place, where the Pacific Ocean mid-oceanic ridge ends in the Gulf of California and is continued by the San Andreas fault zone. Extending from the Red Sea via the Gulf of Elat 600 km north to the Zagros-Taurus Mountains of Turkey, the transform-fault zone is known as the Gulf of Elat–Dead Sea Rift; with the Red Sea rift it is splitting the continental mass into the Arabian and African plates. From place to place along its length the Gulf of Elat–Dead Sea rift is marked by major grabens formed primarily by strike-slip faulting, the most southerly and most spectacular of which is occupied by the Gulf of Elat. Three in-echelon grabens formed by strike-slip faulting make up the three deep basins at the Gulf (Figs. 3.5 and 3.6). Avraham (1983) believes that a transition from the Red Sea spreading center to the strike-slip faulting of a transform-fault zone occurs within the Gulf of Elat, the central and northern Gulfs being more typical transform-fault basins differing substantially from the spreading-center character of the southern one-third of the Gulf. To the north other transform-fault zone grabens are occupied by the Arava Valley, the Dead Sea and the Sea of Galilee (Fig. 3.7).

Work by Avraham, Almagor and Garfunkel (Ben Avraham et al. 1979) has refined our understanding of the origin of the Gulf and the surrounding region. The following explanation is based primarily on their studies.

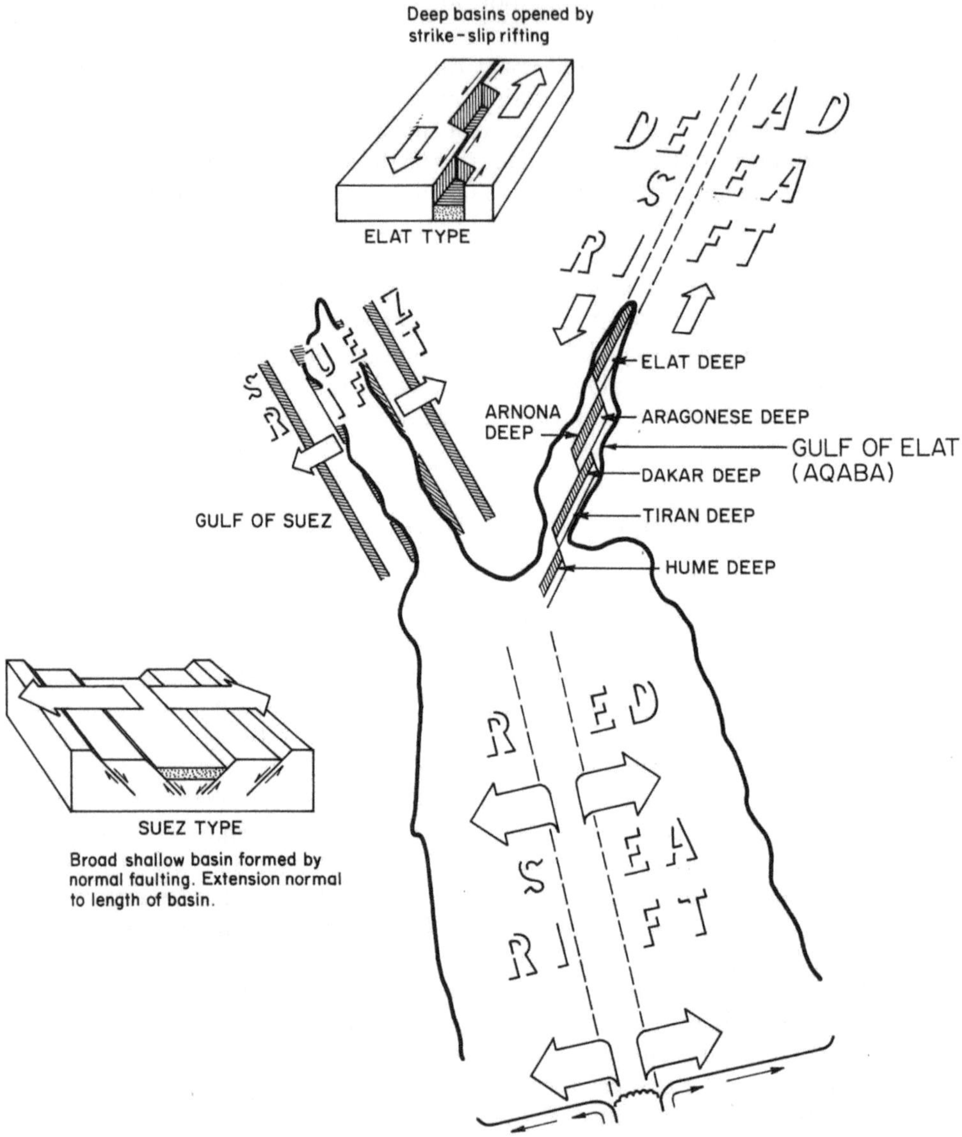

Fig. 3.5. Tectonic framework of the Red Sea and the Gulfs of Elat and Suez. The Red Sea is a spreading center, an extension of the mid-oceanic rift system along which new crust is forming and oceans are growing. The Gulf of Elat is formed by strike-slip faulting typical of its transform-fault zone, but owes its great depth to rhomb-shaped grabens formed as illustrated in *inset*. The Gulf of Suez occupies a graben formed by normal faulting and has been characterized as a failed arm or aulacogen. (Figure devised by M. Raymond Buyce, based on Ben Avraham et al. 1979)

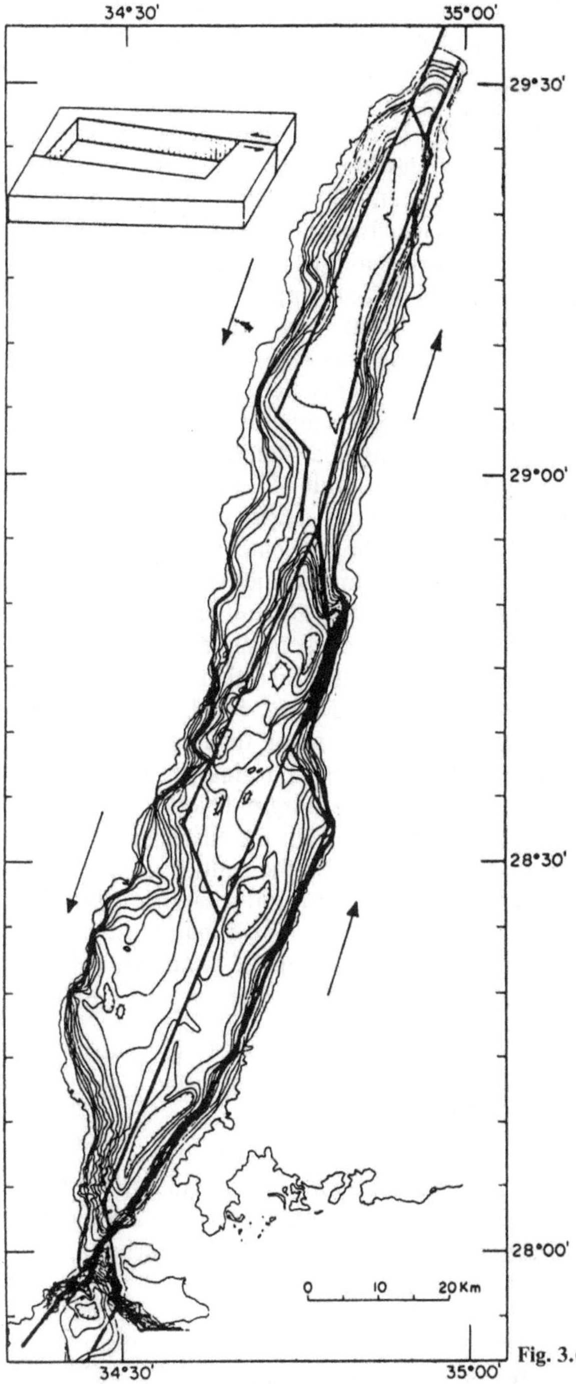

Fig. 3.6

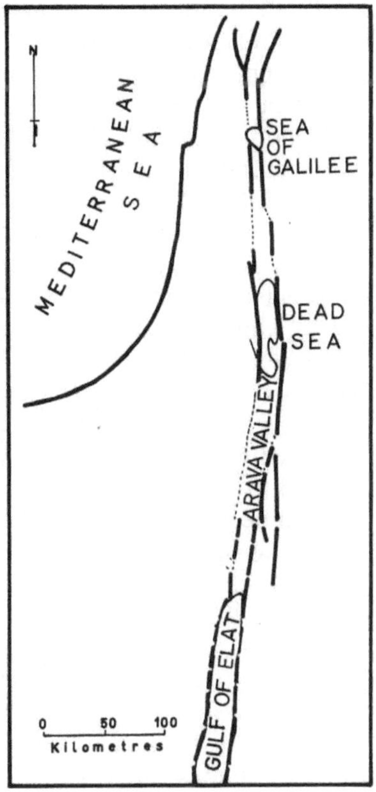

Fig. 3.7

Fig. 3.6. Generalized model for structure of Gulf of Elat; the three basins are interpreted as three in-echelon rhomb-shaped grabens produced by strike slip (see *inset*) (Ben Avraham et al. 1979)

Fig. 3.7. Gulf of Elat-Dead Sea Rift – a series of grabens marking the transform-fault zone of strike-slip faults 600 km from the Red Sea, north beyond the Sea of Galilee

In all, about 105 km of left-lateral displacement has occurred on the Gulf of Elat–Dead Sea rift (Freund 1965, Freund et al. 1968, 1970, Quennell 1959). Movement took place in two stages, as did the opening of the tectonically related Red Sea and Gulf of Aden (Girdler and Styles 1974, Laughton et al. 1970, LePichon et al. 1978, Phillips and Ross 1970). In the first stage, 65 km of left slip occurred possibly during the Miocene or somewhat earlier (Freund et al. 1968, Garfunkel et al. 1973). The Gulf of Suez graben was downdropped by normal faulting at this time (Fig. 3.5). Over the lands which now border the southern Gulf of Elat, especially on the eastern side, evaporites and other marine sediments of Miocene age accumulated (Bentor et al. 1974, Bramkamp et al. 1963, Goldberg 1958). Uplift of lands adjacent to the rift zone could have created as much as a kilometer of relief which was subsequently eroded nearly to base level, exposing large areas of Precambrian basement rock. Along the eastern side at the southern Gulf, the Miocene evaporites and other marine sediments survived the erosion.

The final stage of rifting, involving 40 to 50 km of left-lateral movement, took place 4 to 5 million years ago, shaping the present Gulf of Elat. At that time the present deep basins formed as three in-echelon rhomb-shaped grabens formed by the strike-slip faulting (Figs. 3.5 and 3.6). Strike-slip movement on the rift is still active as shown by geologic, geomorphic and seismic studies (Ben-Menahem et al. 1976, Dubertret 1932, Freund and Garfunkel 1976, Garfunkel 1970, Zak and Freund 1966). The marginal faults, active in the first stage of movement, have been relatively inactive since that time, as shown by the fact that overlying thin terrestrial sediments which are widespread on both sides of the Gulf are essentially undisturbed. Thus, as in the case of the Red Sea spreading, the tectonism here is not continuous but episodic in nature. Between the marginal faults and the younger faults responsible for shaping the Gulf lie the marginal blocks. Relatively narrow marginal blocks and very steep slopes prevail on the eastern side of the Gulf, whereas relatively wide marginal blocks underlie the gentler slopes of the Western side. Faults on the marginal blocks parallel the major faults of the Gulf and are still active. Even the marginal faults are active locally but, for the most part, where they are dramatically exposed in areas flanking the basin, it is as fault-line scarps which show no evidence of recent movement.

3.4 Climate

Climatic conditions strongly effect sedimentation in and marginal to the Gulf of Elat with respect to the influx of terrigenous sediment, the accumulation of supratidal evaporites and microbial mat carbonates, and intrabasinal carbonate deposition associated with fringing coral reefs and pelagic organisms.

3.4.1 Temperature, Precipitation and Evaporation

The Gulf of Elat lies in an area of exceptionally intense evaporation. The climate is hot and dry, and rainfall is scarce. Figure 3.8 shows the mean air temper-

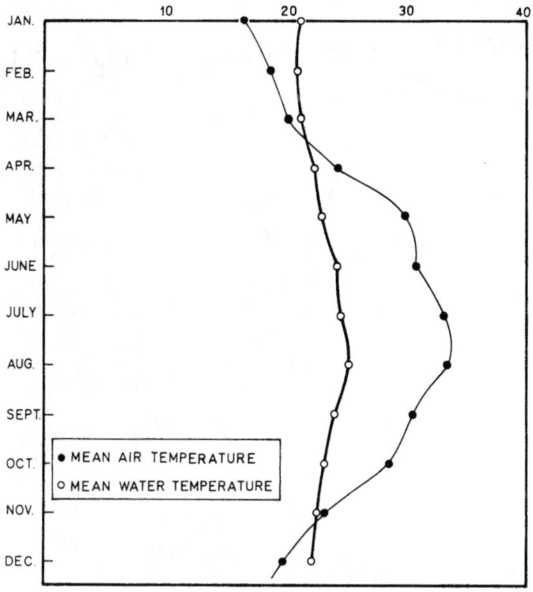

Fig. 3.8. Mean air and water temperatures, northern Gulf of Elat. (Data from Ashbel 1963)

ature at the City of Elat and the water temperature in the Gulf of Elat at Elat. The mean annual air temperature is 25.7 °C. Figure 3.9 is a rainfall map of the Gulf of Elat and contiguous area and shows the extreme aridity of this region. All of the rainfall in the Gulf area occurs within just a few days, mostly in December; during some years none falls. Measured during a 20-year period, the average annual rainfall was about 25 mm for the Gulf of Elat (Red Sea and Gulf of Aden Pilot 1967, Lebedev 1970).

An evaporation map (Fig. 3.10) of the Mediterranean–Red Sea–Gulf of Elat area shows that at Elat and along the northern Gulf of Elat evaporation is three times greater than along the coastal plain of the Mediterranean. Climatological normals for the area (Meteorological Service, 1952) document the large deficit between the amounts of precipitation and evaporation. Rainfall at the northern end of the gulf is normally 22 mm per year. Evaporation is 179 mm, eight times precipitation, leaving a deficit of 157 mm per year.

Inflow of waters from the Red Sea balances the deficit. Neuman and McGill (1962) report that the inflowing current from the Red Sea is a less saline surface flow, whereas the more saline (denser) waters resulting from the large evaporation–precipitation ratio sinks and forms a counterflow to the south. The salinity and water temperature distribution with depth in the nearby Gulf of Suez (Fig. 3.11, after Mohamed 1940) is most likely similar to that in the Gulf of Elat, and is consistent with an inflow of less saline surface waters and a deeper, more saline outflow. A sill at the Straits of Tiran (depth 252 m, J. Hall 1975) separates the Gulf from the Red Sea and restricts deep circulation.

A similar situation exists between the Red Sea and the Indian Ocean, where the evaporative loss of water from the Red Sea is being constantly replenished by

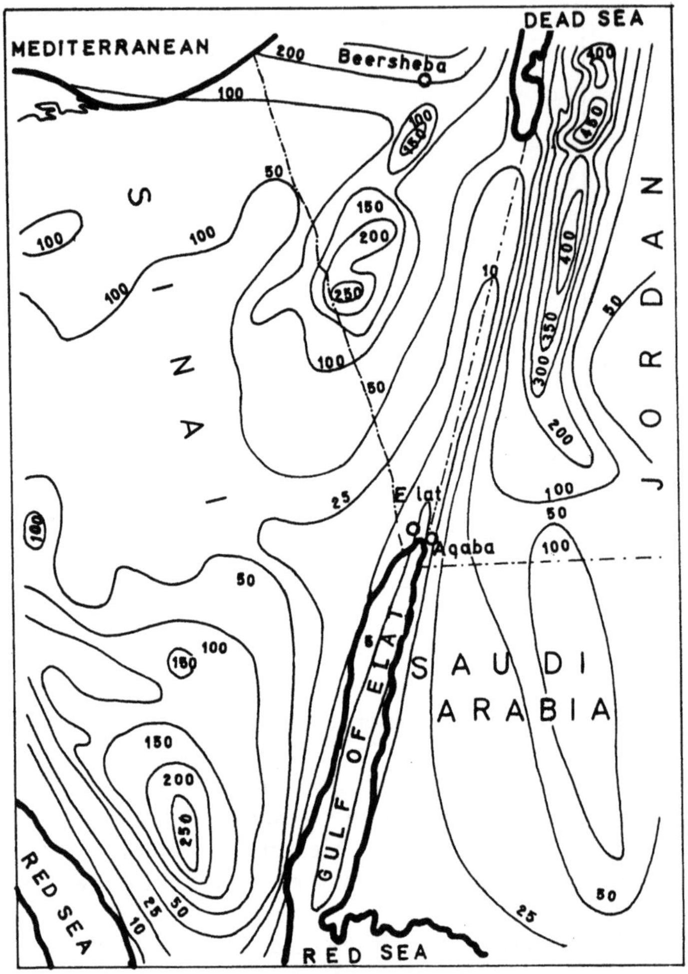

Fig. 3.9. Rainfall map (mm/year) of Gulf of Elat and contiguous area, showing extreme aridity in area of Gulf. (After Ashbel 1951, p. 42)

inflow from the larger body of water. A threshold at 100 to 125 m depth limits circulation between the Red Sea and the Indian Ocean. As a result of replenishment by less saline waters from the south and increase of salinity by evaporative loss across the entire water surface, the salinity increases in the Red Sea from south to north (U. S. Navy Hydrographic Office, 1960, pp. 58, 59) and from south to north in the Gulf of Elat (40‰ to 42‰, Fig. 3.12).

3.4.2 Winds and Waves

Northerly and northeasterly winds predominate in the Gulf of Elat. However, on some winter afternoons, especially in February, the winds change and blow

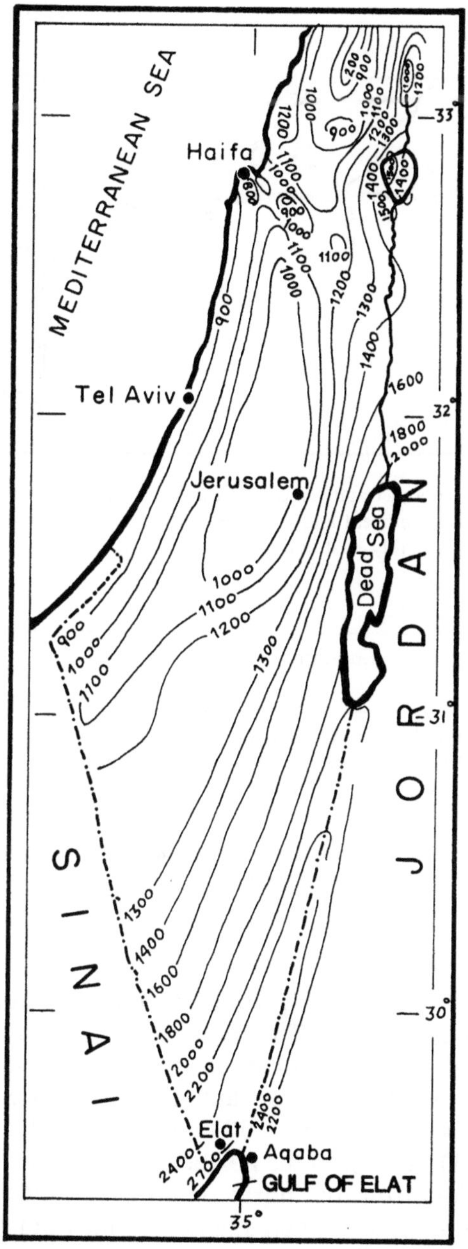

Fig. 3.10. Map of annual evaporation (piche × 0.5) of northern Gulf of Elat and adjacent areas. (After Ashbel 1951, p. 60). Although the numerical values on Ashbel's map differ from those of the Meteorological Service (1952), the trends of the lines are the same

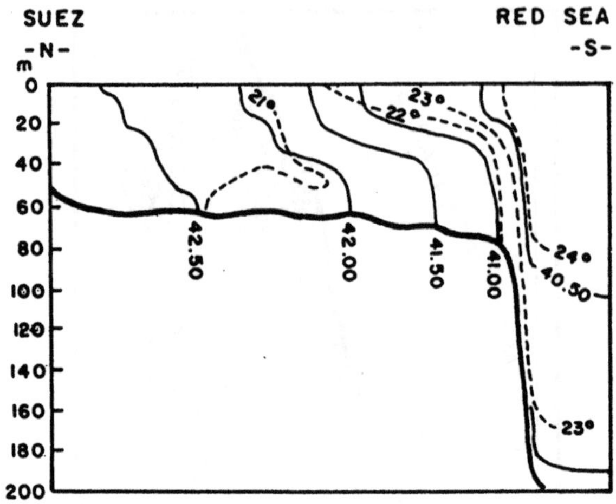

Fig. 3.11. Salinity distribution and water temperatures, Gulf of Suez. Patterns suggest a deep south-flowing cool and saline (dense) current and a fresher warmer north-flowing surface current. Similar conditions probably exist in the Gulf of Elat. (After Mohamed 1940)

Date	Location	Salinity ‰
April 30, 1963*	Straits of Tiran (southern Gulf)	40.2
April 30, 1963*	28°42′N 34°42′E (approx. central Gulf)	40.6
March 25, 1964	6 km south of Elat (northern Gulf)	41.9

Fig. 3.12. Salinity of surface seawater in southern, central, and northern Gulf of Aqaba. Chemical analyses by the Geological Survey of Israel. (*) Samples collected by Y. Nir of the Geological Survey of Israel (G.M. Friedman 1968)

from the south. The strong southerly winds of winter are said to be the most dangerous for navigation on the Gulf. The annual mean wind speed at Elat is approximately 10 knots. High wind speeds have been recorded from the entire Gulf and particularly strong winds blow at Elat and at Tiran, especially at night. On 17 May 1971, a southerly wind raged for 30 min at 150 km/h (approx. 80 knots) at Elat (Sneh 1978, Ph. D. thesis). In the northern Gulf, winds in excess of 34 knots almost invariably blow from the north or northeast (Meteorological Service, 1956, pp. 50, 51); however, south and southeast winds can attain speeds of 7 on the Beaufort scale (28 to 33 knots). For the period of November 1957 through August 1961 only 14 of 2,802 observations of wind frequency in the northern Gulf showed wind speed in excess of 20 knots from the south during the winter months, 2 of 2,806 observations during spring months, and none during summer and autumn (for 2,819 and 2,817 observations, respectively) (pers. comm. Uri Mane, Israel Meteorological Service; 1 July 1964).

Measurements of wind direction and force in the northern Gulf made by the Meteorological Service (1956) through two periods of 3 yearch each (1945–1947,

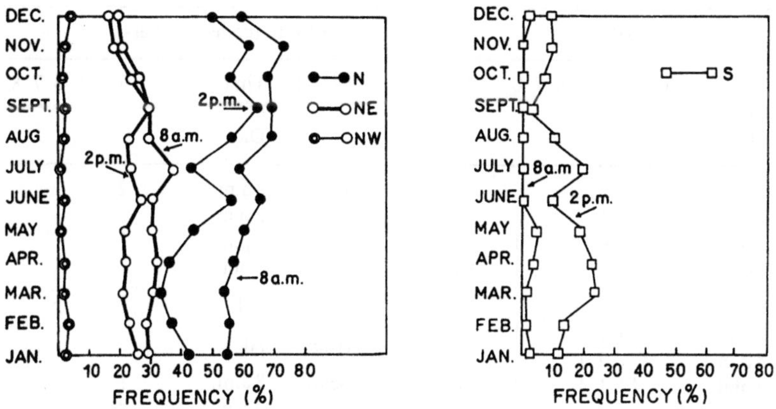

Fig. 3.13. Wind frequency and wind direction, northern Gulf of Elat near City of Elat. (After Meteorological Service 1956)

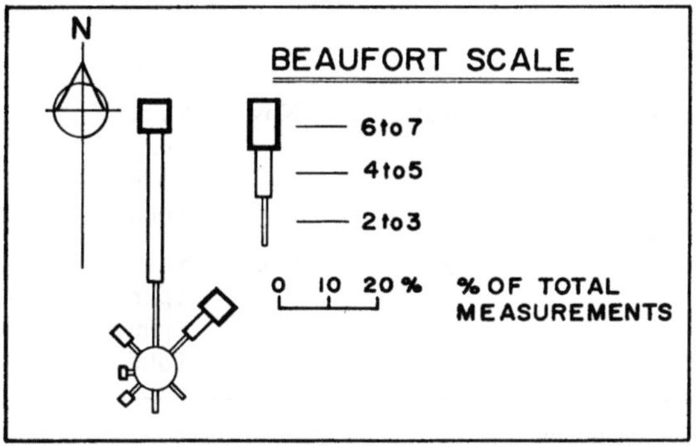

Fig. 3.14. Wind direction and wind velocity, northern Gulf of Elat near City of Elat. (After Meteorological Service 1956)

1950–1952) are summarized in Fig. 3.13. Morning (8 a.m.) and afternoon (2 p.m.) measurements are shown. Because of their close similarity to the morning measurements, evening (8 p.m.) measurements made on the same days have been omitted. The rose diagram (Fig. 3.14) shows the annual wind frequencies and strengths.

In the Gulf, the sea is usually calm; only rarely do wave heights exceed half a meter. Figure 3.15 summarizes data on sea state. These data show that despite the usual calm, on occasion, in the months of January and February, waves 9 to 14 m high have been recorded (G. M. Friedman 1968).

Winds and waves are important factors in sedimentology of the Gulf of Elat. The chimney effect of northerly winds blowing steadily for days on end along the Arava graben toward the Gulf of Elat results in southward transport of sand and its eventual dumping into the Gulf (G. M. Friedman 1968). Commonly, this sand movement can be observed along the southern Arava Valley. Winds also create waves and drive longshore currents, both of which affect sedimentation. Currents redistribute terrigenous debris brought into the gulf by wadis (after the detritus

Sea State (code numbers)[a]									
Month	Time of Day	2†	3	4	5	6	7	8	No. of obs.
Jan.	8 a.m.		1						
	2 p.m.		3	1					
	8 p.m.		3	4				1	
	Total		7	5				1	992
Feb.	8 a.m.		5	1		1			
	2 p.m.		6	1				1	
	8 p.m.		1	2	1				
	Total		12	1	1			1	904
Mar.	8 a.m.								
	2 p.m.		5						
	8 p.m.		5	1					
	Total		10	1					992
Apr.	8 a.m.		2						
	2 p.m.		3	1					
	8 p.m.		4						
	Total		9	1					960
May	8 a.m.		2						
	2 p.m.		4						
	8 p.m.		1						
	Total		7						992

Sea State (code numbers)[a]									
Month	Time of Day	2†	3	4	5	6	7	8	No. of obs.
June	8 a.m.		2						
	2 p.m.		2						
	8 p.m.		3						
	Total		7						960
July	All		3						960
Aug.	All		3						960
Sept.	8 a.m.		1						
	2 p.m.		4						
	8 p.m.								
	Total		5						960
Oct.	8 a.m.								
	2 p.m.								
	8 p.m.		1						
	Total		1						992
Nov.	8 a.m.								
	2 p.m.		1						
	8 p.m.		1						
	Total		2						960
Dec	8 a.m.		2						
	2 p.m.		4	1					
	8 p.m.		1	1					
	Total		7	2					960

Code number[a]	Wave height (m)
2	<0.5
3	0.5 – 1.25
4	1.25– 1.50
5	1.5 – 4.0
6	4 – 6
7	6 – 9
8	9 –14

Data from Uri Mane, Israel Meteorological Service (pers. comm., July 1964)

[a] For code number 2 (wave height of <0.5 m) deduct the combined monthly total for state of sea code numbers 3 to 8 from the monthly total number of observations. Code number 2 data have not been further subdivided into observations for time of day (G. M. Friedman 1968)

Fig. 3.15. Sea state in Gulf of Elat near City of Elat for period 1958–1961

reaches the seaward side of the reefs) and most likely sweep some of the fine terrigenous and pelagic sediments from the submarine slopes into the deeps. The shape of wave fronts as they are refracted by reefs controls the orientation of spurs and grooves on the reefs (Sneh and G. M. Friedman 1980). Waves and currents impinge on the subaerial fans often enough to restrict the supratidal environments that include sabkhas to the lee (southern) side of the distal fans.

3.4.3 Tides and Currents

3.4.3.1 Tides

Tides in the Gulfs of Elat and Suez are semidiurnal (two highs and two lows daily). The tidal range in the Gulf of Elat varies from 70 cm in the north (at Elat) to 90 cm in the south (at Sharm el Sheik) with no nodal points. The tidal regime of the Gulf of Suez is different in that a nodal point exists and near its northern extremity its tidal range is much greater (approx. 2 m). Tides are important to sedimentation even in areas beyond the high-tide line. The groundwater levels are very dependent on the tidal range in sea level because the materials underlying the supratidal zone are very permeable. The large tidal range in the northern Gulf of Suez has resulted in less-well-developed evaporites on the sea-marginal sabkhas. With the groundwater level maintained at about half-way between the high and low tide levels, the groundwater level is commonly too deep for capillary movement of the water to the surface for evaporation (see Chap. 5).

3.4.3.2 Currents

The predominant northerly and northeastern winds generate waves that strike the Gulf's shoreline diagonally, giving rise to a longshore current that flows to the south. The annual mean wind speed of 10 knots is sufficient to create a substantial longshore current. The common storms on the Gulf accompanied by winds of up to 45 to 80 knots provide considerable bursts of energy to these currents. Because some of the high winds are southerly, especially in winter, one can expect to find the normal longshore current's flow to be temporarily reversed. Currents in the southern gulf (e.g., near the Straits of Tiran) are most affected by strong southerly winds.

Previously mentioned was the current system resulting from the predominance of evaporation over precipitation; the evaporative loss generates a replen-

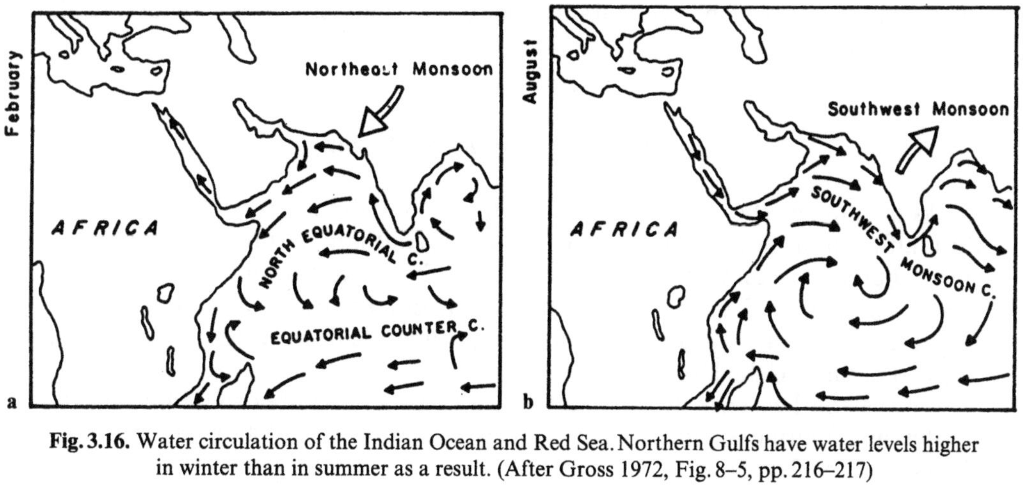

Fig. 3.16. Water circulation of the Indian Ocean and Red Sea. Northern Gulfs have water levels higher in winter than in summer as a result. (After Gross 1972, Fig. 8–5, pp. 216–217)

ishing surface flow of waters from the Red Sea northward into the Gulf, and the density increase accompanying the increased salinity drives a deeper current that flows to the south. Oceanic currents in the Indian Ocean (shown in Fig. 3.16) change the level of the Red Sea and the Gulfs seasonally. The effects in the Gulf of Elat and Gulf of Suez are sea levels that are about 30 cm higher in winter than in summer (Sneh 1978).

3.5 Environments of Deposition

The depositional environments of the Gulf of Elat can be divided into those that are subaerial and those that are submarine. The various environments are restricted to narrow zones parallel to the coast by the prevailing steepness of slope from the bordering mountains falling away sharply to the shores and continuing to plunge abruptly below the surface of the waters into the deep basins. The zones are broadened enough for a suite of environments to form in the areas where the accumulation of terrigenous detritus has provided more gentle subaerial and submarine slopes on the fan deposits. The subaerial environments include fans, dunes, sabkhas (with or without hypersaline pools), berms and beaches. Submerged reefs that fringe the shore are separated from it by narrow lagoons. Seaward from the reefs are the deep water environments of the marginal slopes and basins.

3.5.1 Subaerial Environments

Terrigenous detritus dominates the deposits of the subaerial environments. Considered first is the source in the rocky desert highlands. Discussed next are the fans that are volumetrically predominant among the coastal environments and also provide the more gentle slopes on which the other environments form. Finally considered are the dunes, sabkhas, and beaches that are important not for their aerial extent, which is quite limited here, but as models for more widespread, ancient environments.

3.5.1.1 Sources of Terrigenous Sediments

As mentioned above, a considerable volume of sediment is being supplied to the Gulf from the adjacent uplifted lands. In the southern part of the Sinai peninsula, abundant sediments are eroded from the Precambrian Arabo-Nubian massif, composed of igneous and metamorphic granitic rocks of the Upper Gattarian series, and the volcanic rocks including andesitic flows, tuffs, agglomerates and serpentines of the Dokhan series.

On the southeastern side of the Gulf, a considerable area of Miocene evaporites and marine sediments supply clastic debris (Bramkamp et al. 1963). Weissbrod (1969) reports that sediment brought to the Gulf from wadis draining through a sedimentary terrain in the central and northern Sinai is almost exclusively derived from Paleozoic-Mesozoic Nubian sandstone.

Some of the gravels and sands being deposited on alluvial fans today are second cycle. Deposits of alluvium which now lie above sea level are comprised of some which accumulated about 4,000 years B.P. when sea level worldwide was 1 to 3 m higher (Fairbridge 1961, Jelgersma 1971, Stoddart 1971), and of other deposits which have been tectonically uplifted. Along the southern Gulf, coral reefs at several levels are also being dissected by erosion and reflect similar sea-level changes and/or tectonic uplift. At the southern tip of the Sinai, lime-grainstones of the Quaternary Khashabi Formation (Bentor et al. 1974) are supplying sediment to the gulf.

3.5.1.2 Subaerial Fans

Impressive evidence of stream erosion and deposition exists in the Gulf of Elat area in spite of its being so arid that in some years no rain falls. When the infrequent rain does fall catchment areas of up to tens of square kilometers of rocky desert highlands funnel flashfloods through intermittent stream valleys called wadis out to the narrow coastal margins of the Gulf. Along the entire western coast, the periodic flashfloods build fans of alluvium ranging from 4 up to 15 sq. km in areal extent. The sinuous western coastline is the direct result of the fans building out into the Gulf as much as several kilometers; below the surface of the waters submarine cones of alluvium extend to depths of 150 m and to widths of 1 to 1.5 km (Avraham et al. 1978). With areal extents of the subaerial fans smaller than their catchment areas these fans are definitely not behaving according to the assumptions of Bull (1964). Sneh (1978, Ph. D. thesis) attributes their abnormality to the steepness of slope along the Elat coast where the fan deposits accumulate. On the extremely steep slopes of the eastern coast, no subaerial fans are actively forming. In places, however, deeply dissected alluvial deposits more than 10 km wide line the coast (Avraham et al. 1978). On the western coast, small fans at the mouths of short wadis have steep slopes (up to 10° to 15°) and are composed of larger-sized particles; boulders 1 m and larger predominate in a matrix of cobbles and pebbles with very little sand. Larger fans forming alone or anastomosing at the mouths of more than one wadi (e.g., Wadi Kid and Wadi Adawi) have gentle slopes (1° to 2°) and finer-grained particles; sand grains and layers of sand are more common in the gravels.

Submarine canyons exist that cut down 130 m into cones of alluvial deposits that lie offshore from the fans. The fans themselves have been partially eroded, as shown by the presence of terraces. These are the effects of unusual conditions that prevailed during the late Würm glaciation about 11,000 years ago. Sea level is considered to have been 130 m lower worldwide at that time; a considerable volume of water went into the formation of glacial ice (Gvirtzman et al. 1977, Milliman and Emery 1968). At the very time that the base level of erosion was lowered, the climate is considered to have been wetter (Herman 1965). The net effect of the lowered base level and increased precipitation was a rejuvenation of the cutting power of running water along the Gulf. Previously deposited subaerial fans were down-cut, leaving terraces that are apparent today. Cones of alluvium that were previously and are now offshore from the subaerial fans were subaerially exposed during the late Würm glaciation and canyons were cut into them; the canyons now extend to depths of 130 m.

Recent tectonic events have also affected the fans. The lands on which the fans were deposited have been elevated, resulting in further erosion and terrace formation. Locally, modern fault movements can be seen to have cut across and displaced the fan deposits as well.

Minor dune fields exist on the distal portions of most of the fans as do salt-encrusted flats called sabkhas where evaporites, carbonates associated with microbial mats, and terrigenous muds accumulate.

3.5.2 Coastal Sabkhas

Two types of sabkhas were identified along the northern gulfs of the Red Sea (Sneh 1978): (1) sabkhas associated with sea-marginal hypersaline pools, (2) sabkhas on the supratidal area on the lee side of a distal alluvial fan.

On the Gulf of Elat three hypersaline pools exist: (1) the solar pool just south of Elat; (2) the Dahab pool; and (3) the Nabq pool (part of the sabkha now called Gavish Sabkha). Another hypersaline pool occurs at Ras Muhammad on the Red Sea coast of the Sinai (see description of the Ras Muhammad pool in this volume). Each of the four pools is separated from the sea but receives seawater through cracks and other pores in the intervening rocks and by capillary action. Some pools are inundated periodically by spring and/or storm tides. Very irregularly, sheetfloods reach the pools adding freshwater along with fine-grained terrigenous detritus mud, silt, and sand. Microbial mats flourish in the pools and carbonates accumulate. Water levels in the pools are thought to fluctuate with seasonal sea-level changes (see Chap. 5).

Supratidal sabkhas associated with alluvial fans form on the gentle slopes (1° to 2°) of the distal fans on the southern side where the area is protected from waves and currents that come predominantly from the north. Each sabkha is near the sea, but only rarely in direct contact with it (during spring tides or during some storms). Optimal development of supratidal sabkhas requires: (1) protection from waves; (2) free movement of saline groundwater to the sabkha sediments; and (3) terrigenous mud to aid in capillary action and to serve as a host for evaporites (Chap. 5). Saline waters of the Gulf reach the sediment of the sabkha by free movement through the underlying porous sediments and sedimentary rocks; lateral movement occurs as groundwater flow and vertical movement toward the surface by capillary action. The tidal range along the entire coast of the Gulf of Elat is small enough (70 to 90 cm) for the groundwater level under each sabkha to be within the range of capillary action. This situation results in well-developed evaporites in the sabkha sediments of the Gulf. The situation is different in the northern Gulf of Suez area where the tidal range is too great (2 m) to maintain the groundwater level within the capillary zone; the sabkha sediments of this area include only poorly developed evaporites (for further details see Chaps. 9 and 11).

3.5.3 Submarine Depositional Environments

Because of the steep slopes of the Gulf of Elat the most extensive submarine environments are in the deep waters of the marginal slopes and deep basins. Where the normally steep slopes have been modified by the buildup of terrigenous detritus, offshore of the subaerial fans, fringing reefs dominate the shallow submarine environments. Narrow lagoons separate the reefs from shore and the normal fore reef, reef flat, and reef front environments are well developed.

One of the unusual aspects of the submarine sedimentation in the Gulf of Elat area is the deposition of terrigenous clastic debris side by side with the accumulation of significant buildup of carbonates. This carbonate/terrigenous clastic relationship may be characteristic of tectonically active areas where a new ocean basin is forming in a climate suitable for carbonate-synthesizing organisms. In most regions, any influx of terrigenous debris is extremely deleterious to the organisms responsible for the carbonate buildups. In the Gulf of Elat, however, fan deposition does not seem to effect adversely reef-building organisms. Fringing reefs grow just seaward of the fans in all but the canyon areas that extend through gaps in the reefs and across the submarine cone extensions of the fans. Perhaps it is the coarseness of the terrigenous debris that makes the detritus less of a problem for the reef-building organisms (G. M. Friedman 1982). Turbidity currents in the submarine canyons may carry the bulk of the final materials washed down the wadis directly into the deep waters where the silt and mud can pose no threat to the reef builders. The long periods of quiescence between flashfloods also favor survival of the reefs.

Discussed below are the zonation of sedimentary deposits and organisms; reefs and the factors that control their location and major features; the origin of the carbonate sand and, finally, the marginal slopes and basins that make up the sea floor beyond the reefs.

3.5.3.1 Zonation of Sedimentary Deposits and Organisms

The sedimentary deposits and organisms form a series of zones which vary as depth changes. The author described (1968) a typical traverse from shore into the Gulf at a location 5–6 km southwest of the town of Elat. In the traverse at right angles to the shore one passes through a systematic series of zones characterized by distinctive sediments and organisms. Along the shore are beach rock, fossil reef, and fan debris. Several zones exist in the lagoon and reefs. Still others can be recognized seaward of the reefs (Fig. 3.17).

The most common material along the shore is beach rock that extends from the intertidal zone to about 1 m above the high-tide mark. Locally, especially in the uppermost intertidal zone, a fossil reef occurs (G.M. Friedman 1965, 1969).

The width of the lagoon varies, but the position of the reef on its seaward side is roughly at the same bathymetric level. Where the lagoon is narrower, its profile is steeper. Near the strand line in the lagoon, rounded pebbles of Precambrian igneous or metamorphic origin from the nearby fans are abundant. Toward the reef, these pebbles become progressively less common. In the shallow water of the

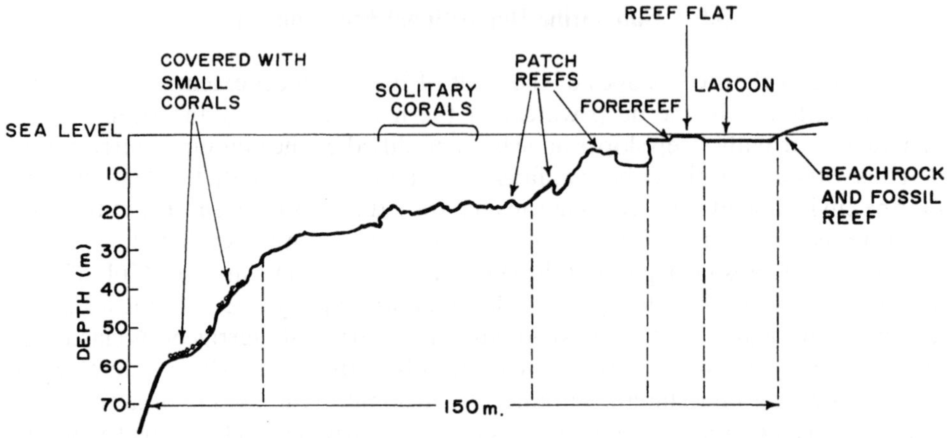

Fig. 3.17. Generalized profile perpendicular to the shore in the Gulf 5 to 6 km southwest of Elat. It is based on the writer's observations out to about 50 m from shore, and for the next 100 m on W. Halpert's diving experience in the deeper water toward the edge of the shelf (G. M. Friedman 1968)

lagoon brown algae, *Padina pavonica* Thivy, are common. These algae deposit an external skeleton of calcium carbonate and are dominant only where the water is very shallow (approx. 40 cm below low tide). In progressively deeper water, they become less abundant, and sparse coral patches begin to develop. Reefward of the *Padina pavonica* zone many individual coral groups appear. Commonly, in the shallow water of the lagoon, the black sea urchin *Diadema saxatile* is abundant. Close to the reef one finds boulder-sized rubble of dead coral, in blocks 1 to 2 m in diameter, that are coated with coralline algae and also support live coral groups. The coral *Stylophora pistillata* (Esper) is abundant here. Progressively reefward, patch reefs begin to develop which ultimately merge with the reef core. Near the reef core, large (15–20 cm long) broken coral fragments are abundant. Contiguous with, and derived from, the reef are thick blankets of skeletal sand. These blankets occur on both lagoonal and seaward sides of the reef as well as in grooves between spurs. On the lagoonal side, the top of the reef stands about 2 to 3 m above the sandy bottom, and on the seaward side about 3 to 4 m.

The mechanism of transportation of the boulder-sized blocks, 1 to 2 m in diameter, of coral rubble found in the lagoon presents a problem. These blocks are as big as or bigger than those described by Ball et al. (1967) from the leeward side of the Florida Reef Tract. These authors concluded that this boulder-sized rubble must have been transported by momentary currents of high velocity which developed when storm waves broke on the reefs during hurricanes. In Florida, during Hurricane Donna in 1960, when winds attained a speed of 140 mph and tides were about 4½ m above normal spring high tides, storm-generated currents succeeded in moving these large blocks. At all other times these large blocks do not move. Other documentation of the formation of coarse reef rubble comes from the Australian Barrier Reef (Fairbridge and Teichert 1948) and the Jaluit Atoll in the Pacific Marshall Islands (McKee 1959). By analogy, it stands to reason that the coarse boulder rubble in the Gulf of Elat reef tract must have resulted from storm waves. The fact that live coral groups are growing on many of these boulders indicates that the last great storm in the Gulf of Elat occurred a long time ago.

The central part of the reef is a reef flat that consists primarily of colonies of dead corals that are covered by coralline algae. In the reef flat, coralline algae are so abundant as to confirm the concept of Odum and Odum (1955) that in terms of biomass the algae contribute much more material to the reef than do all the reef animals. Although the reef is for the most part covered by coralline algae, abundant live corals such as *Stylophora pistillata* (Esper) are found on the lagoonal side.

Spurs and grooves dominate the fore reef. Most of the living corals reside on the distal extremeties of the spurs. A number of species of *Millepora, Acropora*, and *Alveopora* thrive on the seaward side of the reef. Sea urchins (*Diadema*), sponges, and sea anemones are abundant, and on the marine side, crinoids (*Lamprometra klunzingeri*) are common. The living parts of the fringing reef and the patch reefs extend into deeper water than the dead parts.

Although large chunks (5 to 20 cm long) of coral debris lie scattered along the seaward side of the reef, the most abundant reef-derived sediment is sand. At a depth of 15 m and a distance of 40 m from the main reef, this sand is dark-colored, a possible result of reducing conditions at and below the water–sediment interface at that depth. This contrasts with oxidizing conditions measured for surface seawaters (pH 8.20 to 8.60). Skeletal sands persist all the way to the bottom of the Gulf, but where the water is deep they contain abundant admixtures of terrigenous debris derived by flash floods from the wadis and by wind from the Arava Valley. Rapid transport of the skeletal sand is indicated, otherwise organisms would have broken down the sand into mud.

3.5.3.2 Reefs

Where the reefs occur and the relationship of their distribution and the distribution of major features on the reefs to an earlier erosion pattern, and in the case or spurs and grooves to the prevailing wave-front orientations, are discussed below. The relationship of the origin of carbonate sand sediment to the reefs is also considered.

Location. Corals and other reef-forming organisms grow along the entire coastline of the Gulf of Elat and form continuous fringing reefs along most of it (Fig. 3.18). The steepness of the slopes along the Gulf restricts the reefs to areas close to shore (commonly within 10 to 200 m) where the waters are shallow, and restricts the basinward expansion of the reefs (width 10 to 100 m) because the slope at the shelf edge is so steep that the volume of reef talus is not sufficient to provide the shallow foundation necessary for seaward expansion. Individual corals have been observed growing at depths down to 250 m in the clear waters of the Gulf (Roberts and Murray 1983), but they do not form reefs at depth. A narrow and shallow back-reef lagoon lies between the reef flat and shore. Shorelines of the Gulf lacking continuous fringing reefs include the northernmost 10 km of the western shore where patch- and discontinuous-fringing reefs occur

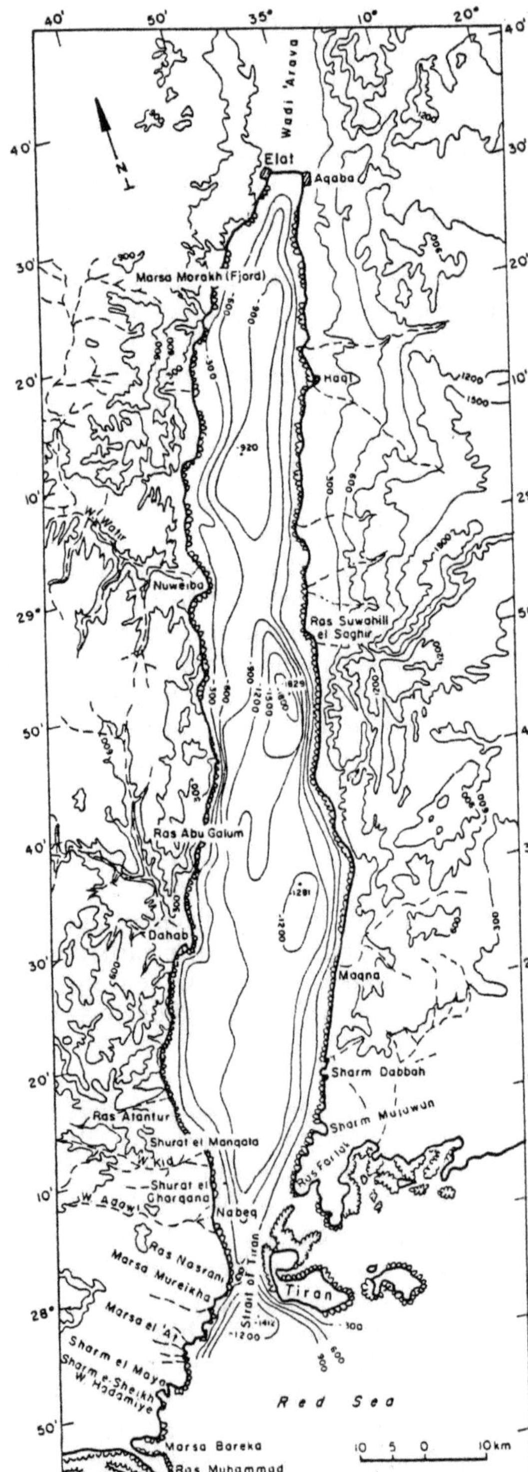
Wadi 'Arava
Elat
Aqaba
Marsa Morakh (Fjord)
Haql
W. Watir
Nuweiba
Ras Suwahill el Saghir
Ras Abu Galum
Dahab
Maqna
Sharm Dabbah
Ras Atantur
Sharm Mujuwan
Shurat el Manqata
W. Kid
Ras Fartak
Shurat el Gharqana
W. Agawi
Nabeq
Ras Nasrani
Marsa Mureikha
Strait of Tiran
Tiran
Marsa el 'At
Sharm el Maya
Sharm e-Sheikh
W. Hadamiye
Red Sea
Marsa Bareka
Ras Muhammad
10 5 0 10 km

Fig. 3.18

(G. M. Friedman 1968), and the northernmost 60 km of the eastern shore, where fringing reefs are also discontinuous (Gvirtzman et al. 1977).

Where the submarine slopes dip more gently off both the western and eastern shores at the southern end of the Gulf, near and beyond the Straits of Tiran, the corals form more extensive reefs (up to 100 m wide), and between the Tiran Islands and the southernmost shore on the east they have grown to form complex patterns of patch reefs, fringing reefs, and barrier reefs (Gvirtzman et al. 1977).

3.5.3.3 Factors Controlling Location of Reefs and Reef Features

Gvirtzman et al. (1977) explained the factors controlling the distribution of the fringing reefs and of features, such as furrows and pools, on the reef flats and particularly the distribution of embayments breaching the reefs (sharms). The researchers demonstrated that the primary factor controlling the distribution of reefs and reef features is the fossil fluvial pattern of erosion and deposition formed during lowered sea level of the late Wurm glaciation. During the glacial low stand of sea level, low temperatures (14°–17 °C) and high salinities (more than 50%) prevailed in the Red Sea and the Gulf of Elat (Reiss et al. 1980) killing most of the planktonic microfauna and presumably all of the corals as well. The lowered base level resulted in intensive erosion that, according to Gvirtzman et al. (1977, p. 487): "... resulted in the incision of deep channels and canyons into the preexisting reefs and superposition of braided streams, alluvial fans and other fluvial forms on the reef flats."

Gvirtzman et al. demonstrate that this fossil pattern is preserved in the featurs seen on the modern reefs (maximum age 10,000–13,000 years) that reestablished themselves, spreading back into the Red Sea and gulfs from the Indian Ocean during the Holocene transgression. A similar idea was espoused by Bloom (1974). Purdy (1974) suggested that modern reefs worldwide form on earlier karst landscapes. However, Gvirtzman et al. (1977) saw no evidence of an earlier karst in the Red Sea and the Gulfs of Elat and Suez.

The fringing reefs along the Gulf are interrupted locally by substantial embayments known by the local Arabic names "sharm", "marsa" or "khor". Gvirtzman et al. (1977) studied the sharms and listed their characteristics (p. 483):

"a) they commonly breach the fringing reefs which follow the shoreline;
b) they always form the outlets of ephemeral streams (wadis);
c) they usually form drowned canyons 30 to 40 m deep, which continue seaward;
d) both the drowned canyons and the continental wadis are partially filled with unconsolidated and unsorted fluvial deposits."

Examples of west coast, Sinai, sharms given (p. 483) are Sharm el Sheikh, Sharm el Moya, Marsa el 'At and the embayment at Dahab; sharms on the Saudi

Fig. 3.18. Map of the Gulf of Elat (Aqaba) showing the distribution of the coral reefs. Height and depth contours in 300 m intervals. Bathymetry based on British Hydrographic Office charts, harbors and anchorages in the Red Sea, chart 3034-1954, chart 3595-1968, London (Gvirtzman et al. 1977)

Arabia coast, the embayments of Haql, Maqna, and Sharm Dabbah. The origin of sharms as canyons eroded during the lowered sea level during the Würm glaciation followed by Holocene flooding appears to be well established. The steep seaward slopes of the canyon bottom at Dahab floored along its axis by fluvial sediments and lacking in any living reefs may show that these canyons are still active (Gvirtzman et al. 1977). The erosive activity is not likely to have cut any of the modern reefs, however, for the corals would not have been able to even get reestablished within these relatively deep and active channels. The embayment of Marsa Murakh called the "Fjord" does not appear to fit this theory of origin, as it is substantially larger than the size of its corresponding wadi would warrant. A local tectonic, fault, origin is postulated for this feature (Gvirtzman et al. 1977).

Reefs in the Gulfs of Elat and Suez are all north of 27° N latitude; they are the northernmost reefs of the Indian Ocean. To the south, in the Red Sea where coral reefs are not at the extreme northern end of their range, reefs colonize even the interior banks of the sharms. In the Gulf of Elat, however, the interior banks of the sharms are essentially coral-free and the reefs flourish only where they can be exposed to the open Gulf.

The furrows, pits, and pools that commonly cut the reef flats and their seaward rims are, like the sharms, also directly attributable to a preexisting fluvial surface (Gvirtzman et al. 1977). The furrows vary in depth from 0.5 to 3 m, in length from 10 to 300 m, and in width from 5 to 200 m (p. 483). The walls of the furrows are vertical or overhanging and they are floored with carbonate sand or gravel derived from the reef. Furrows have cut deeply into the underlying fossil reef rock, demonstrating their erosional origin (as opposed to the development of spurs and grooves by selective growth of reef organisms). Furrows may be straight or meandering, branching into two or forming radial patterns, but always occur offshore from a subaerial fan and most furrows can be matched to channels on the surface of the adjacent fans. Ponds also occur offshore from subaerial fans at wadi outlets but appear to have been closed off from contact with the sea by growth of the coral reefs across former connecting channels. Both furrows and ponds are considered to have originated as extensions of the erosional channels on the fans. As in the case of sharms, furrows and ponds reflect the surface cut during the low stand of sea level during the last glaciation (Gvirtzman et al. 1977). Gvirtzman et al. (1977) note, however, that the large pool in the Ras Abu Galum area does not relate to any recent wadi or fan (p. 485). Small pits on the reef flat probably formed as a result of collapse of underlying subaqueous caves (Sneh 1978, Ph. D. thesis).

In contrast to the erosional features discussed above, the spurs and grooves so common to the reefs in the Gulf are formed primarily by the selective growth of corals and other reef-building organisms (Goreau 1959, Shinn 1963, Sneh and G. M. Friedman 1980). A system of alternate spurs (ridges) and grooves (depressions) occurs on the reef flats, predominantly on the seaward side but extending entirely across the reef locally. Spurs and grooves are easily distinguished from other reef-flat irregularities by their characteristic parallel oriented pattern.

The "spur-and-groove" system (Shinn 1963) or buttress zone (Odum and Odum 1955) is a morphologic characteristic found in the zone of prevailing waves and swells on the seaward sides of most reefs. Spurs, finger-like projections extending from the reef into the prevailing seas and swells, are separated by grooves that are often misnamed surge channels (Shinn 1963). The grooves are quite irregular in detail. Near the reef core, grooves may be only ½ m in width, but at their most seaward extremeties they widen to about 10 m.

Although the reef is mostly dead with the deceased corals thickly coated with coralline algae, vigorous coral growth is occurring on the seaward sides of the reef and on the ends and forward edges of the spurs. Goreau (1959) suggested that spurs and grooves originate by this selective growth of corals and not by erosion. Shinn's (1963) model for the formation of spurs and grooves includes three stages: (1) oriented colonies starting on randomly spaced centers grow seaward into the prevailing seas and swells and develop linearity; (2) carbonate sand, trapped in the grooves between the spurs, makes the grooves unfavorable for colonization by corals; (3) the corals die from crowding and subsequently become completely masked with calcareous algae. In the Florida reef tract, Shinn (1963) has demonstrated that spurs and grooves do in fact originate by coral growth, not by erosion. Shinn's model apparently also applies to the Gulf of Elat reef tracts studied.

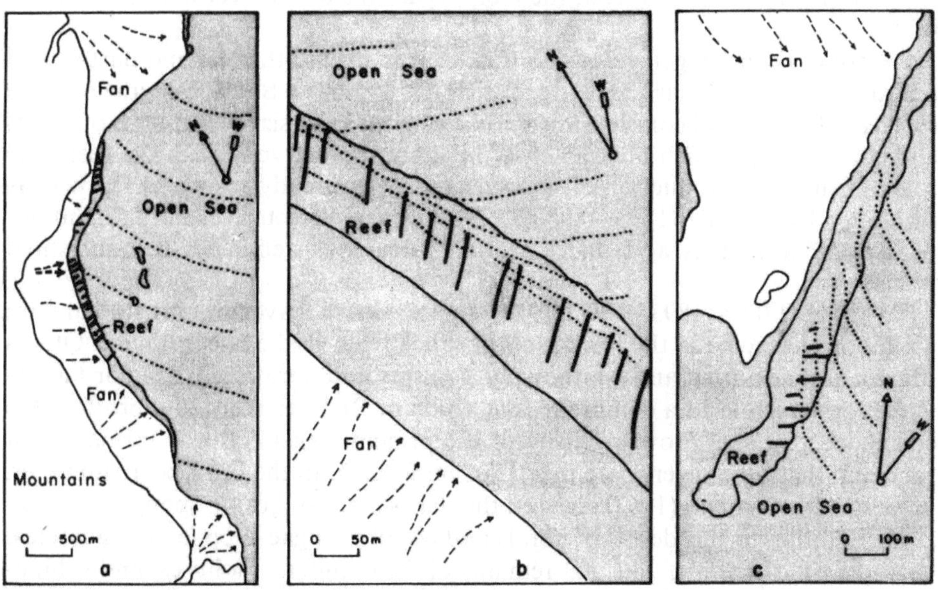

Fig. 3.19 a–c. Fringing reefs from the Gulf of Elat showing spurs and grooves. **a** El Hibeiq, **b** Dahab (north), **c** Dahab (south). Spurs are marked by *heavy lines*, wave fronts by *dotted lines*, and streams on land by *dashed lines*; all traced from aerial photographs. The wind direction is drawn normal to the wave crest and is valid only for the day of the photograph. Wind direction is schematically marked *W* (Sneh and G. M. Friedman 1980)

A detailed study on the origin of spurs and groove systems in the Gulf of Elat and Gulf of Suez was conducted by Sneh and G. M. Friedman (1980). Spurs were shown to form with their long axes normal to the refracted wave fronts impinging on the reefs (Fig. 3.19). No correlation could be found between the orientation of spur-and-groove systems and the trends of streams on land, and no evidence was seen to support the influence of a preexisting karst surface as postulated by Purdy (1974). Growth was again seen to be the predominant factor in spur and grove formation; in only two out of eight localities were clear abrasive patterns seen that suggest an erosive origin for the grooves. In all others, the seaward growth of corals, as well as abrasion, adequately explains their origin (Sneh and G. M. Friedman 1980). Because corals build spurs into the prevailing seas and swells, the spurs form only a small angle with the trends of reefs that, in the Gulf of Elat, are nearly parallel to prevailing winds. By contrast, in the Florida reef tract, where wind- and wave-approach directions are almost normal to the reef trend, the spurs stand at an angle of about 90° with the reef trend (Shinn 1963, Sneh and G. M. Friedman 1980).

3.5.3.4 Carbonate Sands – Relationship to Reefs

The increasing abundance of carbonate sand toward the reef indicates that the skeletal sands are derived from the reef; organisms on the reef make sand from it.

Among the many genera of fish that swarm around the reef and dart in and out around it, parrot fish and trigger fish are common. Some fish graze on the surfaces of corals and coralline algae. By biting off sand-sized fragments of corals and coralline algae, fish literally chew up a substantial volume of the reef. The role of fish in the production of skeletal sands was already known to Darwin, as recounted by Walther (1893–1894, p. 927). Darwin dissected several fish and noticed coral fragments and finely ground calcareous debris in their intestinal tracts.

Walther (1888 a, 1893–1894, p. 927) suggested that scavenging crustaceans are an important factor in the formation of sands from Red Sea coral reefs. Others suggested that boring and benthonic organisms and abrasion are major factors in reef destruction. In a comprehensive study of Red Sea reefs, Nesteroff (1955) concluded that reef-browsing fish are the principal agents in the formation of sand and that other agents are much less important. In the Caribbean, however, Goreau and Hartman (1963) stressed the importance of boring sponges.

In addition to this debris of corals and coralline algae supplied by biological degradation of the reef, skeletal remains of foraminifers, molluscs, and echinoderms which live and die in and around the reef community, are important constituents of reef-derived sands. In the Caribbean, Goreau (1963) has shown that small delicate (green) algae are important contributors to the sediment. In the Gulf of Elat, skeletons of green algae were not observed in the sediment samples studied, and it is therefore inferred that, in this case, green algae are a relatively unimportant source of sediment.

The mineralogy and geochemistry of the carbonate synthesized by corals and by other reef organisms and the size frequency, mineralogy and geochemistry of the reef-derived sands were studied by G. M. Friedman (1968). The mineral composition and chemical makeup of the carbonates synthesized by the reef organisms are identical to those from reefs in other areas, such as the Bahamas and Bermuda. Analysis of the mineral content and chemical makeup (especially their high magnesium content) enabled G. M. Friedman (1968, p. 944) to determine that coral fragments predominate over molluscan fragments. Coralline algae also synthesize high-magnesian carbonate and are major contributors to the reef-derived sands. Many of the sands studied contain significant components of terrigenous debris derived from the Precambrian crystalline rocks and, in some cases, from ancient carbonate reef terraces as well. The terrigenous constituents were reflected in the amounts of insoluble residue and in their chemical makeup. A deep-sea core (v. 14–126) collected by the Columbia University R/V *Vema* in 1960 contains carbonate material ranging from 20% to 75% of the total. Aragonite is sporadic or absent and the unusual combination of high-magnesian and low-magnesian calcite was considered to be attributable to the faunal assemblage (Sanders and G. M. Friedman 1967). Size-frequency analyses of the reef-derived sands showed them to be less well sorted than nearby carbonate beach sands. The fines (62 μm to 125 μm) had been washed out of the carbonate beach sands, presumably by wave action.

3.5.3.5 Sea Floor – Marginal Slope and Basins

Deposition of sediment within the Gulf beyond the immediate effects of the reefs can be considered in terms of two areas: the deep basins and the marginal slopes including the wide gently sloping terrane of the western half of the southern Gulf. Seismic studies (Avraham et al. 1979) indicated a minimum of 2 to 3 km of turbidites, probably of predominantly terrestrial origin, and pelagic deposits have accumulated in the deep basins. The actual thickness of sediments in the northern and southern basins is about 5 km below the sea floor and somewhat less in the central basin (Avraham 1983). When flashfloods wash large volumes of terrigenous debris down the wadis and into the canyons cut in the submarine cones at their mouths, it is likely that turbidity flows are generated that carry the debris into the deep basins. Sediments that accumulate initially on the margins of the deeps are also likely to be swept into the deeps periodically by currents, such as those caused by wind-generated downwelling and/or by flows generated by the tectonic movements that frequently shake the region. The deep-basin sediments commonly meet the basin-margin deposits in a well-defined sharp contact often at the foot of a conspicuous escarpment, evidence of the recent and continuing activity of the faults that define the deep basins (Avraham et al. 1979). Even within the deep basins folds and faults have been detected that are forming as sediment is being deposited (Fig. 3.20). By way of contrast, the major faults defining the valley of the Gulf of Elat show only local movement subsequent to the first phase of tectonism in the Miocene (see Sect. 3.3).

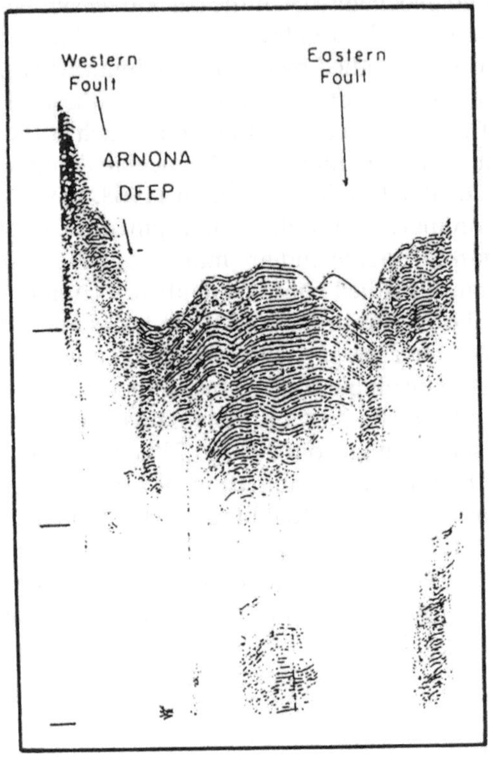

Fig. 3.20. Seismic section across the Gulf of Elat showing folding and faulting synchronous with deposition (Ben Avraham et al. 1979)

Marginal slope deposits were considered by Ben Avraham et al. (1979) to be composed of coarsely stratified deposits. Five deep-sea cores were taken in 1977 from the marginal slope along the western side of the Gulf of Elat by the Woods Hole Oceanographic Institute's R/V *Atlantis II*. Reiss et al. (1980) reported on paleoclimatology of the Gulf as indicated by the fossil assemblages, briefly described the composition and grain size of the sediments, and estimated the accumulation rates of the sediments. Two types of sediments were found to occur on the marginal slope:

a) grey-green partly silty calcareous mud
b) silty clayey sands, mostly quartzose and occasionally arkosic.

The two cores from the northern Gulf taken at depths of 550 and 858 m adjacent to the Elat Deep contained turbidites of alternating silty muds and quartzose sands. The core taken in the southern Gulf at a depth of 963 m on the slope of a coastal bulge attributable to the alluvial buildup of Wadi Dahab contained, not surprisingly, a "thick arkosic sequence in the lower part of the core" (Reiss et al. 1980, p. 295). Using the dates established by fossil content and the thicknesses of the intervening sequences, the accumulation rates of the lutite sediments were estimated at a few centimeters per 1,000 years, and of the arkosic sands near Wadi Dahab more than 40 cm/1,000 years (Reiss et al. 1980, p. 305).

If deep-basin fill accumulated at the rate of 50 cm/1,000 years, the time required to accumulate the measured 5,000 m of sediment partially filling the basin would be 10 million years [(5,000 meters) (0.5 m/1,000 years)$^{-1}$ (1,000 years) = 10,000,000 years)]. As stated above, the strike-slip faulting that formed the major marginal faults of the Gulf of Elat occurred in the Miocene (over 7 million years ago) whereas the in-echelon deep basins formed by faulting about 4 to 5 million years ago. Thus, the measured accumulation rate of turbidites suggests that, even if all of the sedimentary fill of the basins is of turbidite origin, the sediments most likely started accumulating after the first phase of faulting in the Miocene but prior to the second phase of tectonic activity.

3.6 Further Considerations

3.6.1 Plate Tectonics

The Red Sea and its northern gulfs, including the Gulf of Elat are part of a rift system that is forming a new ocean – but not as new as was once thought. The Red Sea has been forming not for only 1 million years but since the Miocene, more than 7 million years ago. An additional complication is that the sea-floor spreading that is forming the ocean has not been continuous but has occurred in two phases. Sea-floor spreading is supposed to be slow, but steady, with no lapses if the patterns of magnetism recorded in the rocks of the sea floor are to be interpreted and correlated with elapsed time, as they are at present. The Red Sea is either not a typical example of the formation of a new ocean or the use of the magnetic strips parallel to sea-floor-spreading centers has to be reconsidered. Further work needs to be done to resolve this.

3.6.2 Age of Formation of the Gulf of Elat

It has been thought that the Gulf of Suez formed in the Miocene with the first stage of the formation of the Red Sea and that only the marginal faults of the Gulf of Elat formed at that time. The faults that actually bound the modern Gulf and that lie inside the marginal faults (separated from them by the marginal blocks) are considered to have formed only 4 to 5 million ears ago, with the second stage of tectonism. The depth of sediment infill of the Gulf of Suez has been known for some time to be on the order of 6,000 m and most likely to include Miocene-age sediments (evaporites?). The Gulf of Elat, with its much deeper water (1,800 m as opposed to 50 m in the Gulf of Suez) was only known to contain "more than" 2,000 to 3,000 m of sediment infill. We now know that there are 5,000 m of sediment infill in the northern and southern ends of the Gulf of Elat and somewhat less in the central area (Ben Avraham 1983). One is forced to wonder if the explanation of the similar depth of fill in the two Gulfs is due to a similar age of formation or due to different sedimentation rates in basins of different ages. Sediments could well have begun accumulating in both the Gulf of Suez and the Gulf of Elat during the Miocene. Using the sedimentation rate of Reiss et al. (1980), the 4 to 5 million years since the last phase of tectonism would only account for about half of the amount of sediment infill of the Gulf of Elat. Considerable accumulation of sediments in the Gulf of Elat prior to 5 million years ago certainly seems likely. Resolution of the issue awaits the sampling of the lower layers of the sediment infill.

3.6.3 Gulf of Elat – Dead Sea Grabens – Type of Faulting

The basin occupied by the Gulf of Elat and the deep basins within the Gulf itself have been ascribed to an origin by strike-slip faulting as has been proposed for the Dead Sea graben (Freund and Garfunkel 1976). There is no universal agreement on this point; some researchers believe that normal faulting was also a factor and may have been predominant. Ben Avraham (1983) has reported that the faults in the Gulf of Elat area do not fit the Dead Sea model and that the Gulf of Elat seems to be an area of transition from the Red Sea spreading-center style of tectonism to the transform-fault style of the Dead Sea graben. The southern Gulf of Elat is very different from the central and northern Gulf, the southern being more like the Red Sea in that the movement seems to be more extensional, perpendicular to the trend of the basin, whereas in the center and north the transform-fault style of strike-slip movement, like that of the Dead Sea graben, seems to predominate. Some boundaries of the basins are not faulted and others have a normal fault component. More such work on the nature of the structure of the area will obviously be needed.

3.6.4 Potential Upwelling and Downwelling

In the northernmost Red Sea just south of the Gulf of Elat the abundance of diatoms in sea-bottom sediments of Late Würm glaciation age was interpreted by Reiss et al. (1980) as evidence of possible upwelling associated with predominant northerly winds. It would be interesting to know if the periodic southerly winds have resulted in periodic downwelling in the northern Red Sea. Similar wind patterns in the Gulf of Elat might be associated with similar upwelling and downwelling.

3.6.5 Reversal of Wind Directions: Longshore Current Effects

Predominant northerly winds drive longshore currents south along the Sinai coast of both the Gulfs of Elat and Suez. In the Gulf of Suez it is only in the northern half of the gently sloping coast that spit complexes bend to the south in response to the southflowing longshore currents. In the southern Gulf of Suez fans along the coast are symmetrical in shape in response to currents and waves approaching from the south as well as from the north. In the Gulf of Elat the only major spit complex occurs on the south-central part of the Sinai (west) coast at Dahab. Presumably, the shore dips too steeply in the northern Gulf of Elat for spits to form. In the southern Gulf of Elat, although slopes are more gentle, southward extending spits do not occur. The lack of spits along the southern shore may be related to the periodic storm winds that blow from the south. Waves and currents driven to the north by these high winds might counterbalance the effects of the predominant northerly winds preventing the formation of spits (in a manner similar to that interpreted for the Gulf of Suez).

4. Coastal Evaporite Systems

BRUCE H. PURSER

Sedimentological study of modern coastal evaporites began in the early 1960's with the discoveries of dolomite, sulfates, and other minerals within the coastal sabkhas of the Arabo-Persian Gulf, demonstrating that evaporite formation was not limited to subaquatic environments. These initial studies of Wells (1962) and colleagues of Shell Research (The Netherlands) and of Shearman (1963) and other researchers of Imperial College (London) have been followed by numerous, often very detailed studies concerning the sedimentology, hydrology, mineralogy, and ecology of littoral evaporitic environments, notably in Abu Dhabi, the Caribbean region, Mexico, Australia and N. Africa. Especially detailed studies of coastal evaporite environments have been carried out along the Sinai coast by Gavish (1974a, b, 1980) and colleagues, the essential results of which are presented in this volume.

This contribution outlines the main aspects of littoral evaporites and, in particular, tries to place the Gavish Sabkha with respect to other areas composing the exceedingly varied complex of coastal evaporite systems. As will be shown, the Gavish Sabkha, in common with many sedimentary environments, is an interesting blend of physical, chemical, and biological factors peculiar to this locality, plus – and more importantly – a certain number of fundamental features common to virtually *all* coastal evaporite systems. Thus, in utilizing the Gavish Sabkha as a possible "model" for interpreting ancient littoral evaporites, one must keep in mind these more fundamental aspects; no single area, whether it is Abu Dhabi or the Sinai coast, should be used as an evaporite "model".

4.1 General Aspects

For a sedimentologist it is very important to distinguish between standing-water (subaqueous) evaporites, which, theoretically, may occur at any depth, and evaporites formed within the sediments essentially by capillary evaporation. Obviously, if we confuse these two basic types, then the potential value of coastal evaporites as paleogeographic markers will be compromised severely or totally lost.

We must try to limit the definition of sedimentary sabkhas to areas of capillary evaporite formation within coastal or inland sedimentary plains. Although, at first glance, this may seem obvious, in fact several important complications arise.

1. Sedimentologists must decide where to place the limit between sabkhas and tidal flats for both are affected by capillary evaporation. It has been convenient to separate the two types partly on a geographic basis, with sabkhas considered to be characteristic of arid, Middle Eastern regions. However, it should not be forgotten that even within the relatively humid climates of the Caribbean, including Florida, near-surface interstitial waters in tidal flats locally attain salinities exceeding 120‰ (Shinn 1964) and thus are potentially capable of precipitating gypsum. This is true also for the relatively temperate Coorongs of S. Australia (von der Borch 1965). That gypsum does not appear to be preserved in these tidal-flat sediments may possible result from excessive seasonal precipitation.

2. In both arid and humid regions, capillary evaporation leads to increased salinity of interstitial waters and thus, in certain cases, may favor the formation of aragonite, dolomite and, eventually, sulfate and other salts. In other words, certain aragonites and dolomites are the product of evaporated interstitial fluids. However, in view of their very frequent formation from nonevaporated marine and continental waters, it would be dangerous to use aragonite and dolomite as criteria for sabkha-type environments; presumably, we must use sulfates as the key minerals, also realizing that even gypsum may be formed (in limited quantities) under nonevaporitic conditions by bacterial decomposition of organic matter in seawater. It is also worth remembering that sulfates are susceptible to bacterial destruction (G. M. Friedman and Sanders 1967), this being the case locally in the Gavish Sabkha.

3. Many areas, including Gavish Sabkha, periodically have standing waters capable of precipitating subaqueous evaporites. However, these waters may dry out seasonally and subsequent capillary evaporation may favor the precipitation of interstitial evaporites; saline-pond and true sabkha conditions may be related intimately.

Finally, dense brines formed by capillary evaporation theoretically may flow vertically and laterally (reflux). These brines may mix with waters of other origins (continental) and precipitate their salts within sediments that were deposited neither on or even below the evaporitic surface. Thus, a problem exists concerning the localization of evaporitic phenomena with respect to the actual evaporitic environment.

Thus, the precise definition of true sabkha environments is not a simple matter of climate. In an attempt to impose a minimum of order within what clearly is a variable system, the author suggests that a sedimentary sabkha (as opposed to a geographic sabkha) is an environment whose sedimentary substratum is affected by capillary evaporation which leads to the precipitation of sulfate and, eventually, other more soluble minerals, whose traces are preserved permanently. Thus, certain coastal evaporite systems, probably including the Gavish Sabkha, are not permanent sedimentary sabkhas for they precipitate part of their salts in the form of subaqueous evaporites from seasonal standing waters. By contrast, if traces of gypsum or other more soluble minerals, either as such or as pseudomorphs, can be demonstrated to occur within tidal flats of more humid climates, then these environments, logically, should be regarded as sabkhas.

4.1.1 The Originality and Economic Importance of Coastal Evaporitic Systems

Both salinas (standing water) and sabkha (capillary) evaporites, although superficially unimportant today, seem to have many analogs in ancient sedimentary basins. Recognition of fossil littoral evaporites, especially true sabkhas, has been made possible mainly as the result of recognition of specific properties of modern equivalents. Precise descriptions and the understanding of evaporitic processes and products of Gavish Sabkha and other modern environments will help further to advance our knowledge of ancient analogs.

Together with earlier published studies of modern littoral evaporitic systems, the study of the Gavish Sabkha confirms the dynamic nature of this particular environment. Perhaps more than any other coastal environment, saline sabkha systems illustrate how rapid lateral sedimentary accretion is followed, almost immediately, by formation of a suite of evaporite minerals. Because of the progressive changes in the compositions and saturations of the interstitial waters as the coastal plain and its evaporite constituents prograde rapidly across the sedimentary basin, these essentially diagenetic products (evaporites) evolve quickly (before deep burial) as brines of marine origin are replaced progressively by saline, or fresh, continental waters.

Several major economic implications are associated with ancient coastal evaporite systems, which are being recognized with increasing frequence in the geological record.

1. Many serve as hydrocarbon reservoirs; this is particularly true for many dolomites and breccias associated with sabkha-type evaporites. The Upper Jurassic "Arab Zones" of the Arabo-Persian Gulf (Powers 1962, Wood and Wolfe 1969) are a good example. Where sabkha anhydrites are well developed, these may serve as impermeable cap rocks as does the Upper Jurassic "hith" anhydrite of Southeast Arabia.

2. Many metal-sulfide deposits are associated with paleosabkha conditions (Renfro 1974), especially where interstitial waters have been derived partly from continental groundwaters. The latter carry dissolved metal ions toward littoral evaporitic plains where the presence of abundant organic material, mainly of microbial origin (see Chap. 11), provokes reducing conditions. Surface evaporation causing the upward capillary flow of interstitial waters favors precipitation of metal sulfides within the organic-rich sediments. These stratiform metal deposits thus are often associated with sabkha-type sulfates and dolomites.

4.1.2 Objectives of this Contribution

This contribution treats coastal evaporite systems, especially sabkhas. These are discussed mainly from a sedimentary angle to demonstrate that marine sabkhas have several, quite different, origins of which the classical Abu Dhabi system is but one. The hydrology and mineralogy of sabkhas are also treated in some detail. However, our knowledge of these important processes is dominated by the results of numerous studies carried out on the sabkhas at Abu Dhabi. This may give a somewhat unbalanced picture of sabkha systems as a whole.

Finally, it should be noted that both sabkhas and salinas (playas) occur both in continental and in littoral environments; only the latter, coastal-type evaporites, will be examined.

4.2 The Sedimentary Origins of Sabkhas

4.2.1 General Geological Settings

Because coastal plains, including those located in arid climates, are developed best where littoral sediments accumulate rapidly, such plains tend to form today, as in the past, on shallow, gently sloping coastal shelves. Lateral sedimentary accretion, the essential aspect of sabkha formation, is favored by shallow offshore waters, by eustatic and tectonic stability, and by the abundant supply of sediment.

In general, modern cratonic areas, such as the Arabian shores of the Arabo-Persian Gulf, and the coasts of Australia, provide the most suitable settings for wide gently sloping shelves. However, many exceptions to this generalization are known. On the contrary, tectonically unstable areas, such as the Persian coast of the Gulf and, to a somewhat lesser extent, the narrow coastal shelves of the Red Sea and its gulfs, are somewhat less favorable for extensive progradation of littoral plains. Thus, the Gavish Sabkha is of very moderate dimensions relative to the wide coastal plains of Abu Dhabi; the littoral platforms of the Red Sea basin are narrow and generally fault-controlled, those of Abu Dhabi are exceptionally wide and have been only very moderately affected by Neogene tectonic movements.

Sabkhas and coastal plains, in general, tend to form along protected shorelines. As such, many occur along leeward coasts or around lagoons and wide embayments protected by nearby sedimentary or structural offshore barriers. This is the case both in the southwestern parts of the Arabo-Persian Gulf (Qatar and Abu Dhabi), and on the west coast of Australia (Shark Bay). Small littoral sabkhas also are forming today along the leeward shores of islands such as Marrawa (Abu Dhabi) and Bonaire (Netherlands Antilles); important, but essentially nonevaporitic, tidal flats are significant features of Bahamian platforms.

Coastal sabkhas are conditioned also by abundant sediment supplies. As is demonstrated in the following pages, these may be related to several, quite different mechanisms that may involve either the landward transport of marine carbonates or the seaward transport of terrigenous detritus.

4.2.2 Sabkha Morphology and Sedimentation

A review of most modern sabkha and related coastal evaporite systems suggests that they are created by one or other of three principal sedimentary mechanisms or, more often, by the combination of these mechanisms. The mechanisms involve supply of marine, eolian or fluviatile sediments.

4.2.2.1 Sabkha Formation by Longshore or Onshore Transport of Marine Sediment

a) Longshore Transport. This form of transport may result in the accumulation of coastal spits or barriers. Progradation of their downcurrent (distal) end leads to the progressive formation of lagoons favoring quiet-water sedimentation, a good example of which has been demonstrated by Shinn (1973 a, b) in northeast Qatar (Fig. 4.1 A). A similar situation exists within the coastal complex of Abu Dhabi (Fig. 4.2) where the presence of extensive coastal lagoons is the result of lateral accretion of spits which tend to link adjacent offshore islands, forming a protective coastal barrier system (Purser and Evans 1973). These sand ridges do not in themselves constitute sabkhas, nor do they normally include evaporite minerals. However, their presence is vital for the accumulation of fine sediments along low-energy shorelines, thus leading to the creation of extensive evaporitic sabkhas.

The formation of a littoral sand ridge encloses a related depression, generally occupied by a bay or lagoon. When the ridge has isolated these waters completely from the adjacent ocean, this depression may be the site of extensive, subaquatic evaporite formation (Fig. 4.3). This clearly has been the case of Sabkha el Meleh on the Tunisian coast (Perthuisot 1975). Here, some 30 m of Early Quaternary halite and gypsum completely fill a coastal depression whose present surface now

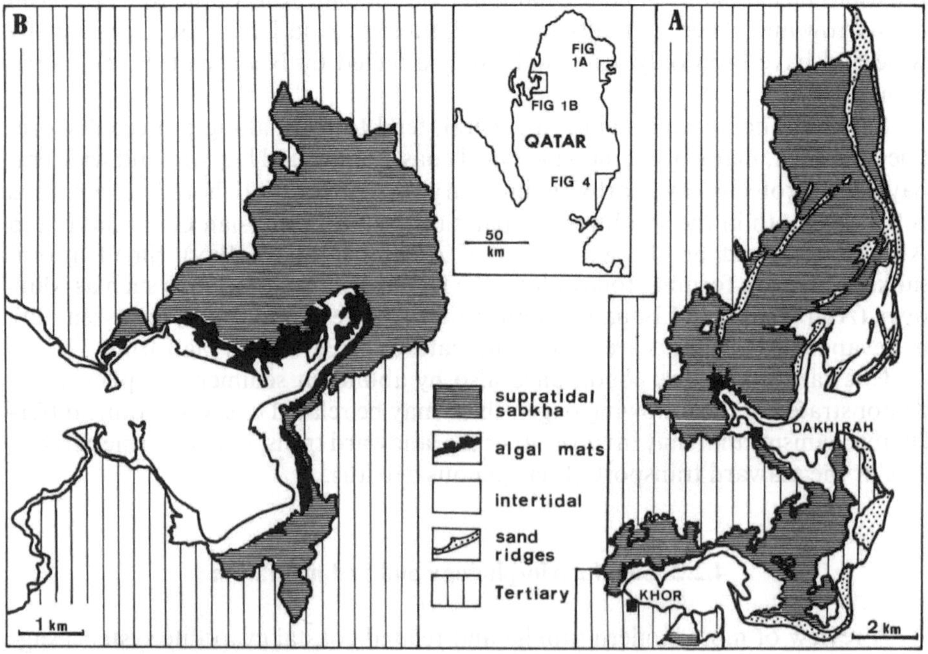

Fig. 4.1. A Sabkha formation patterns in the Bay of Quatar (Persian Gulf). Fig. 1A in inset relates to **A**; Fig. 1B in inset relaTES TO **B**; Fig. 4 in inset relates to Fig. 4.4. Sabkha formation related to coastal barriers, NE Quatar. (After Shinn 1973a). **B** Sabkha Faishakh. (After Illing et al. 1965)

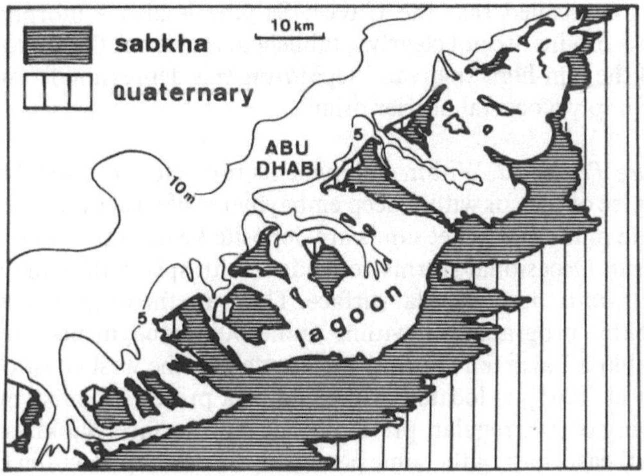

Fig. 4.2. Central part of the Trucial Coast showing distribution of sabkhas and coastal lagoons

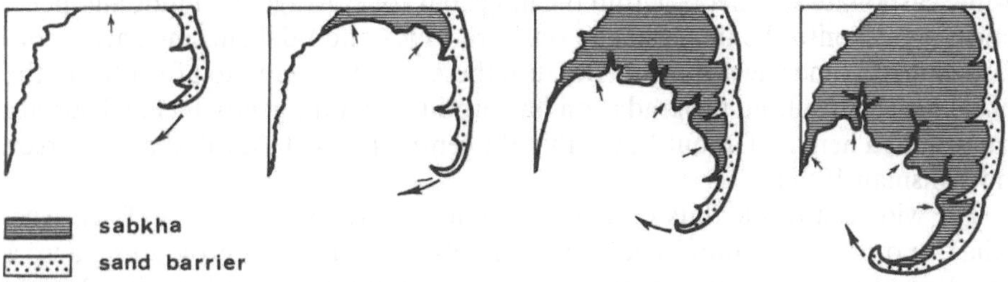

Fig. 4.3. Schematic formation of sabkhas by longshore and onshore transport of marine sediment

constitutes a true sabkha; as seems frequently to be the case, subaqueous evaporite sedimentation leads to progressive filling and emergence favoring the onset of capillary evaporation and true sabkha conditions.

Other complexes of this type occur in northwest Mexico (Baja California) where both the Ometepec and Mormona evaporitic lagoons are separated from the adjacent sea by low sedimentary ridges (Phleger and Ewing 1962, Pierre 1982).

Both lagoons are the sites of true sabkha evaporite formation; flooded periodically, they progressively dry and subsequent capillary evaporation predominates during most of the year, especially around the peripheral parts of these depressions.

The Coorongs of South Australia also are separated from the sea by a series of coastal dune ridges. Continental groundwaters feeding these sedimentary depressions evaporate to dryness during summer when capillary evaporation leads to the formation of interstitial brines (von der Borch 1965). However, in that no permanent sulfate minerals appear to be generated by these evaporitic processes, one should not consider the Australian Coorongs as being true sabkhas.

Finally, it is recalled that the Gavish Sabkha is also a morphological depression whose origins are not clearly established. However, the obvious sedimentary origin of the 2-m-high "plateau" separating this depression from the Gulf of Aquaba may imply a coastal barrier origin.

b) Onshore Transport. Within embayments protected by coastal barriers (see above), along lee coasts, or within deep embayments of structural or erosional origin, carbonate muds and pellet sands are deposited within the intertidal zone by tidal movement. Occasional storms coincident with spring tides may also transport sediments onto the supratidal surface. These are the main processes responsible for sabkha progradation within numerous embayments around Qatar Peninsula. Sabkha Faishakh (Fig. 4.1 B), situated on the west coast of the peninsula (Illing et al. 1965), is located at the end of a protected embayment formed by the drowning of an irregular, pre-Holocene land surface. Sabkhas on the less-protected northeast coast of the same peninsula are developed within bays created by coastal spit formation (Fig. 4.1 A). The vast coastal sabkhas of central Abu Dhabi (Fig. 4.2) whose width locally exceeds 10 km, are prograding across the adjacent lagoon at rates exceeding 2 m/year, mainly in response to the accumulation of fine carbonate sediments within the intertidal zone. Aragonite muds and pellet sands, driven onto the tidal flats by onshore winds and tidal currents, are a constant source of marine sediment. Because these processes tend to affect the entire shoreline, the resulting progradation has produced a flat, gently inclined surface that attains a height of about 1 m within the central parts of the sabkha (Patterson and Kinsman 1981).

The wide evaporitic flats bordering the inner parts of Shark Bay (West Australia), and notably around Hamelin Pool, also appear to be prograding across the adjacent lagoon mainly as the result of the accumulation of marine-derived carbonate sediments (Logan and Cebulski 1970).

This type of marine sedimentary accretion is important for evaporitic mineral formation, for two reasons:

1. It results in a coastal plain situated at, or very near, sea level and thus is susceptible to marine flooding. These marine waters, when present, constantly provide the dissolved ions necessary for efficient evaporite-mineral formation.
2. These sediments are essentially carbonates commonly aragonitic in composition, which provide a ready stock both of Ca and of CO_3 necessary for the formation of dolomite and other evaporite minerals. This is not the case for siliciclastic sediments delivered to coastal sabkhas by other processes.

Tidal channels: Sabkhas formed mainly by marine processes tend to be modified by numerous tidal channels whose presence imparts both a morphological and sedimentological heterogeneity to the sabkha. Sabkhas are not lenses of homogeneous sediment favoring a simple hydrodynamic regime.

Active channels may be floored by coarse, permeable sands, which, after they have been buried by a prograding sabkha, may constitute drains for interstitial waters eventually permitting the seaward flow of brines from the inner parts of sabkha. Although not demonstrated (despite numerous sabkha studies), they may be important for dolomitization and, in ancient sequences, for the emplacement of metals and hydrocarbons.

4.2.2.2 Sabkha Formation by Eolian Supply

Because sabkhas generally occur in desert climates, they are affected strongly by wind. This is an important environmental factor not only because wind stimulates evaporation but also because it tends to supply littoral plains with detrital sediment. That virtually all modern sabkhas are influenced to a variable degree by eolian sediment supply is illustrated by the following examples.

a) Umm Said, Southeast Qatar. The most striking example of eolian accumulation is that studied by Shinn (1973b) at Umm Said (Fig. 4.4), where a sabkha 7 km wide and 40 km in length has developed along the lee coast of the Qatar Peninsula.

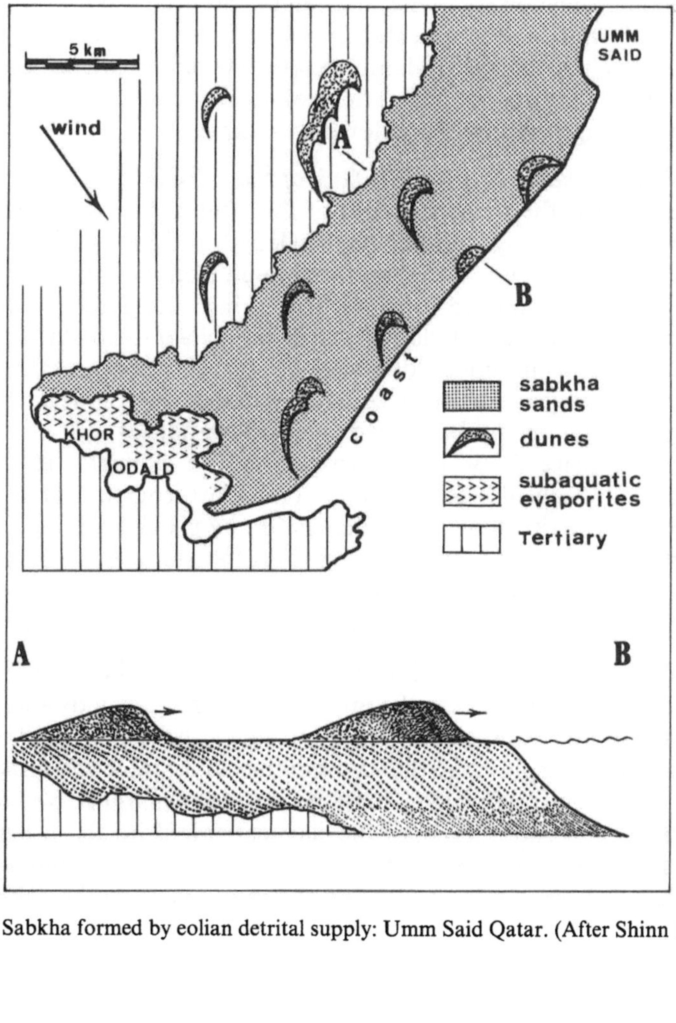

Fig. 4.4. Sabkha formed by eolian detrital supply: Umm Said Qatar. (After Shinn 1973b)

Prevailing northwest shemal winds are piling terrigenous carbonate and silicate sands into spectacular barchan dunes that migrate obliquely across the peninsula, to its leeward, southeast coast (Fig. 4.6 A). Dune sand is constantly delivered to the shoreline where it is deposited and reworked by marine currents. Seaward progradation results in a flat, sandy plain situated only slightly above sea level. The constituent terrigenous sediments are characterized by seaward-dipping cross-strata that reflect progradation of subaquatic avalanche slopes.

Because the surface of the sandy coastal plain at Umm Said is situated just above sea level, it is very susceptible to marine flooding. Constant evaporation increases the salinity of interstitial waters from which gypsum precipitates. Wells drilled by Shell Research have demonstrated that the density of interstitial brines tends to increase downward and seaward (De Groot 1973), constituting one of the rare demonstrations of the theoretic reflux model.

The southern extremity of this terrigenous sabkha complex is prograding across the closed lagoon of Khor Odaid located on the frontier between the states of Qatar and Abu Dhabi. Extremely shallow along its inner margin, this highly restricted lagoon is precipitating subaquatic gypsum (Perthuisot and Jauzein 1978) within microbial mats. Thus, sustained progradation of the adjacent sabkha will result in a vertical sequence grading from subaqueous marine sediment and evaporites at the base to laminated quartz sands with sabkha-type evaporites at the top. This may be a natural evolution of shallow coastal evaporitic environments, as suggested by the following examples.

b) Sabkha el Mellaha, Ras Gharib, Gulf of Suez (Fig. 4.5). This structural depression located near the western shores of the Gulf of Suez (Egypt) includes a highly saline lake measuring about 1,500 m in length and 500 m in width. Situated approximately 10 m below sea level and separated from the sea by a ridge 1 km wide, this depression is supplied constantly by springs along its eastern (seaward) margin (Fig. 4.6B). Seawater obviously is delivered via a hydrostatic head to the salina (pers. obs. and unpublished studies by the University of Assiut). Lake waters are precipitating aragonite, gypsum, and halite which, at least locally, constitute relatively pure, massively bedded layers of evaporites.

The salina occupies the southeastern parts of an elongate (15 km) sabkha whose surface is strongly affected by winds. Transport from the northwest has resulted in the asymmetric filling of the depression by progradation of the sabkha along its northwestern edge. Only the extreme southeastern end of this small basin remains as a residual salina. Initial filling by subaquatic evaporites is being completed by rapid progradation of wind-blown terrigenous detritus (Fig. 4.5).

c) Abu Dhabi Sabkhas. The Trucial Coast is essentially a windward shore oriented obliquely with respect to the prevailing northwest winds and, unlike the nearby leeward shore of southeast Qatar Peninsula, is not particularly susceptible to eolian accretion. Despite this, the vast sabkhas at Abu Dhabi, although composed of marine-derived carbonate along their seaward margins, consist predominantly of terrigenous sands; most of the classical anhydrite and other evaporite mineralogy occur within this terrigenous framework (Figs. 4.15 and 4.16).

The terrigenous sediments composing the inner parts of the sabkha clearly originate in two ways. An important part is supplied by sheetflooding from the

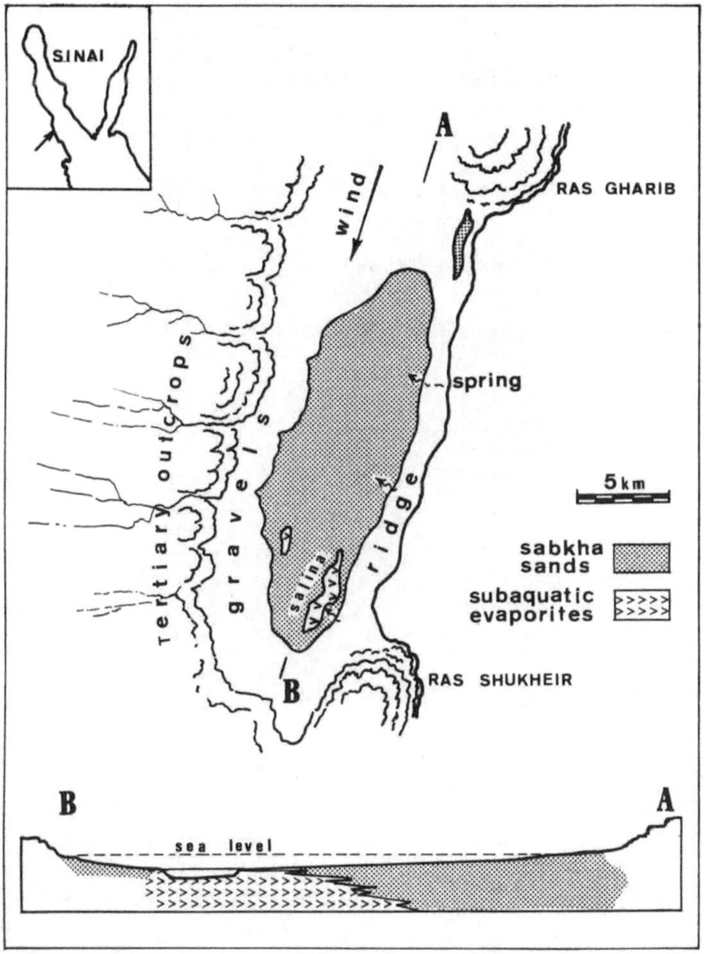

Fig. 4.5. The sabkha at Ras Gharib, Gulf of Suez showing filling from N by eolian sediment supply

adjacent continent. However, occasional offshore winds blowing from the adjacent desert supply terrigenous detritus to much of the sabkha. It is also of interest to note that Pleistocene bedrock underlying much of the sabkha consists of eolianite. Although possibly never a true sabkha deposit, these sands nevertheless receive much of the brine generated near the modern sabkha surface. Because they are very permeable, these Quaternary eolianites may serve as conduits for a refluxing system.

4.2.2.3 Sabkha Formation by Fluviatile Supply

Although sabkhas generally occur in arid climates, many are affected to some degree by nonmarine waters; deserts are characterized by periodic heavy rains followed by rapid, nonchanneled, surface runoff. This results in an efficient (but

Fig. 4.6. A Salina at Ras Gharib, W Gulf of Suez, showing peripheral barrier ridge (horizon), saline spring waters and crusts of halite. **B** Sabkha at Umm Said, SE Qatar, formed by progradation due to sediment supply by eolian dunes. (Photo E. Shinn)

little-studied) transport process that brings great quantities of detritus to topographic lows including both inland and coastal sabkhas. The landward fringes of virtually all littoral sabkhas around the Arabo-Persian Gulf are supplied by this process.

In addition to nonchanneled supply of continental detritus, important coastal sabkhas may be formed on fluvio-marine deltas developing in arid climates.

a) The Estuarine Shatt al Arab Sabkhas, Mesopotamia. Despite arid climates, major river systems do occur in arid areas. This is the case for parts of the Nile, the Senegal River and, especially, the Tigris–Euphrates system in Mesopotamia. These twin rivers, supplied by snows and rains in remote temperature climates, carry an important charge of fine sand and clay to the lakes and marshes of the Lower Mesopotamian Plain. Although much of the detritus is deposited as lacustrine deltas, an important part is delivered to the Shatt al Arab estuary which is supplemented by an important Iranian river, the Karun, bringing detritus from the adjacent alpine chain. Rapid fluviatile supply to the adjacent shores of the Arabo-Persian Gulf has created a vast, lobate deltaic plain (Fig. 4.7 A) composed essentially of finely laminated clays and silts (Purser et al. 1982). Because the delta front is located on a protected, leeward coast, the resulting deltaic plain lacks a frontal sand barrier and fine sediments are ubiquitous.

Interstitial waters within the fine sediments of the deltaic plain are highly saline. This situation undoubtedly reflects the somewhat unusual sedimentary origin of this flat, desiccated surface. Unlike most deltaic plains, the constituent sediments are not supplied by lateral fluviatile flooding but are transported to the front of the delta by tides whose amplitude is in the order of 4 m. Thus, although sediment supply is fluviatile, seaward accretion is related closely to marine hydrodynamics. This marine influence is important for evaporite formation. Interstitial waters, even close to the estuary, are highly saline and this favors the formation of gypsum and other evaporite minerals.

With the exception of date palms and other cultivated vegetation which occupy a 3-km-wide zone fringing the estuary, the arid plains that constitute the Shatt al Arab delta are virtually devoid of vegetation (Fig. 4.7 B). Their salt-encrusted, cinder-like morphology is almost identical to that of the Gavish Sabkha. The intimate association of evaporite minerals forming from highly saline brines, and low-salinity (< 1‰) estuarine waters of the Shatt al Arab, is instructive; it demonstrates that deltas accreting from protected shorelines in arid climates may form extensive sabkhas, composed of laminated gypsiferous clays that are not uncommon in the geological record.

b) The Fluvial-fan Sabkhas of Southeast Iran. The Zagros alpine chain in Iran is drained by numerous, semi-permanent streams many of which discharge onto the northeastern shores of the Arabo-Persian Gulf. The Mehran River (Fig. 4.8), studied by Baltzer et al. (1982), is forming a fan whose radius is in the order of 15 km and whose littoral fringes exhibit all the typical features of coastal sabkhas. This simple, relatively symmetric cone differs from the more important Shatt al Arab delta in that continental sediment supply is essentially nonchanneled; an axial talweg exists but this is of very modest dimensions. Aerial photos and mor-

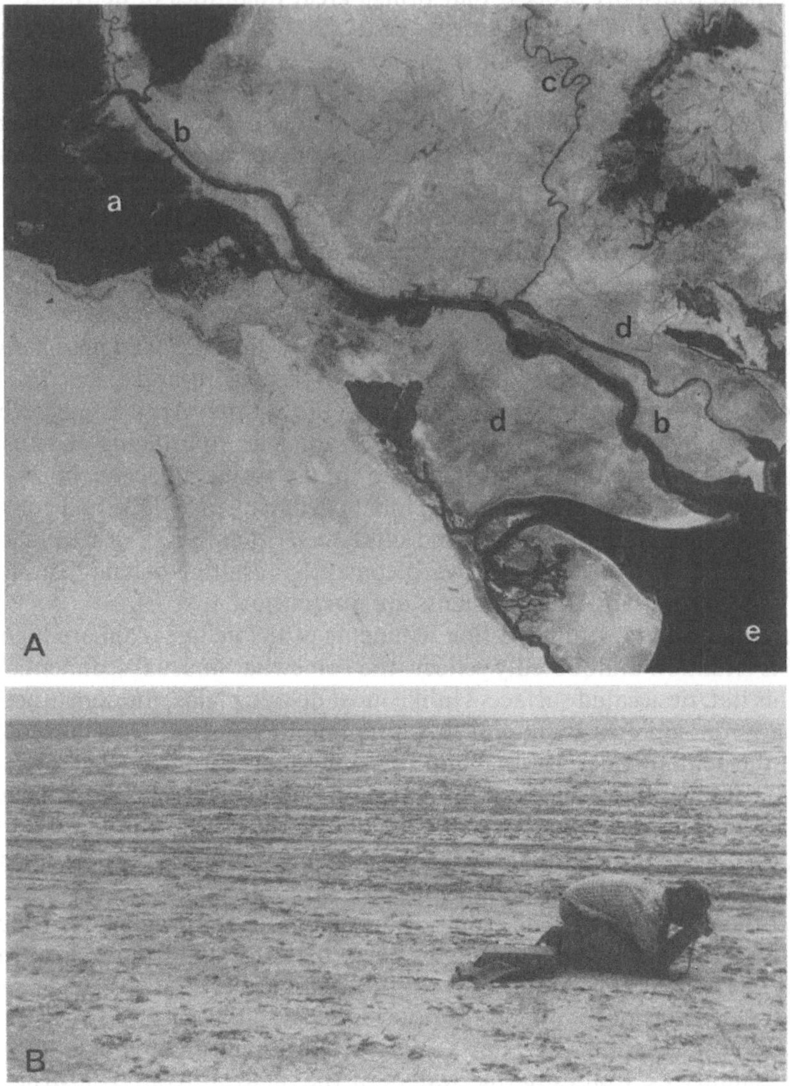

Fig. 4.7. A Satellite photo of the Shatt al Arab delta showing: *a* Lakes and swamps of the Lower Mesopotamian plain; *b* Shatt al Arab estuary; *c* Karun River; *d* estuarine sabkhas; *e* Arabo-Persian Gulf. **B** Surface morphology of the estuarine sabkha

phology clearly suggest that surface runoff during violent rains affects the entire surface of the cone. Sheetflooding results in a granulometric evolution: the finest silt and argillaceous mud fractions are being deposited around the periphery of the cone whose surface is subhorizontal and covered locally by microbial mats. The brown, laminated clays underlying this arid, littoral plain contain numerous mm-sized crystals of gypsum and up to 30% dolomite. Hydrology undoubtedly is dominated by meteoric waters during periods of violent rain, but these are in-

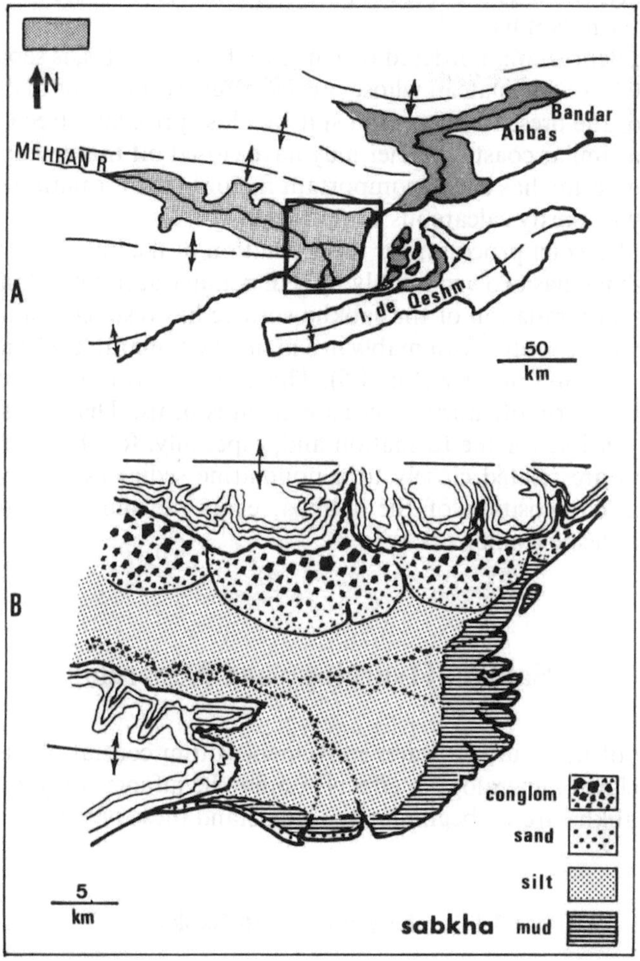

Fig. 4.8. Localization and sediment distribution on a fluviatile-fan sabkha, Mehran R., SE Iran

frequent and the highly saline interstitial waters probably reflect flood recharge by marine waters whose access is facilitated by exceptionally high tidal ranges (5 m) characteristic of this area.

This Iranian example also is instructive. It shows that evaporites, admittedly in small quantities, may form in tectonically active, perialpine settings providing that these are located within an arid climate. It also confirms that sabkhas may constitute an integral part of fluvio-detrital systems.

4.2.2.4 The Sedimentary Origins of the Gavish Sabkha

The examples discussed show that coastal sabkhas may be formed by at least three basic sedimentary processes involving marine, eolian or fluviatile supply.

Most frequently, sabkhas are formed by a combination of these processes, as is the case at Gavish Sabkha.

This ovoid depression, bordered by true sabkha evaporites, is separated from the sea by a 400-m-wide plateau whose precise origin seems to be in doubt. However, in view of the presence of coastal spits in close proximity (see Chap. 9) it is possible that a similar coastal barrier may have closed off this embayment. That marine sedimentation has been an important factor in the formation of this ridge is indicated by its partly calcareous composition.

Whatever the exact processes for the formation of the initial depression, it is clear that its shape has evolved mainly by sedimentary accretion along its southern periphery. Progradation of the sabkha surface has resulted mainly from eolian supply from the south, presumably in a manner comparable to that at Sabkha el Mellaha in the Gulf of Suez (Fig. 4.5). This terrigenous input is possibly augmented by surface runoff during occasional rainstorms. That the sedimentary processes responsible for the formation and, especially, for the lateral filling of this depression are related closely to a nonmarine sediment supply is demonstrated by the composition of the sabkha; evaporite minerals are dispersed through a siliciclastic framework.

4.3 Sabkha Hydrology and Mineralogy

The study of near-surface water movement within coastal plains in general and of their related mineralogical effects is still in its infancy; only at Abu Dhabi and Gavish Sabkha are we beginning to understand these processes.

4.3.1 Arabo-Persian Gulf Sabkhas

4.3.1.1 Hydrology of the Sabkhas

The interstitial waters from which evaporite minerals are precipitated have three potential origins: (a) the sea, (b) continental groundwaters, and (c) rain (presumably a negative factor in mineral precipitation). The hydrodynamic properties of these waters depend mainly on morphology both of the sabkha itself and of the adjacent continental terrain. They depend also on the width of the coastal plain. Obviously, marine waters tend to be greater factors along the littoral periphery. As the sabkha widens by sedimentary accumulation, its central and landward parts become less susceptible to marine incursions.

a) Marine Water Supply. In coastal systems it is clear that marine waters constitute the main, but not the only, source for evaporite minerals. It is also evident that the formation of important quantities of minerals requires important quantities of water and therefore an efficient hydrodynamic system. The principal elements of such a system include surface flooding and interstitial flow.

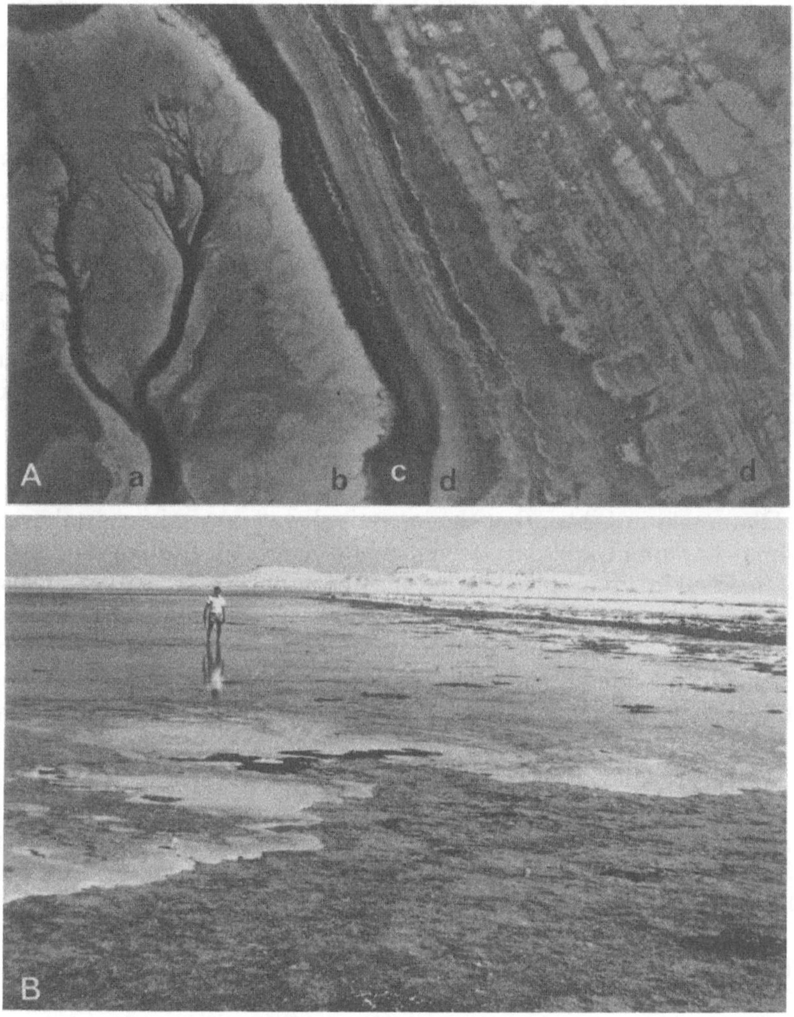

Fig. 4.9. A Aerial photo showing the tidal zones at Abu Dhabi. *a* Lagoon with subaquatic channels; *b* intertidal zone; *c* dark-colored microbial mats marking limit between intertidal and supratidal zones; *d* supratidal sabkha showing accretion lines. **B** Surface flood-recharge, the essential hydrodynamic process on the Abu Dhabi sabkhas

Surface flooding: Because the surface of the sabkha is situated near sea level, it is particularly susceptible to flooding during periods of exceptionally high tides (Fig. 4.9), especially if these coincide with strong onshore winds. This process was termed "flood-recharge" by Butler (1969). Many of the sabkhas of Abu Dhabi and Qatar are inundated by a thin pellicule (several cm) of water several times per year. Where the sabkha is narrow (several km), as at Sabkha Faishakh (West Qatar), these marine waters may cover the entire surface. However, where sabkhas are excessively wide (10 km), as in central Abu Dhabi, only the outer half of the plain is affected regularly. This periodic surface flooding furnishes the ions necessary for mineral formation and probably constitutes the essential supply mechanism for sabkhas lacking a periphereral sedimentary barrier (see following paragraph).

Interstitial flow: Seawater is supplied to sabkhas via subsurface flow when the groundwater table within the sabkha is situated at, or near, sea level. This is the case where sabkhas fringing salinas are separated from the sea by a sedimentary barrier, as at the Gavish Sabkha and at Ras Gharib.

G. M. Friedman and Sanders (1967), and others, have stressed the importance of upward capillary flow of interstitial fluids stimulated by surface evaporation. These vertical capillary movements, in turn, may lead to lateral migration of fluids by a mechanism termed "evaporative pumping" (Hsü and Siegenthaler 1969). Although probably not operating where sabkha water tables rise landward (as in Abu Dhabi), it may be an important mechanism where the water table is horizontal.

b) Continental Water Supply. Geochemical studies have demonstrated, both at Abu Dhabi and in Libya, that the interstitial waters within coastal evaporite systems may be supplied, in part, from the adjacent land. If these waters are sufficiently enriched in dissolved salts, they may be important sources of coastal evaporite minerals. As with their marine equivalents, these waters may be delivered to the sabkha system either by surface runoff during occasional heavy rains, or, more frequently, by interstitial groundwater flow. Such groundwater flow is especially efficient where the hinterland is relatively high or where rainfall is frequent. It is aided by the fact that the water table within a coastal sabkha, such as at Abu Dhabi, is gently inclined toward the sea.

c) Hydrology Within the Sabkha. Evolution of interstitial brines: our knowledge of sabkha hydrology is based almost exclusively on studies carried out in Abu Dhabi. Those of Kinsman (1969a), Butler (1969), Hsü and Siegenthaler (1969), Bush (1973), McKenzie et al. (1980), Patterson and Kinsman (1981) constitute a solid base from which emerge certain aspects of interstitial brine evolution (Fig. 4.10).

Interstitial waters, as already noted, are delivered to the sabkha from opposing sources – the sea and the land.

Seawaters: Because of evaporation within the adjacent lagoons, the initial salinity of seawater may be concentrated to about 60‰. These are delivered to the outer half of the sabkha essentially by flood-recharge, this supply naturally being most frequent along the marine margin. Following flooding, waters sink into the sabkha rapidly, elevating the water table, which is lowered subsequently by capillary evaporation. This evaporation provokes not only a rapid increase in salinity but also stimulates upward movement of water within the phreatic zone by evaporative pumping. This vertical movement is important because it results in constant supply of ions necessary for mineral formation.

Salinities of interstitial waters increase from 70‰ to 365‰ across the sabkha; they attain a maximum within the central parts. This progressive concentration is also reflected both in the ^{18}O of the waters which increases from $+1$ in the lagoon to about $+5$ in the central parts of the sabkha (McKenzie et al. 1980), as

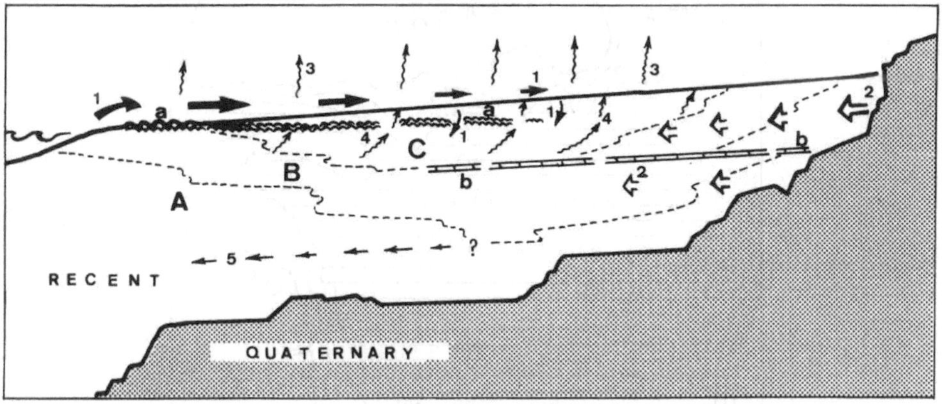

Fig. 4.10. Profile showing water movements within the Abu Sabkha: *1* flood recharge supply; *2* continental groundwater supply; *3* surface evaporation; *4* evaporative pumping; *5* possible reflux in buried channel sands. Salinity zones: *A* moderate, *B* high, *C* very high

well as by the Cl/Br and the K/Br ratios of the brines (Patterson and Kinsman 1977).

Progressive increase in salinity (chlorinity) toward the central parts of the sabkha (Fig. 4.11) is an expression of water supply from both sides, and its contemporaneous evaporation; the central parts represent areas renewed the least frequently from the periphery and whose stagnating waters therefore undergo greatest evaporative concentration.

Continental groundwaters are delivered to the landward parts of the sabkha. They have been identified not only by the decrease of salinity within the sabkha brines but also by high K/Br ratios which are in the order of 150 (verses 15 for marine waters) (McKenzie et al. 1980). The input of relatively low-salinity groundwater which, like its marine equivalent, progressively evaporates, results in the high-chlorinity plateau being located in the central parts of the sabkha (Fig. 4.11).

The high-salinity (chlorinity) plateau shown by Patterson and Kinsman (1977) to exist in the middle parts of Abu Dhabi sabkha also represents the zone of mixing of highly evaporated marine and continentally derived waters. However, as the sabkha progrades, this plateau moves laterally toward the sea. The rate of migration, as calculated by Patterson and Kinsman, is in the order of 30 to 100 cm/year.

Seaward migration of interstitial brines is implied by the gently inclined (seaward) nature of the water table (Patterson and Kinsman 1977). However, as shown by McKenzie et al. (1980), this reflux movement is very slow and probably not sufficient to explain extensive dolomitization. Furthermore, at any given point on the sabkha, salinity tends to decrease downward probably as the result of the related effects of surface evaporation and upward movement of phreatic waters in response to evaporative pumping. This downward decrease in salinity, also noted by Illing et al. (1965) in Sabkha Faishakh (West Qatar), tends to limit the possibilities of an evaporative reflux system, especially within the fine sabkha sediments which are most affected by capillary evaporation.

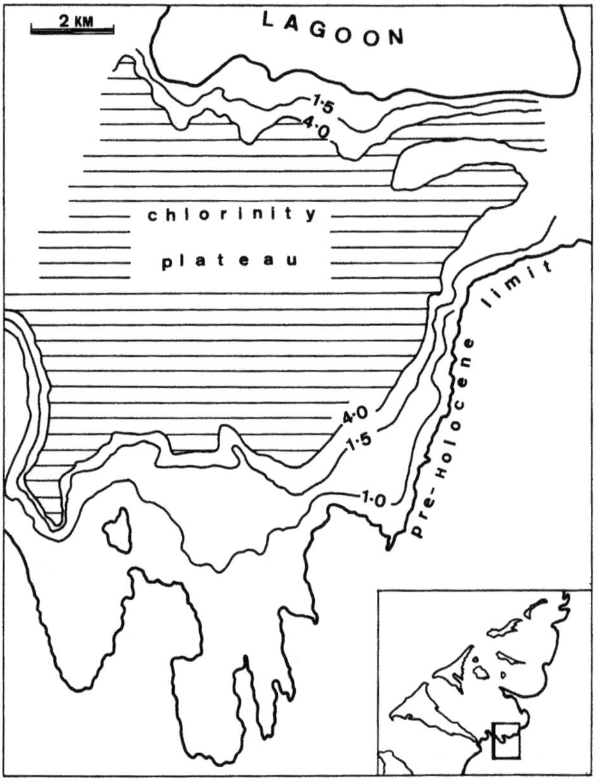

Fig. 4.11. Isochlorinity distribution on Abu Dhabi sabkha showing central plateau and lateral decrease both toward lagoon and continent. (After Patterson and Kinsman 1977)

Lateral movement of interstitial waters is severely limited by the impermeable nature of these sediments which, at least within the seaward parts of the sabkha, are composed essentially of carbonate mud. As already noted, however, sabkha sediments are not homogeneous. Effective permeability barriers exist both in the form of buried microbial mats (Bush 1973) and as cemented layers (Hsü and Schneider 1973, McKenzie et al. 1980), that act as aquacludes subdividing the sabkha sequence into at least three separate hydrostatic zones. Continental groundwater supply may be concentrated within the lowest level and it reaches the upper horizons (via evaporative pumping) only where the impermeable crust has been broken.

Although studies suggest very slow return of brines toward the sea, the relatively minor quantities of interstitial halite within the sabkha could suggest flushing of brine. Furthermore, the sabkha sequence overlies a very permeable Pleistocene eolian sand which may act as an efficient reservoir. Finally, the presence of numerous buried channels, partly filled with coarse sediment, could also act as local conduits for dense, possibly dolomitizing brines.

Reflux of brines within the sandy sediments underlying the sabkha at Umm Said (Southeast Qatar) seems to be demonstrated by De Groot's (1973) studies

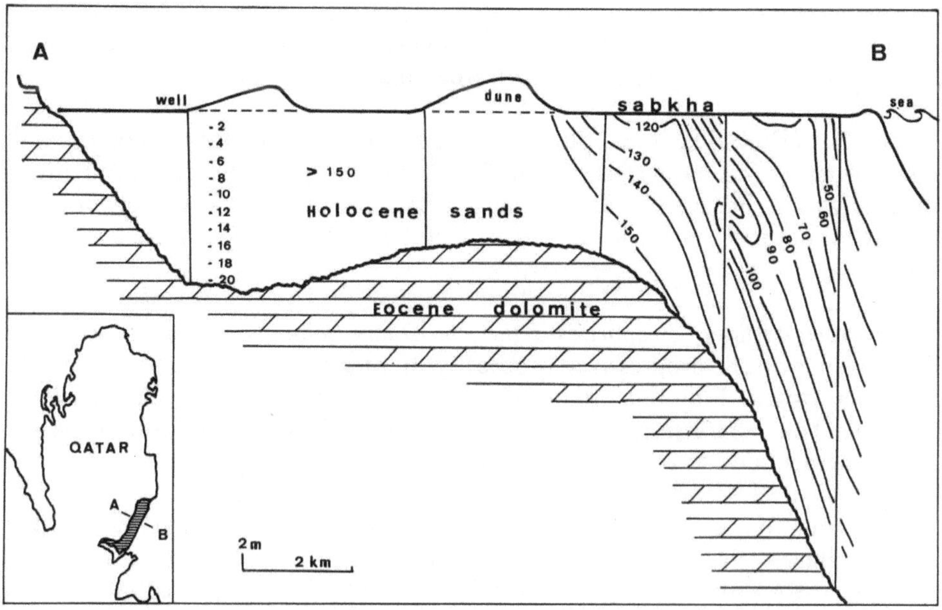

Fig. 4.12. Isochlorinities at Umm Said sabkha. The disposition suggests a possible reflux system. (After De Groot 1973)

that reveal a slight downward and seaward increase in salinity of the sabkha brines (Fig. 4.12). The sedimentary composition of this sabkha is quite different from that at Abu Dhabi; it is composed of homogeneous, permeable, essentially terrigeneous sand. Although minor amounts of Ca-rich dolomite muds have been identified, they occur frequently near the surface and the importance of the reflux mechanism vis-a-vis dolomite formation remains to be demonstrated.

4.3.1.2 Mineralogy of the Arabo-Persian Gulf Sabkhas

Although evaporite minerals do precipitate from standing waters occurring in local depressions on the sabkha (Fig. 4.14 A) and, especially, within certain closed lagoons such as Khor Odaid (Fig. 4.4), the vast majority of the Arabian evaporites are precipitated from interstitial waters within the sabkha; they are largely secondary in origin.

a) Evaporation Rates. Annual rainfall in the southwestern parts of the Gulf is in the order of 40 to 100 mm/year, whereas net seawater evaporation is about 1,240 mm (Butler 1969). However, because the minerals formed within the sabkha sediments are a response to capillary evaporation, it is clear that these rates will be greatly inferior to those cited for surface-water evaporation.

Experiments by Schneider (in McKenzie et al. 1980) have suggested capillary evaporation rates to be between 200 and 500 mm/year.

Furthermore, as noted by McKenzie et al. (1980), the high humidity, especially at night, (60%–70%) tends to reduce evaporation and may even contribute to the surface-water supply; it results in the diurnal dissolution of surface halite crusts.

Finally, it should be noted that evaporation rates depend not only on ground temperatures (which range from 10° to 53 °C), but especially on speeds and directions of winds. Because winds are most persistent during the winter months, highest evaporation rates do not necessarily coincide with the hottest summer months.

b) Precipitation of Evaporites in the Abu Dhabi Sabkhas. Precipitation naturally depends on the progressive evaporative of interstitial waters that are mainly of marine origin. Flooding frequency is related to proximity to the lagoonal shoreline. Thus, chlorinities of interstitial waters and the distribution of the related minerals, tend to be zoned, both laterally and vertically, with respect to the shore (and, to a lesser degree, to the continental edge of the sabkha).

This zonation, established by Kinsman (1969a), Evans et al. (1969), Butler (1970b), Bush (1973), Patterson and Kinsman (1977), and others, in the wide sabkhas situated immediately southwest of Abu Dhabi Island, is essentially as follows (Fig. 4.13):

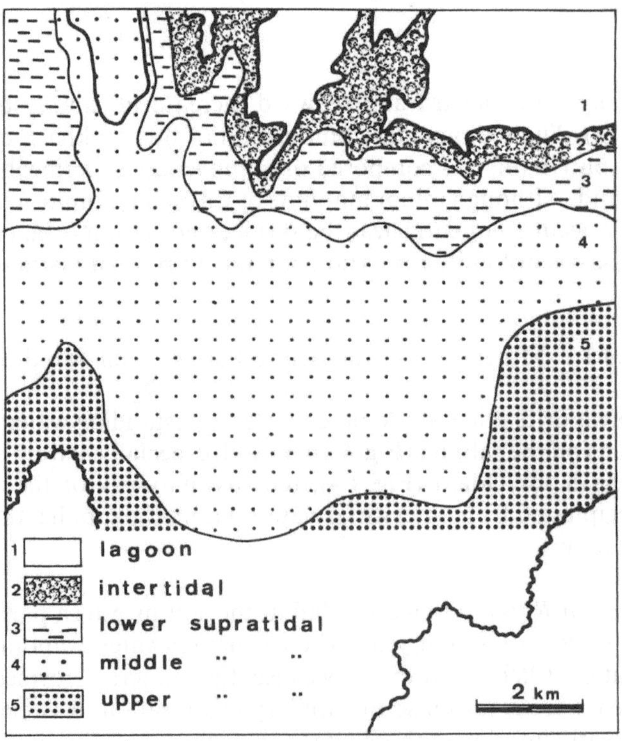

Fig. 4.13. Schematic distribution of the diagenetic zones in the sabkha SE of Abu Dhabi Island. *1* Lagoon; *2* Intertidal; *3* Lower Supratidal; *4* Mid Supratidal; *5* Upper Supratidal

Lagoonal environments: The saline (60‰–70‰) waters of the lagoons lose calcium carbonate, presumably to physico-chemical precipitation, before delivery to the sabkha (Kinsman 1969). Thus, a part of the lagoonal aragonitic muds must be of evaporitic origin.

Intertidal zone: In addition to further precipitation of aragonite, both in the form of mud and as beach rock cements, the upper parts of the intertidal zone are characterized by the formation of dolomite and small (5 mm), flattened crystals of gypsum. Both these minerals occur mainly within, or immediately below the microbial mat that marks the transition between the intertidal and supra-

Fig. 4.14. Subaquatic evaporites in pond on Abu Dhabi sabkha: *a* oriented gypsum crystals; *b* white halite incrusting gypsum produced by progressive evaporation of pond; *c* residual pond waters. **B** Sabkha evaporites within microbial mat, Abu Dhabi. This mush of small gypsum crystals in upper parts of the intertidal zone occurs in laminated dolomite *(beige)* and microbial mat *(black)*

tidal zones. Factors favoring the formation of calcium-rich, poorly ordered dolomite are not yet established with certitude. They may be related to:

- the presence of organic matter within the fine, aragonitic muds;
- the increased Mg/Ca ratios (about 10) caused by the precipitation of gypsum;
- or to the removal of sulfate ions whose presence tends to inhibit dolomite precipitation (Kastner 1982).

Lower supratidal zone: The top 30 cm of sediments are a complex mixture of aragonite and dolomite mud, and small gypsum crystals which form a compact

Fig. 4.15 A, B. Typical anhydrite structures in mid-supratidal zone Abu-Dhabi sabkha. **A** Nodular and bedded anhydrite. (Photo G. Butler). **B** Bedded anhydrite and nodules coalescing to form typical chicken-wire structures

mush situated within the microbial mat (Fig. 4.14 B). In addition, larger (up to 20 cm) gypsum crystals are dispersed within the underlying lagoonal muds at depths of about 1 m (Fig. 4.15). The presence of undeformed microbial laminae and other inclusions within many of the larger gypsum crystals shows that these form by replacement of carbonate sediment.

Within this same zone, at distances of approximately 1 to 2 km from the permanent lagoon, the first traces of anhydrite are noted. According to Butler (1969), most of the anhydrite forms by secondary replacement of metastable gypsum; this is demonstrated by the presence of gypsum pseudomorphs within the anhydrite. However, part of the anhydrite seems to be formed also as a by-product either of dolomitization (Ca2+) or by dissolution/reprecipitation of aragonite.

Middle supratidal zone: This zone corresponds to maximum chlorinity (Fig. 4.11), which coincides both with maximum evaporation of brines and with the mixing of marine and continentally derived waters. The salt-encrusted surface is developed on terrigeneous, eolian-derived sands within which most preexisting gypsum is replaced by anhydrite (Fig. 4.15). The anhydrite, whose microstructure consists of a felt of small (2–100 microns) fibrous crystals, exists firstly in the form of nodules, each nodule having replaced a gypsum crystal. Nodules tend to coalesce and progressively form the characteristic chicken-wire structure as anhydrite replaces the intervening sedimentary framework. Nodules coalesce forming layers of pure white anhydrite measuring up to 50 cm in thickness. These soft layers tend to be deformed, especially toward the inner parts of the supratidal zone. Deformation structures include enterolithic folds, microdiapirs and polygonal structures (usually visible on the sabkha surface).

Underlying lagoonal or intertidal aragonite muds are replaced, almost entirely (80%), by diagenetic dolomite having $\delta^{18}O$ in the order of $+2.5$–$+3.7$. These lagoonal muds also include large, discoidal gypsum crystals, as in the lower supratidal zone.

Uppermost supratidal zone: The innermost parts of the sabkha, composed almost entirely of terrigeneous sediments, are supplied mainly by continental groundwaters. This zone includes abundant anhydrite nodules, generally grouped into distinct layers (Fig. 4.16). However, the main peculiarity of this zone concerns the presence of gypsum which either replaces the anhydrite, or occurs as intergranular cement.

Halite is a common mineral over most of the sabkha but is limited, almost entirely, to the surface. Crusts rarely exceed a few cm in thickness and these generally dissolve following each marine flood. This mineral does not normally constitute a typical element of the sabkha evaporite series.

c) Precipitation of Evaporites in Sabkha Faishakh, West Qatar. Studies by Illing et al. (1965) have demonstrated that gypsum and dolomite are forming under conditions comparable to those existing at Abu Dhabi. However, despite an almost identical climate, Sabkha Faishakh is characterized by only rare traces of anhydrite and this mineral, probably the most characteristic evaporite element in Abu Dhabi, is lacking in Qatar. This difference may possibly be the result of the somewhat smaller size of sabkhas in Qatar (<5 km), this favoring their more frequent marine flooding and generally lower interstitial salinities.

Fig. 4.16. Complete sequence in middle parts of the supratidal zone, Abu Dhaba sabkha, showing: *a* massive, partially dolomitized, lagoonal muds; *b* scattered, large crystals of gypsum in lagoonal muds; *c* laminated microbial mat and dolomite; *d* desiccated microbial mat; *e* gypsum mush being transformed into white anhydrite; *f* diapiric anhydrite; *g* nodular anhydrite; *h* terrigenous sand; *i* hammer (scale)

4.3.2 Divers Coastal Evaporite Areas

4.3.2.1 North African (Tunisian and Libyan) Sabkhas

Virtually all of the numerous sabkhas along the North African coast are located behind littoral sand ridges. Many, including the Sabkha el Meleh at Zarsis in Tunisia (Perthuisot 1975), were forming already during the Pleistocene as subaquatic lagoonal evaporite basins closed by coastal spits. As such, their hydrology must have been quite different from that of the Abu Dhabi sabkhas; parental waters were delivered through a permeable coastal barrier or via narrow lagoonal channels. Progressive filling of these basins with massive, subaquatic halite and other evaporite minerals, has led progressively to true sabkha conditions, notably around their peripheries.

On the edge of the Gulf of Sirte in Libya, a double barrier ridge limits a coastal sabkha 20 km wide whose surface is permanently dry (Fig. 4.17 A). Studies of sulfur isotopes (Rouse and Sherif 1980) show that gypsum within the lower parts of the evaporite sequence (below the sabkha) have been derived from seawaters ($\delta^{34}S$ in the order of $+20$) possibly prior to complete closing of the coastal bar-

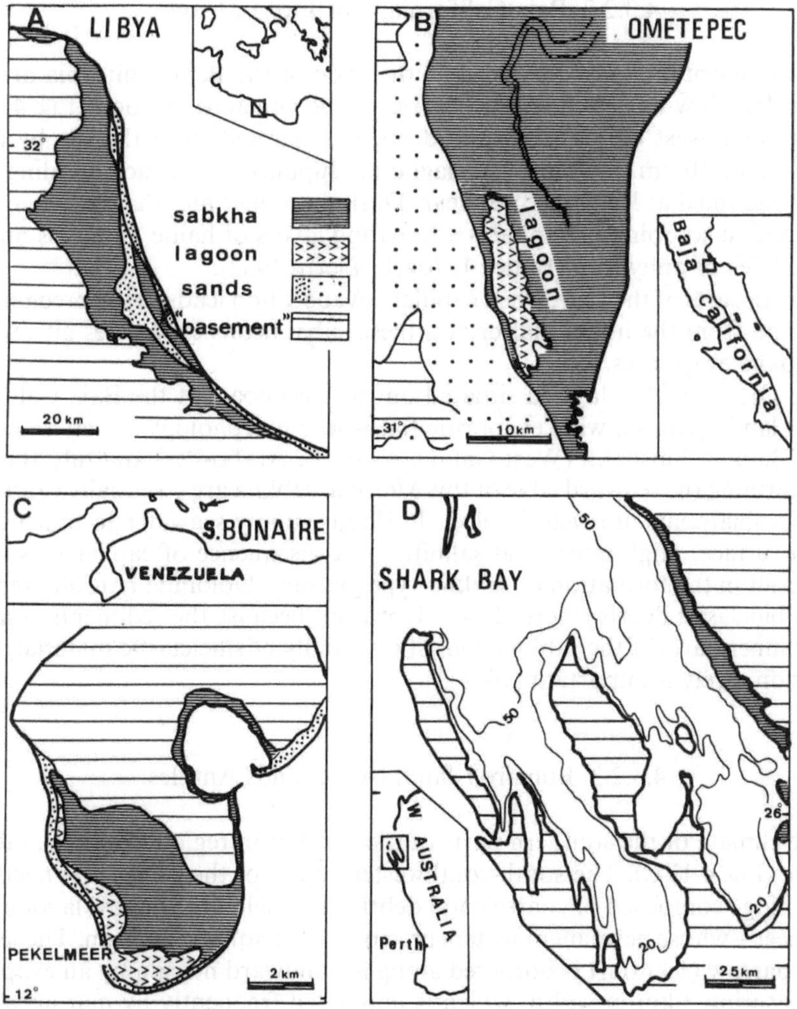

Fig. 4.17 A–D. Diverse coastal evaporite systems. **A** Libya; **B** Baja California; **C** Bonaire Island; **D** Gladstone Embayment, Shark Bay

rier. The near-surface parts of the sabkha, however, include gypsum whose ^{34}S-values are about +15, suggesting a nonmarine origin. These values correlate with the isotopic properties of the brackish (3‰ salinity) waters in springs located along the inner edge of the sabkha. The relatively constant, low isotopic values extend across the sabkha to the peripheral barrier. This implies that continental groundwaters are being supplied to the entire sabkha, and possibly even to the sea.

This Libyan example is interesting in that it may well be a modern analog for certain ancient coastal evaporite systems, such as parts of the Messinian of the Mediterranean (Pierre 1982) and the Eocene evaporites of the Paris Basin whose sulfur isotopes also seem to reflect, in part, a continental origin.

4.3.2.2 Baja California, Northwest Mexico

The evaporitic plains situated on either side of the Baja California are bordered by shallow bays and coastal barriers. The Ometepec lagoon (Fig. 4.17 B), situated southwest of the Colorado delta, is separated from the sea by a low plateau some 10 km in width. This lagoon is supplied by surface flooding from the sea and filled at least once per year. During summer months the water evaporates and in so doing precipitates alternating layers of halite and gypsum with rare nodules of anhydrite (Butler 1970a, b, Pierre 1982).

During parts of the year, when capillary evaporation leads to the precipitation of gypsum from the interstitial brines, these subaqueous evaporites give way to true sabkha evaporites.

The Ojo de Liebre lagoon situated on the west coast of the Baja California extends landward as a wide evaporitic flat whose morphology closely resembles that of Sabkha Faishakh (West Qatar). As in the Arabo-Persian Gulf, the sediments forming the seaward edge of this Mexican sabkha are essentially calcareous and microbial mats are well developed. Floods supply seawater to much of the sabkha surface. High interstitial salinities, a consequence of capillary evaporation, result in the formation of sabkha-type gypsum. Dolomite partially replaces certain bioclastic debris (Pierre 1982). However, because the sediments comprising the inner parts of the sabkha consist essentially of siliciclastic material, dolomitization rarely is important (10%).

4.3.2.3 Bonaire Island, Netherlands Antilles

The climate of the southern parts of the Caribbean region is arid, notably at Bonaire (Fig. 4.17 C). The south-southwestern coast of the island is fringed by a barrier ridge composed of coarse coral debris that shelters a shallow lagoon – the Pekelmeer – whose permanent waters precipitate subaquatic gypsum. The lagoon is very narrow (1 km). It is bordered along its landward margin by an evaporitic flat measuring 3 km in width, which is inundated frequently by marine waters. Evaporation of interstitial waters and the formation of highly saline brines lead to the precipitation of gypsum and, especially, to the development of lithified crusts composed essentially (80%) of fine, diagenetic dolomite (Deffeyes et al. 1965, Lucia 1968).

Although not demonstrated by coring, it is probable that the Bonaire sabkha is prograding laterally over the adjacent lagoon. This will result in a characteristic coastal evaporite sequence with subaquatic (lagoonal) gypsum at the base and sabkha evaporites with dolomite at the top, demonstrating again the close relationships between the two basic evaporite systems.

4.3.2.4 Shark Bay, West Australia

The southern parts of Shark Bay include a number of highly restricted areas (Fig. 4.17 D). In the Gladstone Embayment studied by Davies (1970 a), salinities of lagoonal waters increase to at least 269‰ within the interstitial waters of the supratidal zone.

The formation of brines is a direct expression of the semi-arid climate (rainfall 220 mm/year) and a low-relief hinterland. Both elements compare closely with the environments of the southern parts of the Arabo-Persian Gulf.

Evaporation of groundwaters within the upper intertidal zone leads to the formation of a mush of small (5 mm), flattened gypsum crystals similar to those formed in a comparable part of the sabkha at Abu Dhabi. Finer aragonitic sediments in the supratidal zone include somewhat larger (3 cm) gypsum crystals associated with traces (up to 12%) of dolomite whose precise origins have not yet been established.

In common with other sabkha-type settings (Abu Dhabi, Bonaire, Sinai, etc.), the inter- and supratidal zones in Gladstone Embayment are characterized by numerous lithified aragonitic crusts whose formation is also the result of intense evaporation.

The wide intertidal flats surrounding the more protected parts of Shark Bay have much in common with the sabkha environments of the Arabian coasts. However, despite the precipitation of interstitial gypsum, no anhydrite has been found. Furthermore, gypsum formation is limited to capillary evaporation. Despite the isolated setting of these lagoons, subaquatic marine evaporites are rare or absent. These marginal evaporite conditions therefore probably reflect the semi-arid nature of the climate with total annual evaporation rates of about 780 mm (Logan and Cebulski 1970) compared with evaporation in the order of 1,200 mm/year under the more arid climate at Abu Dhabi (Bush 1973).

4.3.3 Hydrology and Evaporitic Minerals at Gavish Sabkha: Comparison with Other Regions

The hydrology and related evaporite mineral formation at the Gavish Sabkha have been studied by Gavish (1980; see also Chap. 9). Marine waters are supplied through the peripheral sand ridge via a series of springs. Virtually no surface flood-recharge comparable to that at Abu Dhabi takes place.

Within the Gavish Sabkha there are two major hydrographic environments: (1) subsurface groundwaters, from which evaporites precipitate as a result of capillary-water evaporation; and (2) surface waters located around the periphery of this depression from which subaqueous evaporites are precipitated. However, much of the resulting gypsum is destroyed subsequently by bacterial action (see Chap. 11 and G. M. Friedman and Sanders 1978).

Although the hydrologic system at the Gavish Sabkha has much in common with those of other sabkhas and salinas separated from the sea by peripheral sedimentary barriers, and notably at Ras Gharib, Gulf of Suez (Fig. 4.5), nevertheless it differs from the classical Arabian sabkhas. These Arabian sabkhas are supplied both by surface flood-recharge from the sea and by interstitial groundwaters derived from the adjacent continent.

In the Gavish Sabkha dolomite constitutes up to 10% of the fine fraction. Its somewhat greater concentration within this depression relative to the adjacent desert sediments suggests that it cannot be totally detrital in origin. However, this mineral is relatively unimportant volumetrically. This may result in part from the limited volumes of aragonite mud within the essentially siliciclastic sediments. It is recalled that dolomite tends to form by diagenetic replacement of aragonitic mud which is volumetrically important in the sabkhas of the Arabo-Persian Gulf.

In the Gavish Sabkha, gypsum is formed both within the sediments in a manner comparable to that in Abu Dhabi, and on the surface as crusts peripheral to the surface waters of this seasonal salina. This latter situation exists very rarely in Abu Dhabi. Finally, gypsum must also be precipitated from the highly saline (350‰) surface waters of the salina but most such gypsum is destroyed subsequently by bacterial action. Relatively pure gypsum attaining thicknesses of nearly 1 m have been encountered both on the surface (see cores 4 and 5 in Chap. 9) and at depths of 120 cm (core 7). The authors of Chapter 9 believe that virtually all this gypsum in core 7 could have resulted from a refluxing mechanism. Massive development of gypsum in Gavish Sabkha has no true equivalent in the Arabian sabkhas of the Gulf, where gypsum generally is scattered through a carbonate (dolomite) framework. Where crystals constitute a mush within the microbial mat, they rarely exceed 30 cm in thickness, and generally are transformed rapidly into anhydrite.

Anhydrite, the most typical mineral in the Abu Dhabi sabkhas, occurs in relatively small quantities in the Gavish Sabkha where it is disseminated as fine crystals within the lutitic fraction. Despite the highly saline brines and elevated summer temperatures no nodules or chicken-wire structures have been found.

Halite seems to be well developed in Gavish Sabkha where it constitutes surface crusts up to 20 cm thick. It has not been found in depth. Its better development relative to the Arabian sabkhas must be related to the locally favorable hydrographic conditions within Gavish Sabkha. The presence of exceedingly saline waters (350‰) and their position close to the surface may favor greater capillary evaporation, leading to development of relatively thick crusts of halite. At Abu Dhabi, the water table within the supratidal zone is situated generally at depths often exceeding 1 m.

4.4 Conclusions

The characteristic morphologic features of coastal evaporite bodies depend not only on climatic factors but also on the processes responsible for the development of the sedimentary framework within which evaporite minerals are pre-

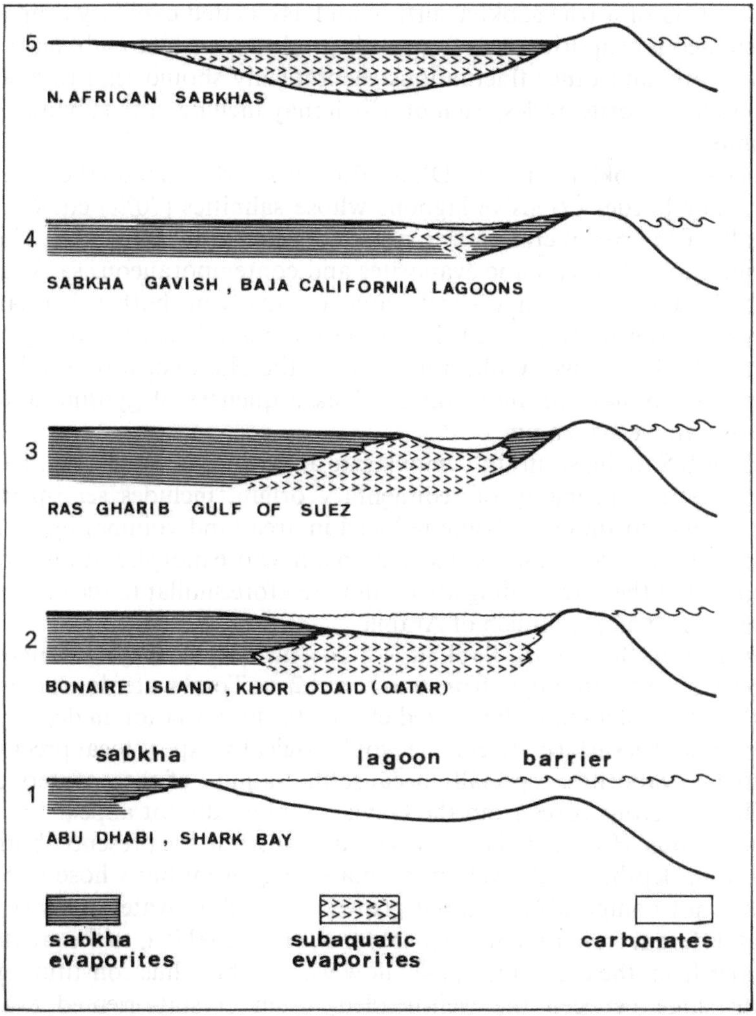

Fig. 4.18. Schematic comparison of the principal types of coastal evaporite systems. *1* Abu Dhabi, Shark Bay: restricted lagoon (lacking evaporites) and well developed peripheral sabkha. *2* Bonaire Island, Khor Odaid (Qatar): coexisting salina (well developed) and peripheral sabkha. *3* Ras Gharib, Gulf of Suez; residual salina with peripheral sabkha which is rapidly prograding over subaquatic evaporites. *4* Gavish Sabkha and Baja California lagoons: traces of a morphological depression remain but this is only temporarily flooded; evaporites are mainly sabkha-type, but thick layers probably due to very shallow water table and excessive water loss. *5* N. African sabkhas: subaquatic evaporites are followed by sabkha which completes filling of basin

cipitated. Detrital sedimentation, the evolution of parental brines, and the chemical precipitation of evaporite minerals are all closely interrelated.

It has been shown that sabkha formation may result from the combination of three major processes involving marine, eolian, or continental supply of detritus. Marine supply may lead to the formation of coastal barriers favoring restricted lagoons and the subsequent development of subaqueous evaporites. However, progressive filling, mainly by peripheral progradation, generally leads to the formation of a true sabkha surface and associated capillary evaporation. Thus, subaqueous evaporites very frequently grade upward into sabkha-type evaporites. Eustatic and other fluctuations theoretically should lead to a series of superimposed evaporite cycles, each of which may include both subaqueous and sabkha units.

The classical sabkhas at Abu Dhabi do not overlie subaquatic evaporites. However, they border extensive lagoons whose salinities (70‰) could increase dramatically if the peripheral barrier system developed to perfection. The eventual precipitation of subaquatic evaporites and contemporaneous sabkha accretion could lead readily to a vertical sequence involving both subaquatic and sabkha-type evaporites (Fig. 4.18). This situation possibly exists already in other parts of the Arabo-Persian Gulf, notably near the Hasa coast of Saudi Arabia where certain sabkhas are underlain by thick sequences of gypsum and halite (D.J. Shearman, verbal comm.).

The Gavish Sabkha seems to demonstrate a transitional stage (Fig. 4.18); this littoral depression, probably of sedimentary origin, includes semi-permanent standing waters but these are being reduced in area (and volume) by peripheral sabkha accretion. It will soon become a flat plain whose morphology will be comparable to that of the surrounding areas, and therefore similar to the monotonous sabkha surfaces typical of much of Arabia.

In conclusion, the Gavish Sabkha is a coastal evaporite system involving a water table whose depth ranges from 0 to about 2 m. Total water loss at any given point within the system must be related closely to this variation in depth. In that the system includes surface waters one would logically expect local precipitation of subaquatic evaporites, especially because the salinity of these waters exceeds 300‰. The numerous cores from the Gavish Sabkha do not appear to confirm these expectations. However, these cores do demonstrate the presence, both at the surface and in depth, of relatively pure, massive gypsum but whose unoriented crystals are not comparable to those grown in standing water. This situation, somewhat different from that of other sabkhas, may be related to the greater total loss of water from the exceptionally shallow water table, thus constituting an intermediate stage between the well-bedded, often crystal-oriented evaporites formed in subaquatic environments, and the dispersed, generally nonorientated evaporite facies developed mainly by capillary evaporation within sabkhas.

5. Hypersaline Sea-marginal Flats of the Gulfs of Elat and Suez

AMIHAI SNEH and GERALD M. FRIEDMAN

The northern gulfs of the Red Sea as a whole may be regarded as a model for sedimentation within a hot and arid climatic zone and adjacent to a newly forming ocean. The climatic conditions that prevail enable considerable quantities of carbonates and evaporites to form and the active tectonism results in the unusual juxtaposition of abundant terrigenous clastics. We studied the characteristics and distribution of the various sea-marginal environments of deposition and of the strata deposited in each to provide a better understanding of ancient strata formed under similar conditions. The Gulfs of Suez and Elat (Aqaba) were compared to show the similarities and differences between adjacent marine embayments with the same climate but different topographics (i.e., the broad gentle coastal plain and shallow water setting of the Gulf of Suez versus the very steep narrow coast and extremely deep Gulf of Elat). Significant differences were also found to exist due to the large (2 m) tidal range in the northern Gulf of Suez compared to the areas of lesser tidal ranges, the Gulf of Elat and the southern Gulf of Suez.

5.1 Methods

The authors have been engaged in geologic studies of the Pleistocene, Holocene, and modern sediments of the northern gulfs of the Red Sea, since 1967.

The areas chosen for detailed study (Figs. 5.1, 5.2 and 5.10) are on the west coast of the Gulf of Elat and east coast of the Gulf of Suez and were selected after studying aerial photographs of the entire length of the shores of the Sinai Peninsula and on the basis of reconnaissance field trips. Sampling of sediments was carried out along traverses normal to the shoreline from the land seaward, down to about a 10–15 m water depth. In each of the areas selected for detailed study a series of samples was taken along at least one traverse; normally several traverses were run. At some stations trenches were dug to provide three-dimensional representation. Echo-sounding profiles to establish the sea-bottom morphology were taken with a Barograph echo-sounder installed on a small boat (operation of the equipment and measurements by Y. Nir of the Geological Survey of Israel), and the location of the stations along the profiles were established by triangulation from shore.

The investigation of the samples included detailed petrographic analysis. All samples were studied with binocular microscopes using standard staining techniques for identification of aragonite, calcite, dolomite, and high Mg-calcite (G. M. Friedman 1959).

Particle size frequency distributions of grab samples were determined with 1/4 phi interval sieves. An attempt was made to characterize various subenvironments on the basis of statistical analysis of the grain size distributions following Folk and Ward (1957), G. M. Friedman (1962), McKinney and G. M. Friedman (1970), and Visher (1969).

To determine the effects of the seasons, the areas selected for detailed study were investigated both during the summer and winter and aerial photographs of the summer and winter shorelines were also compared.

5.2 Geological and Climatic Setting

The geological and climatic setting of the Gulf of Elat is given in Chapter 3 and will be recapped only briefly here with the addition of information concerning the Gulf of Suez.

The northern gulfs of the Red Sea (Fig. 5.1) are two large elongated embayments which are part of the rift system dividing the African and the European-Asian plates.

The Gulf of Elat Rift is narrow, 10–20 km wide, and is steep sided. The modern gulf which is about 1,800 m deep occupies the full width of the latest phase of rifting, and its slopes are, in fact, fault planes. The modern Gulf of Suez occupies the central trough of the Suez Rift and its depth is only 50 to 70 m. A relatively wide gently sloping, coastal plain exists along this gulf. The climate is hot and dry, and rainfall is scarce. Average annual rainfall ranges from 5 to 25 mm, all of which fall in winter within a few days (G. M. Friedman 1968). The area is one of intense evaporation (179 cm in the northern Gulf of Elat; Meteorological Service 1952) with evaporation rates several times greater than precipitation. The mean highest daily temperatures in the northern Gulf of Elat are 22 °C in January and 40 °C in August (Red Sea and Gulf of Aden Pilot 1967). Salinity in the Gulf of Elat ranges from 40‰ at the entrance to the Gulf to 42‰ in the northern Gulf (G. M. Friedman 1968, Table 2, p. 903). Salinity of the waters of the Gulf of Suez varies from 40.5‰ to 42.5‰ (Mohamed 1940).

Northerly and northwesterly winds dominate the Gulf of Suez and northerly and northeasterly winds the Gulf of Elat. Stormy southern winds are much less frequent and occur mainly in February in both Gulfs. The winds generate waves that approach the shorelines of the Gulfs diagonally, thus creating a longshore current directed southward. Storms are common along the Gulf of Elat, especially during the night (G. M. Friedman 1968, Meteorological Services 1956, 1967, Red Sea and Gulf of Aden Pilot 1967).

Although south-flowing currents, generated by the prevailing northern winds, exert a major force that affects the sea-marginal depositional environments, other northward-moving currents in the southern parts of both Gulfs counteract this influence. These northward-flowing currents include currents resulting from salinity differences; a warm less saline surface water current flows into the Gulfs

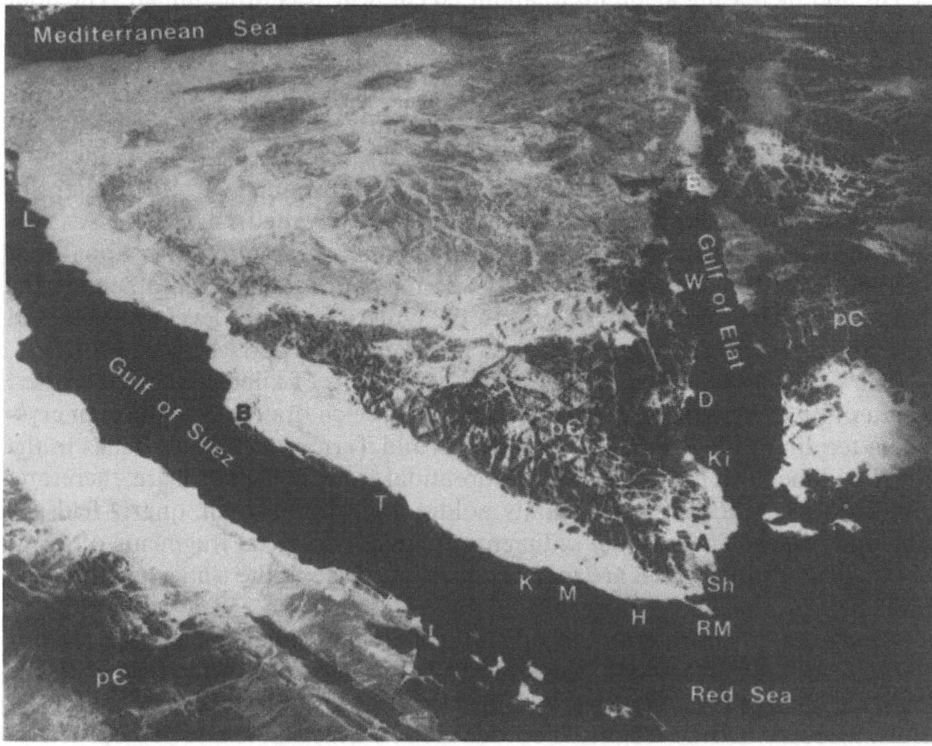

Fig. 5.1. View north across the Sinai Peninsula and the northern gulfs of the Red Sea (Gulf of Suez and Gulf of Elat) taken from Gemini XI. *1* Fault line (transform fault) from Dead Sea (D. S.), *upper right* of photograph, down to Elat *(E)*, continuing southward along the east coast of the northern part of the Gulf of Elat. *2* Absence of coastal plain along the shores of the Gulf of Elat from Elat *(E)* down to Sharm el Sheikh *(Sh)* and seaward protruding fans (*W* Watir fan; *D* Dahab fan). *3* Cuspate spits south of Suez *(S)* and north of Ras Muhammad *(RM)*. *4* Elongated faulted blocks on the Gulf of Suez arm south of the Belayim lagoon *(B)* and on the western side of the Gulf. *5* Precambrian basement *(pE)* in southern Sinai, on the western side of the Gulf of Elat, and on the eastern side of the Gulf of Suez. Note locations of Ras el Haddan *(H)*, Ras Kanisa *(K)*, Ras el Millan *(M)*, Et Tur *(T)*, Ras Lahata *(L)*, Wadi Kid *(Ki)*, and Marsa el At *(A)*, which are cited in the text. Photo from NASA

from the Red Sea replacing waters lost by evaporation and lost by an outflowing deeper density current of more saline cooler waters (Mohamed 1940). Stormy winds from the south at the tip of the Sinai Peninsula drive currents northward in both Gulfs; in the Gulf of Suez, where a nodal point exists at Et Tur, the ebb-tidal current in the southern Gulf also flows to the north.

The physiographic configuration of the Red Sea, and the Gulfs' long, narrow and almost closed embayments, dictate the nature of the tides which have several nodal points along the axis of the Red Sea. Tides are semi-daily and their characteristics differ in the two Gulfs. In the Gulf of Suez, a nodal point occurs near Et Tur (Fig. 5.1) about 80 km north of the southernmost limit of the Gulf, and 220 km south of the antinodal point of Suez (Fig. 5.10). The tidal range in the Gulf of Suez, near its northern limit, is about 2 m, decreasing southward to 0 at

Et Tur and increasing again up to about 60 cm near Ras Muhammad. The tidal range in the Gulf of Elat is about 70 cm at Elat and 90 cm near Sharm el Sheikh (Fig. 5.1). No nodal point exists along the Gulf of Elat.

Water circulation in the Red Sea and its northern gulfs is affected by the surface water circulation in the open Indian Ocean. In summer, the southwest Monsoon Current replaces the Northern Equatorial Current, as surface winds blow toward Central Asia. When this occurs, a southward-bound current develops in the Red Sea. In winter, the reverse takes place. Air flowing from Central Asia toward the Indian Ocean results in the reappearance of the Northern Equatorial Current. When this current reappears, water from the Indian Ocean enters the Red Sea and hence, its northern gulf. In the Gulf of Suez, sea level was observed to be about 30 cm higher in winter than in summer (see also Morcos 1970).

The Gulf of Suez has a coastal plain several tens of kilometers wide crossed by intermittent braided stream courses (wadis) which drain a Precambrian crystalline terrain in the south, and Cretaceous and Tertiary carbonate rocks in the north. Sediments of the intertidal and supratidal zones in the north are, therefore, composed of carbonate sand, mainly oolites and, in the south, quartz-feldspar sand and pellets. The subtidal sediments are mainly skeletal fragments of disintegrated coral reefs, which are sparsely distributed along the whole length of the Gulf.

Along the northwestern Gulf of Elat terrigenous sediments are derived primarily from the Paleozoic to Mesozoic Nubian Sandstone, while further to the south and along the northeastern gulf Precambrian granitic rocks provide the detritus. Miocene marine sediments border the southeast shore and provide terrigenous sediments there. Wadis draining the rugged mountainous deserts bordering the entire gulf deposit sediments in alluvial fans or in fan complexes along the western (Sinai) coast; the fans reach their largest areal extent along the southwest coast where the widest and most gently dipping coastal area exists. Sabkhas (see definition in Chap. 1) have formed on the southern (lee) margins of most of the larger fans and the fans are fringed by narrow reefs. Areas up to 2 km wide of highly dissected alluvial deposits border the eastern coast locally; no active fans exist along the east coast. Reefs also fringe nearly the entire eastern coast.

5.3 Sea-marginal Environments of Deposition

Coastal environments comparable to those of the northern gulfs of the Red Sea are known from several arid regions. In Baja California, deposition of evaporites, mostly aragonite, gypsum and halite, occurs in shallow pans which are flooded during high spring tides (Kinsman 1969 a, Phleger 1969, Shearman 1970). The sabkha environment of the supratidal flats of the Trucial Coast, especially of Abu Dhabi, is known in detail and dealt with in Chapter 3 (Kinsman 1966, 1969a, b, 1970, Evans et al. 1969, Butler 1969, Bush 1973, Hsü and Schneider 1973, G. M. Friedman and Sanders 1978, p. 335–339). Supratidal flats constitute a 10-km-wide sea-marginal belt of terrigenous, microbial, and carbonate sediments. The sediments host evaporite minerals, including gypsum and anhydrite.

All of the sea-marginal environments of the Gulfs of Elat and Suez are closely related to the deposition of terrigenous detritus by the intermittent stream courses (wadis). The alluvial fans, especially in their distal portions, are the substrate for most of the other sea-marginal environments, especially along the Gulf of Elat. Along the more gently sloping coastal areas of the Gulf of Suez the fans spread more widely and where the distal fans reach the sea cuspate spit complexes form at the mouths of the wadi channels. They protrude seaward about 1 km. Shepard (1973) defined cuspate spits as sandy cusp-shaped projections of the shoreline. Shepard's definition applies to humid regions, which usually have ponds and lagoons as part of the complex and are associated with a river system. In arid areas, the cuspate spits occur with fans and supratidal sabkhas. Glennie (1970) described a system of coastal sabkha–spit complexes in an arid area in the Persian Gulf, where the associated environments are alluvial fans, dune fields, coastal sabkhas, sand spits, and tidal creeks.

In the Gulf of Suez the spit complexes exhibit several interrelated environments which max include dunes, berms, back-berm channels, and seasonally-flooded sabkha/lagoons or supratidal sabkhas. A sand-wave belt characterizes the intertidal environment and extends into the subtidal environment which is also characterized by reefs. The gentle submarine slopes of the shallow Gulf of Suez make it possible for wide reef belts to form to considerable distances from the shoreline.

The sea-marginal environments along the west coast of the Gulf of Elat are part of fan complexes that also protrude seaward, some of them for several kilometers. Fans, dune fields and distal-fan supratidal sabkhas with berms and back-berm channels are the associated environments of the complexes. Locally hypersaline pools are part of the sabkhas. Narrow belts of reefs fringe the fan complexes, the steepness of the submarine slope sharply limiting their distance from shore and their width.

5.4 Gulf of Elat – Sea-marginal Environments

5.4.1 Fans

Fans and fringing reefs extend as belts parallel to the main fault scarp along almost the entire western coastline of the Gulf of Elat, from Elat in the north to Sharm el Sheikh in the south (Figs. 5.1 and 5.2). Most of the fans are small, on the order of 4 km^2, but some are quite large, exceeding 15 km^2. Networks of braided streams lace the fans. The small fans dip at angles of about 2.5°, whereas the larger ones dip at only 1° to 2°. At the mouths of very short wadis, fans are steep, slopes range from 10° to 15°. The catchment-basin areas that serve the fans are on the order of tens of kilometers, and hence they are much larger than the fan areas themselves. Flashfloods occur rarely (once in several years).

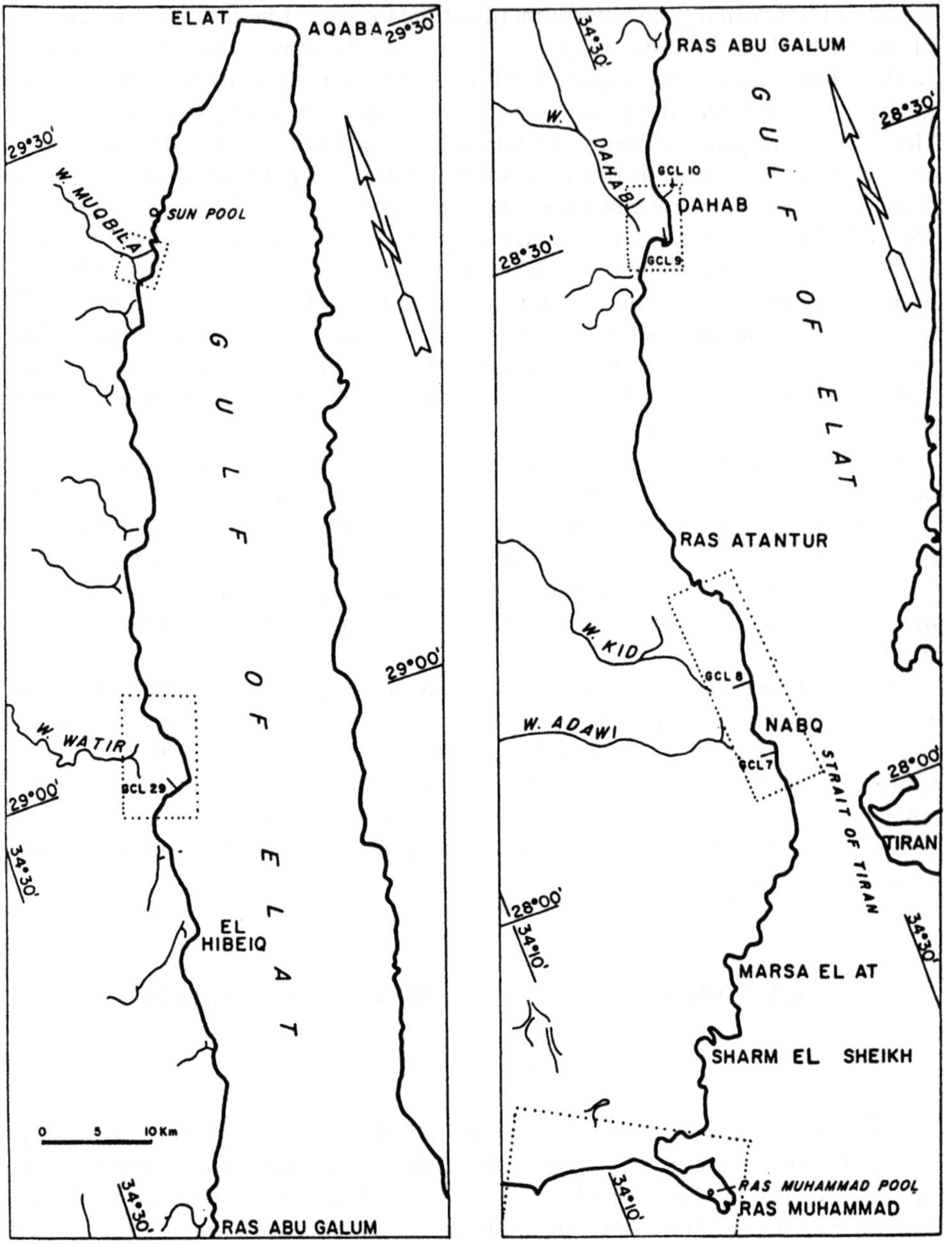

Fig. 5.2. Location map; Gulf of Elat. *Dotted rectangles* indicate study areas

5.4.2 Sabkhas

Coastal supratidal sabkhas are located on the southern flanks of the fans that have prograded across the shelf. There the sabkhas are sheltered from the intensive process of winnowing which results from prevailing southwestward-attacking waves. Waves approaching from the northeast have built pointed spits to the southwest; the spit complex at Dahab is the largest (Figs. 5.3 and 5.4).

The area of the mouth of Wadi Watir illustrates a typical example of fans and coastal supratidal sabkhas (Fig. 5.5). One large trunk stream, that of Wadi Watir, emerges from the edge of the mountains and spreads out into a large fan. Light-colored sand and gravel underlie this fan. Several small fans composed of darker-colored and coarser-grained sediment have grown across the parts of the large fan (Fig. 5.5). A radiating network of incised channels covers the surface of the large fan and dunes have spread across parts of its distal side. The crests of these dunes strike approximately east–west. Reefs fringe the distal margin of the fan, growing most vigorously into the waves and currents moving southward along the coast. Thus, despite the tendency of the waves to destroy such exposed fans and sabkhas, both may persist because of the protection by the reef (G. M. Friedman and Sanders 1978, p. 302–304). As mentioned above, the close juxtaposition of flourishing reef growth and copious terrigenous sediment deposition is a phenomenon that seems to be characteristic only of this particular climatic and tectonic setting (hot and dry, adjacent to a newly forming ocean). The long lapses of time between periods of sediment influx and the predominance of coarse-grained material deposited near the reef-building organisms may explain their ability to thrive. The sabkha surface, characterized by its hummocky topography, is developed on the southern flank of the fan.

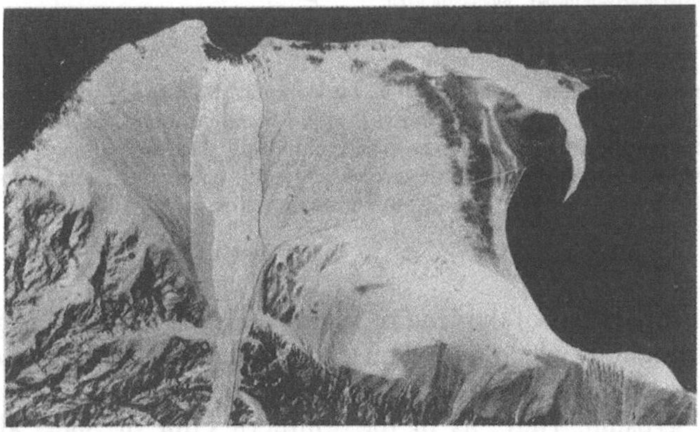

Fig. 5.3. View vertically downward from an airplane showing Dahab fan complex of the Gulf of Elat. Bedrock *(at bottom)* consists of Precambrian rocks composing Arabo-Nubian Shield. Seaward side of large fan underlain by light-colored sediments is fringed. Dry bed of incised straight channel (about 2-km long), which crosses fan from apex to periphery, displays well-developed pattern created by a braided stream. Drowning of seaward end of this dry channel has created small embayment in periphery of fan. Radiating network of faintly incised channels covers surface of large fan. Several small fans (*upper left* and *lower right*) have grown across parts of large fan. Waves approaching from *upper left* have built pointed spit on right side of large fan

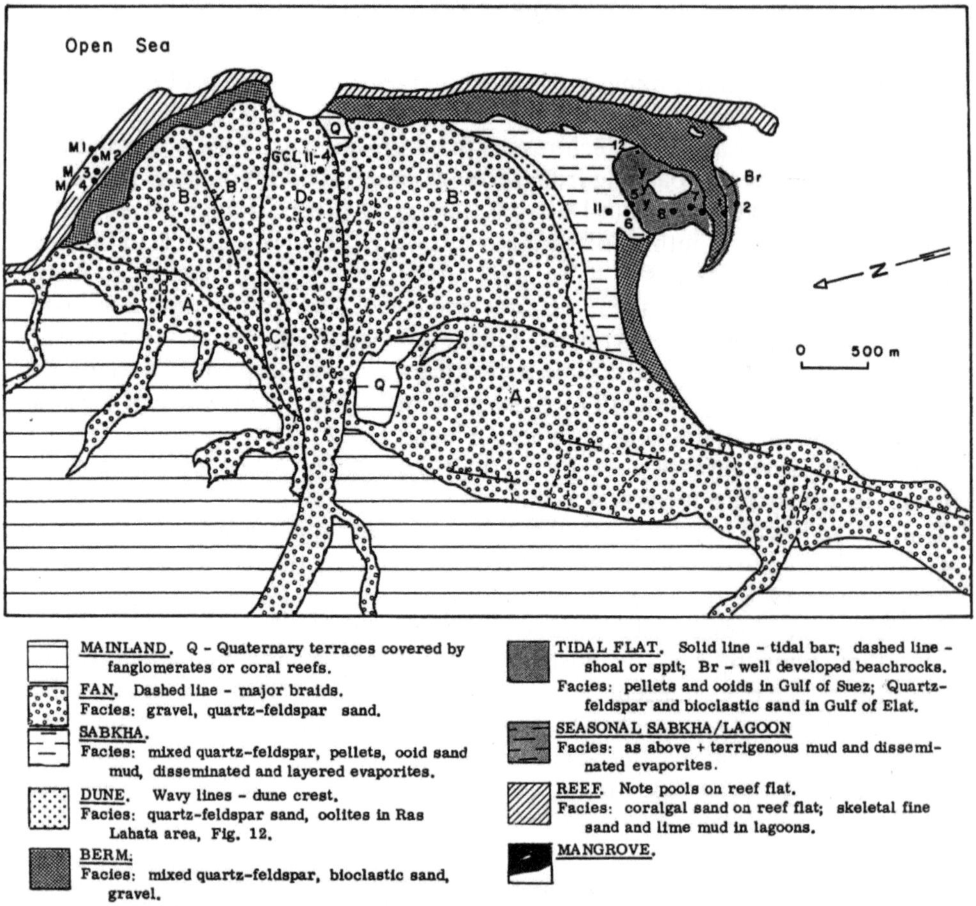

Fig. 5.4. Sedimentary environments in the Dahab fan complex. Map shows locations of sampling stations. *A* Small, steep fan; *B, C* old fan surfaces; *D* modern main drainage course; during heavy flash floods waters entering the fan area spill over to the south; *solid lines* separating *B, C* and *D* are cliffs; *B'* scarplet of northern bank of intermediate surface below *B; y* oyster reef. Note faults on fan surfaces parallel to mountain scarp

Pits dug into sabkhas of the Gulf of Elat reveal interbedded terrigenous sands and muds throughout which abundant gypsum crystals are dispersed. Mud enhances capillary migration and serves as a good host for brines, thus aiding precipitation of gypsum. Here, as in the southern Gulf of Suez, abundance of evaporites seems dependent on the occurrence of mud. Some of the mud is deposited in back-berm channels as described below. In the example of Wadi Kid the uppermost 40 cm below the surface of the sabkha consists of brown, very fine-grained to silty sand with abundant disseminated gypsum crystals and intercalated layers of very fine-grained sand that have been cemented by gypsum. Twenty centimeters of moist brown mud underlies the silty sand (Fig. 5.7). At the southern, protected side of the Dahab fan and spit complex (Figs. 5.2–5.4) the

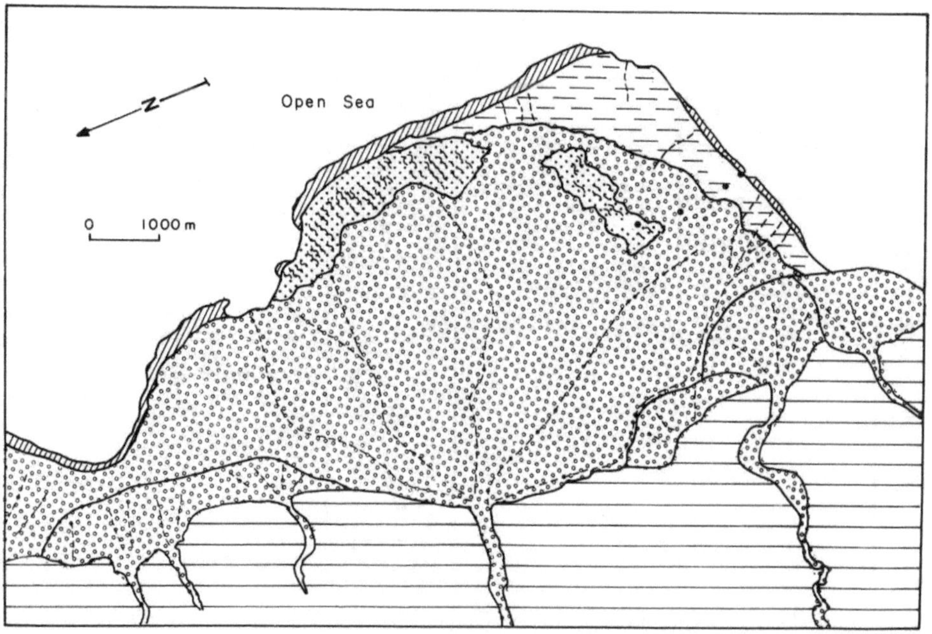

Fig. 5.5. Sea-marginal environments in area of Wadi Watir, Gulf of Elat. *Black dots* indicate location of pits. For legend see Fig. 5.4

Fig. 5.6. Oblique aerial photograph of fans and sabkhas of Wadi Watir area, Gulf of Elat, View due east showing bedrock of Arabo-Nubian shield, fault scarp, and fans of several sizes. One large fan, that of Wadi Watir, underlain by light-colored sand and gravel. Several small fans composed of darker-colored and coarser-grained sediment have grown across parts of large fan

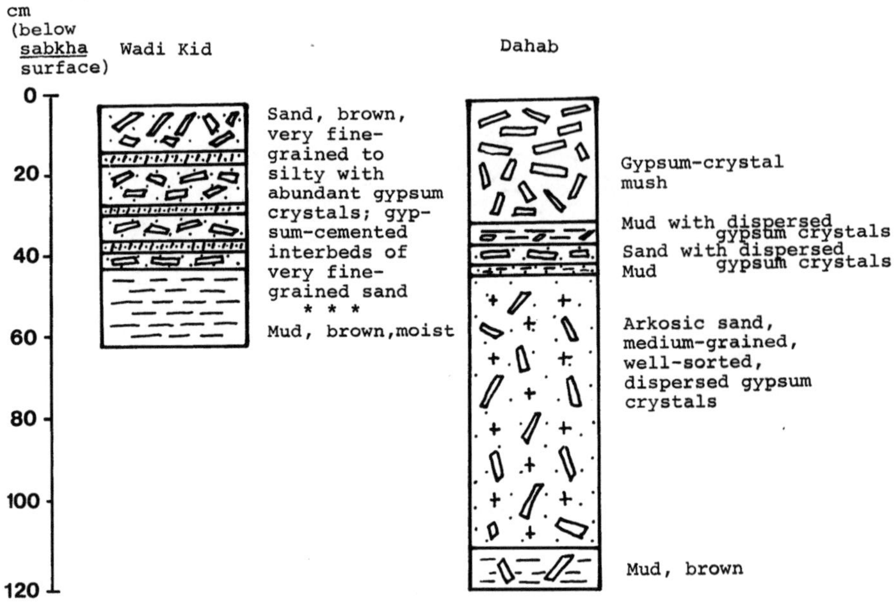

Fig. 5.7. Vertical sequences in coastal sabkhas of Gulf of Elat, as revealed in pits dug in sabkhas of the Wadi Kid and Dahab areas

halite-encrusted surface of the sabkha is underlain by an upper 30 cm of gypsum crystal mush, followed below by a few thin streaks of mud and 70 cm of well-sorted, medium-grained feldspathic sands throughout which gypsum crystals are dispersed; 10 cm of brown mud underlies the sands (Fig. 5.7).

The gypsum crystals, when developed within a clayey host, form a structure which resembles "birds eyes". The sabkha sequence at Marsa el At is very similar. The upper 13 cm are loose, wet sands with gypsum hardpans underlain by brown mud which includes sand layers down to 40 cm, below which coarse sand occurs.

5.4.3 Sea-marginal Pools

Sea-marginal hypersaline pools are located in three places along the west coast of the Gulf of Elat and at Ras Muhammad (Fig. 5.2). These are: (1) the solar pool south of Elat (see Fig. 15.2 b in Chap. 15), (2) the Dahab pool, (3) the Nabq pool (part of the *sabkha* now called Sabkha Gavish, Fig. 5.8), and (4) the Ras Muhammad pool (see Fig. 10.3 in Chap. 10). The Dahab pool displays deposition of pelletal lime mud. All four pools are separated from the sea, and seawater passes the dividing barrier through cracks. Nevertheless, during severe storms, at least Sabkha Gavish is flooded by seawater. Microbial mats are common to all of the pools. The Ras Muhammad pool, located at the southernmost tip of the Sinai Peninsula is more closely decribed in Chap. 10. The Solar Lake has been treated in several comprehensive surveys (Krumbein et al. 1977, G. M. Friedman et al. 1973 a, Krumbein and Y. Cohen 1974, see also Chap. 15).

Fig. 5.8. Puffy microbial mat surface, sea-marginal hypersaline pool, Nabq (Gavish Sabkha)

5.4.4 The Berm and the Intertidal Zone

In the Gulf of Elat the intertidal zone is relatively narrow; most of it is covered with boulders, cobbles and gravel cemented into beach rocks in the uppermost intertidal zone. The intertidal flat at the mouths of the wadis (e.g., the main wadi of the Dahab fan) consists of well-sorted medium sand; beach rocks are extensively developed. The berm-top sediments are generally poorly sorted and coarser grain sizes occur.

The barrier bar at the southern flank of the Dahab fan demonstrates beach rocks on both of its slopes. These beach rocks consist of igneous pebbles. The beach rocks display large-scale cross strata, dipping seaward. There seem to be two sets of beach rocks: the first one parallels the shoreline of the barrier, whereas the second is drowned in the sea and meets the barrier at an angle. Both beach rocks are in the process of disintegration. The berm on the open-sea side of the barrier is characterized by coral and mollusc fragments of cobble size and by igneous pebbles. Sandy supratidal flats are covered by oyster-reef mounds, by sea grass, by well-sorted sands comprising low tidal bars and by pseudo-oncolites (Fig. 5.9). Similar oncolites from the Ras Muhammad area were studied by Lewy (1972, see also Sneh and G. M. Friedman 1973). In both places they occur in the upper intertidal zone. The "oncolites" are several centimeters in size, consisting of fine and medium-sized skeletal particles. Their internal structures show one or two discontinuous, concentric, carbonate laminae. Some are rounded. Rounded oncolites were found rolling on the intertidal flat and in the uppermost subtidal zone. Some column-shaped ones were found embedded in the indurated host sediments. The rounded oncolites are probably products of the disintegration of the column-shaped ones.

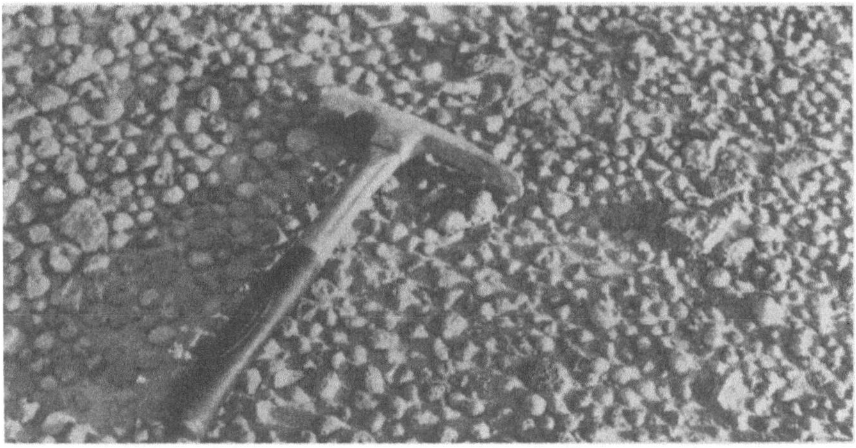

Fig. 5.9. Pseudo-oncolites on intertidal flat; Dahab fan complex

5.4.5 Back-berm Channels

Fine-grained terrigenous mud accumulates in channels cut across sabkhas during flashfloods. The mouths of the channels, although initially connected to the sea, are soon blocked by a berm trapping a shallow, elongate pool in which terrigenous, organic-rich mud settles out. During periodic storms in the Gulf the back-berm channel is flooded by marine water which quickly becomes hypersaline; at other times flashfloods provide more freshwater and terrigenous mud.

5.4.6 Dunes

Dunes partially cover the sabkhas and berms along the Gulf of Elat, some are stabilized by vegetation, others migrate. The dune sands reach the flat by way of the sea. Observation made on the following day after a flashflood in Wadi Watir (Figs. 5.1, 5.2, 5.5 and 5.6) revealed that the flood delivers the sands to the waters beyond the wadi's mouth; part of the sand probably continues to roll below wave base and moves away under the power of the longshore current; however, prevailing winds drive much of the sand back up onto the land surface where it forms the dunes.

5.5 Southern Gulf of Suez – Sea-marginal Environments

The Gulf of Suez has gently sloping broad coastal plains and its submarine slopes are equally gentle, reaching a maximum depth of only 50 to 70 m. The gentle slopes contrast sharply with the steep slopes of the Gulf of Elat. Tides in the Gulf of Suez are also different; the tidal range at Suez (Fig. 5.10) is nearly 2 m

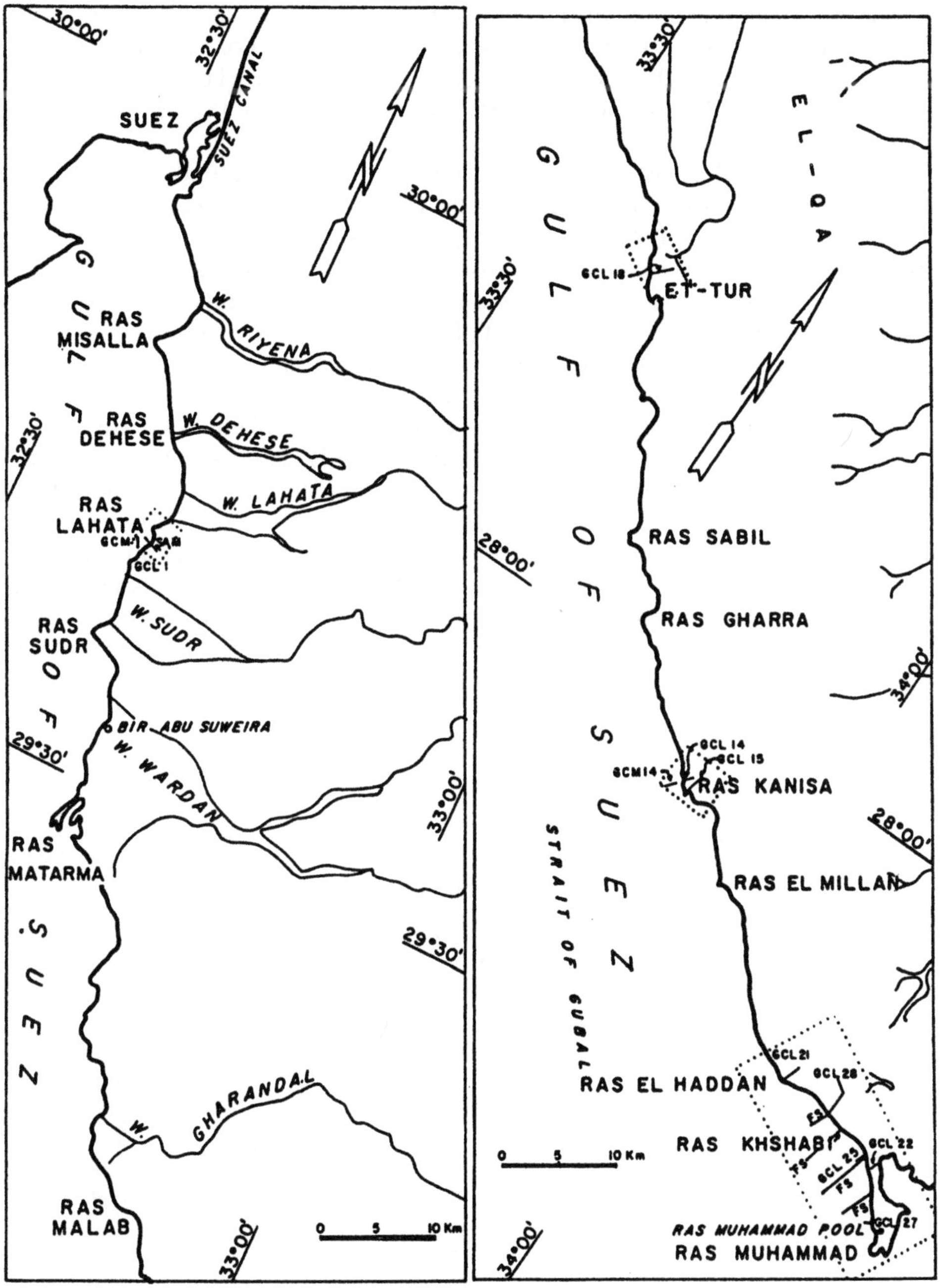

Fig. 5.10. Location map; Gulf of Suez. *Dotted rectangles* indicate study areas

– twice the 90-cm maximum of the Gulf of Elat. At Et Tur, where a nodal point exists, there is essentially no tide and the southernmost Gulf of Suez has a 60-cm tidal range. Because of the gentle slopes both periodic sea-level changes, (1) tides and (2) the seasonal change (30 cm) affect large areas along the sea margins of the Gulf of Suez.

Large intertidal areas exist where sand waves are associated with cuspate spits. The cuspate spit complexes provide the settings for lagoons which are seasonally flooded in winter, only to be exposed as sabkhas in summer during the lowest sea level. Some sea-marginal environments differ between the southern and northern Gulf areas. All of the sea-marginal environments are considered first with emphasis on the southern Gulf of Suez. In a separate section concerning the northern Gulf the differences between the sea-marginal environments there and those of the southern Gulf will be highlighted.

The sea-marginal environments at Ras Kanisa (Figs. 5.10 and 5.11) are typical of the southern Gulf of Suez and hence will be briefly described. From land to sea, the observed environments and underlying sediments include (1) a distal fan, composed of coarse sand and gravel laced by braided streams, (2) supratidal sabkhas underlain by terrigenous sediment composed of particles of quartz, feldspar and rock fragments, terrigenous mud, pellets and ooids, and disseminated as well as intercalated evaporites, mainly gypsum. Dolomite has been reported from the sabkha of Ras el Haddan (Gavish et al. 1969), (3) berms and sand waves (tidal bars) which are associated with cuspate spit complexes whose sediments are

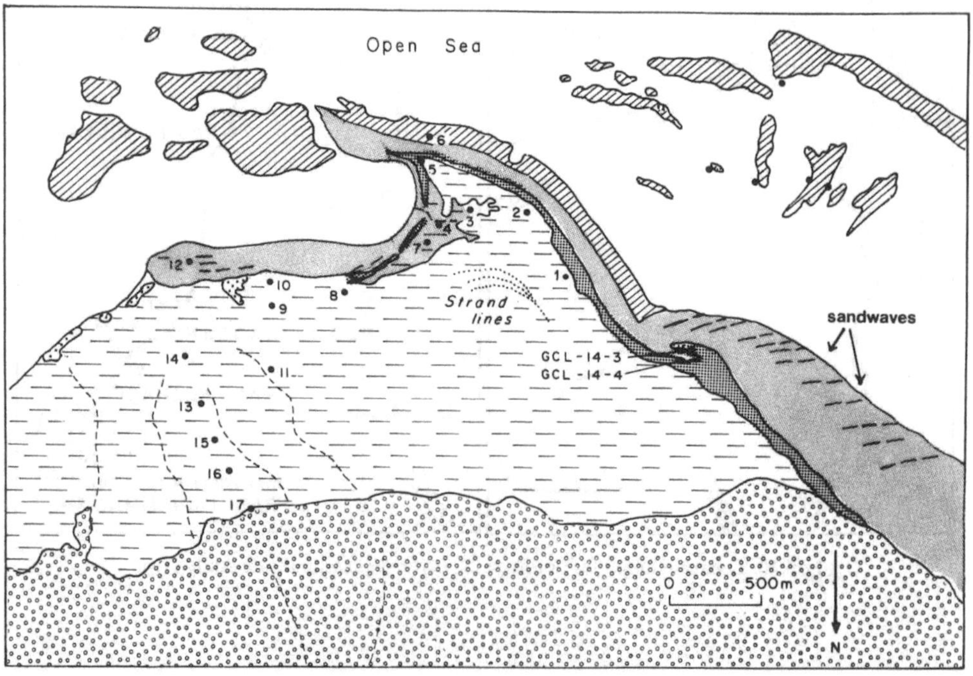

Fig. 5.11. Sea-marginal environments in Ras Kanisa area, southern Gulf of Suez. Map shows location of sample pits. For legend see Fig. 5.4

mostly sand and composed of mixed quartz, feldspar and rock fragments, pellets and ooids, (4) a lagoon, underlain mostly by skeletal sand, lime mud, ooids and pellets, fringed by (5) reefs. Sand waves and berms stretch out as spits to the southeast in response to the southward-directed longshore current. Here, as in the Gulf of Elat, a distinctive aspect of sedimentation in the gulfs is seen: reefs not only survive but, in fact, thrive in close juxtaposition to the transport and deposition of copious amounts of terrigenous sediment. The reefs even protect the fans along their distal margins. Thus, despite the tendency for the waves to destroy a fan, it persists because of protection by the reef (G. M. Friedman and Sanders 1978, pp. 302–304).

Supratidal sabkhas of the southern Gulf of Suez were studied at four locations (see Fig. 5.10): Et Tur, Ras Kanisa, Ras el Millan, and Ras el Haddan. Sand waves (tidal bars) of the intertidal zone were studied at two locations, Ras el Haddan and a few kilometers south of it.

5.5.1 Fans

Along the southern Gulf of Suez the Precambrian mountains are 5 to 8 km from the Gulf and the fans spread out shoreward from there. It is on the distal portions of small coalescent fans that the sea-marginal environments of the supratidal zone form. The areal extent of the very thin gently dipping fan deposits exceeds that of their catchment area in the highlands, as is common for alluvial fans (Bull 1964).

5.5.2 Supratidal Sabkhas and Dunes

Supratidal sabkhas occur almost continuously from Et Tur (Fig. 5.10) southward and extend landward up to 1 km. Nearly all the sabkhas are located at the distal margins of fan aprons. The sabkha surface is generally elevated 50 to 100 cm above the adjacent tidal flat; eolian sand accumulates there and, most likely, growth of evaporites within the sabkha sediments swells the surface. The sabkha is slightly hummocky with local flat surfaces. At the boundary of these slightly raised sabkhas and the fan margins a shallow topographic depression occurs. During flashfloods, the depressions hold freshwater and the suspended load of fine terrigenous sediment is deposited. The surface of this area is characterized by deep mud cracks when the water dries up.

Small dunes stabilized by plants are common on the sabkha flats. At Ras el Millan, the small dunes are salt-encrusted. Large dunes of the dikaka type, which are scrub-covered accumulations of dune sand (Glennie and Evamy 1968, Glennie 1970), are likewise common. All the sabkha flats described are commonly wet and display halite crusts on their surface. In some places, the sabkha surface is crinkled, but no specific pattern was observed. The sea-marginal environments of Ras el Millan and Ras el Haddan are comparable to those of Ras Kanisa.

Vertical sediment sequences below the surface of the supratidal sabkha vary depending on the proximity of the sample locations to the fan (Fig. 5.12). Near the boundary with the fan or with the berm only the uppermost layers contain

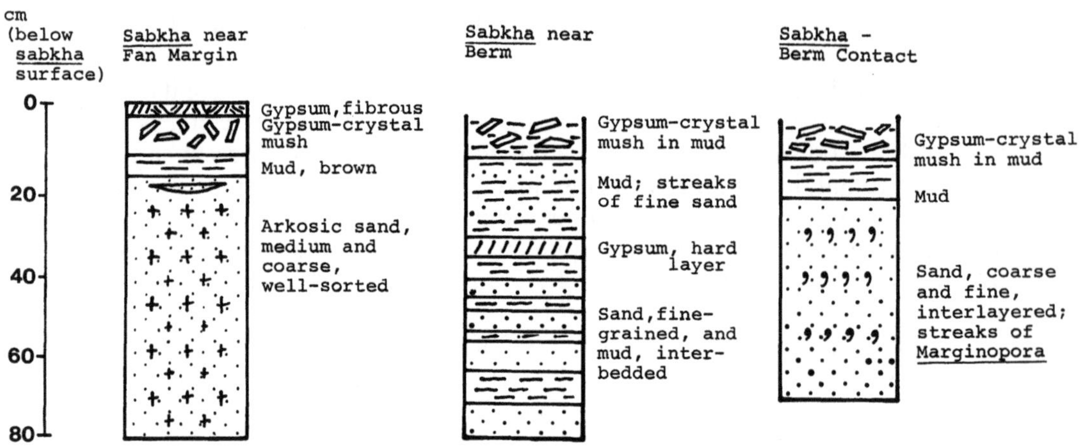

Fig. 5.12. Vertical sequences in coastal sabkha of southern Gulf of Suez and their variation depending in proximity to the fan, as revealed in pits dug in sabkha of Ras Kanisa (Fig. 5.11)

evaporites and mud; coarser sediments underlie them. Near the fan margins at Ras Kanisa, the section, from top to bottom, consists of 2 cm of fibrous gypsum, 5 to 10 cm of gypsum–crystal mush, and 5 cm of brown mud. Well-sorted medium and coarse feldspathic sand comprises the section downward to a depth of 80 cm. Closer to the berm, the section includes 10 cm of gypsum–crystal mush in mud, 20 cm of mud and some fine sand streaks, and another 5 cm hard layer of gypsum which underlies the mud and overlies a continuous section of interbeds of fine sand and mud. Where present, mud enhances capillary migration and serves as a good host for brines, thus aiding the precipitation of gypsum. At the boundary with the berm, the upper 10 cm of the sabkha also consists of gypsum–crystal mush overlying a mud layer. However, the lower section down to a depth of 70 cm consists of interlayers of coarse angular sand and of fine sand displaying faint cross-laminae. 5 cm streaks of *Marginopora* mush suggest that, at the time of deposition, the area was subjected to storm surges or to seasonal storm tides.

At Et Tur, the tidal range is negligible and the groundwater table is at a depth of 20 to 30 cm. As a result, the sabkha section lies entirely within the capillary zone. Braided streams lace the sabkha, dissecting the surface into longitudinal bars which are flat-topped and covered with a thin halite crust (Fig. 5.13). Scrub is absent; troughs are 20 to 50 cm deep. The bottom 50-cm section of a pit dug to a depth of about 100 cm shows a trough sequence consisting of 5-cm-thick interlayers of (1) ripple cross-laminated fine and medium sand, (2) mud, and (3) hard layers of gypsum (Fig. 5.14). This trough sequence is overlain by a longitudinal bar sequence (Fig. 5.6) about 50 cm thick, and consisting of fine and medium sand displaying horizontal stratification. Gypsum polygons are common on the *sabkha* surface.

A 300-m-wide zone adjacent to the shoreline is characterized by dikaka accumulations of dune sand. These are about 4 m high and their internal dune structure has been completely destroyed by the plant roots. Sea grass transported by the winds is an important constituent of these dunes.

Fig. 13

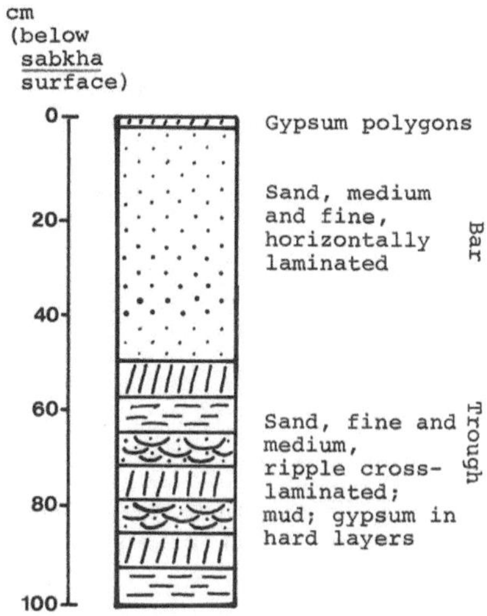

Fig. 14

Fig. 5.13. Shallow troughs and flat-topped bars on sabkha surface; note halite crust covering bar. Et Tur area

Fig. 5.14. Vertical sequence in braided streams of coastal sabkha in southern Gulf of Suez, as revealed in pig dug in Et Tur area

5.5.3 Back-berm Channels

Temporary spits are developed in front of wadi-fan channels which downcut into the sabkha surfaces at the distal margins of the Wadi Sudr, Wadi Wardan and Et Tur fans (Fig. 5.10). When flashfloods occur, sandy sediments are transported seaward through wadi channels and wadi mouth bars, and small "deltas" develop. However, in a rather short period of several months, reworking by waves takes place and a small spit is formed. The sediments migrate southward with the longshore current and are rearranged in sand waves (tidal bars). Eventually a berm bar is formed at the mouth of the wadi channel. The construction of these bars is very well demonstrated by landward dipping large-scale foresets (Fig. 5.15). Once the bar is established to its full height, it begins to prograde, accreting sediments on its seaward side; most probably during storms. Large-scale seaward dipping foresets characterize this stage; coastal erosion is minor. Following the bar building, a layer of black, mucky, organic-rich clay, several centimeters thick, develops on the bottom of the blocked wadi channels. The banks

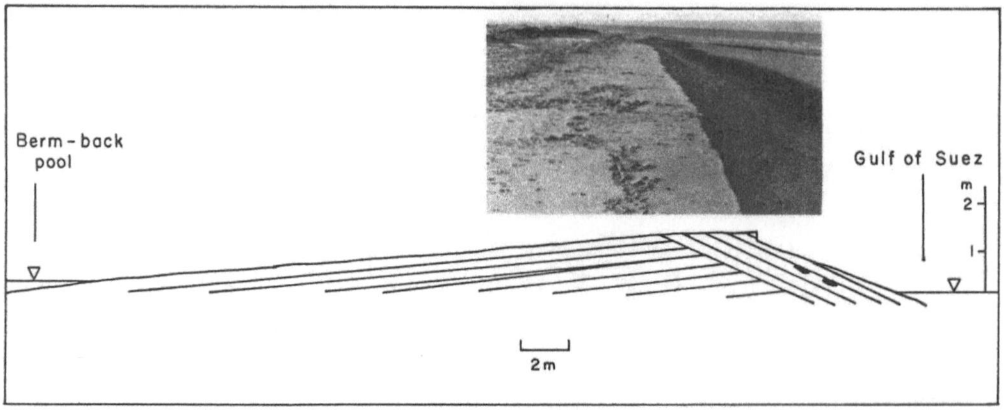

Fig. 5.15. Photograph and cross-section diagram of wadi-mouth berm, Wadi Wardan, south of Ras Lahata, northern Gulf of Suez, and diagram of typical internal structure. *Black spots* indicate interstratified tar pollution

of the channels, which are 50 to 70 cm high, have sections which display both interlayers of mud and sand, and 20-cm-thick black, organic-rich mud beds. During storms the channels are flooded with marine water and become hypersaline pools. Infrequent flashfloods provide more terrigenous mud and fresh water to the blocked channels. Evaporite precipitation does not occur.

Unlike the sabkha profiles described, that of Ras el Khashabi at the southern edge of the Gulf is poor in evaporites, and consists mainly of quartz and feldspar grains, pellets and some ooids. Wadi-derived mud is practically absent, as the sabkha is located on a coastal strip devoid of wadi fans.

5.5.4 Berms

A berm is developed along the beach about 1 m above high-tide level; when associated with the spit complex, it constitutes the sandy spit ridge behind which the sabkha of the complex develops. The berm belt, reaching widths of tens of meters at most, is characterized by loose sand and gravel surfaces.

Mounds, probably constructed by callianassids, composed of sand-sized particles are very common and reach up to 50 cm in height. The sand particles are loosely lithified, forming aggregates of about 1–2 mm in size.

Small dunes of about 1 m in height and generally fixed by halophytes are another common feature that develop on the berm. Strandlines with accumulations of gravels and shells as well as beach rocks are sporadic on the seaward slope of the berm. An accumulation of shells at the toe of the berm was observed in several places. The sedimentary structures within the berm down to at least 50 cm are completely obscured by burrowing.

The sands are poorly sorted and consist of quartz and feldspar, ooids, pellets, some skeletal fragments and relatively abundant magnetite.

The berm of the Ras Lahata spit complex in the northern Gulf of Suez is a 150-m zone that separates the intertidal zone from the sabkhas. Three subzones are recognized in this environment: (1) beach rocks – they cover the slope of the berm, dipping seaward by 10° and are situated in the uppermost intertidal zone forming a narrow discontinuous strip, (2) berm-top terrace this subenvironment consists of loose oolite sediments and abundant mollusc fragments; and (3) "strand-plains"; the landward side of the berm is a strand-plain that consists of ooids and abundant oolitic limestone flat pebbles, at least 60 cm thick.

5.5.5 Cuspate Spit Complexes

Several spit complexes exist along the northern 90 km and the southern 90 km coastal strips of the Gulf of Suez (Fig. 5.10). The wadis which drain the mainland in the northern part of the Gulf are associated with the spit complexes that protrude seaward about 1 km. The sea-marginal environments are all associated within the spit complexes. Glennie (1970) described a similar system of coastal sabkha-spit complexes in an arid area in the Persian Gulf, where the associated environments were alluvial fans, dune fields, coastal sabkhas, sand spits, tidal creeks, and deltas. In the northern Gulf the main current along the beach runs southward, and the spit complexes are located slightly south of the mouths of the wadis, and they tend to develop lengthy southward-bending barrier bars. This is true for the Ras Misalla spit south of Wadi Riyena, for the Ras Dehese spit complex south of Wadi Mirba, where oolite environments were investigated by Walther (1888 b), for two small spits south of the mouth of Wadi Lahata (Ras Lahata spit complex in this study), for the Ras Sudr spit south of Wadi Sudr, for the Ras Matarma spit south of Wadi Wardan, and for the Ras Malab spit south of Wadi Gharandal (Fig. 5.10). The shoreline from Et Tur in the north to Ras Muhammad in the south also displays a series of cuspate spits, i.e., Et Tur, Ras Kanisa, Ras el Millan and Ras el Haddan spits. The spits south of the tidal nodal point at Et Tur are symmetrically triangular, and at Ras el Haddan, the southernmost cuspate spit, the head of the spit bends slightly toward the north. Although currents, generated by the prevailing northern winds, exert a force that tends to bend the spits southward, other northward-moving currents, unique to this region, counteract the southward-moving currents. These northward-moving currents are described above. The coriolis force may likewise have an effect, as it deflects currents entering the Gulf toward the eastern shore. Tidal channels cut through the southern flank of many of the spits, whereas at Ras el Haddan they cut through the northern flank.

5.5.6 Seasonal-flooded Sabkhas

The seasonal-flooded sabkha is part of the spit complex. It is an environment that reflects the seasonal changes of sea level which characterize the Gulf of Suez. During the winter, the sea level is higher by 20 to 30 cm and certain sabkha areas of the spit complexes are covered by either shallow lagoons or by tidal channels.

During the summer, the sea level falls and the lagoons and tidal channels become part of the supratidal zone.

This seasonal change in conditions was observed in several small spits along the northern flank of the Ras Kanisa spit complex (Fig. 5.12) and to some extent in five other spits north of Ras el Haddan. It was also observed in the northern Gulf in Ras Lahata and Ras Malab spits, where they are protected from the sea by a sandy barrier. The environmental changes are reflected in the difference in sedimentary products. Mud cracks, halite crust, and other subordinate evaporite minerals, mud streaks, and high concentrations of organic matter (i.e., black mud layers in Ras Malab) are characteristic of the seasonal sabkha lagoon, when it is subaerially exposed in the summer, whereas accumulations of shells of gastropods *(Cerithium)* and other organisms or just skeletal fragments and trails, pellets, and ooids are characteristic of its winter lagoonal or tidal-channel phase. The main seasonal sabkha/lagoon of the Ras Kanisa spit complex is paved by a black crust of microbial mats. Products of the seasonal sabkha/lagoon environment are similar to those described in berm-top spillover ponds by G. M. Friedman and Sanders (1978); but whereas berm-top ponds are filled with seawater during storms only, the seasonal sabkha/lagoon is constantly covered with seawater all winter.

5.5.7 Intertidal Zone: The Tidal Flat and Sand Waves (Tidal Bars)

The intertidal zone along the southern shoreline of the Gulf of Suez is a prominent 400-m to 800-m wide belt easily distinguishable from other environments. The belt includes the area exposed during the lowest tide possible, at the ebb of spring tide during the summer. Rippled flats and shallow channels, only a few centimeters deep, characterize the higher landward intertidal zone, whereas sand waves (tidal bars) are the major features along the rest of the belt (Fig. 5.16).

Fig. 5.16. Sand waves (tidal bars) meeting shore at 30°; note rippled surfaces; Ras Khashabi

Sand waves of the southern Gulf were studied in detail at Ras el Haddan and a few kilometers farther south. In the first location, the irregularity of the shoreline caused by the spit of Ras el Haddan has a pronounced effect on wave refraction. The wave energy is concentrated on the spit and spreads out to the south, resulting in the deposition of a series of sand waves which display a pattern similar to that of the refracted waves themselves. About 500 m seaward of the shoreline, beyond the influence of the spit, the direction of the sand waves changes. Each one of the sand waves has a flat top, 3 to 10 m wide, rising 20 to 40 cm above the wave trough, which is also about 3 to 10 m in width. The sand-wave slopes are asymmetric; a gentle slope, 10° to 20°, characterizes the seaward side of the sand wave, whereas the landward slope has a gradient of 30°. Superimposed small-scale ripples run along the top of the sand waves with their strike intersecting that of the sand waves. While trend of the sand waves is determined as a result of a long-term consistent longshore current, that of the ripples results from wind-driven water waves perpendicular to the wind direction. Very commonly, linear structures a few meters long and only 0.5 cm high form in the wave troughs during the ebb current. A few kilometers south of Ras el Haddan a series of parallel sand waves (Fig. 4.16) strike the shoreline at a 30° angle. Each of these sand waves starts as an extension of a very small spit. As at Ras el Haddan, there is another set of sand waves with a different strike direction lying farther seaward.

An attempt to reveal the sand waves' internal structures failed; nevertheless, the surface features of the sand waves indicate that initially they had been formed as mega-ripples, about 10 cm high, and were later truncated and mantled by a train of small-scale ripples. Similar sand waves first described by Bucher (1919, pp. 173, 181) have been identified as mega-ripples. They are common in the intertidal oolite shoals in the Bahamas (Imbrie and Buchanan 1965) and in the North Sea (Hülsemann 1955). The sand waves migrate over slightly indurated surfaces that in places (e.g., at Ras Sudar and a few kilometers north of Ras Muhammad) are covered with only a thin (1 cm) sheet of loose sand (G. M. Friedman 1975, p. 392, G. M. Friedman and Sanders 1978, p. 156). Indurated surfaces in this area occur mainly on intertidal flats close to the berm.

The typical intertidal sediment consists of quartz and feldspar grains, ooids, pellets, skeletal fragments and heavy minerals. The sands are moderately sorted and negatively skewed, suggesting high-energy conditions, and hence winnowing. The sands are fine- and very fine-grained (Fig. 5.17); the mode (most abundant size class) is between 2.50∅ and 3.75∅. Quartz and feldspar grains constitute 45% to 85% of the total grain population. They are generally subangular to subrounded. A larger percentage of quartz and feldspar grains are present but many are superficially coated with micrite, forming ooids. Pellets are abundant – 10% to 45% – and they appear in a variety of colors ranging from light-gray to dark tones, and occur in several shapes, e.g., spheres, generally cryptocrystalline, cylindrical tubes and cubes (Fig. 5.18). Generally, the pellets are aggregates of aragonite and containing quartz silt and magnetite grains. Some pellets were apparently elongate initially, polished by movement, and then broken into more equant particles; two opposing sides are relatively rough, implying separation after the polishing phase. Many pellets are bored and the cavities are filled with aragonite mud (Fig. 5.19). Skeletal particles consist of forams (*Marginopora, Peneroplis, Spiro-*

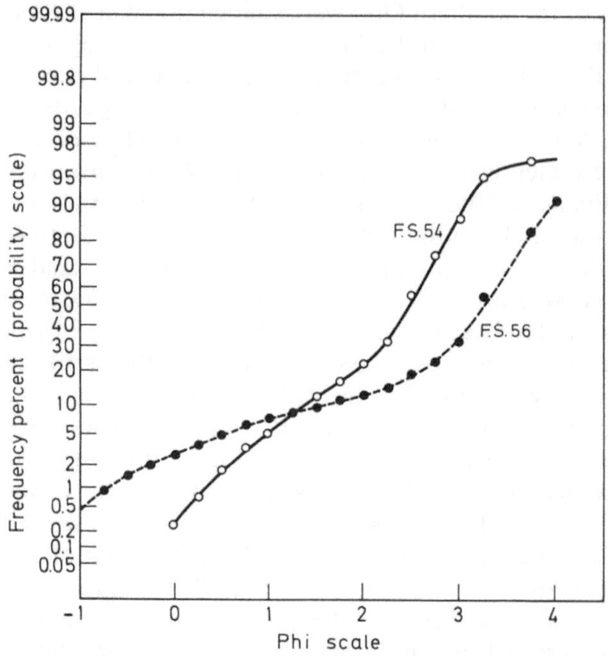

Fig. 5.17. Grain-size analyses from tidal flat between Ras el Haddan and Ras Khashabi

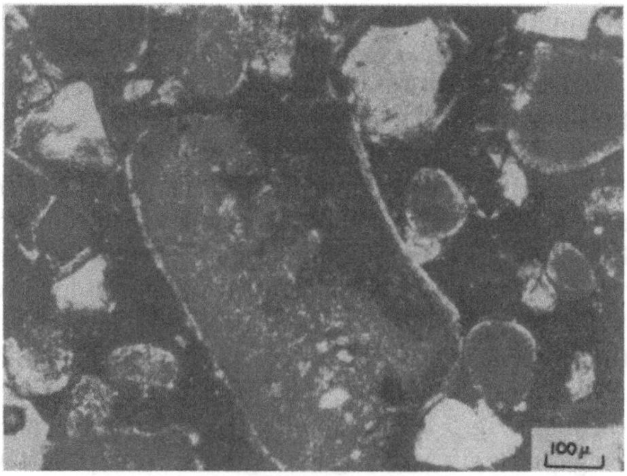

Fig. 5.18. Large fecal pellet consisting of cemented quartz and carbonate grains; note pellet has a rectangular cross-section; uneven surfaces of the two short sides compared to the other two sides suggest that the grain is the disintegration product of a tube-shaped fecal pellet; Ras el Haddan. Thin-section photomicrograph; cross-polarized light

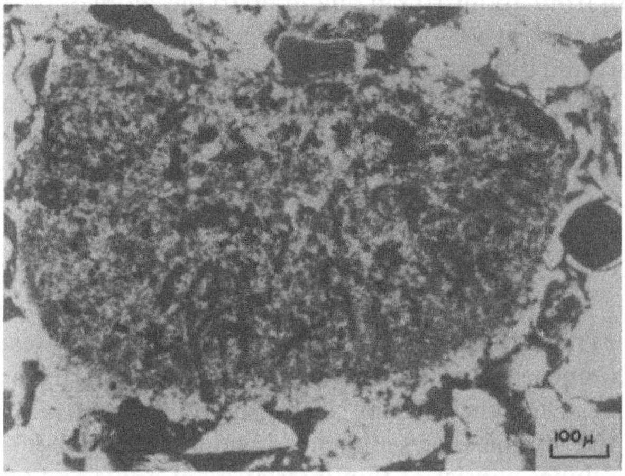

Fig. 5.19. Large fecal pellet with bore holes (dark linear areas) filled with cryptocrystalline carbonate (aragonite mud); Ras el Haddan tidal flat. Thin-section photomicrograph, cross-polarized light

lina, miliolids), mollusc fragments and some ostracodes. Fragments of corals and of coralline algae are nearly absent. In three traverses across the intertidal zone we observed progressive decrease in the quantity of skeletal particles, from the outer intertidal zone landward. The different foram species are distributed randomly; their distribution is dependent on hydrodynamic conditions. *Marginopora,* for example, is concentrated at the foot of the steep slopes of the asymmetric sand waves. Another example is the continuous band of mollusc shell *(Mactra)* accumulation along the high-tide line. Magnetite is abundant in all intertidal samples and mica also is generally present.

5.6 Northern Gulf of Suez: Sea marginal Environments

Sea-marginal environments similar to those of the southern Gulf stretch along the coast of the northern Gulf, especially along a 90-km costal strip between Suez and Ras Malab (Fig. 5.10). As mentioned above, important differences exist in some of the environments between the southern and northern Gulf. In the north, southerly directed currents dominate, there is a very high tidal range, terrigenous materials are derived from ancient carbonates, and the distal fans differ as described below.

5.6.1 Fans

As already noted, sabkhas of the southern Gulf spread across the distal margins of fans. By contrast, fans are less extensive among the sea-marginal environments of the northern gulf. Intermittent braided stream courses (wadis) gradually increase in width from about 10 km inland up to the shore of the Gulf. The slope of these wadi channels is very gentle (approx. 0.25°). Sediment delivered to the Gulf is moved around by waves and currents. As the main current along this coast flows to the south, cuspate–spit complexes are directed southward of the wadi mouths. Examples include Ras Misalla, Ras Dehese, Ras Lahata, Ras Sudr, Ras Matarma, and Ras Malab (Fig. 5.10).

5.6.2 Supratidal Sabkhas and Dunes

Coastal dunes partially cover the supratidal sabkha. They reach up to 8 m in height and are composed of ooids. These ooids were transported by the wind from the intertidal zone. The dunes display steeply dipping cross-strata formed by their advance across the sabkha. The sediments down to a depth of 80 cm in the landward side of the dunes consist of thin interlayers of ooids and terrigenous mud. The ooids below the floor of the sabkha are derived from the coastal dunes and were redistributed across the floor of the sabkha. During and following flash-floods the dunes act as a barrier by ponding water across the sabkha as a result of which terrigenous mud is deposited from the suspended load. The water of the flooded sabkha drains into the Gulf through low saddles in the coastal dunes and/or through percolation. In places the sabkha extends seaward beyond the coastal dunes. There the sabkha surface is always wet and covered by crusts of halite, which commonly cement the ooids, or by thin brown mud patches. The berms and back-berm channels present are essentially the same as those in the southern Gulf. Below the surface, down to a depth of about 60 cm, the sediments consist of brown mud layers, loose ooids commonly mixed with mud, slightly indurated chips of ooids and flat pebbles which are disintegration products of indurated thin ooid layers. The flat pebbles are about 1 cm thick and up to 5 cm in diameter; generally, they are horizontally stratified. Granulometric analyses of the sabkha sands reveal that only about 15% of the total population is in the very fine-grained range, and is not composed of ooids; 8% is coarse to very coarse and represents aggregates of ooids. At the lateral stratigraphic contact with the dune field, porous silts have accumulated and form a low 50-cm step.

Gypsum is generally absent below the surface of the sabkha. As explained below, the depth of groundwater, here 1 m deep, is related to the tides, and is too great for a capillary rise of brines to cause gypsum precipitation.

Cuspate–spit complexes and seasonal-flooded sabkha/lagoons of the northern Gulf were discussed above with their southern Gulf counterparts.

5.6.3 Intertidal Zone: Tidal Flat and Sand Waves (Tidal Bars)

The intertidal zone in the northern Gulf of Suez in front of and between spit complexes is a prominent belt of 400 m. Sand waves, several tens of centimeters high, characterize the intertidal and the upper subtidal zones, extending over a belt 600 m wide; they are less developed opposite the mouths of the wadis. The sand waves display a pattern which includes three basic types. The first and more

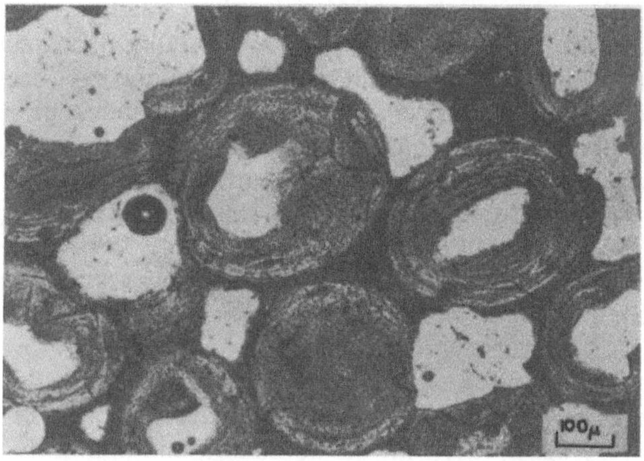

Fig. 5.20. Oolites from Ras Lahata beach rocks, uniform in size (well sorted). Thin-section photomicrograph, plane-polarized light

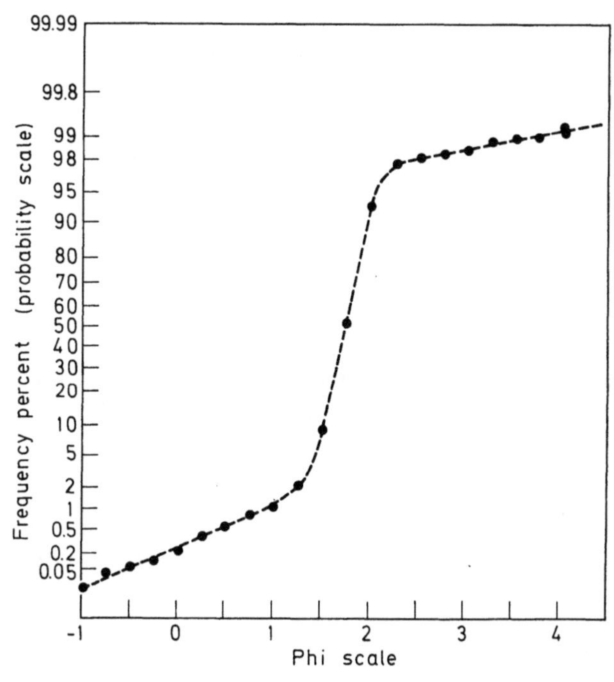

Fig. 5.21. Grain-size analysis of a sand-wave (tidal bar) sample from the Ras Lahata area showing it to be medium-grained with the mode between 1.50 ∅ and 2.25 ∅

common type consists of sand waves that run in a north-northwesterly–south-southeasterly direction and meet the upper intertidal zone at an angle. The second type consists of sand waves that parallel the shoreline; this type occurs only in the upper intertidal zone. The third type of sand waves includes the subaqueous extensions of small spits, which tend to run in a southwesterly direction in shallow water; the avalanche faces dip toward the southeast. However, at water depths of 3 to 5 m the spit extensions are no longer stable and they change to a southeasterly trend.

The major sedimentary constituents of the sand waves in the northern Gulf are ooids. Those between Ras Dahese and Ras Sudr were first investigated by Walther (1888 b). The ooid sands are extremely well sorted (Fig. 5.20). More than 90% of the total population is of medium sand size. The sands are negatively skewed and only 2% of the grains are finer than 2.2∅ (Fig. 5.21). The ooids are of two colors, very light gray, which is dominant, and light-gray. Skeletal fragments which are generally bigger than the ooids are very rare, but become more common in the transition to the subtidal zone. Mollusc fragments as well as echinoid spines and forams are common. Brown algae, including *Padina*, and sargassum are common.

The ooid sand waves migrate over a lithified pavement which consists of aragonite-cemented oolites. The oolites of this pavement are very well sorted and of medium grain size. The lithified oolites floor the intertidal zone up to 400 m away from the shoreline. Here they build a low, 20-m-wide ridge south of the Ras Lahata spit complex parallel to the shoreline, which could be interpreted as a beach rock formed before the area was covered by sand waves. G. M. Friedman (1975) discussed the origin of the cement, claiming that cyanobacteria may be responsible for the precipitation of the aragonite cement.

5.7 Effect of Tidal Range on Precipitation of Evaporites in Supratidal Sabkhas

Pits dug below the surfaces of supratidal sabkhas reveal that continuous layers of gypsum are absent from the northern Gulf of Suez and that crusts of halite are only poorly developed. By contrast, in the southern Gulf of Suez and in the Gulf of Elat, gypsum is abundant and forms hard layers, strata of gypsum-crystal mush, cement in what would otherwise be loose sand, or occurs as dispersed crystals through sand (Figs. 5.7, 5.12 and 5.14). The presence or absence of gypsum may be related to the tidal range. Gypsum is a supratidal evaporite mineral which forms at or below the surface of the sabkha. In the northern Gulf of Suez, where the precipitation of gypsum is at a minimum, such as in the area of Ras Sudr (Fig. 5.10), the tidal range varies from approx. 60 cm to more than 2 m. Measurements of tidal changes and depth to the groundwater table near this area were taken at four stations over a period of a full month's cycle in the summer of 1971 and in the winter of 1971–72. One station was located at the shoreline and

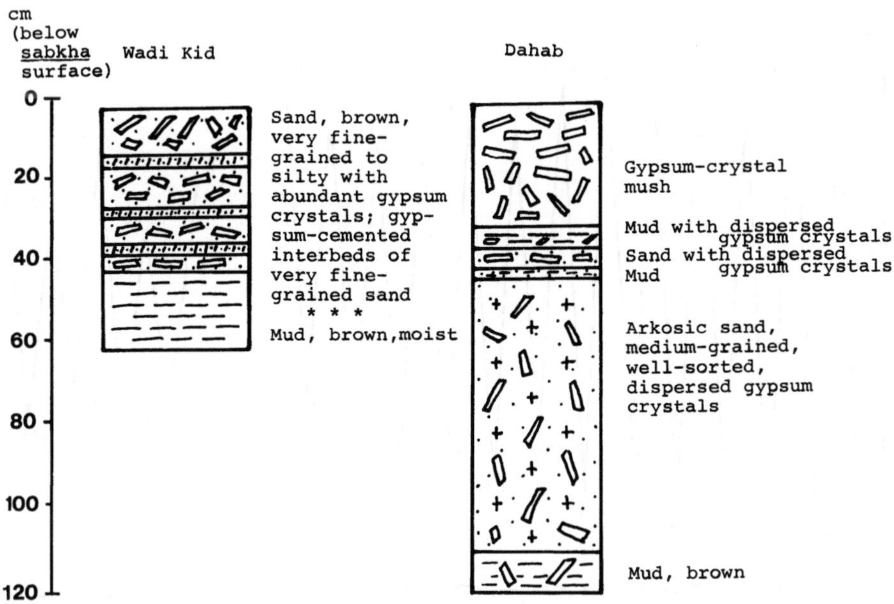

Fig. 5.22. Vertical sequences in coastal sabkhas of Gulf of Elat, as revealed in pits dug in sabkhas of the Wadi Kid and Dahab areas

three other stations were at 100-m intervals progressively landward across the sabkha (Fig. 5.23, Table 5.1). Seawater from shore penetrates landward unimpeded through relatively permeable sands. The tidal variation controls fluctuation of the groundwater table below the surface of the sabkha. Daily fluctuation is of the order of 1 to 3 cm (Fig. 5.23); monthly fluctuation 16 to 21 cm. Depth to groundwater is shallower in the winter than in the summer (Table 5.1); seasonal fluctuations range from 16 to 21 cm.

During the summer, when evaporation is at a maximum, the depth to the groundwater level is 83 cm or more (Table 5.1, Fig. 5.24). Such a depth is too great for a capillary rise of brine to cause precipitation of gypsum. Although depth to groundwater is less in winter (Table 5.1), the prevailing cooler winter temperatures are not as favorable for gypsum precipitation.

By contrast, in the southern Gulf of Suez, as in the Gulf of Elat, the maximum tidal range is 90 cm, and generally the range is 50 cm or less. Hence, the depth to the groundwater table is generally on the order of a few tens of centimeters. Here, the supratidal sabkha sediments are also entirely within the capillary zone and so remain moist throughout the entire year, producing favorable conditions for the precipitation of gypsum. Precipitation may be explained by capillary upward migration of brines accompanied by evaporation (Friedman and Sanders 1967, West et al. 1979). Gypsum is precipitated within a framework of sedimentary particles, as in the continental sabkhas of the Arava Valley north of the Gulf of Elat (Amiel and G. M. Friedman 1971, G. M. Friedman and Sanders 1978, pp. 213–215, Gavish 1974a).

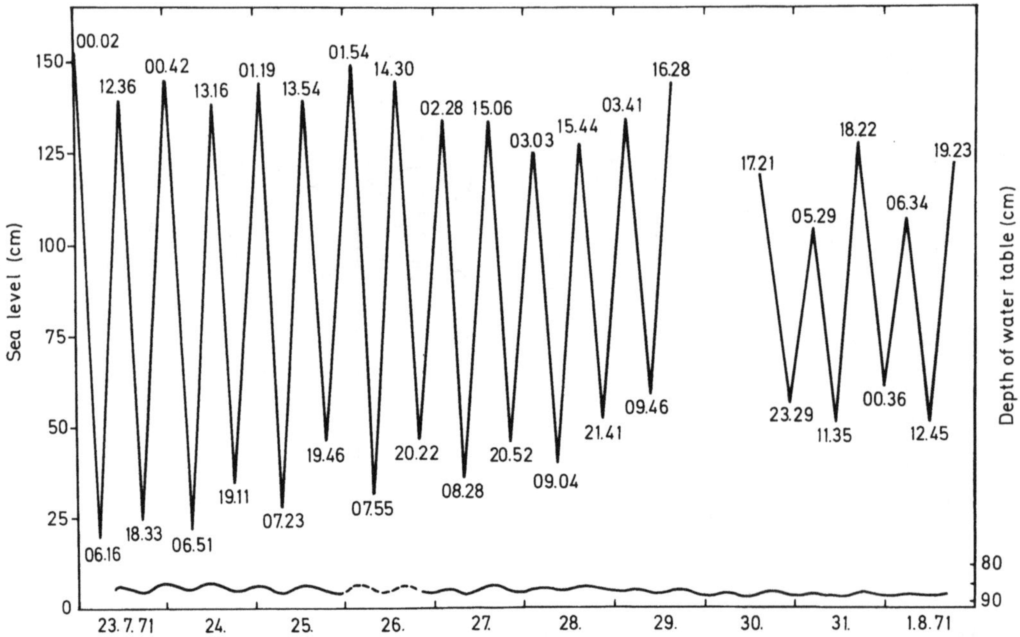

Fig. 5.23. Tidal variations of shoreline and their effect on groundwater fluctuations below the surface of the sabkha at a station 100 m from shoreline. Site of measurement: mouth of Wadi Wardan near Ras Sudr, northern Gulf of Suez (Fig. 5.10); time of measurement: between July 23, 1971 and August 1, 1971 (dates shown on *horizontal axis*). *Vertical axis*, on *left side* of graph, relates daily tidal fluctuations at shoreline to an arbitrary datum for sea level. For each daily cycle the times for high tide and low tide are provided. Graph *at bottom* of figure shows daily changes in depth to groundwater table below surface of sabkha; depth to groundwater is shown on *right vertical axis* of graph. Tidal fluctuation controls fluctuation of groundwater table, but daily variation in groundwater table ranges from only 1 cm to 3 cm. (A. Ecker of the Geological Survey of Israel operated the meters and prepared the tidal curves)

Table 5.1. Depth to groundwater table below sabkha near Ras Sudr (month of Wadi Wardan)

Distance of station from shoreline (m)	Depth to groundwater table (cm) Summer	Winter
100	85	68
200	83	62
300	115.5	99

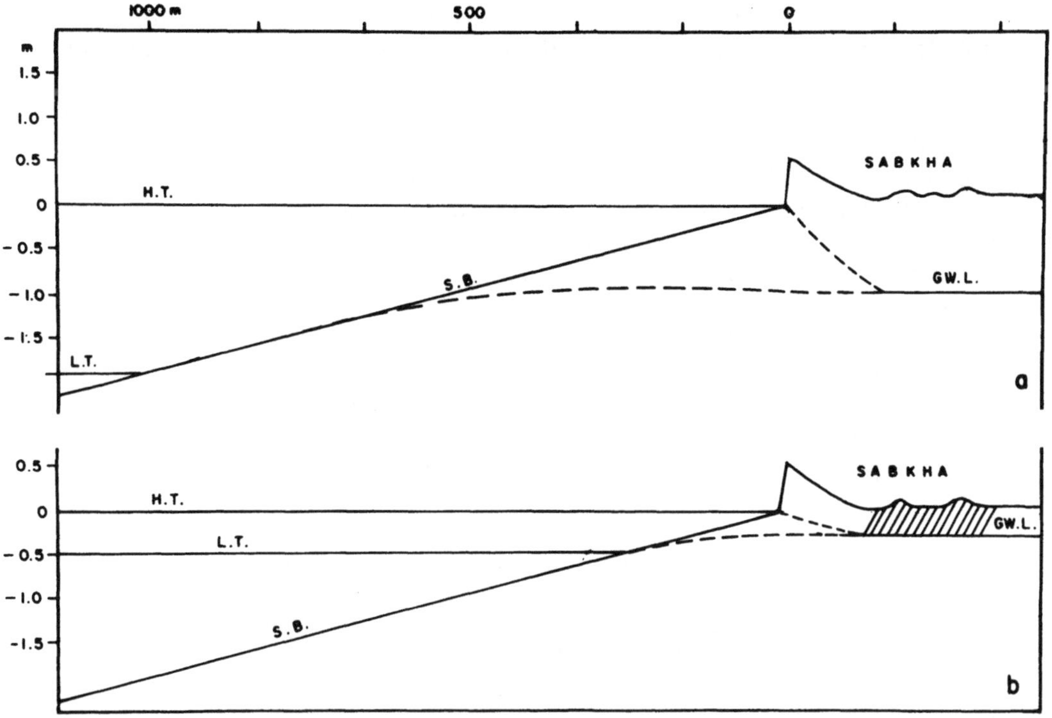

Fig. 5.24 a, b. Coastal sabkhas and tides. *H.T.* high tide; *L.T.* low tide; *S.B.* sea bottom; *G.W.L.* groundwater level below surface of sabkha. Measurements made in July 1971. Relationship between tidal range and level of roundwater below surface of sabkha: **a** northern Gulf of Suez, **b** southern Gulf of Suez. Depth to groundwater level in northern Gulf, where tidal range is up to 2 m is approx. 1 m. By contrast, depth to groundwater level in southern Gulf, where tidal range is up to 0.5 m, depth to groundwater is approx. 0.5 m

5.8 Requirements for Formation of Supratidal Sabkhas with Evaporites

Both the Gulf of Elat and the Gulf of Suez lie in a climatic and geographic zone in which supratidal sabkhas develop. The necessary environmental conditions for sabkha development include (1) a hot arid climate, (2) proximity to the sea with a continual source of saline water, (3) addition of mud to the depositional site. When mud is added to a very permeable sandy section, it enhances potential capillary migration and also serves as a good host for brines. However, certain other environmental conditions are necessary requirements for the proper development of supratidal sabkhas in the Gulf of Elat and Suez. With respect to the second requirement listed above, for an adequate quantity of saline water to be supplied to the supratidal flat in this area the tidal range must be small enough to maintain the groundwater level within the zone of capillary action. In the northern Gulf of Suez, for example, where the tidal range is high, the sabkha sequence is poorly developed and essentially lacking in evaporites. This condition obviously does not hold for other regions where the physical conditions are favorable for seawater to become supplied to the supratidal flats periodically and trapped there until it evaporates (Posnjak 1940). A requirement for sabkhas along the gulfs that is unrelated to the factors listed above is that the environments must be protected from the waves and nearshore currents, which are very strong on occasion. The distal fans are protected by a belt of fringing reefs. The supratidal sabkhas of the Gulf of Elat are located on the southern flanks of these fans and are protected from being reworked by waves and current generated by the prevailing northern winds both by the northern and central portions of the fans and the fringing reefs. In the Gulf of Suez, the protected sites are within the cuspate–spit complexes. The required addition of mud to the site restricts the sabkhas of the gulfs to locations at the toes of fans (including spit complexes), where transport and deposition of terrigenous mud occurs. In the southern Gulf of Suez, in places where the fans contribute only small quantities of mud, the authors found the coastal supratidal sabkhas to be poorly developed.

5.8.1 Sequence of Sea-marginal Environments Through Time: Gulf of Suez

In the Gulf of Suez the coastal accretion in the spit complex areas results in regressive sedimentary sequences in which sabkha deposits are underlain by intertidal deposits. In the Ras Kanisa area, 50 cm of pellets and ooid sands with interlayers of mud and gypsum of the sabkha type are underlain by *Marginopera* sands which are of intertidal origin. Migration of the spit-barrier bars in this area has a southward and a landward component, resulting in the landward superposition of the barrier over the protected seasonal sabkhas. In small spits in the Ras Kanisa area, stratified coarse *Cerithium*-bearing sand deposits of the barrier, about 1 m thick, are underlain by at least 20 cm of thick organic-rich black mud. The black mud crops out at the bottom on the seaward side of the barrier; this results from the migration of the barrier landward.

The sabkha deposists are overlain by eolian dunes, which migrate mainly over the sabkha surface, on the berm-top flats and on the distal margins of the fans. The sand particles that comprise the dunes derive from the sea. Sands added to the Gulf by flashfloods along wadi channels are, in large part, driven back onshore by prevailing winds, as observed in the Gulf of Elat and described above. Dune sands accumulate in several spit complexes south of the mouth of a main fan-wadi.

Study of successive aerial photographs dating from 1956 to 1978 and periodic field observations since 1967 show that accretion activity along the Gulf's coast takes place mainly on the spit complexes. When spits extend far enough into the subtidal zone, they bend southward and continue to develop as hook-shaped spits. During storms, the process of spillover produces southward, landward-dipping cross-strata. The sandy spit ridge stabilizes and becomes a barrier bar protecting the subsequently developing lagoons, seasonal-flooded sabkha/lagoon and eventually the supratidal sabkhas. Each one of the spit complexes does not reflect the whole set of environments mentioned above, but rather it displays a stage of development within the sequence. The Ras Matarma spit complex, and

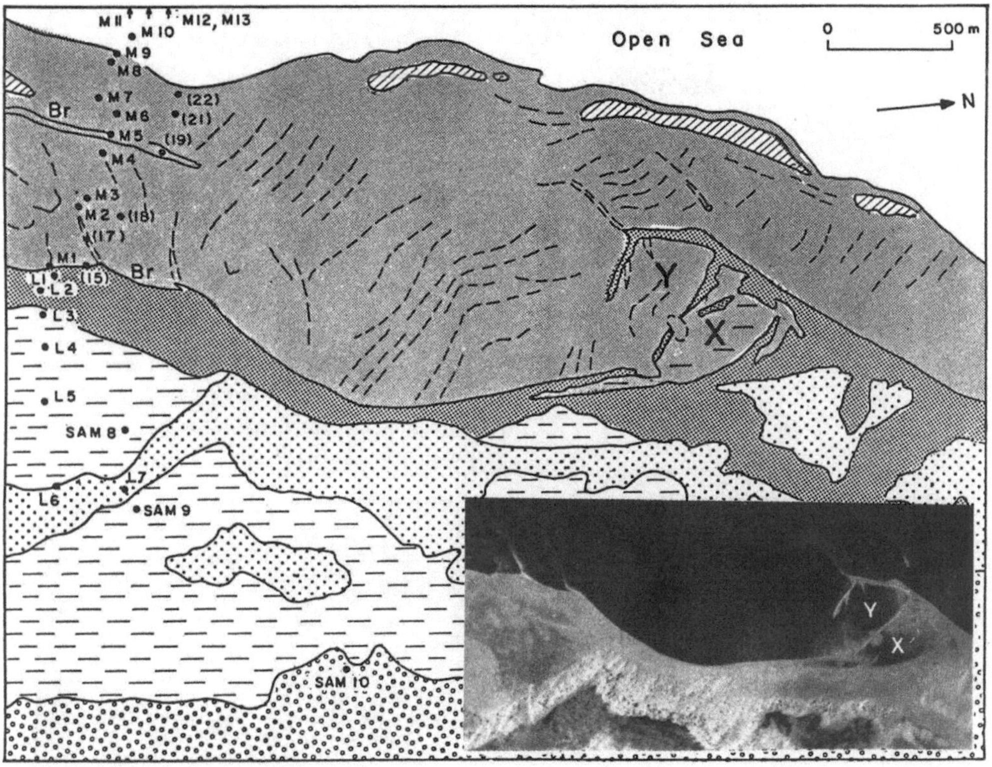

Fig. 5.25. Ras Lahata cuspate–spit complex in the northern Gulf of Suez, an example of a spit-barrier bar redeveloping seaward from the bending point to form a second lagoon (lagoons marked *X* and *Y*); diagram of sedimentary environments (see Fig. 5.4 for legend) and aerial photograph of the same area

the Ras el Millan spit complex display the lagoonal stage; tidal-flat conditions prevail in the lagoon and very shallow tidal channels migrate laterally over the tidal flat. The Ras Malab, as well as many of the small secondary spits, display seasonal sabkha/lagoons in the protected zone behind the spit-barrier bars. The Ras Lahata spit complex in the northern Gulf of Suez is an example in which the spit-barrier bar redevelops from the bending point seaward, thus forming a second spit complex (Fig. 5.25). Seasonal sabkha/lagoon conditions exist in the first complex, close to the mainland, whereas in the second spit, situated farther out, a lagoon exists, protected behind the barier bar.

In Ras Matarma, the spit complex actually displays two lagoons separated by a sandy ridge. Hence, the evolutionary trend seems very similar and, therefore, the explanation given by Sass et al. (1972) based on Ras Matarma observations, i.e., that the lagoons developed in eroded areas, may not necessarily be the only one, though indeed backward erosion by tidal channels was also found to play a role in the development of spits. Though rapid accretion takes place, beach rocks are formed around the intertidal borders of the spits.

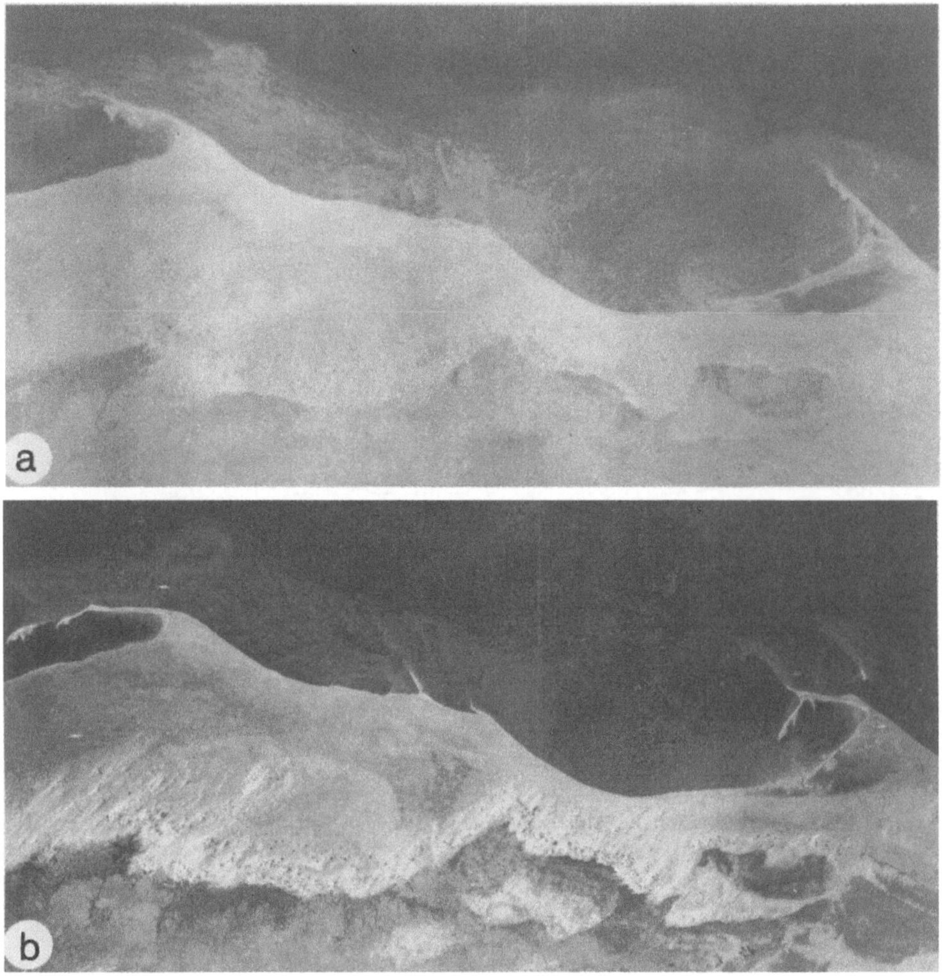

Fig. 5.26 a, b. Ras Lahata spit complex in *right side* of photographs, scale 1:10,000 in both **a** 1956 and **b** 1971. Note construction of additional spits; note also development of a spit shown *on left side* of photograph

An estimate of the rate of migration can be made by considering two examples. Figure 5.26 presents examples of accreted spits from the Ras Lahata area. The outer spit was constructed approximately 200 m southward in less than 15 years between 1956 and 1971, as was another spit farther south. It seems more likely, however, that the spits were accreted under stormy conditions and hence, although they were accreted over a 15-year period, their formation was not gradual. Photographs taken of the Ras el Millan area show similar rates of spit growth.

5.9 Conclusions

- Sea-marginal flat environments, including sabkhas, in the gulfs are intimately associated with terrigenous deposition from wadis. In the Gulf of Elat the suite of environments is associated with distal fan complexes, and in the Gulf of Suez with cuspate–spit complexes.
- Vertical sequences below the surface of supratidal sabkhas are dependent on their location relative to the fan or berm. Only the uppermost layers (10 to 20 cm) include evaporites and muds in areas near the fan or berm boundary in the Gulf of Elat and southern Gulf of Suez.
- Requirements for supratidal sabkha formation include (1) hot and arid climate, (2) a sea-marginal location with adequate supply of saline water to the sabkha to develop evaporites, and (3) a supply of terrigenous mud. Requirements for supratidal sabkha formation in this region also include (4) location on the lee side of fan or spit complexes where protection is afforded from prevailing winds and waves, and where wadis provide terrigenous mud which aids in capillary action and serves as a host for precipitation of evaporites, and (5) a low tidal range, so that the groundwater level can be maintained within the zone of capillary rise to permit adequate supply of saline water for evaporite formation.
- Coastal accretion in the Gulf of Suez occurs by a sandy spit ridge or a sand wave becoming a barrier bar protecting a lagoon, followed by a seasonal-flooded sabkha/lagoon, and eventually by a supratidal sabkha. A regressive sedimentary sequence is laid down with supratidal sabkha deposits underlain by intertidal deposits. Subsequent landward migration of the barrier may result in barrier sands overlying the supratidal deposits in some cases.
- Spits studied in aerial photographs developed (migrated) at a rate of several tens of meters during an active season.

6. Anchialine Pools – Comparative Hydrobiology

Francis D. Por

The term "anchialine pools" was first used by L. B. Holthuis, in his H. Steinitz Memorial Lecture of April 1973 at the Steinitz Marine Biology Laboratory in Elat. It was first published in October of the same year (Holthuis 1973). "Anchialos" meaning in Greek "near the sea", the new term stands for "pools with no surface connection, containing salt or brackish water, which fluctuates with the tide".

The term anchialine pools is used here in an enlarged sense as a replacement for such terms as "sea-marginal pools" (Gavish 1975, Y. Cohen et al. 1977a), "seepage pools" (Por 1974a, Por and Dor 1975a, b), or „ufernahe Seen" (Krumbein and Y. Cohen 1974). The term is used mostly in hydrobiological context, while the terms "sabkha" and "salina" are widely distributed in geoscience and ecology. Some of the individual approaches to the consequences of climatic conditions, erosion and sedimentation, tidal influences and tidal pumping as related to sea-marginal flats and pools have been discussed in Chapters 3, 4 and 5.

6.1 What Is an Anchialine Pool?

Anchialine pools are water bodies of several tens, up to a few hundreds of meters in diameter, situated close to the seashore. Their most important feature is the fact that they have a permanent subterranean connection with the open sea and no permanent, or seasonal subaerial surface contact. Likewise, they are not connected with any permanent flow of freshwater on their landward side. Finally, these pools have an active and permanent subterranean outflow. In conclusion, the hydrography of the anchialine pools is defined by discrete, subterranean water exchange.

Anchialine pools are found first of all on emerged coral shores, rich in crevices. An alternative substrate are the lavaflow shores – such as in Hawaii or Ascension Island – which contain also a network of cracks and holes. Anchialine water bodies cannot become established on sedimentary bottoms, and probably only exceptionally on massive eruptive or metamorphic rock.

Anchialine pools cannot exist in areas of active fluviatile deposition or long-shore sedimentation. They can survive occasional short-time floods or storm overspill from the sea, but not as a permanent or seasonal feature of their hydrography.

A secular stability of the environmental conditions is therefore an important characteristic of the anchialine pools: rather than exhibiting unpredictable salinity fluctuations like the surface-supplied nonmarine nearshore water bodies, their salinity signature is a stable and seasonally predictable one. This is defined chiefly by geomorphological factors, such as distance from the sea and the texture of subterranean crevices available for in- and outflow of water and indeed by the size and depth of the water body and its surrounding topography. The annual average evaporation must by far exceed the annual average rainfall over the pool: therefore anchialine pools are characteristic for arid or semiarid shores.

In accordance with the secular seawater level changes, small-scale uplift or sinking movements of the shore, and the changes of the climate in the area, anchialine pools are stable ecosystems. Every anchialine pool has its individual hydrography – as far as salinity is concerned they may be situated anywhere on the range extending from polyhaline to euhaline, metahaline and hypersaline conditions. Unlike many open-supply estuarine environments, the small anchialine pools tend to develop stable or predictable vertical stratification.

6.2 Anchialine Pools as Mature Ecosystems

Our pools are islands of homeostasis and successional maturity unlike the open-supply estuaries with their imprevisible parameters. Making amends for the absolute limits set by high salinity, the anchialine pools are inhabited by a high diversity of species, which exhibit clear biotic zonation patterns within the pool and a well-established niche partitioning. To the contrary, the open-supply estuarine water bodies are inhabited by a small diversity of opportunistic species. As such, the anchialine pools deserve first of all interest because of their being islands of ecological climax-like stability in the domain of the brackish and hypersaline waters, which are in their wast majority dominated by impoverished opportunistic biota.

Owing to the exclusivity of the subterranean marine contact, the anchialine climax-like biota are protected from the penetration of these aggressive estuarine opportunistic biota: Here belong the schools of omnivorous or predacious euryhaline fish, penaeid shrimp or portunid crabs, that usually feed in the lagoons but reproduce in the open sea. To the same complex belong also the extremely widespread pioneering species of sessile animals, such as barnacles and mussels. The pools are exclusively inhabited by algae and animals that can reproduce within the pool itself, or its subterranean afferent system. The lack of competition on the part of the most aggressive component of the inshore and estuarine fauna is a most important consequence of the anchialine situation. In pools of the hypersaline range, aerial transport or wind-borne settling becomes important.

Finally, because of the smallness and partly also because of frequent vertical stratification, anchialine pools have practically no pelagic community. Primary production is a benthic one, or else the trophic basis is supplied by anemotrophy, i.e., wind-swept import of fine terrestrial organic matter. A need for good recycling of the autochthonous benthic photosyntetates is also conductive to intricate food webs.

How do the anchialine pools integrate in the continuum of the different types of aquatic environments? On the one hand they can be delimited from the submarine caves, the "marginal caves" of Riedl and Ozretic (1969), and from lagoons that have a mixed supply of subterranean and subaerial seawater.

Examples for submarine caves are many: these have no elements of trophic autarchy, no algae, no grazers. They are inhabited by blind decapods, mysids and amphipods (see also Stock and Vermeulen 1982). This does not imply, however, that – as we shall see below – some animals cannot be common to both caves and anchialine pools.

Lagoons with mixed supply can be considered to be the mangrove lagoons of Southern Sinai (Por et al. 1977) or the lagoon of El Qardud, likewise in Sinai (Fig. 6.1): these are inhabited, especially in their hydrographically more extreme recesses where the estuarine migrators cannot penetrate, by elements of the anchialine flora and fauna.

Finally, toward the extremely widespread littoral sabkha environments (sensu latu), and the littoral hypersaline rock pools, the anchialine pools are delimited by the predictability of their environment and consequently by the diversity of their biota.

Fig. 6.1. Lagoon of El Qardud in Sinai, seen from the west. The semi-closed metahaline area is indicated by an *arrow*

6.3 World Distribution of the Anchialine Pools

A first listing of anchialine pools of the polyhaline marine type from the Pacific Ocean results from the publications of Holthuis (1963, 1973) and Wear and Holthuis (1977): they are reported from Hawaii, Fiji, Funafuti, Ellice, Tokelau and Loyality archipelagos of the Central Pacific, and Halmahera and the Philippines in the Eastern Pacific. In the Indian Ocean, marine anchialine pools were reported by Fricke (1976) from Aldabra and Europe Island. In the Red Sea, marine salinity pools have been found on Entedebir Island in the Dahlak Archipelago (O. H. Oren 1962 b, Por 1968 a), on the Ras Muhammad peninsula of Sinai and the adjacent Sinafir Island (Por and Tsurnamal 1973, Por and Dor 1975). Sinai has the only well-investigated pool of metahaline salinity, the Di Zahav pool (Por and Dor 1975), and three hypersaline pools, Solar Lake, Haniqra Pool and the Gavish Sabkha, extensively discussed in this volume.

From the Central Atlantic there is a report of a marine-type pool on Ascension Island (Chace and Manning 1972). The islands of Curaçao and Bonaire (Netherlands Antilles) are rich in various types of high-salinity anchialine pools (see, among others, Kristensen 1970, 1971, Lucia 1968). Cuba, the Bahamas and the Islands of Bermuda have many anchialine-like littoral water bodies, although from the literature it is somewhat difficult to clearly identify them (see Sket and Iliffe 1980, Hart and Manning 1981).

Conclusions are as yet highly premature; however, it seems to be evident that anchialine pools are a circumtropical feature and that they exist everywhere where the shore has a suitable structure and where evaporation exceeds precipitation on a yearly average. The higher frequency of anchialine habitats in the Indopacific might be an objective situation, especially due to the many oceanic islands that have no riverine systems.

6.4 The Acting Environmental Factors

In order to understand the genesis of the different types of anchialine pools, a listing of the main environmental factors is needed (see also Fig. 6.2). The factors are listed in an approximate order of importance and marked by letters for easier use:

1. The seawater seepage into the pool *(SE)*
2. Evaporation *(E)*
3. Pool water reflux into the substrate *(RE)*
4. Evaporative pumping by the pool brine *(EP)*
5. Influx of freshwater (rain and/or runoff) *(F)*
6. Surface-to-depth relation of the pool water body (S/D)
7. Exposure to wind

An attempt will be made to organize all these factors into a mathematical formula, giving the salinity of the pool waters, where PS is expressed in per mil salin-

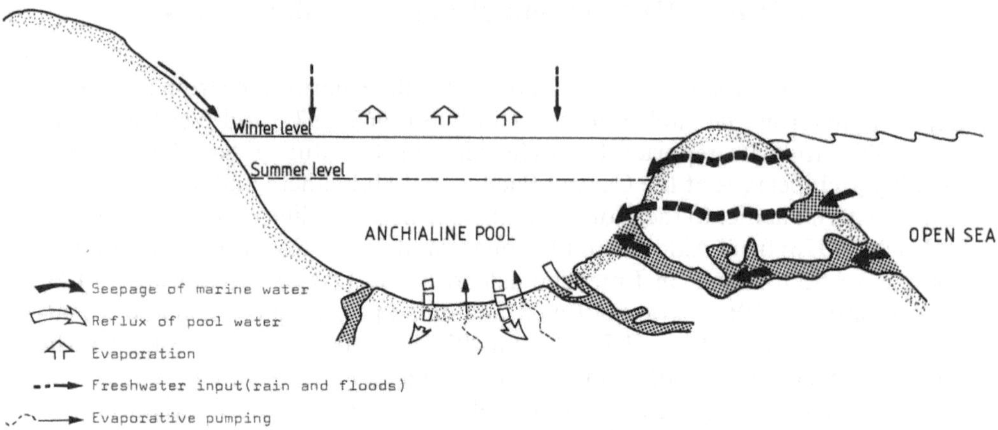

Fig. 6.2. A schematic cut through an anchialine pool. (See discussion in text)

ity. PS = 1/1 meaning seawater salinity; PS = ±2/1 would express a more or less metahaline salinity, etc.

1. Seepage from the sea depends on structure of the separating bar and the distance from the sea. In cases of cave-like connections with a nearby sea, tidal movement can be found in the pool; the effect of tide will be the less felt, the farther the distance separating the pool from the sea; also there will be a concomitant retardation of the tidal cycle. Seepage can be seasonally variable, since it is determined also by the pressure of the water column in the open sea: this may be reduced, when, as in the high summer of the Northern Red Sea, the level of the sea decreases. However, in general, seepage is a remarkably constant factor. Salinity of the seepage water is considered to be marine salinity. I do not foresee that pools will be found that are supplied by brackish waters.

2. Evaporation with its seasonally changing values is the main factor responsible for the seasonal variation in the hydrographic parameters of the pools. Evaporation is much more effective in shallow and wind-exposed pools; however, the seasonality is still the most important element here.

3. Reflux, or escape of pool waters into the substrate is probably the less-known factor, since reflux usually occurs in the deeper parts of the pool and the nature of the aquifer to which it leads in unknown. Even with little hope to be able to measure the volume of reflux, it can be reasonably quantified from the following formula:

$$PS\,(‰) = \frac{E}{SE - RE} \quad \text{(Por 1975a)}. \tag{1}$$

However, this simple formula has to be complicated in most of the cases.

4. Evaporative pumping (sensu Hsü and Siegenthaler 1969, Gavish 1975a; see also Chapts. 4 and 9) acts in pools with concentrated brine that absorbs less saline water from the surrounding sediments. This factor should be included into our formula:

$$PS\,(‰) = \frac{E}{(SE + EP) - RE}. \tag{2}$$

It would be very interesting to calculate experimentally, what is the absolute value of salinity and the differential values between pool and aquifer water, where evaporative pumping sets in.

5. Although anchialine pools exist in areas where evaporation predominates over precipitation, pools receive relatively small and unpredictable inputs of rainwater and freshwater runoff from the surrounding hills. Nonetheless, this relatively secondary factor has also to be included into our formula:

$$PS\,(‰) = \frac{1}{F} + \frac{E}{(SE+EP)-RE}\,. \tag{3}$$

6. The surface-to-depth relation of the water body is the last of the factors influencing the saline regime of the pool. Thus:

$$PS\,(‰) - \frac{1}{F} + \frac{E}{(SE+EP)-RE} + \frac{S}{D}\,. \tag{4}$$

7. Wind has been mentioned as a factor increasing evaporation. It is directly responsible for the vertical mixing of the pool waters: wind protection is favorable to the establishment of a halocline in the pool, with the ensuing heliothermal heating of the hypolimnion and appearance there of anaerobic conditions.

6.5 Typology of the Anchialine Pools

There is a wide range of variation in salinities among the different anchialine pools. However, they can be classified into three categories of increasing salinity, each category characterized by different sets of biota. Possibly, the three types can be seen as evolving one into another due to subgeological, secular changes in the pool environment.

6.5.1 Polyhaline-Euhaline Anchialine Pools

These pools are well supplied with seawater, usually showing marked tidal fluctuations, with active reflux; they are as a rule deep, narrow and with shaded areas. As a consequence, PS ‰-1/1; i.e., pool salinities are equal or slightly lower than the open-sea salinity. These pools exhibit great seasonal stability in salinity and other hydrographic factors.

Without doubt, the characteristic biota of these pools are the red-colored shrimp (Holthuis 1963), caridean decapod crustaceans of deep-red color belonging to the families Palaemonidae, Hippolytidae, Atyidae, Alpheidae and Procarididae. Several genera can be considered to be exclusive to the anchialine environment: *Procaris, Halocaridina, Antecaridina, Calliasmata, Metabetaeus, Barbouria* and eventually *Typhlatya*. According to Holthuis (1963, 1973) there are some 11 species typical for the pools; this figure may be already minimalistic.

Some of the species have an extremely wide geographic distribution: *Antecaridina lauensis* is reported from Sinai, Entedebir Island (Southern Red Sea), Europe Island (Indian Ocean) and Fiji and Hawaii (Pacific); *Calliasmata pholidota* is known from Sinai and from Hawaii. Of the two known species of the anchialine genus *Procaris* one *(P. hawaiiana)* is found in Hawaii, the other *(P. ascensionis)* on the Atlantic Ascension Island. It is clear that these species must have a continuous distribution in marine caves or unknown anchialine pools between the two extreme geographical sites.

So far only one Atlantic species, *Barbouria cubensis,* shows a widespread distribution in its zoogeographic area. But new species have been reported from this region recently: such is the case of the new genus and species *Sommersiella sterreri* and of *Typhlatya iliffei* (Hart and Manning 1981).

An extremely interesting characteristic of the red shrimp fauna is the coexistence of several species in the same pool. In Hawaiian pools, situated in the same area, five species (*P. hawaiiana, C. pholidota, Halocaridina rubra, A. lauensis* and *Metabetaeus lohena*) occur in different combinations of twos and threes. On Ras Muhammad, *C. pholidota,* and *Periclimenes pholeter* live together in the same pool (Holthuis 1973, Por and Tsurnamal 1973), while another species, *A. lauensis,* lives in another rocky crack, a few hundreds of meters distant. In Tucker's Town Cave on Bermuda, *S. sterreri, T. iliffei, B. cubensis* and *Automate dolichognatha* are all found together (Hart and Manning 1981) in a pool of 7/15 m.

The red shrimp are usually found together with some representatives of the epigeic shrimp fauna, such as *Palaemon debilis* in the Indopacific. In Sinai, the anchialine *P. pholeter,* representative of the big genus *Periclimenes,* does not have the typical deep-red color of the other anchialine shrimp. It is therefore possible that stages of the evolutionary process leading to the red shrimp can be still found in situ.

Holthuis (1973), Wear and Holthuis (1977) and Fricke (1976) give interesting observations about the behavior of the red shrimp. It seems that these species commute, sometimes in a tidal rhythm, between the well-lit open pool waters and the dark and shaded cracks and subterranean waterways. The big amount of red-colored carotenoids can eventually be related to the frequent changes between the well-lit tropical pool and the dark "marginal cave" environment. Wear and Holthuis (1977) found that *Ligur uvae,* a widespread Pacific red shrimp, becomes translucent to chalky in color during the night, owing to the condensation of the carotenoid pigment in the chromatophores.

The rest of the biota of the polyhaline-marine pools did not reveal till now anything characteristic. To my knowledge, the only other new species described from such waters is the gobiid fish *Cabillus anchialinae* (Klausewitz 1975). Por (1968 a) described several new species of benthic copepods from Devil's Crack on Entedebir Island (Southern Red Sea), but it is doubtful if these species are typical for the anchialine habitat. The only eventual exception is the benthic calanoid *Pseudocyclops steinitzi.* A new, yet undescribed species of this genus has been found also in the cracks of Ras Muhammad (Sinai).

In general, the biota of these type of pools is similar to that of the shaded rocky shores. Among these, mention should be made of the crustose red algae, siphonocladacean green algae, sponges, bryozoans and solitary ascidia.

6.5.2 Metahaline Anchialine Pools

At PS per mil values of 1/2–2/1, i.e., when salinities in the pool are up to about twice seawater salinity, one can speak of pools with metahaline salinities. The metahaline salinity range is 41‰–70‰ in the Red Sea area (Por 1972) and for the Caribbean it should be considered, following Kristensen (1970), to be 38‰–70‰. This salinity range is still inhabited by typical marine biota.

This is probably the least frequent of the three types of anchialine pools. It needs a particular combination of environmental factors to maintain the annual salinity fluctuations within the limits of the metahaline values. The only better investigated metahaline pool is that of Di Zahav on the Sinai coast (Por and Dor 1975), where salinity fluctuates seasonally between 45‰ and 60‰. On the island of Curaçao, two water bodies, Lagun Blancu and Lagun Sant Juan, can also be characterized as belonging to this category (unpubl. pers. data). Recently also an association typical for metahaline pools was found in the nearly closed to the sea corner of the lagoon of El Quardud in Sinai (Fig. 6.1) – no doubt a seepage pool in statu nascendi.

Two features are characteristic for the metahaline pools of Sinai and of Curaçao: the thick sedimentary carpet of fine calcareous mud that covers most of the pool bottom and the boulders of serpulid reef.

The calcareous muddy bottoms are inhabited by a very diversified meiobenthos (for Di Zahav, see Por 1979). Probably among the only macrobenthic inhabitants of these bottoms are the upside-down medusae of genus *Cassiopeia* (*C. andromeda* in the Red Sea and chiefly *C. xamahana* in the Caribbean).

The biogenous serpulid roocks (built in the Di Zahav pool by *Vermiliopsis pygidilis*) are settled by a diversified and zonally disposed flora of siphonocladacean green algae, the colonial ascidian *Ecteinascidia* and several species of sponges. In Di Zahav, the boulders harbor also keyhole limpets of genus *Diodora,* and sphaeromid isopods of genus *Cymodocea.*

The metahaline pools present also a peculiar beach rock-like gravelly bottom, consolidated by sponges (in Di Zahav, chiefly *Hymeniacidon sanguinea*) which give footing for the red alga *Laurencia papillosa.*

The metahaline waters on Curaçao are characterized also by the abundant growth of the extremely euryhaline green alga *Batophora oerstedti.*

Among the fishes, gobiidae are still present, but the predominating taxa are already the cyprinodonts [*Aphanius dispar* in Di Zahav and *Rivulus marmoratus, Poecilia sphaenops* and *Cyprinodon dearborni* in Netherlands Antilles (Kristensen 1970)].

Cyanobacteria are present in the wet littoral of the metahaline pools but never forming mats. They are probably efficiently grazed by the amphibious snails that are numerous in this environment: *Pirenella conica* in Sinai and *Batillaria minima* on Netherlands Antilles, both belonging to the family Potamidiidae.

6.5.3 Hypersaline Anchialine Pools

Above the metahaline limit, i.e., PS per mil 2/1, the hypersaline pools are characterized by an extremely impoverished general diversity of species. Restriction is due to high absolute salinity, modified ionic composition, anoxyc and occasionally high temperature conditions in the hypolimnia, seasonal drying out or considerable size reduction. Eventually, antibiotic effects of the massive cyanobacterial populations have also to be taken into account. Solar Lake, the first known pool of this type has been first mentioned by Por (1968 b).

Since this type of pools forms the subject of the other contributions in this volume, their discussion here will be very succint.

Hypersaline anchialine pools are probably a very frequent coastal feature in the dry-tropical zone. In Sinai, three such pools have been investigated, namely Solar Lake, Gavish Sabkha and Haniqra Pool on Ras Muhammad. There are several such pools in Curaçao, though severely modified by salt exploitation, the best known being Lagun Jan Thiel (Kristensen 1970). On Bonaire, the lagoon called Pekelmeer is well known from the geological side (Lucia 1968).

Hypersaline anchialine pools are first of all characterized by the predominance of cyanobacterial mats and photosynthetic bacteria, accompanied by a reduced diversity of very euryhaline diatoms (see Chap. 14). The fauna is characterized first of all by the rich diversity of aquatic and amphibious coleopterans (Por 1975 b; see also Chap. 15). Ephidrid brine flies may appear, although they are not as dominant as in other types of salt pans. The presence of *Artemia,* too, is not always a dominant one. On Curaçao, two cyprinodont fish species may occasionally appear in these pools, although salinity is in general above their resistance limit.

The meiobenthos is first of all characterized by a fairly rich ciliate fauna (Wilbert and Kahan 1981), a species of rhabdocoelan flatworms, several nematods and two species of harpacticoid copepods (Por 1975 b).

Because of the extreme salinity fluctuations and seasonal drying out in some of these pools, it is believed that only animals that have resting or anabiotic stages, or else can leave the water or survive in the wet pool bottom, can be permanent elements of the pool fauna.

As shown in the above-mentioned paper (Por 1975 b), a wet salt-marsh fauna develops, feeding on the big amount of cyanobacterial material, left to dry during the seasonal level decrease of the pools. Mention should be made here first of all of the oniscoid isopods, pseudoscorpions and beetles. The fact that hypersaline pools are frequently above the salt tolerance limit of 100‰ of *Pirenella conica* (Taraschewski and Paperna 1981) is probably the reason of the lurious growth of the cyanobacterial mats.

7. Botanical Studies on Coastal Salinas and Sabkhas of the Sinai

Michael Evenari, Yitzchak Gutterman, and Eliezer Gavish†

Salinas and salt marshes (hydrosalinas) of the world have often been studied by biologists (Chapman 1974, 1975, 1976), but the many salinas and sabkhas of the Sinai Peninsula have never been investigated in detail, and only a few papers dealing with them have been published (Migahid et al. 1959, Danin 1981, 1983).

When we first visited the Sinai in 1967, we were struck by the pronounced difference in the floral composition of the salinas of the southeastern (Gulf of Elat), southwestern (Gulf of Suez) and northern (Mediterranean) coast of the Sinai. We decided to describe these differences and, if possible, to find out the reason for them. At the same time, we wanted to investigate the zonation of the salina vegetation and the relationship between the root systems of the dominant plants of each salina and soil structure, chemical composition and soil water of each habitat. The study of this relationship seemed to be of special interest since "little is known on properties of soil and water in natural saline habitats" (Poljakoff-Mayber and Gale 1975).

While working in the Sinai, we soon noticed that investigating salt plains was of more than scientific interest. The salinas of the peninsula, especially those on the coast of the Gulf of Elat have a strange beauty lying between the blue-green water of the Red Sea and the somber dark-hued monumental massif of the magmatic mountain ranges of the central Sinai. Because of this beauty, a favorable climate during winter and excellent opportunities for swimming and snorkeling, they became, soon after 1967, when good access roads were built along the coast, great tourist attractions. Tourist resorts were established and the whole coast was overrun by visitors. This led to deterioration and even partial destruction of many natural special environments. For some, the Egyptian proposal to turn "parts of the Gulf of Aqaba and the Gulf of Suez, Tiran Island, ... the Bardawil and Tina sabkhas" (Kassas 1965, 1967, 1981) into natural reserves, may have come already too late. Since our investigation was carried out when the salinas and sabkhas were still undisturbed, this chapter documents their original state which may be of some importance for later generations.

An investigation like the one we undertook may also be of some practical value. It has been stated lately that coastal deserts with large amounts of available saline water "constitute a substantial natural resource that is presently under-exploited" (Norlyn and Epstein 1982). Regarding the so-called biosaline concept (Aller 1982), some authors even speak of "saline silviculture" (Teas 1979, 1982).

By selection and management of their native flora or by planting them with highly salt-tolerant cultivated plants, some of the Sinai salt plains as part of a

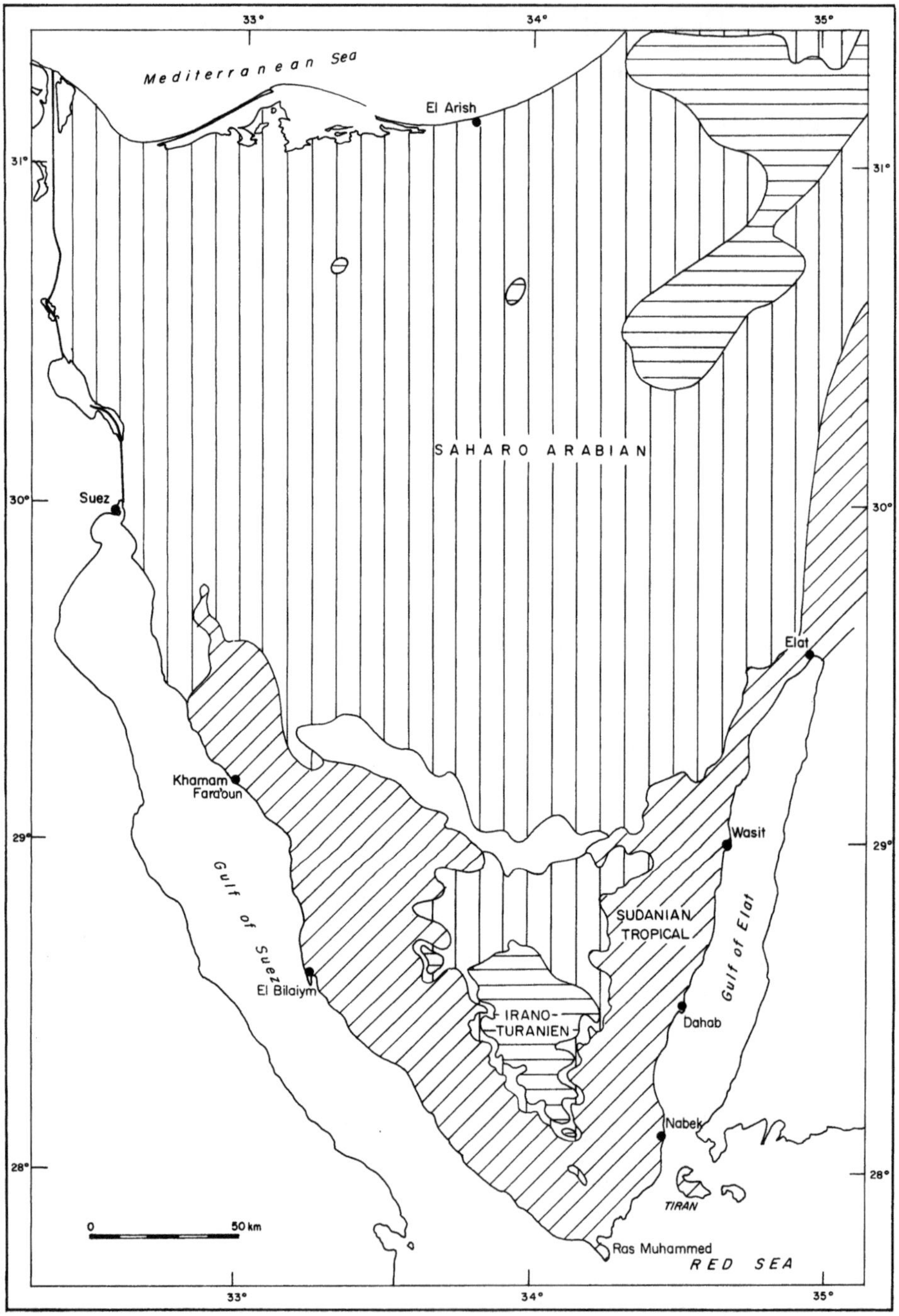

Fig. 7.1. Map of Sinai indicating the location of the investigated sabkhas and the phytogeographical regions of the peninsula. (After Orshan-Danin, unpublished)

coastal desert could become economically useful (Boyko 1966, Mudie 1974, Epstein et al. 1979, Rains 1979, Somers 1979, Tewari 1967, Felger and Mota-Urbina 1982). Such a hydrohalophyte which could be used for practical purposes is, e.g., *Juncus arabicus*, growing in the Sinai salinas, since Bloch et al. (1954) have shown that *Juncus maritimus*, a species closely related to *J. arabicus*, is good raw material for the production of cellulose.

Since it was not possible to study all of the many salt plains and sabkhas of the Sinai, only a limited number of the most typical ones could be investigated. They are: Wasit-Nueiba, Dahab, Nabq, Island of Tiran, Ras Muhammad, El Belayim and Ras Sheratib, Chamam Faràoun (Fig. 7.1). In order not to overload the chapter, we present only data from Nabq, Ras Muhammad, El Belayim, Ras Sheratib and Chamam Faràoun.

7.1 Methods

In studying the zonation of the salinas and sabkhas there was no need to use complicated methods to determine the dominant plants of each zone since each zone was visibly and exclusively dominated by one and sometimes two species.

Table 7.1. Name, chorotype and photosynthetic type of the plants, investigated in the salinas of the Sinai

Name	Chorotype	Photosynthetic type
Acacia raddiana Savi	S (Afr.)	C_3
Acacia tortilis (Forssk) Hayne	S (Afr.)	C_3
Aeluropus lagopoides (L.) Trin.	EM–ESA	C_4
Arthrocnemum fruticosum (L.) Moq. = *Salicornia fruticosa* (L.) L.	M	C_3
A. macrostachyum (Moric.) Moric et Delponte = *A. glaucum* Ung.-Sternb.	M–Sa	C_3
Avicennia marina (Forssk.) Vierh.	T	C_3
Calotropis procera (Ait.) Ait. f.	S (SA)	C_3
Halocnemum strobilaceum (Pall.) M.B.	M–IT–SA–ES	C_3
Hammada salicornica (Moq.) Iljin = *Haloxylon salicornicum* (Moq.) Bge et Boiss.	S (ESA)	C_4
Hyoscyamus muticus L.	SA–S	?
Juncus arabicus (Aschers. et Buchenau) Adamson	SA–IT	C_3
Limoniastrum monopetalum (L.) Boiss.	WM–SA	C_3
Limonium axillare (Forssk.) O. Ktze = *Statice axillare* Forssk.	SA–NS	C_3
Nitraria retusa (Forssk.) Aschers.	SA (S)	C_3
Salvadora persica Baker	S	C_3
Tamarix nilotica (Ehrenb.) Bge	S-Afr.	C_3
Zygophyllum album L. f.	SA (M)	C_3
Z. coccineum L.	SA	C_3
Z. simplex	S	C_4

S = Sudanese, ES = East Sudanese, NS = North Sudanese, Afr = African, M = Mediterranean, EM = East Mediterranean, WM = West Mediterranean, T = Tropical, SA = Saharo-Arabian, ESA = East Saharo-Arabian, IT = Irano-Turanian, WIT = West Irano-Turanian

The fact fits rather well into a general rule applicable to deserts: the more extreme the habitat, the more pronounced the dominance of a few and sometimes only one species.

The extent of each zone was determined by a survey on the ground – aided by detailed maps and aerial photographs.

When possible, the roots of the plants were excavated down to the extreme tip of the main, or one of the main, roots and, in most cases, the excavations were continued until the water table was reached. Only the larger roots could be excavated and even those could often not be fully laid open. During the excavations the roots were drawn to scale on millimeter paper. But although in the drawings the root system is represented more or less exactly, the above-ground parts are drawn schematically. During the excavations soil profiles were made and soil and water samples were taken at various depths, their conductivity and pH determined and analyzed for their chemical composition and particle size (for more details see Gavish 1974b). The values of these parameters are represented in drawings, which also indicate schematically the distribution of the main roots and rooting depth of the plants investigated.

Table 7.1 gives the names and the phytogeographic distribution (chorotype) of the dominant plants of all the salt plains we investigated according to Feinbrunn-Dothan (1978), Greenberg-Fertig (1966), Taeckholm (1974) and Zohary (1966, 1972, 1976). The synonyms of the plant names are given also to avoid confusion (e.g., Limonium–Statice, Hammada–Haloxylon, etc.). Table 7.1 also indicates to which photosynthetic type the plants belong, i.e., if they are C_3, C_4 or CAM plants. These data are taken from Winter (1978), Ziegler et al. (1981), Winter (1981), Winter et al. (1976, 1983), and Shomer-Ilan et al. (1981).

7.2 Climate

Climatic data of the Sinai Peninsula are very scanty and exist only for a few localities (Ganor et al. 1973). From Table 7.2, it can be seen that the mean annual maximum and minimum temperatures of the coast of the Gulf of Elat are higher than those of the Gulf of Suez. The northern coast of Sinai is considerably cooler than both other coasts. The mean relative humidity of the Gulf of Elat is considerably lower than that of the Gulf of Suez, whereas the Mediterranean coast has the highest mean relative humidity.

The temperature of the seawater of the Gulf of Elat is in winter by 3°–4 °C and in summer by about 0.5°–1 °C higher than that of the Gulf of Suez (Table 7.3; Markovitz 1973, Oren 1962a). The winter temperatures of the seawater along the Mediterranean coast of the peninsula are lower than those of the seawater of its eastern and western coast, but in summer they are higher. The seawater of the Gulf of Suez is more saline than that of the Gulf of Elat, but both have a higher salinity than the water of the Mediterranean. The temperature of the lagoon of Bar Dawil on the northern coast of Sinai is much lower in winter and considerably higher in summer than that of the water of the north, east and west coast, and it has a very high degree of salinity.

Table 7.2. Meteorological data of various localities of the Sinai

Locality	Mean annual temperature (°C)	Mean annual maximum temperature (°C)	Mean annual minimum temperature (°C)	Mean annual relative humidity (‰)	Mean annual rainfall (mm)
Elat	25.0	31.2	18.8	39	25
Sharm el Sheikh	25.7	30.4	21.0	38	–
Et Tur	22.8	28.5	17.2	57	9
Abu Rodes	22.4	26.5	18.4	52	–
Abu Zneima	23.6	28.9	18.2	–	20
Suez	22.5	29.0	16.0	53	24
El Arish	19.6	25.2	14.0	68	97
Gaza	20.1	25.3	14.9	70	371

Table 7.3. Winter and summer temperatures and salinity of the water of the Gulf of Elat, Gulf of Suez, the Mediterranean and the lagoon of Bar Dawil

Locality	Mean temperature in winter (°C)	Mean temperature in summer (°C)	Mean salinity (‰)
Gulf of Elat	21	26	40.6
Gulf of Suez	18	26	42.0
Mediterranean coast of Sinai	17	28	39.1
Lagoon of Bar Dawil	14	32	70.0

7.3 Plant Communities of the Coastal Salinas and Sabkhas

7.3.1 Nabq

Of all the salinas and sabkhas on the Gulf of Elat, Nabq, 175 km south of Elat, is the most interesting one because of the presence of mangroves (Walsh 1974).

In Nabq, the alluvial fans of a number of smaller and larger wadis which are offsprings of two large wadis, Wadi Samra and Wadi Kida, unite. The alluvial material, brought down from the mountains, does not reach the seashore. The soil along the coast consists therefore only partly of wadi alluvium and contains a high proportion of carbonates – the source of which is the sea. It is pumped inland mainly by evaporative pumping (see Chaps. 4 and 9).

The external belt of the saline lagoon of Nabq is formed by the mangrove *Avicennia marina*. *Avicennia* does not form a continuous belt but appears in the Nabq area in four main stands, the largest of which is situated near the ship wreck (Fig. 7.2). *Avicennia* forms there a dense "forest" of trees touching each other (Fig. 7.3 a). Further north only single *Avicennia* plants are found, the most northern one of which is shown in Fig. 7.3 b). Most of the Avicennias grow in seawater, rooted in mud, but some grow on sand off the shore; they are at low tide about 15 m, and at high tide about 1 m distant from the shore. One such plant was ex-

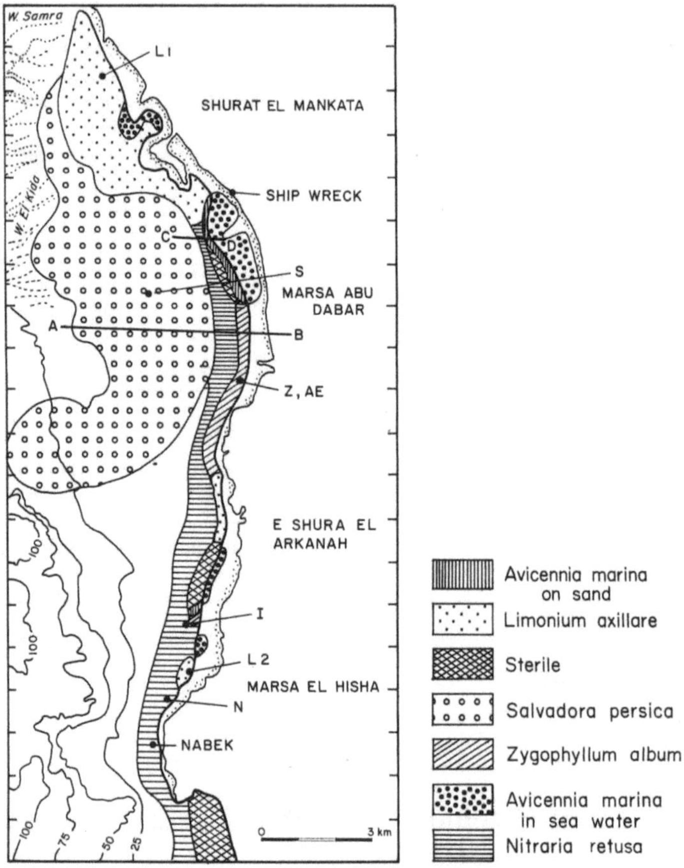

Fig. 7.2. A situation plan of the Nabq area indicating the various vegetational belts and the location of the plants and the root system which was excavated

cavated. The upper 50 cm of the soil consisted of sand with little clay. The main roots were found at a depth of 50–70 cm in a layer of clay and silt. Groundwater reached by some roots was found at 90 cm. It had a temperature of 26 °C, a pH of 6.7 and a conductivity of 22 mmho/cm. Plants growing on the shore had no pneumatophores.

The vegetation of the salinas along the shore of Nabq is dominated by *Zygophyllum album, Nitraria retusa* and *Limonium axillare*. The first two species occur also in Wasit-Nueiba and Dahab, whereas *Limonium* is not found north of Nabq. Further inland lies an area populated exclusively by *Salvadora persica.*

Close to the Gavish Sabkha depression (see Chap. 9) all macrophytes are practically excluded by the extreme cementing and enrichment with bitter salts. Although many seeds are transferred into the gypsum and salt crusts surrounding the Sabkha and on the central elevated part, even after heavy rains with freshwater supply no seedlings were observed. The Gavish Sabkha thus is a microorganisms- and mainly prokaryote-dominated environment.

Fig. 7.3. A A dense stand of *Avicennia marina* at Nabq. **B** The solitary northernmost *Avicennia* plant in the Gulf of Elat

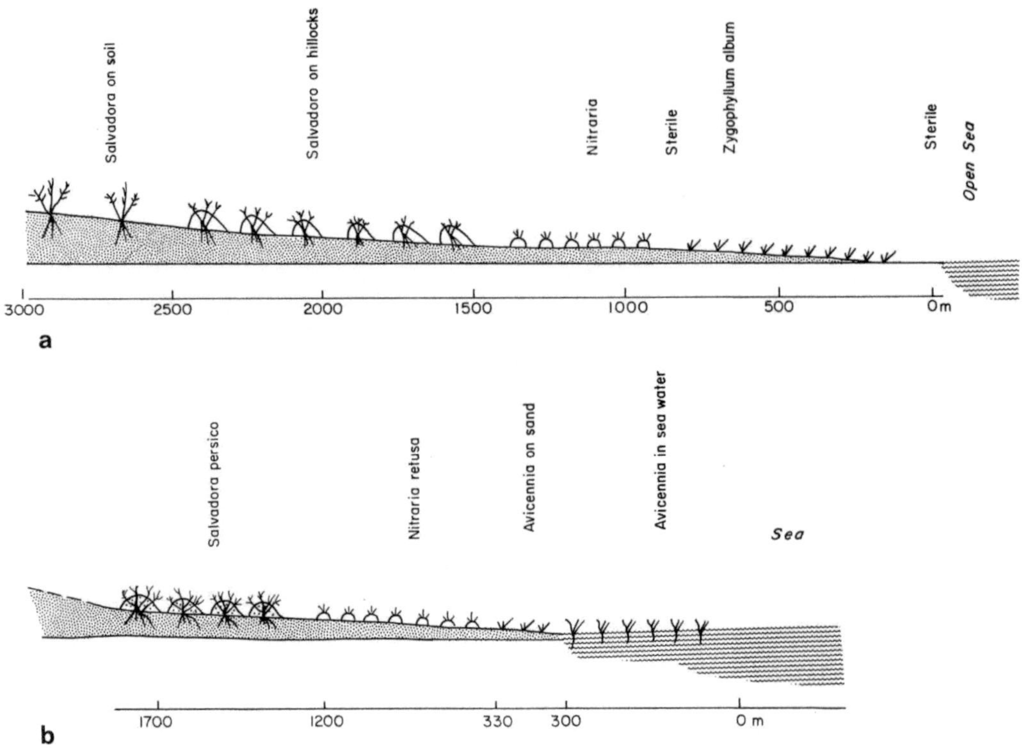

Fig. 7.4. a A schematic vegetational transect along *A–B* of Fig. 7.2. **b** A schematic vegetational transect along *C–D* of Fig. 7.2

Fig. 7.5. A photograph of the belt of *Zygophyllum album* of the transect *A–B* of Fig. 7.2

Figure 7.4a (transect A–B of Fig. 7.2) shows the belt about 30–50 m inland of the sterile strip along the shore. It is a pure stand of *Zygophyllum album*. The plants from Nabq form rebkas and sometimes ridges carrying a number of shrubs. Figure 7.5 depicts part of this belt and shows that the phytogenic hillocks consist of fine sand blown unto the plants and retained by it, whereas the larger sterile patches between the plants consist of coarse sand grains which are too large to be transported by the winds.

Inland from the *Zygophyllum* belt and separated from it by a small sterile zone is a *Nitraria retusa* community in which the plants grow also on phytogenic hillocks of various configurations.

The vegetation of the belt between the *Nitraria* community and the foot of the mountains is a pure stand of *Salvadora persica* consisting of two sections. In the one nearest the sea the plants from large phytogenic hillocks (Fig. 7.6), mostly in the form of rebkas, 5–10 m long and 3–6 m high. In the section nearest the mountain ranges the plants creep on the soil surface. The *Salvadora* community extends inland on the fan of Wadi Kida up to a line on the sloping fan above which the groundwater of the wadi cannot be tapped anymore by the roots of *Salvadora*. The main stem of the plants grows parallel to the soil surface or at a slight angle to it. The wind covers it with fine sand and a rebka is formed with its long axis in a north-south direction. Inside the hillock the main stem forms many side branches and the tip of the branches and their leaves stick out of the hillocks. When the hillocks have reached a certain critical height, the wind starts to blow the sand from the northern side of the hillock unto its southern flank, thus denuding the northern part of the plant. These parts die off but remain lying on the surface of the flat ground giving the landscape a quite bizarre appearance (Fig. 7.7). Since the branches newly covered by sand form roots, the dying of the main stem does not kill the plant. In this way the hillock, together with its plant, moves

Fig. 7.6. Some phytogenic hillocks at Nabq formed by *Salvadora persica*

Fig. 7.7. Some partly denuded *Salvadora* plants at Nabq, moving south

slowly south driven by the northerly winds, and thus *Salvadora* is at Nabq a "wandering tree".

In the center of the *Salvadora* community the distance between the individual plants is comparatively small (5–30 m). Sometimes the hillocks even touch each other (see background of Fig. 7.6). At the edges the stand thins out and the distances from plant to plant increase to 100 m and more (Fig. 7.8 a). The ground between the plants consists of coarse sand covered by smallish stones of various sizes (Fig. 7.7), and looks like a reg in status nascendi. As in real regs, vesicular soils were found below the stones (Evenari et al. 1974).

Figure 7.5 b represents transect C–D of Fig. 7.2. The belt of *Avicennia* (in seawater and on sand) is followed inland by a zone dominated by *Nitraria*, which borders further inland on the *Salvadora* community.

In the whole Nabq area, *Nitraria* is by far the dominant plant of the salt plains, forming a continuous belt 5 km long (Fig. 7.2). *Zygophyllum album* plays only a minor role. The largest area at Shurat El Mankata is dominated by *Limonium axillare* (Figs. 7.2 and 7.8 b).

Figure 7.9 is a more detailed sketch of the vegetation along the C–D transect. *Nitraria* dominates, but near the shore it is accompanied by *Aeluropus lagopoides* growing in patches.

Whereas on the flat areas *Nitraria* is mostly the only plant growing there, the runnels which represent only a very small percentage of this area carry a different vegetation. The runnels are the terminal branches of wadis and are therefore run-on habitats. Their vegetation is a mosaic of *Zygophyllum album, Z. coccineum* with a few patches of *Aeluropus* and here and there a solitary plant of *Nitraria*. At Nabq these runnels are the habitat of *Zygophyllum coccineum*. The runnels

Fig. 7.8. A Solitary plants of *Salvadora* separated by large distances at the edge of the main stand. **B** *Limonium axillare* in Nabq

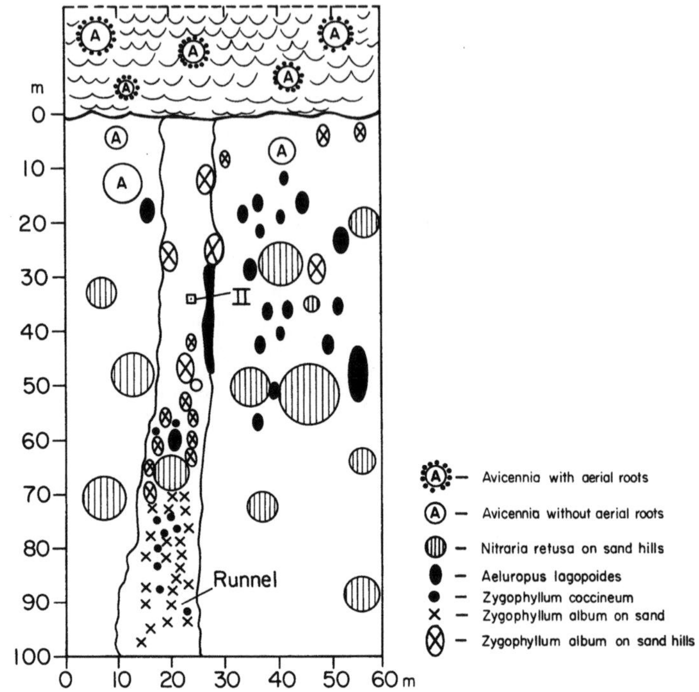

Fig. 7.9. A situation plan of a small strip near the seashore along the transect *C–D* of Fig. 7.2

which were visited a few weeks after a flood had occurred contained many seedlings in contrast to the flat areas outside them where only a few seedlings were found. In Fig. 7.10, which represents a small part of a runnel (2 m^2, 35 m from the coast, area II of Fig. 7.9), the distribution of the seedlings is shown in detail. The mortality of the seedlings of *Z. album* was quite high with only 17 seedlings surviving out of 56 which had germinated, whereas out of 9 seedlings of *Z. coccineum* 8 had survived.

At Nabq two plants of *Salvadora,* six of *Limonium,* five of *Nitraria,* two of *Zygophyllum album* and two of *Aeluropus* were excavated. One of the five excavated *Nitraria* plants (N in Fig. 7.2, 200 m from the coast) is depicted in Fig. 7.11 a. Only one of its branches carried green, live leaves. The other two were blown out by the wind and were dying. The roots grew in the direction of the long axis of the plant's rebka. In the upper 30 cm of the soil, consisting of fine sand with a high conductivity and chloride content, the thick roots did not branch. The main root system had developed at a depth of 50–90 cm in a layer of coarse sand, large stones, much gypsum and carbonate, where conductivity and salinity were very much lower. There were very few roots below 100 cm. The roots penetrated a compact layer of gypsum and carbonate at a depth of 75–90 cm. They did not reach the groundwater table at 150 cm. This groundwater, as that of all other plants excavated at Nabq, had a high conductivity (45–60 mmhos/cm).

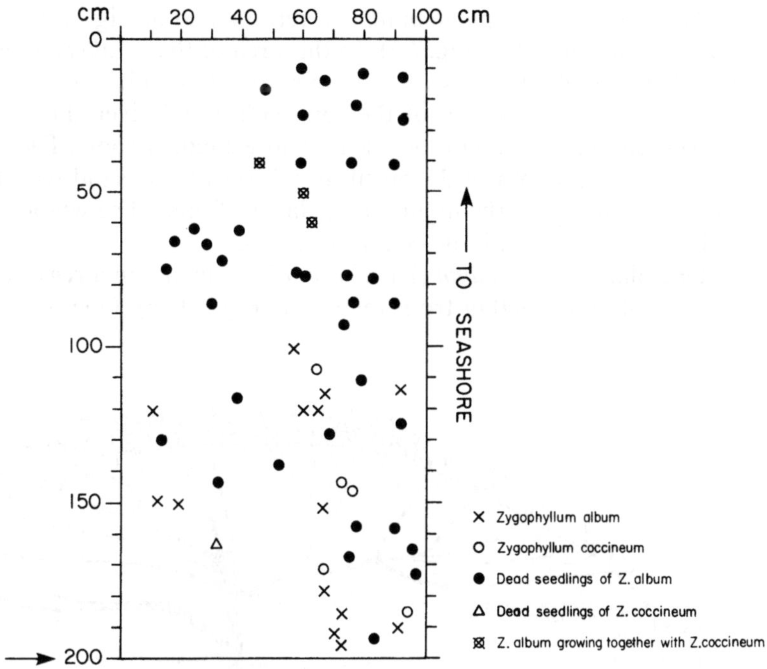

Fig. 7.10. A sketch of area II of Fig. 7.9

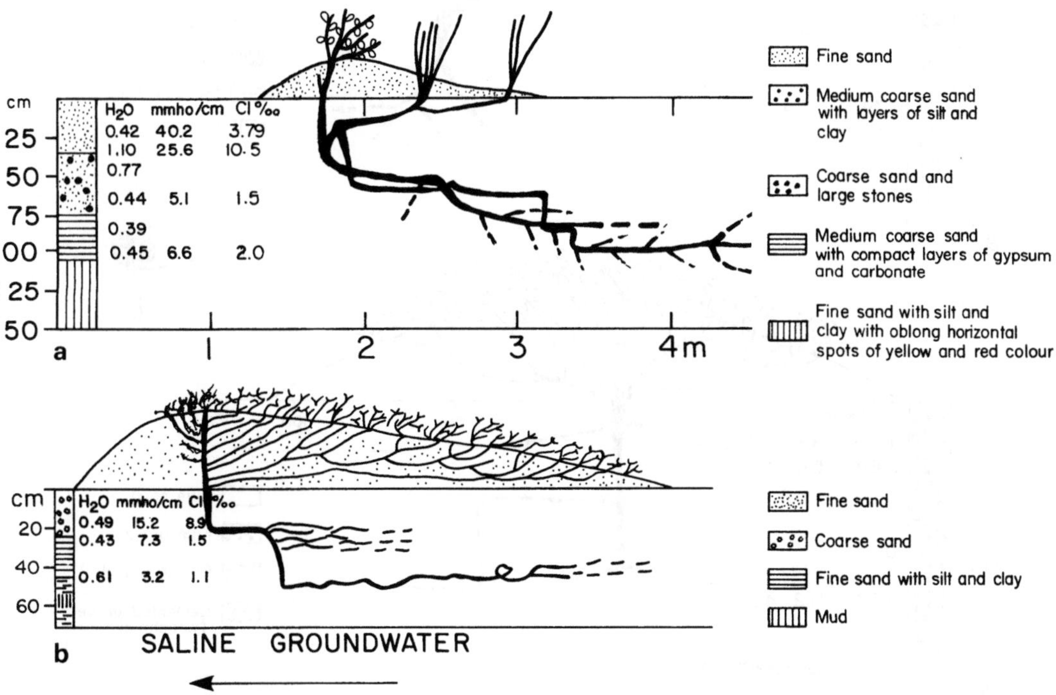

Fig. 7.11. a The root system of *Nitraria retusa* excavated at Nabq in locality N of Fig. 7.2. **b** The root system of *Zygophyllum album* excavated at Nabq in locality Z of Fig. 7.2. Figures for H_2O are % of weight

The plant of *Zygophyllum album,* represented in Fig. 7.11 b, grew 30 m from the shore (locality Z of Fig. 7.2). In this area of the *Z. album* community the distance between plants was with 4–10 m relatively small. It increased further to 10–30 m. At the upper 15 cm of the soil, with its relatively high conductivity and chloride content, the main root carried no secondary roots. The main root system was found at a depth of 25–50 cm in a layer of fine sand with low salinity. No root penetrated into the highly saline layer of mud. The whole root system grew in the direction of the long axis of the rebka.

One plant of *Limonium* (L1 of Fig. 7.2, growing on a reg) was excavated at a distance of about 500 m from the coast (Fig. 7.12 a). One part of the roots had

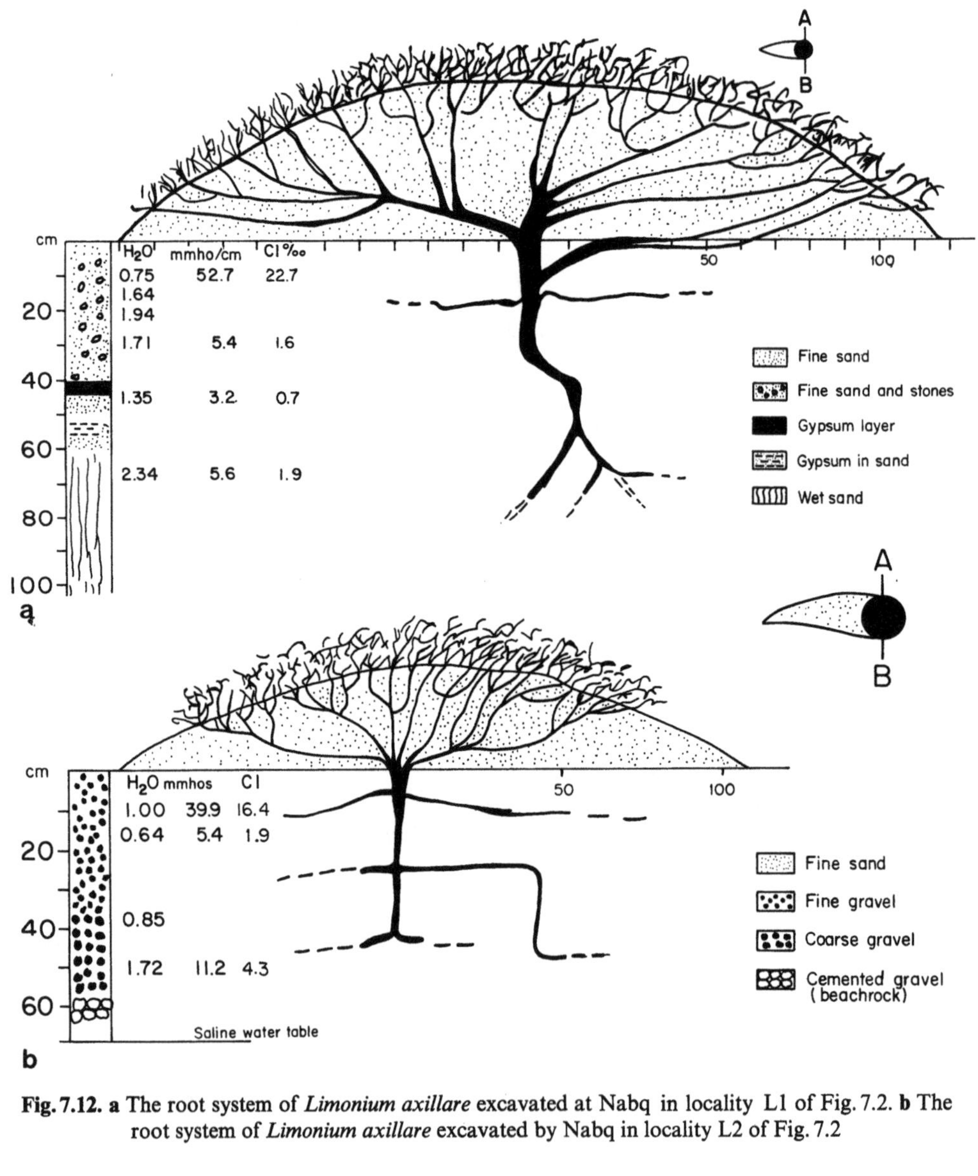

Fig. 7.12. a The root system of *Limonium axillare* excavated at Nabq in locality L1 of Fig. 7.2. **b** The root system of *Limonium axillare* excavated by Nabq in locality L2 of Fig. 7.2

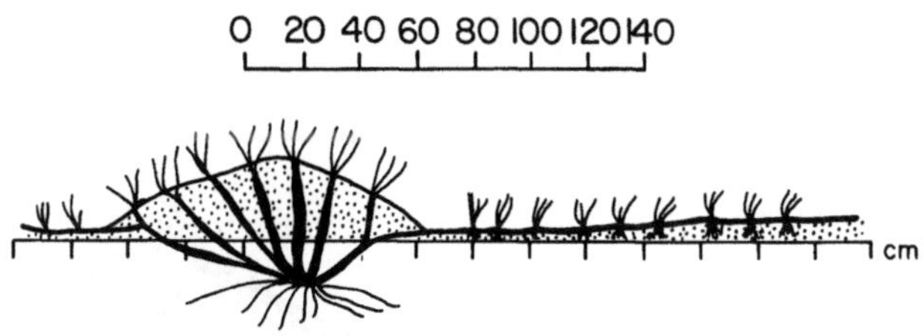

Fig. 7.13. *Aeluropus lagopoides* at Nabq in locality AE of Fig. 7.2

developed in a layer of fine sand containing stones of various sizes about 15 cm below the highly saline soil surface. The main root grew straight downward, penetrated a gypsum layer, branching at a depth of 60 cm in a layer of sand containing much gypsum. We lost it in a stratum of wet sand at a depth of about 80 cm. Although groundwater was not found even at a depth of 150 cm, we suppose that the roots did reach it, since all the soil layers below the upper 10 cm had a low conductivity and chloride content.

The root system of another *Limonium* plant (L2 of Fig. 7.2, growing 12 m from the seashore) is shown in Fig. 7.12 b. The uppermost horizontal root system was found at a depth of 12–15 cm in a layer of fine gravel which, in contrast to the upper 10 cm of soil, had a low conductivity and chloride content. A second horizontal root system grew at a depth of 30–35 cm. The main root branched horizontally in a layer of coarse gravel. Below this layer was cemented beach rock. The highly saline groundwater table was found just above the beach rock. No root touched it.

The lateral horizontal roots of *Limonium* were not excavated in their complete length in the plants represented in Fig. 7.12 a, b. In other excavated plants, they extended to a distance of 2.5–4 m from the main stem.

Aeluropus lagopoides grew in small rebkas and nebkas (Fig. 7.13). Each plant had a number of long rhizomes carrying many secondary rooted plantlets. All the roots of *Aeluropus* were found in the upper highly saline 20 cm of the sandy soil.

Salvadora persica has a very large extended and deep-reaching root system. The plant shown in Fig. 7.14 had a thick dead trunk denuded by the wind. The living trunk with many side branches grew in a very large rebka. The main trunk carried many thick roots. The one which was followed turned horizontally above a stone layer at a depth of 60–70 cm, but some of its branches grew downward, penetrating the stone layer. It was lost at a depth of about 100 cm. It is most probably that these roots reach the nonsaline groundwater which is found in sandy loam at a depth of 5–6 m. The whole soil profile is nonsaline and has a low conductivity and chloride content.

7.3.2 Ras Muhammad

Ras Muhammad is the southern tip of the Sinai Peninsula. The channels inside the coral reefs ("reef cracks", Por and Tsurnamal 1973; Fig. 7.15 a) are bordered by dense stands or *Avicennia marina* (Fig. 7.16). In Sinai, mangroves are found

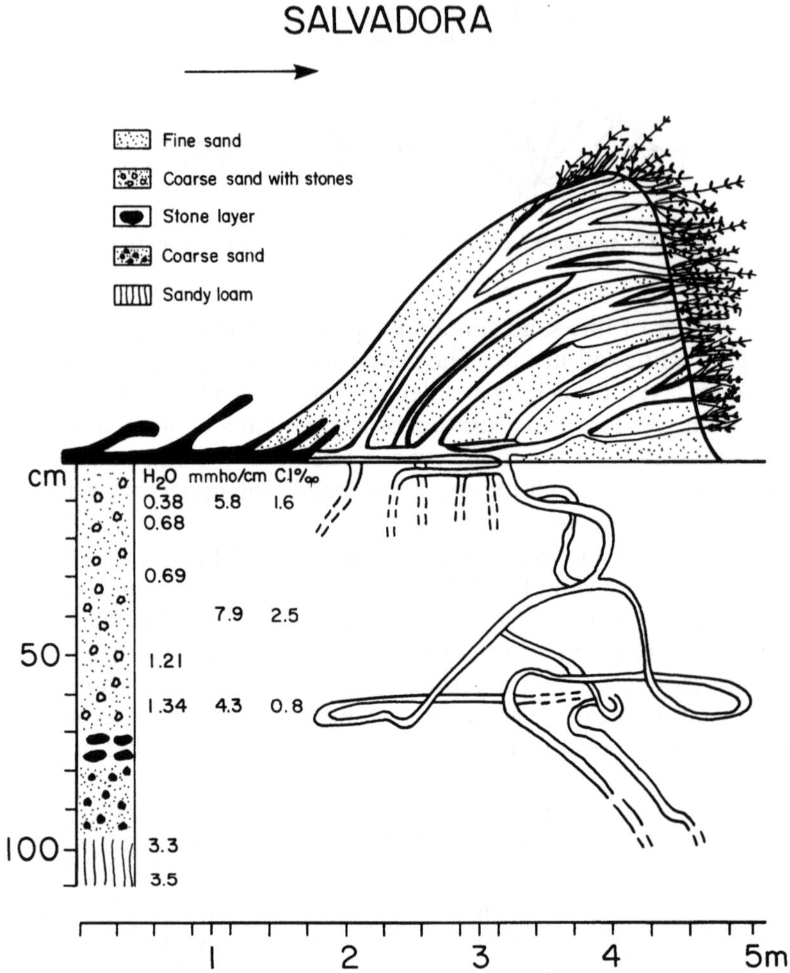

Fig. 7.14. The root system of *Salvadora persica* excavated at Nabq in locality S of Fig. 7.2

only on the Gulf of Elat from Nabq to Ras Muhammad. They are absent on the western shore of the peninsula.

North of Ras Muhammad are two saline ponds (Fig. 7.16 a), the northermost of which was investigated. This pond is about 250 m long and 150–250 m wide (Fig. 7.15 b). Since the level of the water of the pond changes with the tide, it is obvious that an underground communication exists between the pond and the Red Sea. The temperature of the pond's water was 24.5 °C, its pH 8.3, and its conductivity 50 mmhos/cm. At various places seawater seeps continuously into the lake. At these localities the conductivity of the water was only 27.5 mmhos/cm and the pH 7.7.

The vegetation around the pond consists of very dense stands of *Halocnemum strobilaceum* arranged in strips of various lengths (Fig. 7.15 b and 7.17). The largest strip is about 200 m long and 2–15 m wide. At the points of seawater seepage

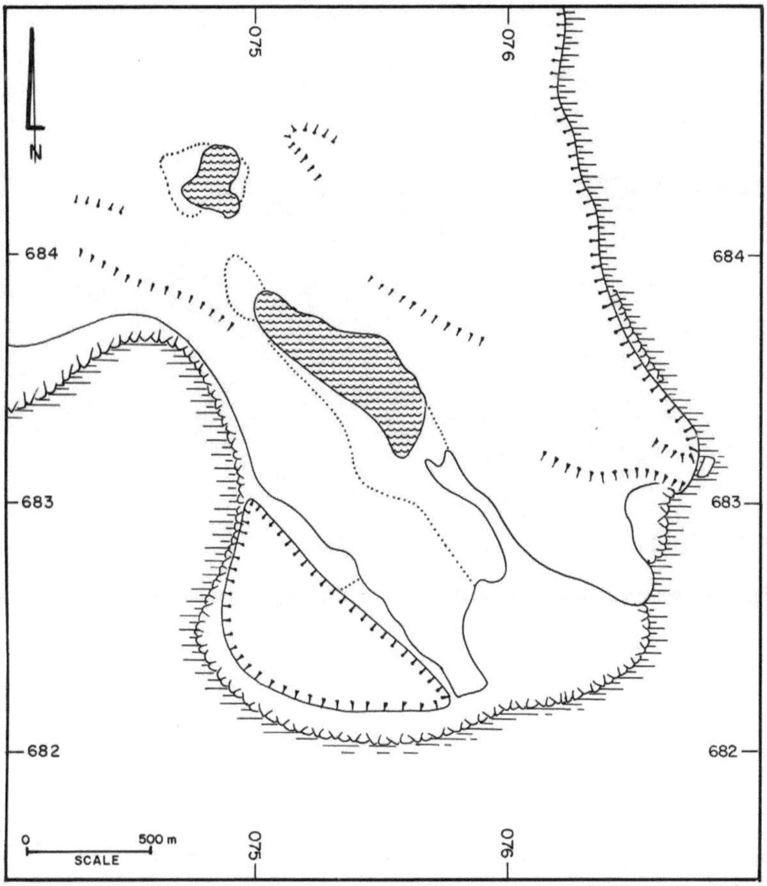

a

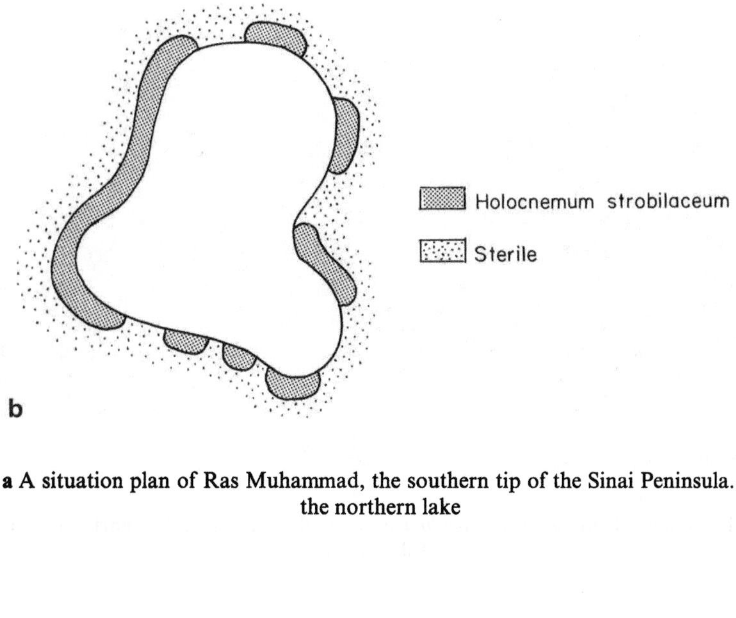

b

Fig. 7.15. a A situation plan of Ras Muhammad, the southern tip of the Sinai Peninsula. **b** A plan of the northern lake

Fig. 7.16. Dense stands of *Avicennia marina* bordering the channels inside the coral reefs

Fig. 7.17. The stands of *Halocnemum strobilaceum* on the shores of the northern pond at Ras Muhammad

into the pond the water was covered by a green algae (*Enteromorpha* spec.). The area outside the *Halocnemum* belt is completely bare. Large parts of the surface are covered by a thick crust of salt.

7.3.3 El Belayim

The lagoon and the lake of El Belayim lie on the coast of the Gulf of Suez 135 km northwest of Ras Muhammad and about 8 km northwest of Abu Durba (Fig. 7.18). The geochemistry and mineralogy of the area has already been investigated by Gavish (1974 b). The lagoon is situated north of the lake and is considerably larger than it. Gebel Abu Durba stretches up to the southern tip of the lagoon. A number of wadis coming down from Gebel Abu Durba form alluvial fans toward the coast. The plain in between Gebel Abu Durba and Gebel Qabeliat is completely void of vegetation (Fig. 7.19 a). It is the side branch of El Qa, a much larger sterile plain. The lake is surrounded by a belt of *Halocnemum strobiliaceum* forming a nearly pure stand containing only a few plants of *Zygophyllum album*. On the western part of the lake the stand is dense and the sterile patches between the plants occupy only a small percentage of the area

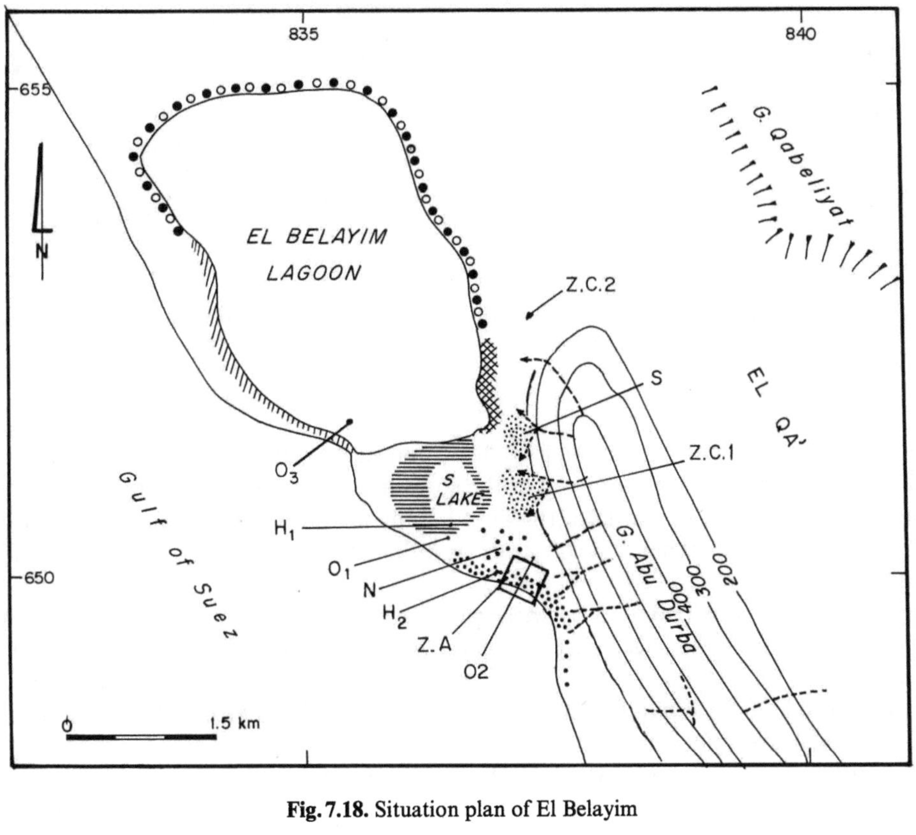

Fig. 7.18. Situation plan of El Belayim

Fig. 7.19. A Vegetationless plain between Gebel Abu Durba and Gebel Qabiliat (in background). **B** *Halocnemum strobilaceum* on the western shore of the southern lake of El Belayim

(Fig. 7.19 b). This stand is about 1 km long and at its widest part 0.5 km wide. In some places the groundwater reaches the soil surface, as can be seen in Fig. 7.19 b. On the eastern part of the lake the stand thins out and is much narrower with many more sterile patches. The phytogenic hillocks formed by the plants are small and low.

The lagoon is surrounded by three plant communities. Its northern and part of its eastern and western side are characterized by a narrow strip of *Halocnemum strobilaceum* and *Zygophyllum album*. At the southeastern edge of the lagoon is a pure stand of *Zygophyllum album*, and at its southeastern side a mosaic of *Halocnemum strobilaceum* and large sterile patches.

South of the lake along the coast *Halocnemum strobilaceum, Zygophyllum album* and *Nitraria retusa* form a community. *Zygophyllum coccineum* and *Ham-*

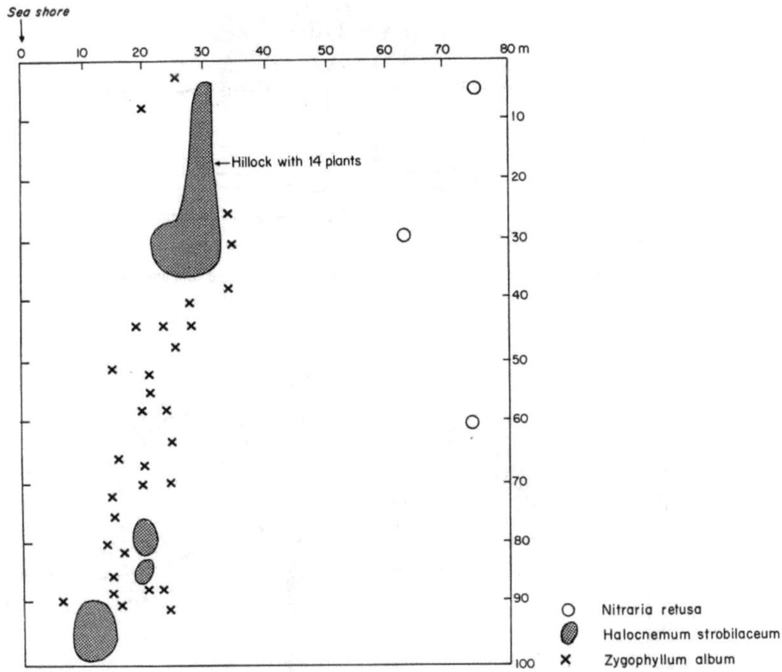

Fig. 7.20. A detailed situation plan of the plants growing in the *square* indicated in Fig. 7.18

mada salicornica are the dominant plants in the runnels on the fans formed by the wadis carrying floodwater from Gebel Abu Durba down to the shore. In some of these runnels *Zygophyllum coccineum* is the only plant to be found.

Figure 7.20 shows the distribution of *Zygophyllum, Halocnemum* and *Nitraria* in a strip 10–80 m broad and 5–10 m off the sea shore (square indicated in Fig. 7.18). The plants nearest the shore are *Halocnemum* and *Z. album. Nitraria* grows farther inland. Figure 7.20 shows also that sometimes many plants of *Halocnemum* thrive on the same phytogenic hillock and that the farther inland, the larger the distance between plants.

In El Belayim two plants of *Halocnemum* (H1 and H2 of Fig. 7.18), one plant of *Zygophyllum album* (ZA of Fig. 7.18), two plants of *Z. coccineum* (ZC1 and ZC2 of Fig. 7.18) and one plant of *Hammada salicornica* (S of Fig. 7.18) were excavated. Three boreholes were also made in sterile areas (01, 02, 03 of Fig. 7.18). Only Fig. 7.21 a, b represent the root system of two excavated plants. In Figs. 7.22 through 7.29 the roots of the various plants are only schematically drawn, mainly in order to show to what depth the roots penetrated.

The root system of *Halocnemum* (Fig. 7.21 a) is typical not only for this plant but also for *Zygophyllum album* and *Nitraria*. The roots do not reach the groundwater. Just above the surface of the groundwater the roots turn horizontal. Since *Halocnemum* H1 grows directly at the shore of the lake, groundwater is found at a depth of only about 40 cm. In the other cases (H2, ZA, N of Fig. 7.18) the groundwater table is reached at a greater depth. It is also typical for all the plants

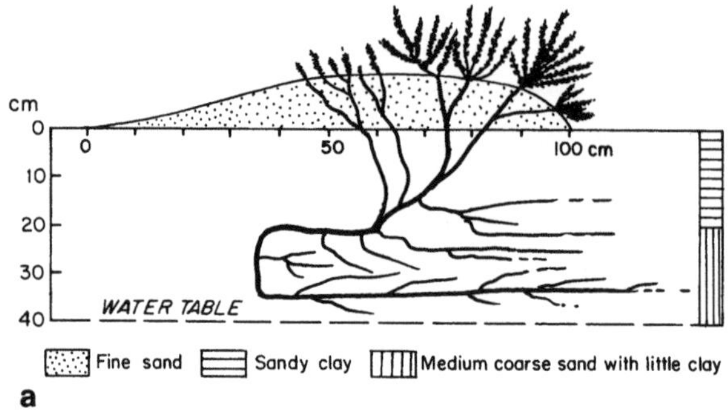

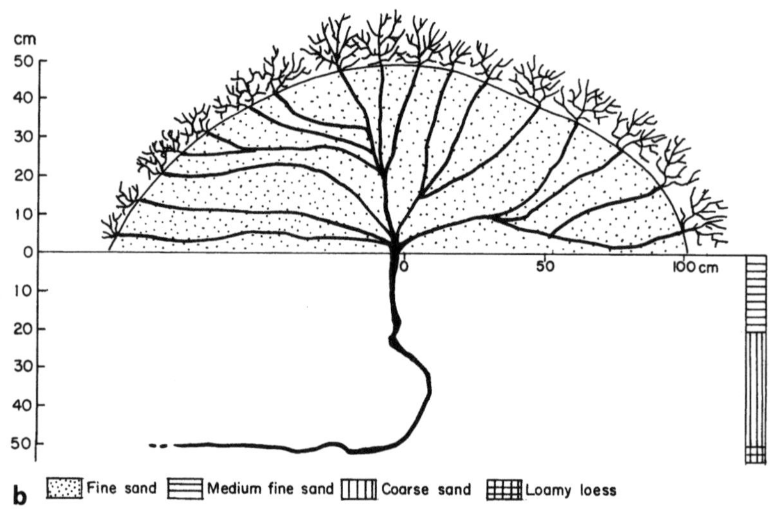

Fig. 7.21. a The root system of *Halocnemum strobilaceum* (H1 of Fig. 7.18). **b** The root system of *Hammada salicornica* (S of Fig. 7.18)

growing in the salinas and sabkhas of El Belayim that the soil around the roots of the plants is full of fungal hyphae. They were observed with the help of a binocular and there seemed to be a connection between the roots and the hyphae, perhaps indicating the presence of a mycorrhiza.

Figure 7.21 b shows the root system of *Hammada salicornica* growing in a runnel of a wadi fan (S of Fig. 7.18). The upper 10–20 cm of the soil were extremely saline, but the 20–40 cm layer of medium and coarse sand contained nonsaline run-on water (see also Fig. 7.27). At a depth of about 45 cm the root turned abruptly at an angle of 45° and grew from there on horizontally. The soil below this level was completely dry (see also Fig. 7.26). The main root did not branch at all and carried only many fine rootlets.

Figures 7.22 through 7.28 represent the mineral composition, pH and conductivity, water content and particle size at various depths of the habitat of a number of plants.

The upper 15 cm of the soil of the habitat of *Halocnemum strobilaceum* (Fig. 7.22 H1 of Figs. 7.18 and 7.21 a) growing in the sabkha around the lake have

TOTAL SALINITY ‰	DEPTH (cm)	‰ Cl	SO$_4$	Ca	Mg	K	Na	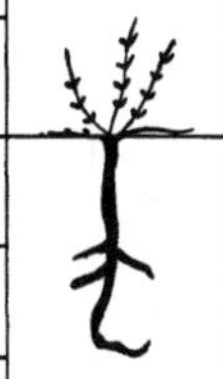	PH	CONDUCTIVITY mmhos / cm	WATER CONTENT OF SOIL (% of weight)	PARTICLE SIZE (mm) > 1.00	1.0-.125	.125 >
23.65	0–15	13.35	1.20	0.45	0.66	0.66	7.33		8.6	37.2	84.3	5	55	40
8.40	15–30	6.53	0.60	0.49	0.25	0.33	3.61		8.7	20.05	15.0	25	67	8
54.94	30–	31.14	3.32	0.66	2.00	0.65	17.17	water table	7.4	67.91	100			

Fig. 7.22. Mineral composition, conductivity, water content, pH, and particle size at various depths of the soil at the habitat of *Halocnemum strobilaceum* (H1 of Fig. 7.18)

the highest conductivity, the highest Na and Cl content and the highest salinity of all the habitats investigated in El Belayim, and also the highest water content. In this layer the main root does not branch and carried no rootlets. At the main rooting depth salinity, conductivity and NaCl contents are considerably lower than in the upper layers. With the exception of the habitat of *Zygophyllum coccineum* (Fig. 7.24) the groundwater has the highest salinity, conductivity, and Na and Cl content of all the habitats investigated in El Belayim and is considerable more hypersaline than the water of the Gulf of Suez.

The soil analysis given for the habitat of *Halocnemum* is typical for all the plants growing in the sabkha bordering the lake and the lagoon.

The plant of *Nitraria retusa* (Fig. 7.23, N of Fig. 7.18) grew in the center of an area of medium fine sand covered by stones forming a kind of desert pavement. This seemed to be a reg in formation. Around the plant a small low hillock of fine sand was collected, which is a phytogenic hillock either in formation or in destruc-

TOTAL SALINITY ‰	DEPTH (cm)	‰ Cl	SO$_4$	Ca	Mg	K	Na		PH	CONDUCTIVITY mmhos / cm	WATER CONTENT OF SOIL (% of weight)	PARTICLE SIZE (mm) >1.00	1.0-.125	.125 >
15.93	0–10	7.86	2.46	1.65	0.86	0.20	2.9		7.15	25.87	1.08	5	87	8
13.70	10–28	7.40	1.37	1.34	0.31	0.08	3.2		7.45	21.32	5.47	11	88	1
4.24	28–57	2.08	0.55	0.27	0.09	0.16	1.09		7.3	7.76	5.75	10	85	5
9.68	57–80	5.21	0.90	0.58	0.28	0.08	2.63		6.75	16.33	11.5	12	85	3
46.72	80–	27.14	2.99	0.64	1.92	0.29	13.74	water table	7.4	62.25	100			

Fig. 7.23. *Nitraria retusa* (N of Fig. 7.18, text as for Fig. 7.22)

TOTAL SALINITY ‰	DEPTH (cm)	Cl ‰	SO_4 ‰	Ca ‰	Mg ‰	K ‰	Na ‰		PH	CONDUCTIVITY mmhos / cm	WATER CONTENT OF SOIL (% of weight)
19.78	0–10	9.55	2.99	1.36	0.30	0.08	5.50		7.5	30.72	5.21
40.20	10–20	20.96	4.13	1.47	0.42	0.25	12.97		7.45	56.59	21.0
7.65	20–25	1.91	2.56	0.76	0.05	1.00	1.37		7.60	10.02	7.80
4.62	25–60	2.26	0.72	0.24	0.05	0.09	1.26		7.75	8.89	9.68
11.76	60–85	6.86	2.09	0.78	0.27	0.16	1.60		8.45	22.96	14.5
80.36	85–	46.26	5.21	0.82	2.61	0.84	24.62	water table	7.0	84.08	100

Fig. 7.24. *Zygophyllum coccineum* (ZC2 of Fig. 7.18, text as for Fig. 7.22)

tion. On this otherwise sterile plain only a few plants of *Nitraria* are found 100–150 m distant from each other. The upper 25–30 cm of the soil are quite saline and dry and contain gypsum and carbonate concretions. The main root does not develop lateral roots in this layer. At a depth of 30–60 cm the sandy soil contains available water and has a low conductivity, salinity and ion content in contrast to the soil above and below. The main roots have developed in this layer. At a depth of about 80 cm the main root turns horizontal and does not touch the highly saline groundwater (compare with Gavish 1974b, Fig. 8, where the same habitat is represented in pit 4).

Zygophyllum coccineum (Fig. 7.24, ZC2 of Fig. 7.18; corresponds with pit 7 of Gavish 1974b) grows in a depression (40–150 m) of the sterile plain not far from the road leading from Abu Durba to Ras Sheratib. In the soil profile layers of sand and clay alternate. The most hypersaline groundwater of all the habitats investigated at El Belayim is found at a depth of 90 cm in a layer of pure sand. The upper 20 cm of soil are highly saline. The main root system has developed at a depth of 20–60 cm where the sand is wet and salinity comparatively low. At a depth of about 60 cm, i.e., 30 cm above groundwater level, where salinity has increased to 12‰ and more, the main roots turn horizontal. Very many fungal hyphi are found down to a depth of 65 cm in sand and clay. From the soil analysis it is obvious that some nonsaline runoff water from Gebel Abu Durba reaches the habitats of *Zygophyllum coccineum* and *Nitraria,* enabling the plants to grow in an otherwise saline environment.

The habitats of *Halocnemum* H1, *Zygophyllum coccineum* ZC2 and *Nitraria* N contain not only a high concentration of Cl and Na but also of SO_4, Mg, K and Ca. This is evident when comparing the relevant data of Figs. 7.22–7.24 with those of Figs. 7.25–7.27.

Zygophyllum album (Fig. 7.24, ZA of Fig. 7.18) and *Halocnemum strobilaceum* (Fig. 7.26, H2 of Fig. 7.18) grow in the same plant community along the seashore,

TOTAL SALINITY ‰	DEPTH (cm)	Cl ‰	SO_4 ‰	Ca ‰	Mg ‰	K ‰	Na ‰		PH	CONDUCTIVITY mmhos / cm	WATER CONTENT OF SOIL (% of weight)	PARTICLE SIZE (mm) >1.00	1.0-.125	.125 >
1.57	0–21	0.64	0.36	0.02	0.05	0.05	0.45		8.4	2.91	1.9	1	95	4
2.39	21–40	1.22	0.26	0.03	0.06	0.08	0.74		8.7	4.53	4.0	5	85	10
1.17	40–60	0.51	0.21	0.04	0.03	0.04	0.34		8.75	2.1	2.85	10	85	7
1.23	60–70	0.60	0.14	0.03	0.03	0.04	0.39		8.7	2.42	2.47	15	82	3
2.89	70–105	1.52	0.29	0.06	0.08	0.07	0.92		8.75	5.49	4.25	20	75	5
43.77	>105	24.02	3.41	0.53	1.63	0.52	13.66	water table	7.65	54.12	100			

Fig. 7.25. *Zygophyllum album* (ZA of Fig. 7.18, text as for Fig. 7.22)

TOTAL SALINITY ‰	DEPTH (cm)	Cl ‰	SO_4 ‰	Ca ‰	Mg ‰	K ‰	Na ‰		PH	CONDUCTIVITY mmhos / cm	WATER CONTENT OF SOIL (% of weight)	PARTICLE SIZE (mm) >1.00	1.0-.125	.125 >
1.77	0–20	0.72	0.33	0.11	0.05	0.06	0.50		7.60	3.23	0.5	3	94	3
1.70	20–50	0.23	0.88	0.05	0.02	0.04	0.48		7.85	2.91	4.25	0	98	2
1.03	50–70	0.44	0.17	0.05	0.02	0.03	0.32		7.90	2.10	2.14	22	78	0
	70–95	0.35	0.23	—	0.02	0.02	0.36		7.85	2.10	2.3	25	73	2
1.78	95–120	0.80	0.29	0.06	0.03	0.05	0.55		7.75	3.55	5.33	5	75	20
43.04	>120	24.02	3.08	0.50	1.60	0.49	13.35	water table	7.3	53.36	100			

Fig. 7.26. *Halocnemum strobilaceum* (H2 of Fig. 7.18, text as for Fig. 7.22)

i.e., a belt about 100 m wide between the shore and the metamorphic rocks of Gebel Abu Durba. The soils of both habitats are down to a depth of 90–100 cm characterized by low conductivity, low salinity and low ion concentration. In both cases the main root system is found at a depth of 30–90 cm, where the total salinity of the sandy soil amounts to no more than 1.17‰–2.89‰. The roots do not reach the highly saline groundwater. The source of the water of low salinity, conductivity and ion concentration of this habitat is the runoff from Gebel Abu Durba. The very high salinity of the groundwater derives from the seawater,

Table 7.4. Conductivity and chemical composition of groundwater of various sampling localities at El Belayim (see Fig. 7.18) and Khamam Fara'oun

Locality	Conductivity (mmhos/cm)	Ph	‰ NA	K	Ca	Mg	Cl	SO_4	Total salinity
H 1	67.91	7.4	17.17	0.65	0.66	2.00	31.14	3.32	54.94
H 2	53.36	7.30	13.35	0.49	0.50	1.60	24.02	3.08	43.04
N	62.25	7.40	13.74	0.29	0.64	1.92	27.14	2.99	46.72
ZA	54.12	7.65	13.66	0.52	0.53	1.63	24.02	3.41	43.77
ZC 2	84.08	7.00	24.62	0.84	0.82	2.61	46.26	5.21	80.36
O_1	53.36	–	12.77	0.52	0.52	1.60	24.17	3.31	42.89
O_2	58.21	7.35	12.94	0.49	0.57	1.68	25.34	3.88	44.90
O_3	59.83	–	14.61	0.59	0.55	1.73	26.54	4.03	48.05
Sea water	51.74	–	13.05	0.49	0.48	1.55	22.99	3.23	41.79
Khamam Fara'oun (Juncus)	37.35	6.75	6.90	0.50	0.82	0.60	13.78	1.92	24.52

which near the shore has a much higher salinity than the water of the Gulf of Suez at a distance from the coast (Table 7.4).

Zygophyllum coccineum (Fig. 7.27, ZC1 of Fig. 7.18) grows in a runnel on the slopes of an alluvial fan at the foot of Gebel Abu Durba. The soil surface down to 10 cm and the soil below 60 cm have à very low salinity, conductivity and ion concentration and a low water content. The main root system of the plant has developed in the upper 30–40 cm of the soil which have a higher salinity and a higher water content than the soil above and below. Along the roots many thickenings were observed which look like the bacterial nodules of the Leguminosae. The upper 10–40 cm of the soil consist of fine wet sand. In our excavations we did not reach the groundwater, which seems to be at a depth of at least 2 m. The vertical distribution of salinity and water content indicates that the habitat of *Zygophyllum coccineum* receives runoff from the mountain.

Total salinity ‰	Depth (cm)	‰ Cl	SO_4	Ca	Mg	K	Na		PH	Conductivity mmhos/cm	Water content of soil (% of weight)	Particle size (mm) > 1.00	1.0–.125	.125 >
2.69	0–12	1.47	0.22	0.31	0.07	0.05	0.57		7.4	5.5	0.86	0	88	12
6.84	12–30	3.67	0.64	0.41	0.15	0.08	1.89		7.35	12.61	3.2	20	70	10
1.81	30–50	0.86	0.29	0.13	0.03	0.04	0.46	No ground water	7.7	2.99	0	30	65	5

Fig. 7.27. *Zygophyllum coccineum* (ZC1 of Fig. 7.18, text as for Fig. 7.22)

TOTAL SALINITY ‰	DEPTH (cm)	‰ Cl	SO$_4$	Ca	Mg	K	Na		PH	CONDUCTIVITY mmhos/cm	WATER CONTENT OF SOIL (% of weight)	PARTICLE SIZE (mm) 1.0	1.0-.125	.125
6.25	0–5	2.91	1.15	0.59	0.15	0.08	1.37	STERILE AREA WITHOUT PLANTS	7.23	10.61	0.23	10	89	10
11.75	5–18	6.10	1.83	1.23	0.36	0.08	2.75		7.45	20.21	0.61	8	89	3
7.39	18–32	1.96	2.94	0.82	0.13	0.05	1.44		7.65	9.70	0.84	8	89	3
4.65	32–50	2.30	0.55	0.17	0.08	0.06	1.49		7.75	8.40	1.57	15	82	3
6.17	50–95	3.08	0.85	0.20	0.13	0.08	1.83		7.65	10.83	6.9 6.6	12	68	20
44.90	95–	25.34	3.88	0.57	1.68	0.49	12.94	water table	7.35	58.21	100			

Fig. 7.28. Soil analysis of the sterile spot O_2 of Fig. 7.18

In three places (01, 02, 03 of Fig. 7.18) holes were dug in localities void of plants, and soil analyses were made. The data for 02 are given in Fig. 7.28. Salinity, conductivity and ion concentration of the soil of 02 do not differ much from the respective figures for the habitat of *Nitraria* (Fig. 7.23). These factors are therefore not responsible for the sterility of 02, the reason for which must be the very low water content of the upper 50 cm which apparently prevents germination and seedling establishment.

7.3.4 Ras Sheratib

Along the seashore of Ras Sheratib, about 10 km north of El Belayim, is a strip 60–150 m wide of a pure stand of *Halocnemum strobilaceum* growing on phytogenic hillocks. The same vegetation continues south and especially north for many kilometers. The special feature of this vegetation is that here the largest and highest nebkas and rebkas were observed in the whole of Sinai, 5–6 m high and 20–30 m long (Fig. 7.29). On each hillock grow a number of *Halocnemum* plants. The largest hillocks are found nearest to the coast at a distance of 1–8 m from the shoreline. Inland the hillocks become smaller until at the inland border of the

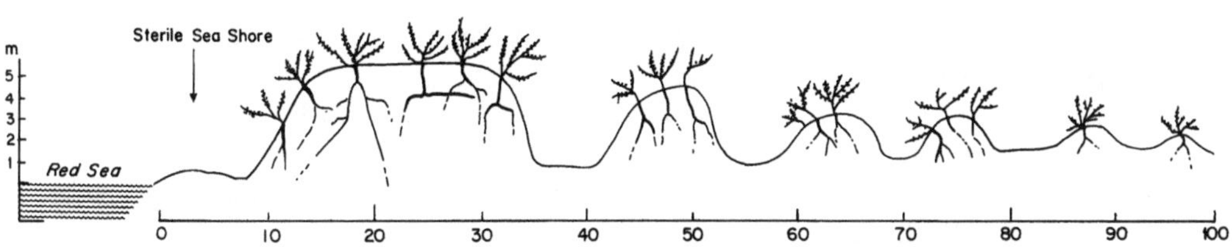

Fig. 7.29. A schematic vegetational transect of a pure stand of *Halocnemum strobilaceum* at Ras Sheratib

belt, 80–110 m from the shore, they are no more than 15–20 cm high, carrying only one plant (Fig. 7.29). As always, the hillocks consist of fine sand. At a depth of 5–10 cm the soil is a mixture of sand and clay. At lower levels layers of sand and clay interchange. At 30 cm and 50 cm layers of heavy wet clay were found. At a depth of 70 cm above the groundwater level there is wet sand. The roots penetrate to this level and turn there horizontal. Salinity, conductivity and ion concentration are more or less the same as described for *Halocnemum* H2 in El Belayim (Fig. 7.26). The plain inland from the *Halocnemum* belt up to the foot of the mountains 7–8 km distant from the coast is mainly sterile with the exception of the runnels which harbor *Hammada salicornica* and *Zygophyllum coccineum*.

7.3.5 Khaman Fara'oun

The hot spring of Khaman Fara'oun lies about 75 km north of El Belayim. The area and its vegetation have been already described by Migahid et al. (1959).

The spring at the foot of Gebel Khaman Fara'oun has a temperature of 70 °C. Its water has a strong sulfureous smell. The spring water flows through a small channel into a basin from which it percolates through many small runnels into the sea (Figs. 7.30 and 7.31). The water of the basin and the runnels contain green algae and cyanobacteria which grow at a temperature of up to 55 °C. Between the basin and the foot of Gebel Khaman Fara'oun is a sandy slope which extends southward (Fig. 7.30). The main vegetation there consists of scattered plants of *Zygophyllum album, Arthrocnemum frutisosum* and, in runnels, *Zygophyllum coccineum* (Fig. 7.32).

Between the basin and the sea is a sandy ridge. The dominant plant there is *Juncus lagopoides* and some trees and shrubs of *Tamarix nilotica*. Further south is a swampy area covered by a dense mat of *Aeluropus*.

One plant of *Tamarix* (T of Fig. 7.30), one of *Zygophyllum album* (Z of Fig. 7.30) and one stand of *Juncus* (J of Fig. 7.30; see also Fig. 7.32) were excavated.

We do not present here the root system of *Zygophyllum* because with the exception of the fact that its roots grew in soil with a temperature of 35°–38 °C, it was not different from those excavated in other localities.

The habitats of the three species excavated were very wet (Figs. 7.33 and 7.34) since they receive water from the brackish spring. The soil layers in which these plants developed their main roots have low conductivity, total salinity and ion concentration.

The soil surface of the habitat of *Tamarix* was covered by a solid crust of salt. The upper 5 cm were extremely saline (Fig. 7.33). The same is true for the habitat of *Aeluropus*. In contrast to this, the habitat of *Juncus* was without a salt crust, apparently because the whole soil profile including the soil surface was completely water-saturated. The upper 60 cm of soil were only moderately saline (Fig. 7.34) in comparison with those of *Tamarix* and *Aeluropus*.

The most remarkable fact about the roots of both plants is that they grew in a medium (black soil overlain by sand) of a temperature of 33°–40 °C. But their

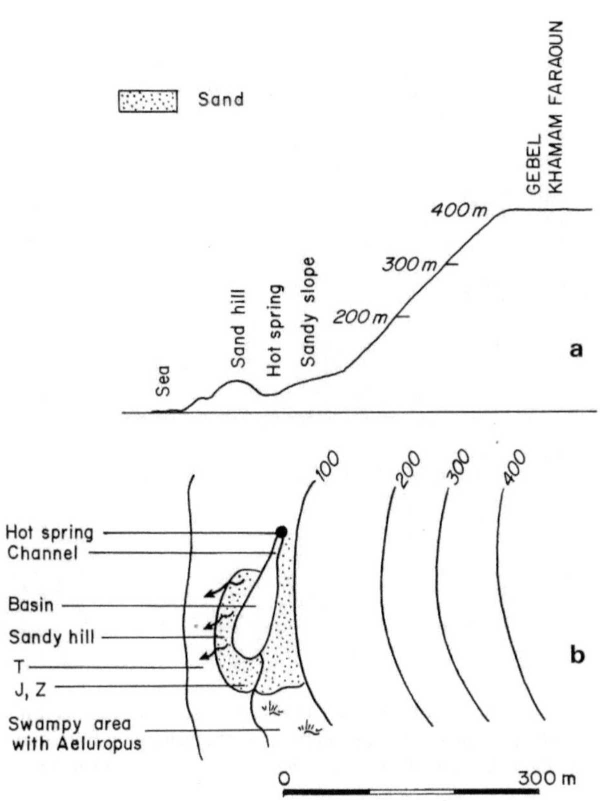

Fig. 7.30. **a** A section profile of Khamam Fara'oun from Gebel Fara'oun to the shore. **b** A situation plan

Fig. 7.31. The beach of Khamam Fara'oun showing the sterile strip along the coast and the runnels through which the water of the hot spring flows into the sea. Plants of *Juncus arabicus* are seen at the left fore- and background. Salt crusts are visible as *white spots*

Fig. 7.32. The salt marsh at Khamam Fara'oun. *At the left* a stand of *Juncus arabicus, at the right* a mat of *Aeluropus lagopodes*. On the sandy slope between Gebel Khamam Fara'oun and the salt marsh are scattered plants of *Zygophyllum album, Z. coccineum* (in runnels) and *Arthrocnemum fruticosum*

TOTAL SALINITY ‰	TEMPERATURE (°C)	DEPTH (cm)	‰ Cl	‰ SO_4	‰ Ca	‰ Mg	‰ K	‰ Na		PH	CONDUCTIVITY mmhos / cm	WATER CONTENT OF SOIL (% of weight)
57.59	29	0–5	33.65	2.41	4.25	1.24	0.78	15.26		7.45	80.85	25.5
4.13	33	5–20	2.24	0.30	0.31	0.06	0.07	1.15		7.35	7.76	17.3
3.78	43	20–30	1.71	0.56	0.50	0.05	0.07	0.89		7.45	7.11	20.2
24.92	48	30–40	13.78	2.35	0.82	0.60	0.50	6.87	water table	6.8	37.35	

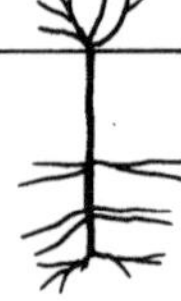

Fig. 7.33. *Tamarix nilotica* (T of Fig. 7.30b)

TOTAL SALINITY ‰	TEMPERATURE (°C)	DEPTH (cm)	‰ Cl	SO4	Ca	Mg	K	Na	pH	CONDUCTIVITY mmhos/cm	WATER CONTENT OF SOIL (% of weight)	PARTICLE SIZE (mm) > 4.0	0–125	125 >
7.88	32	0	2.98	2.00	0.85	0.12	0.16	1.77	7.8	12.12		0	82	18
1.78	38	8	0.98	0.17	0.16	0.02	0.04	0.41	7.9	2.18	SATURATED	0	84	16
2.80		23	1.51	0.16	0.24	0.04	0.05	0.80	7.8	5.20		0	94	6
3.31	42	35	1.97	0.14	0.28	0.05	0.07	0.80	7.7	6.63		0	96	4
24.52	44	62	13.78	1.92	0.82	0.60	0.50	6.90	6.75	37.35				

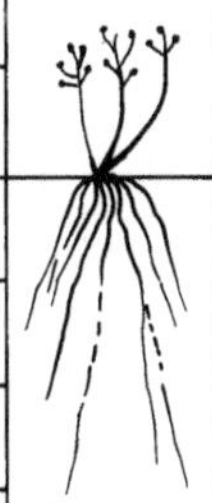

Fig. 7.34. *Juncus arabicus* (J of Fig. 7.30b)

roots did not penetrate the deeper soil layers with temperatures between 40°–48 °C, apparently because of a combination of high temperature and high salinity.

The chemical analyses of the groundwater found in the various plant habitats and in the holes made in sterile areas at El Belayim and Khaman Fara'oun are given in Table 7.4.

The groundwater of Khaman Fara'oun has the lowest conductivity, salinity and ion concentration of all samples taken. Its salinity is far below that of the water of the Gulf of Suez sampled at shallow depth (41.8‰) because of the brackishness of the water of the hot spring. The water of the lagoon (O_3) is considerably more saline than the seawater. The groundwater of the sabkha near the lagoon (ZC2) is with a salinity of 80.36‰ about two times as saline as the seawater and the water of the lagoon. The groundwater of the sabkha around the south lake is also hypersaline (H1). The groundwater of the sterile plain (N) is also considerably more saline than the seawater, but less saline than ZC2, H, and O3. The groundwater of all the other localities along the seashore is constantly, by only 1.5‰–3‰, slightly more saline than that of the seawater. The proportion of all the ions in all the samples is nearly constant. The conductivity shows a close relationship with the total salinity (see also Gavish 1974b).

7.4 Discussion

7.4.1 Distribution of Plants

The plant species found in the salinas of Israel, the salinas on the coast of the Gulf of Elat, the coast of the Gulf of Suez and the northern coast of Sinai, are not identical. What are the possible reasons for the unequal distribution of some of the species? To find an answer to this question, the geographical distribution of the main species found in the salinas of Sinai will be described, taking into ac-

count their distribution in Israel, on the Mediterranean coast of Egypt and on the eastern coast of the bay of Suez, and the Red Sea littoral of Egypt and the Sudan. The species which will be considered are:
Avicennia marina, Salvadora persica, Limonium axillare, Zygophyllum album, Zygophyllum coccineum, Halocnemum strobilaceum, Suaeda vermiculata, Arthrocnemum macrostachyum, Arthrocnemum fruticosum, Nitraria retusa.
In dealing with these species we will also take into account their occurrence in the salinas of Wasit-Nueiba, Dahab, the Island of Tiran and the lagoon of Bar Dawil, salinas which we investigated but did not describe in this chapter.

Mangals of *Avicennia marina* do not occur north of Nabq on the Gulf of Elat, and its last continuous stand on the eastern coast of the Gulf of Suez is at Ras Muhammad. Mangals in general are restricted to the tropical regions of the Old and New World, and air temperatures limit their distribution. According to Chapman (1977) they thrive "when the temperature in the coldest month does not fall below 20 °C and where the range is about 10 °C. Cockayne (1958) states that mangals develop only under conditions of "... a muddy substratum... absence of frost, warmish water during summer, tide erosion of insufficient power to uproot the young seedlings." These conditions are fulfilled concerning the Sinai mangals of *Avicennia marina,* which is chorotypically a tropical element (Table 7.1), but the question remains why, on the eastern coast of Sinai, *Avicennia* is found up to a latitude of 28° 04′ at Nabq and on the western coast its northernmost point of distribution is at 27° 44′ (Ras Muhammad) where it was recorded already by Ferrar (1914) and Zahran (1965, 1967). That the conditions on the western Sinai coast are unfavorable for *Avicennia* is also indicated by the fact that on the Egyptian side of the Gulf of Suez *Avicennia* grows in stands only up to Hurghada (28° 04′ N) and south of it, where it grows together with another mangrove, i.e., *Rhizophora mucronata,* which is completely absent from Sinai (Kassas 1957, Kassas and Zahran 1962, Zahran 1977). South of Murghada in the Sudanese coast of the Red Sea, *Bruguiera gymnorhiza,* another mangrove, joins *Avicennia* and *Rhizophora* in forming mangals. Its northern limit is near Suakin in the Sudan (19° 08′; Zahran 1977). There seems to be, therefore, a gradient of tropicality: *Bruguiera-Rhizophora-Avicennia.* The reason for the expansion of *Avicennia* toward the north in the Gulf of Elat can be seen from Tables 7.2 and 7.3. The mean annual, mean annual maximum, and mean annual minimum air temperatures, as well as the mean annual winter temperatures of the sea-water, are higher on the eastern side of the coast of Sinai than on its western and northern one. As mentioned before, some *Avicennia* plants grow on the sandy shore where they lack rhizophores and not in the muddy underground of the seawater. The same has been observed on the Egyptian Red Sea coast (Ayyad and Ghabbour 1984). It is probable, as stated by these authors, that the silting of the shoreline is responsible for this phenomenon.

Salvadora persica is a Sudanese element which needs tropical conditions and a good supply of nonsaline water. These conditions prevail on the coast of the Gulf of Elat up to Dahab (28° 30′ N), which apparently is its northernmost locality as a stand. Solitary plants grow in the Arava and Jordan Valley in oases with tropical climatic conditions northward up to about Ein Gedi (31° 28′ N) (Zohary and Orshansky 1947). Some plants of *Salvadora* grow also in the Islands of Tiran and Sinafir (Danin 1983), but there are no stands of *Salvadora* in the western Sinai coast.

On the Egyptian coast of the Gulf of Suez the northernmost "patches of almost pure growth (of *Salvadora*) within the channels of the main wadis" (Kassas and Zahran 1962) are found up to a latitude of about 27°–28° N. North of this latitude only isolated plants are found (own observation), whereas its center of distribution is in the south (Sudan).

As already mentioned by Kassas and Zahran (1962), *Salvadora persica* has two growth forms. It grows either as an upright tree or "a cushion-like patch of short branches" (Kassas and Zahran 1962). On the coast of the Gulf of Elat, the tree is found only in its prostrate form. On the Island of Tiran, *Salvadora* is an upright tree, as well as in the Jordan and Arava valleys of Israel. According to Kassas and Zahran (1962) the prostrate form "may be the result of repeated cutting; *Salvadora* is the toothbrush tree of the Orient" (for its use as a toothbrush, see Evenari and Gutterman 1973). According to Gupta and Sakena (1968) the two growth forms of *Salvadora* may be two different ecotypes.

Limonium axillare, another Sudanese element (Table 7.1), is found only south of Nabq and on the Island of Tiran and is missing on the west coast of the Sinai. According to Kassas and Zahran (1962), Zahran (1977) and Ayyad and Ghabbour (1984) *L. axillare* is found only in isolated localities on the eastern coast of the Gulf of Suez and appears in stands only on the Red Sea coast south of Qusei (26° 04′ N).

Zygophyllum album and *Z. coccineum* are common on all three coasts of the Sinai and on the eastern coast of the Gulf of Suez. But there is a difference between the two species. *Zygophyllum album* is a dominant community-building species with a very wide range of ecological conditions (Kassas 1967). On the one hand, it is common down south to the Sudanese Red Sea coast and, on the other hand, it penetrates into the Mediterranean coastal lands, though there it is "rare". Its chorotype, according to Zohary (1972), is SA(M) (Table 7.1), but it is in reality SA(M-S). *Z. coccineum* is not a community-building species. It is purely Saharo-Arabian. In Israel it is "uncommon" (Zohary 1972) and is found only in the Arava and Jordan valleys up north to En Boqeq (31° 12′ N).

Halocnemum strobilaceum is common on the western and northern coast of the Sinai but was not found on its eastern coast and does not occur in Israel. On the eastern coast of the Gulf of Suez it occurs down south to Hurghada (28° 04′ N; Zahran 1967). Chorographically, the plant is a Mediterranean-Saharo-Arabian-East Sudanese element (Table 7.1) and is therefore found on the Mediterranean coast of Egypt. But it is not clear why it is missing on the Mediterranean coast of Israel and on the eastern coast of the Sinai, including the Island of Tiran, and in the Jordan and Arava valleys.

Sueda vermiculata, a Saharo-Arabian-Sudanese element, is found on all the coasts of the Sinai, including the Islands of Tiran and Sanafir (Danin 1983), but in Israel it occurs only in the Arava valley up north to Jotvata (29° 53′ N). It is also found on the eastern side of the Gulf of Suez and on the Red Sea coast of Egypt and Sudan.

Arthrocnemum macrostachyum, A. fruticosum and *Nitraria retusa* are, in contrast to the species mentioned so far, found on all the coasts of the Sinai and in

Israel. *Arthrocnemum macrostachyum* occurs in the Mediterranean and Saharo-Arabian phytogeographic regions of Israel and is found on the eastern side of the Gulf of Suez, especially in its southern part and on the Red Sea coast of Egypt and Sudan (Zahran 1967, Ayyad and Ghabbour 1984). In Egypt, *Nitraria retusa* is confined to its Saharo-Arabian coast (Marsa Alam, 25° 03′ N; Zahran 1967).

From all this it evolves that because of its special climatic conditions the eastern coast of Sinai harbors three tropical-Sudanese community-building species (*Avicennia marina, Limonium axillare* and *Salvadora persica)* which, as far as we know, reach there their northernmost point of distribution. Because of lack of more specific data this discussion does not deal with the effect of specific environmental factors like light and relative humidity of the air on halophytes and their distribution, although it is known that these factors affect halophytes and their salt tolerance (Strogonov 1964, Gale et al. 1970, Gale and Poljakoff-Mayber 1970, Nieman and Poulsen 1971, Gale 1975, Poljakoff-Mayber 1975, Poljakoff-Mayber and Gale 1975).

7.4.2 Root Systems of the Different Species and Their Relation to Soil Conditions

7.4.2.1 El Belayim

In all investigated habitats, the mechanical composition of the soil (particle size, Fig. 7.22–7.27) and, therefore, the matrix potentials are nearly identical. This parameter therefore cannot be the reason for either the different behavior of the root systems or the different distribution of the various species. What differs from habitat to habitat is the water content, the salinity and conductivity of the soil.

Based on the values represented in Figs. 7.22–7.27, we calculated the osmotic and water potentials of the main rooting zones of each habitat.

In the sterile plots (Fig. 7.28, 02) the total salinity and the soil's osmotic potential down to a depth of 95 cm is not higher than in the habitats colonized by plants as, e.g., *Nitraria retusa* (Fig. 7.23). But since the water content is very low, the total soil water potential is lower than −14 bar. This is apparently the reason for the plot's sterility. The same is true for 01 and 03, the other investigated sterile plots.

The main rooting zone of *Nitraria retusa* (Fig. 7.23), only a short distance from 02, has at a depth of 50–80 cm a higher conductivity, a higher water content than 02. Its water potential is with about −6 bar considerably higher than that of 02, enabling the plant to grow in that habitat.

In the main rooting zone of *Zygophyllum album* (60–100 cm, Fig. 7.25) conductivity and water content are low and the water potential is with about −11 bar higher than that of 02 but lower than that of the *Nitraria* habitat. The situation at the rooting zone of *Halocnemum strobilaceum* (H2, Fig. 7.26) growing nearby is similar to that of *Zygophyllum album*.

The other *Halocnemum* plant (H1, Fig. 7.22) investigated grows in the *Halocnemum* belt around the South Lake. Its rooting zone is confined to a depth of 15–30 cm, where conductivity and water content are high. The water potential is with

−7 to −8 bar quite high. The rooting zone contains the highest chloride content of all the habitats investigated. The active roots are constricted to a very limited depth and cannot grow either above or below this zone in spite of the very high water content because conductivity and chloride content are too high. A comparison of the two habitats of *Halocnemum* shows that the species can grow in a comparatively wide range of soil salinity, chloride and water content.

Zygophyllum coccineum is another case where two habitats of the same species can be compared. One plant (ZC1, Fig. 6.27) grows in a wadi fan where in the two different rooting layers conductivity, chloride and water content are low with a total water potential of about −12 bar. The situation of the other *Zygophyllum coccineum* plant (ZC2, Fig. 7.24), growing in a run-on depression, is quite different. In the main rooting zone of 20–60 cm conductivity and chloride content are low and water content high with a water potential of about −4 bar. In spite of a very high water content the roots cannot grow below 60 cm because of very high conductivity and a very low osmotic potential.

With all the differences between the various habitats in El Belayim there are two facts common to all of them. The first one is that no root of any species can grow in layers containing groundwater because they cannot tolerate a total salinity of 40‰ and a conductivity of more than 50 mmhos/cm. The second common fact concerns germination. During our various stays at El Belayim we did not find a single seedling of either one of the species investigated or of any other species. The reason is that the upper soil layer (0–10 to 0–20 cm) has a very low water content and often, in addition, a high conductivity resulting in a water potential which is so low as not to permit germination. Germination and seedling establishment are therefore only possible when in certain years unusually high rainfalls and/or much run-on increase the water potential above a certain minimum, below which either no germination occurs or seedlings cannot establish themselves.

Such a pattern of germination and seedling establishment is typical for many and possibly all desert perennials. It had been observed for *Artemisia herba alba* (Evenari and Gutterman 1976), *Hammada scoparia* (Evenari and Gutterman, unpublished) and *Zygophyllum dumosum* (Evenari et al. 1982). Only in certain specific years with certain rainfall patterns do seeds of these plants germinate and seedlings survive. Therefore, the population of adult plants of these species are composed of individuals of specific age-groups with age-group gaps inbetween.

Once germination and seedling establishment have taken place, the survival of the halophytes studied depends on an equilibrated interplay between osmotic potential and water content in the main rooting zone, resulting in a soil water potential not lower than about −14 bar.

7.4.2.2 Khaman Fara'oun

At this very special habitat the rooting depth of *Tamarix nilotica* and *Juncus arabicus* is limited neither by water content nor by conductivity nor by water potential of the soil but by the temperature which increase with depth. Apparently, the roots of both species cannot function in temperatures higher than about 40 °C to 45 °C.

The habitat of *Juncus arabicus* is characterized by full water saturation of the soil and by low conductivity. Only the water of the water table, which the roots do not reach because of its high temperature, is very saline. The roots spread directly from the soil surface to a depth of about 30 cm.

The main rooting depth of *Tamarix nilotica* is at 5–20 cm, where conductivity is low and water content high. In contrast to *Juncus,* the roots cannot spread in the upper soil layer because of its very high conductivity and salinity.

7.4.2.3 Ras Muhammad

In this locality only *Halocnemum strobilaceum* grows in a belt around the saline pond. The soil is fully water-saturated. The conductivity of the pond water ranges from 27.5 to 50 mmhos/cm. *Halocnemum* grows in areas where conductivity lies around 20 mmhos/cm and the calculated water potential of the soil amounts to about −11 bar, a value similar to that found for *Halocnemum* H2 at El Belayim. No plant whatsoever grows outside the *Halocnemum* belt because salinity there is so high that a salt layer is formed on the soil surface impeding germination and growth.

7.4.2.4 Nabq

In Nabq we investigated several different habitats. *Salvadora persica* (Fig. 7.14) is a nonhalophyte. Its habitat has a very low conductivity and chloride content.

The habitats of two plants of *Limonium axillare* we investigated are quite different from each other. L1 (Fig. 7.12a) grows on a regoid plain. Below the highly saline soil surface, in which no active roots were found, the roots develop in the 20–90 cm layer which has a low chloride content and a low conductivity. In a depth of about 15–20 cm a well-developed horizontal root system was found. At the time of our investigation these roots had no rootlets and were apparently inactive. They seem to function only when a certain minimum amount of rain or run-on water penetrates to this layer. The deeper root system relies on the water contained in the upper layer of the wet sand which, according to our calculations, has a water potential of about −10 bar.

The habitat of the other *Limonium* plants (L2, Fig. 7.12b) is near the seashore. The plant has the same inactive horizontal root system as L1. Its main roots are found in the coarse gravel. These roots turn horizontal and do not penetrate the soil layer below a depth of about 50 cm where the water potential is lower than −14 bar.

The *Nitraria retusa* plant we excavated was part of the large extended *Nitraria* belt along the seashore. No active roots were found in the upper 30–40 cm of the soil where water content is low and conductivity and chloride content high. The main roots develop in layers of coarse and medium sand where the water potential of the soil amounts to about −12 bar. The roots do not enter the layers of fine sand and silt below a depth of 100 cm where the water potential is much lower than −12 bar.

In the belt of *Zygophyllum album* the upper soil layer and the mud layer at a depth of 60 cm are saline. The upper soil layer contains practically no water, whereas the mud is water-saturated and has a much higher conductivity than the upper 15 cm of the soil. The root system of *Zygophyllum album* lies between these two layers (Fig. 7.11 b).

Only in Nabq and on the Island of Tiran were seedlings of *Zygophyllum album* and *Z. coccineum* found. In Nabq the seedlings of both species were observed in runnels inside the *Nitraria* belt (Fig. 7.10). Most of the seedlings of *Z. coccineum* had succeeded in establishing themselves, whereas the majority of those of *Z. album* had died. Runnels are, according to our observations, the typical habitat of *Z. coccineum*. Outside runnels and run-on depressions seedlings of *Z. coccineum* were absent. Some seedlings of *Z. album* were found in the belt formed by this species mostly not far from adult plants.

Aeluropus lagopoides (Fig. 7.14) is rooted in the upper highly saline soil layer in between plants of *Nitraria* and *Zygophyllum album*.

7.4.3 Photosynthetic Pathways

Of all the species investigated none is a CAM plant. All the hydrohalophytes dominant in their respective communities are C3 plants (Table 7.1): *Avicennia marina, Halocnemum strobilaceum, Limonium axillare, Nitraria retusa, Zygophyllum album* and *Tamarix nilotica*.

With the exceptions of *Aeluropus lagopoides* and *Hammada salicornica*, which are C4 plants, most of the halophytes, which in the Sinai are only companion plants in their respective communities, are C3 plants: *Arthrocnemum fruticosum, A. macrostachyum, Limoniastrum monopetalum, Zygophyllum coccineum*.

The glycophytes growing in the vicinity of the Sinai salinas are also C3 plants: *Acacia raddiana, A. tortilis, Calotropis procera* and *Salvadora persica*.

Without discussing the ecological advantages and disadvantages of the three photosynthetic pathways (for such discussion see Ziegler et al. 1981, Hattersley 1983, Pearcy 1983, Pearcy and Calkin 1983, Pearcy and Ehleringer 1984, Evenari 1984), it can be stated that for the dominant hydrohalophytes of the Sinai salinas the C3 pathway seems to be well fitted to their specific ecological situation. The complete absence of CAM plants is most remarkable.

This certainly cannot be generalized for other halophytes growing in different ecological environments.

7.4.4 Biological Passports

In this section a short ecobiological description of the main dominant species growing in and around the salinas of the Sinai will be given. *Avicennia marina:* A tropical community-building hydrohalophyte needing warm seawater and muddy substrate, forming in seawater the outermost belt of the salinas; leaves salt-excreting (Waisel 1972, Waisel and Agami 1979); a mangal pioneer reaching its northernmost point of distribution in the Sinai. *Limonium axillare:* A Sudanese hydrohalophyte forming belts near the seashore; leaves salt excreting; grows in substrate of sand and gravel, forms phytogenic hillocks, penetrating inland to regoid plains; reaches its northernmost point of distribution in the Sinai. *Zygophyllum album:* A hydrohalophyte with an extended area of distribution (Red Sea coast, Sinai, Mediterranean coast) with a large range of ecological conditions; forms phytogenic hillocks; occurs in pure stands in narrow belts and in mixed stands with either *Halocnemum* or with *Halocnemum* and *Nitraria*; has no salt-excreting glands but, like the xerohalophyte *Zygophyllum dumosum* (Evenari et al. 1982) rids itself of surplus salts by dropping first its leaf blades and then its petioles (Levitt 1972, Waisel 1972, Waisel and Agami 1979). *Halocnemum strobilaceum:* A hydrohalophyte with a distribution area extending from Sudan and the west coast of the Sinai to the Mediterranean coast of Egypt; reasons unknown for its absence from the eastern coast of the Sinai and from Israel; grows mostly in pure stands; forms very high phytogenic hillocks; has a relatively shallow root system; grows in substrate of fine sand and mud; its main rooting zone contains the highest sodium and chloride content of all the Sinai halophytes investigated. *Nitraria retusa:* A hydrohalophyte common to the salinas of the whole Saharo-Arabian region; penetrates into the adjacent Irano-Turanian and Sudanese regions; forms in the Sinai extended belts of nearly pure stands, mostly on more inland fringes of belts formed by other species; has no salt-excreting mechanisms; tolerates very high salt concentrations (Zohary 1962, Waisel 1972, Waisel and Agami 1979); has a very extended and deep rooting system where saline groundwater is not too near the soil surface. *Aeluropus lagopoides:* A hydrohalophyte with salt-excreting leaves; occurs in patches together with *Zygophyllum album, Limonium axillare* and *Nitraria retusa;* has very shallow roots; tolerates very high salinity. *Salvadora persica:* A tropical glycophyte reaching its northernmost point of distribution as pure stands in Sinai; grows on wadi fans and colluvium inland from the coastal salinas; forms very large phytogenic hillocks with which it moves ("wandering tree"); has a very extended deep root system possibly reaching the nonsaline water table.

C. The Gavish Sabkha – A Case Study

8. Introduction

Wolfgang E. Krumbein

In this section we deal with special aspects of the Gavish Sabkha, and in two contributions with the very similar environments of the Ras Muhammad Pool (Chap. 10) and the Solar Lake (Chap. 15). Comparisons are made, however, in the individual contributions to other evaporative systems.

Much of the work was stimulated by our late friend Eli Gavish during his productive years of sedimentological studies along the shores of the Gulf of Elat. Another stimulus was initiated by the agreement that the Solar Lake and the Gavish Sabkha were selected as model cases for a Project of the International Geological Correlation Program. Its subject is *Early Organic Evolution and Mineral and Energy Resources*. It is presently chaired by one of the chapter authors (M. Schidlowski). A subproject, chaired by S. Golubic and W. E. Krumbein is entitled *Fossil stromatolitic microbial ecosystems and their modern analog*. Another subproject chaired by G. Eglinton deals with the possibilities of finding marker molecules in ancient oil-producing sedimentary environments by comparison with nonpolluted modern systems. Also, this group had chosen the Solar Lake and the Gavish Sabkha as case study systems. Organic geochemistry, metal deposit formation in the biosedimentary environment and the comparison of recent and ancient productivity and sulfate reduction systems thus were the main purposes of the specialized studies undertaken in the Gavish Sabkha and the Solar Lake.

The Gavish Sabkha is a microbe-dominated system with scarce macrophyte flora at its borders and a very specialized faunal community. Therefore, most of the case studies after an introductory section on the sedimentological setting and parameters of the Gavish Sabkha and similar systems (Chaps. 9 and 10) deal with the microbial communities (Chaps. 9–14). Some of the plant elements and plant relationships of the Gavish Sabkha are already described in Chapter 7 and need not be reevaluated in a special chapter here. The faunal elements of the Gavish Sabkha and the Solar Lake are described and compared in Chapter 15. The remaining special studies deal with the biogeochemical aspects of sabkhas and use the Gavish Sabkha as example. Heavy metal distribution, isotopic ratios produced and incorporated into the sedimentary column, as well as organic markers and fossilization potential of the microorganisms are treated extensively (Chaps. 11 and 16–19). We were not able to analyze carefully for methane production and other volatile gases and we had no chance to further elaborate on the interrelationship of the cycles of phosphorus, nitrogen and sulfur with the general carbon cycle within such environments. Future studies will have to be done to further document the great importance of this intermediate coastal systems, serving as huge nutrient, metal and organic matter traps for the material washed down from the continent and extracted from the evaporating waters of the adjacent sea.

9. Geomorphology, Mineralogy and Groundwater Geochemistry as Factors of the Hydrodynamic System of the Gavish Sabkha

ELIEZER GAVISH †, WOLFGANG E. KRUMBEIN, and JACOB HALEVY

The purpose of this study is to contribute to the understanding of hydrodynamic and diagenetic processes in sabkhas by close observations of variations in mineralogy and brine geochemistry in the uniquely restricted marine system of the sabkha south of Nabq, called the Gavish Sabkha. Also, we give a more specific geological and geochemical description of this particular sabkha than has been described previously (Gavish 1980, Gavish et al. 1978, Krumbein et al. 1979 a).

9.1 Geological Setting

The closest geographically defined place is the coastal alluvial fan oasis of Nabq (Fig. 9.1). The sabkha itself is a round depression in the alluvial plain, measuring about 500 m in diameter. It is separated from the Gulf by about 400 m of beach sand, covered by alluvial fan facies up to about 2 m above mean sea level. The internal morphology of the sabkha is shaped like an inverted dinner plate; the central part is elevated about 50 cm above a fringing water rim (Figs. 9.2 and 9.20). The rim of water, especially on the sabkhas northeastern side, piled up by floods of desert sheetwash, may reach a width of 10–20 m and water depths of 5–50 cm. Water is supplied to the sabkha by flow channels leading from numerous peripheral small springs to the rim. The Gulf waters, with a tidal range of about 60–70 cm, do not flood over the bar and into the sabkha, even during storms. From time to time, however, the sabkha may be immersed by storm floods which, after breaching the lower coastal bar 1.5 km farther south, wind their way northward along the coast.

The climate of this area is arid; annual rainfall is only 10 mm. The mean annual air temperature is about 26 °C and the relative humidity about 50% (Israel Meteorological Service, 1967). Evaporation is about 4.0 m per year (Griffiths 1972). Clay and silt layers in the sabkha and sand deposits at its margin record the desert sheetfloods. Sometime around 1950, and again in 1979 and 1980, such floods occurred.

The unique shape of the sabkha and its location within the coastal plain (Fig. 9.3) raises questions about its age and mode of formation. It is unlikely that the sabkha depression started as a solution sink; the climate is arid and no other such sinks have been observed in this area. A more plausible, but still unproven, possibility is that it occupies a preexisting topographic low in the underlying bedrock; quite possible a former backreef lagoon. Irregular and fast growth of a reef

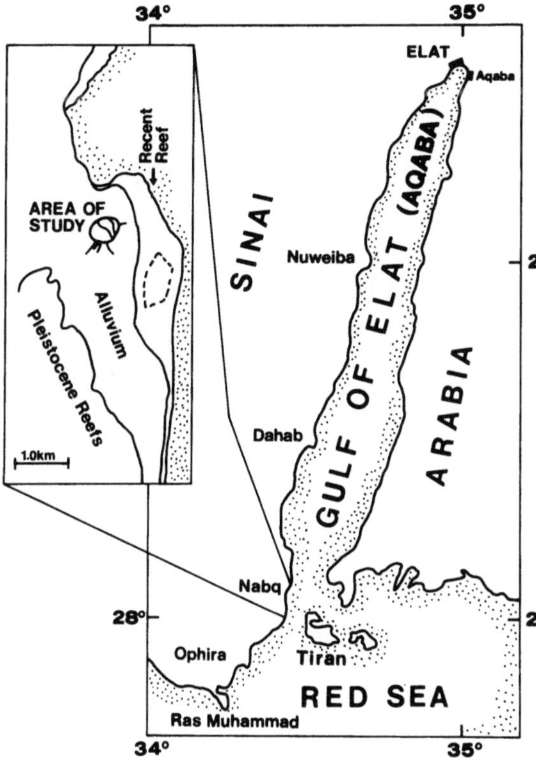

Fig. 9.1. Index map showing the location of the Gavish Sabkha along the Gulf of Aqaba

could have left such a circular depression in its reef flat, since similar depressions are observed in the modern reef flat. Figure 9.4 shows an example of such irregular reef growth along the coast to the north of the sabkha, which may eventually produce a similar depression. Such a reef underneath the sabkha would have to be more recent than the nearby raised Pleistocene reefs and possibly younger than the last postglacial sea rise.

G. M. Friedman (1965, 1972) and Neev and Friedman (1978) have concluded a relative uplift of reefs near Elat occurred about 2,000–4,000 years ago (Chap. 3). Studies conducted by the authors on beach rock along the coast of the Sinai, indicate a corresponding relative drop in the sea level of about 1–1.5 m. It is possible, therefore, that this old reef flat was exposed subaerically about 2,000 years ago, isolating the depression from the sea. Since then it has been filling with clastics and evaporites, forming the present-day sabkha. A similar tectonic event was proposed by Krumbein et al. (1977). Unfortunately, we could not drill deeply enough to prove that the sabkha is indeed underlain by a reef. Therefore, the bedrock under the sabkha was not sampled in any of our numerous cores for sedimentological, mineralogical, geochemical and biological purposes.

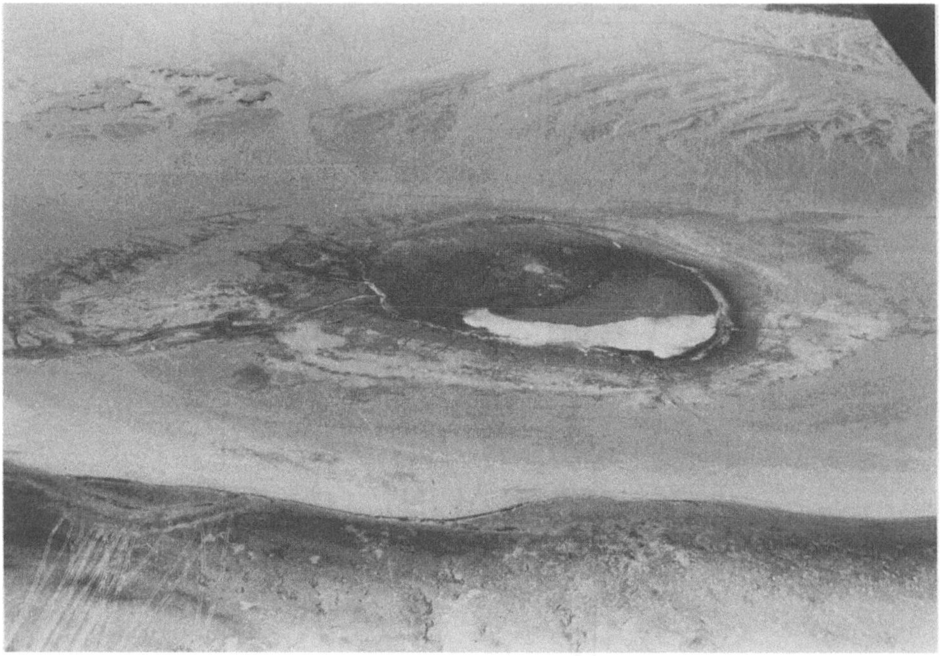

Fig. 9.2. An oblique view of the Sabkha from the air. Note the flow channels east and south of the Sabkha

Fig. 9.3. Aerial view of the Sabkha showing its location on the coastal plain

Fig. 9.4. A round gap in a modern fringing reef north of Nabq, showing the similarity in shape and thus indicating a possible way of genesis (see Fig. 9.3)

9.2 Sampling Procedure

Sampling sites in the Sabkha were located to major sedimentological units along a north–south transect; sediments outside the sabkha were sampled as well (Fig. 9.9). Ten sediment cores were taken with Polyvinyl-Chloride pipes; four of them reaching sediments deeper than 1.5 m. In addition, a number of surface sediments were sampled to check lateral variations. At one location close to the open water rim, 2.8 m of core was recovered. The entire core consisted of sabkha sediments; the underlaying bedrock was not penetrated.

Surface water samples were collected from the open water rim and when possible from the deeper sediments at each core site. The interstitial brines were not extracted directly from the cores, because (a) it could not be done immediately in the field, (b) the up to 20% compaction of the cores during sampling could have mixed the interstitial brines, making any samples representative of one specific core location. Therefore, a special device was constructed for sampling the interstitial water directly from the sediments. This device, a pipe with a built-in tubing and filter, was pushed into the sabkha near the coring sites and the interstitial water was sucked out from a predetermined depth with the help of a special pump. This system was successful only in a few places and at relatively shallow depths. Obviously, the technique needs to be improved. Water samples were also taken from places in the sabkha where cores were not taken.

In the laboratory the cores were subdivided into segments according to apparent differences in texture, mineralogy or color. Each sample was air-dried and analyzed for its bulk mineralogy by X-ray diffraction analysis. The samples were also sieved through a 63 um sieve and the mineralogy of the lutitic fraction was again analyzed for the major ions Na^+, K^+, Ca^{2+}, Mg^{2+}, Cl^-, and SO_4^{2-} by atomic absorption spectroscopy, flame photometry, chloridometer and volumetric methods.

For Ca^{2+} and Mg^{2+} we used atomic absorption spectroscopy, for Na^+ and K^+ flame photometry and for Cl^- and SO_4^{2-} volumetric methods as well as a field reflectometric salinometer.

9.3 Geomorphological and Mineralogical Zonation

Various observations and aerial surveys (S. Cohen 1975, Neev and G.M. Friedman 1978) along the Sinai coast of the Gulf of Aqaba have revealed evidence of several earthquakes and landslides; some have been so recent that no fringing reef has been reestablished and that the color of the mountain slopes has not yet developed the red weathering patina which is so typical for arid desert areas. This is the case at a place known locally as the Spectacles or geographically as Ras Abu Galum. A few hundred meters farther south a round hole exists in the fringing reef (Fig. 9.4), which suggests that an earthquake and landslide could have formed the depression behind the coastal bar south of Nabq.

Alternatively, such holes in the reef may be produced tectonically and by the special features of the growth of a fringing reef when it meets with the clastic input of desert wadis (e.g., at Dahab and Nueba). All metahaline and hyperhaline pools along the Gulf (Por 1972) have one or the other of the above-mentioned origins. In the case of the Nabq oasis area it is most probable that tectonics and desert wadis' terrigenous clastic sedimentation interfered with the elevated reef complexes visible in the southwest of Fig. 9.2 to create the depression. It is quite obvious from this aerial view that several different hydrodynamic and geomorphological factors control the evolution of the Gavish Sabkha.

1. Seawater seepage creates wet spots, and springs seasonally supply a variable amount of seawater according to the difference between the mean sea level and the level of the open water rim of the Gavish Sabkha. The open water rim is clearly visible to the east of the white spot (Figs. 9.2 and 9.3). In summer, evaporation causes the water level to sink a maximum of 10 cm, and more seawater is supplied. In winter, with decreasing evaporation less seawater is supplied even though the winter tides run much higher than the summer tides. It has been observed, however, in the Solar Lake (Krumbein and Y. Cohen 1974, Y. Cohen et al. 1977a) and at the Gavish Sabkha that the water level rise in winter occurs rather quickly, within a few days. This is due to a sudden rise in tidal levels all along the Red Sea and the Gulf of Aqaba in fall. This phenomenon (Fig. 9.5) is not related

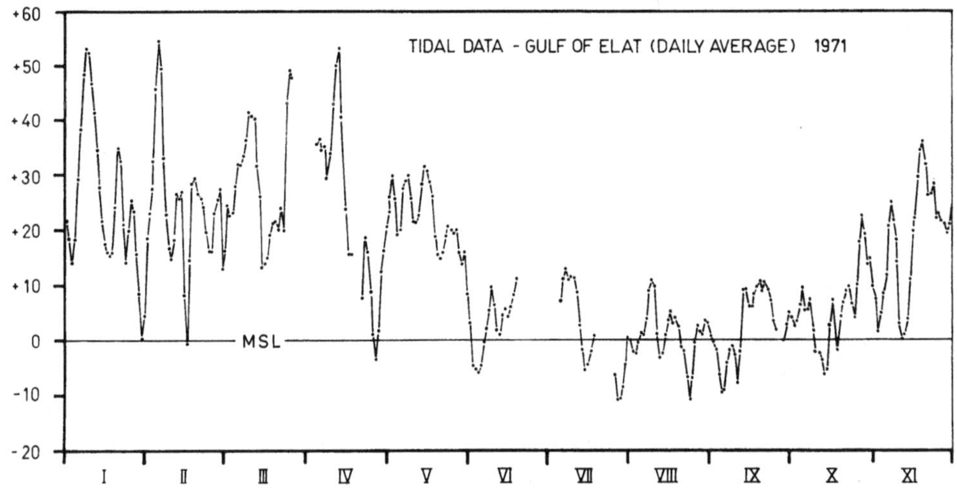

Fig. 9.5. Tidal data from the Gulf of Elat (1971). Note the variation of higher and lower tidal levels as a response to annual fluctuations of the climatic conditions

to the tides themselves, but to the climatic responses (e.g., monsoons) to the sun movement around the equinox.

2. Desert sheetfloods sporadically supply large amounts of rain water. This water creates the small wadi systems which are visible especially in the southern and northeastern parts of the Gavish Sabkha (Figs. 9.2 and 9.3). Such floods have occurred in the 1950's and in 1979 and 1980. In 1975, in the airport area of Ophira (Scharm el Sheikh), a rainfall of 44.5 mm was observed, which apparently had very little actual effect on the Gavish Sabkha, as can be seen in all our cores. On 20 October 1979, 75 mm of rain was observed at Ophira which corresponds to the average rainfall of 7 years in this area. The Sabkha water level was raised by 60 cm on that occasion and it took 5 months till the normal water level of the open water rim was reestablished by seepage reflux and evaporation. Another catastrophic rain was observed on 25 to 27 December 1980, when a 20.5 mm rain was observed at Ophira; even more rain must have fallen at the Gavish Sabkha inasmuch as the water level was raised by about 25 cm at that time, while no water level rise had taken place after the 44.5 mm rain in 1975. Similar floods have been observed also in other parts of the world in the same years (Stolz 1984). This demonstrates the force of the desert sheetfloods and also the patchy distribution of rain and the difficulties in estimating the influence of rainfall on such systems.

3. The tide itself is largely reduced over the seepage distance of a few hundred meters from the sea through the sand bar and the fracture system of the underlying reef. At the Solar Lake, which is much closer to the sea than the Gavish Sabkha, no real tidal influence is observed and the water supply shifts only slightly with the tides. At the Gavish Sabkha between 0 and 2 cm water level changes occur due to the tides, but with considerable delay in response time. On the other hand, the zone above the open water level is wetted more or less during one tidal cycle and the periods of dryness change considerably when periods of

high tide or low tide coincide with the maximal evaporation activity at noon. This in turn has little influence on the mineralogy and microbiology of the sabkha system but influences the faunal behavior considerably (see Chap. 15).

4. Sporadical wetting and drying strongly influence the higher parts of the Sabkha margins and the central part of the Sabkha, inasmuch as wetting enables microbial mats to form new surface layers and drying desiccates and cracks the new crusts. Rise in water level is caused by extremely high tides, spring tides and storm tides, as well as by the desert sheetfloods and seiche produced by frequently occurring southerly and northerly winds blowing strongly for several days. The microbial mats glue the sandy and wind-blown sediments together. Thus sediment accumulation by agglutination develops with a typical upward directed pattern. Wind and slope gravity lead to an undulated, wrinkled surface which later leads to surface forms we will call "Petee" structures: in contrast to the early and exclusively abiogenic sharp-edged triangular desiccation-crack structures usually designated as "Tepee" structures (after the tents of the prairie Indians). Petee structures (biogenic) and Tepee structures (abiogenic) are very difficult to distinguish and even a very experienced ecologist or geomicrobiologist will find it difficult to distinguish between the two possible origins of a surface in a continental or marine sabkha, when he sees it during completely dry conditions, which is the normal situation for 980 out of 1,000 days in a desert environment. Figures 9.6 and 9.7 show Tepee and petee structures respectively. B. Javor and J. Stolz (pers. comm.) as well as one of the authors (W. E. Krumbein) have observed such Petee structures at the Baja California microbial mats during the big rains of 1979 and 1980, which also occurred at the Gavish Sabkha in the same years. In the case of the Baja California mats, prior to the establishment of the final Petees, the whole mat was moved above the ground by the winds like a tablecloth, producing tablecloth folds. In the Gavish Sabkha this was never as impressive, as the stabilizing mats were rather sand-embedded mats than surface covering mats like the *Lyngbia* mats of Baja California.

5. Another distinct geomorphological feature influencing the water regime of the Gavish Sabkha is the formation of carbonate cemented zones which, when close to the surface, may evoke the impression of beach rock. These carbonate cemented zones within the sand dunes form hard layers which then may get exposed to the air by wind and water erosion (Gavish 1974a). The picture of several beach rocks is justified inasmuch as two different beach rock plates have been observed at the Solar Lake, one below the other, separated by microbial mat sediments, thus indicating relative sea level changes by tectonical events and/or eustatic sea level rises. Similar cementation zones occur at the Gavish Sabkha as well, clearly visible in Fig. 9.8. The dark rim at the upper part of Fig. 9.8 coincides with the edge of the most prominent carbonate cemented sand plate. During extremely high tides the inflowing seawater emerges from underneath this plate which is also the only safe ground for vehicles close to the Gavish Sabkha. Therefore, the location of the car tracks at the lower right of the photo also indicates the location of the carbonate cemented zone. A lower (older) carbonate plate is situated further down toward the open water rim visible in Fig. 9.8 by the change from white to dark in the seawater channel leading to the left (northern microdelta).

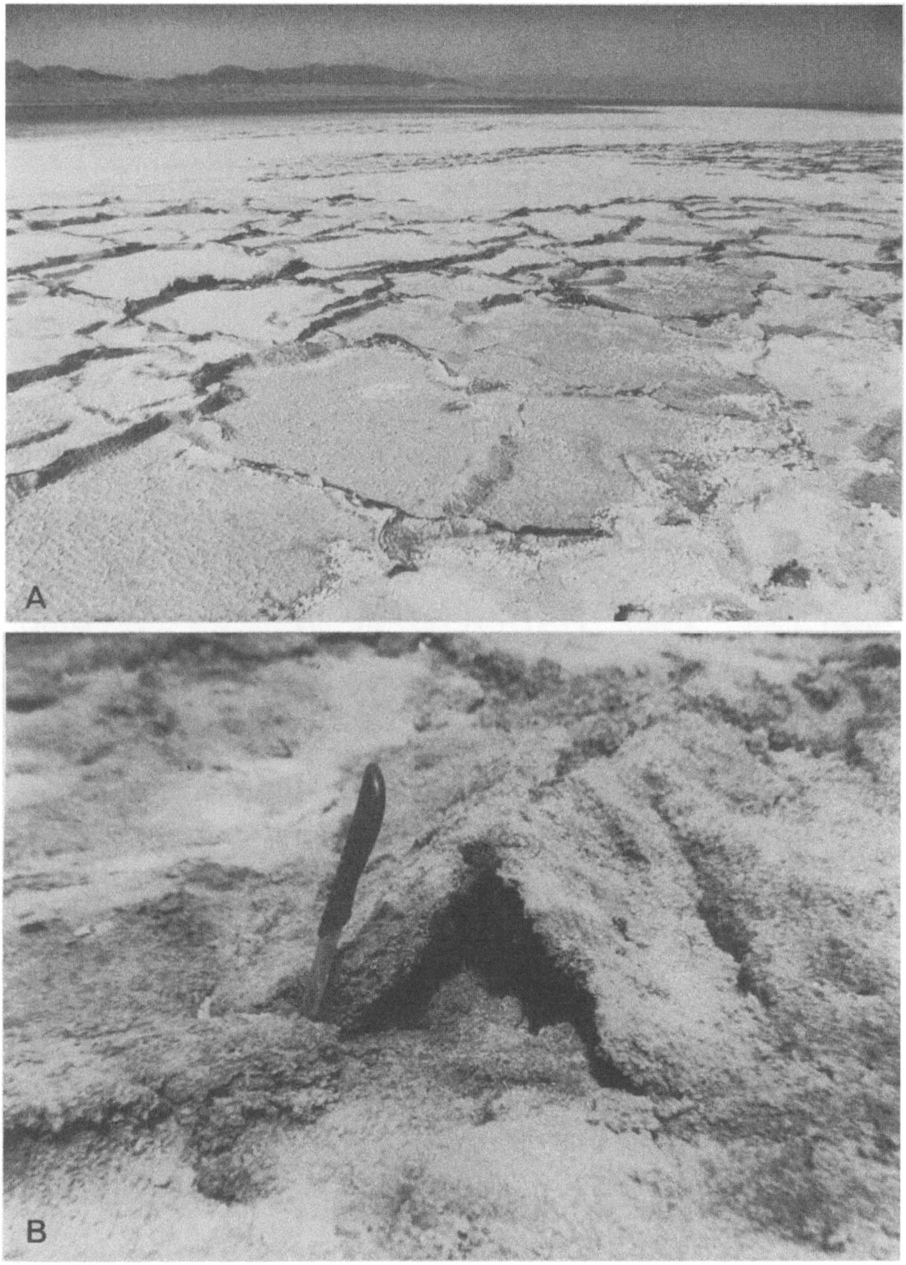

Fig. 9.6 A, B. Coarse halite polygons on the surface of Zone 9. The close-up of **B** shows the characteristic "tepee" structure which originates abiogenically as compared to "petee" structures, shown in Fig. 9.7

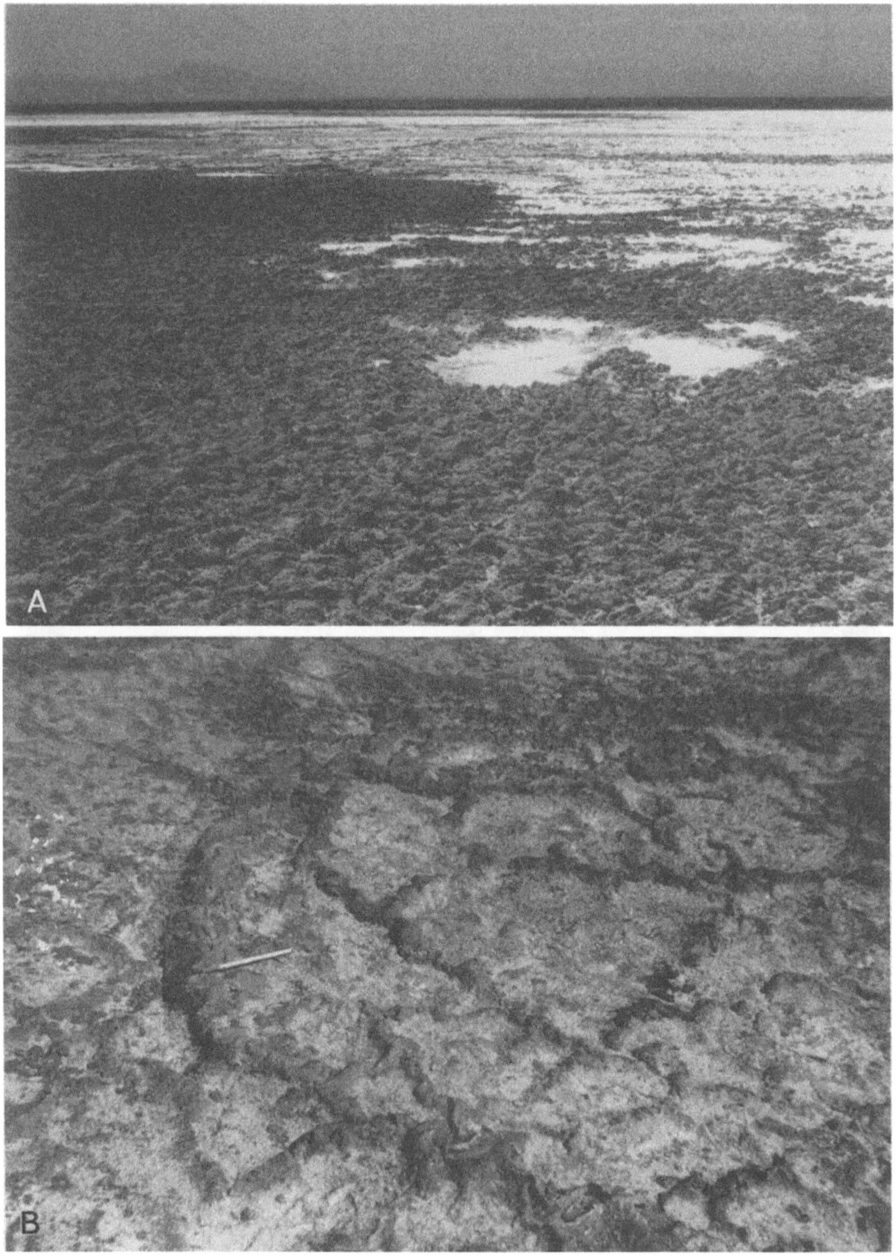

Fig. 9.7 A–D. The surface patterns in Zones 3 and 8 ("Petee" structures) mirror the interplay between wetting and drying on higher parts of the Sabkha margins and the center. Wetting favors the formation of microbial mats and stabilization of the sediments. Drying causes desiccation cracks. Wind and slope gravity lead to the undulated, wrinkled surface. In their final stage, these structures are very similar to "Tepee" structures but, by studying their initial stages, they can be clearly derived from the interplay between biological and physical properties

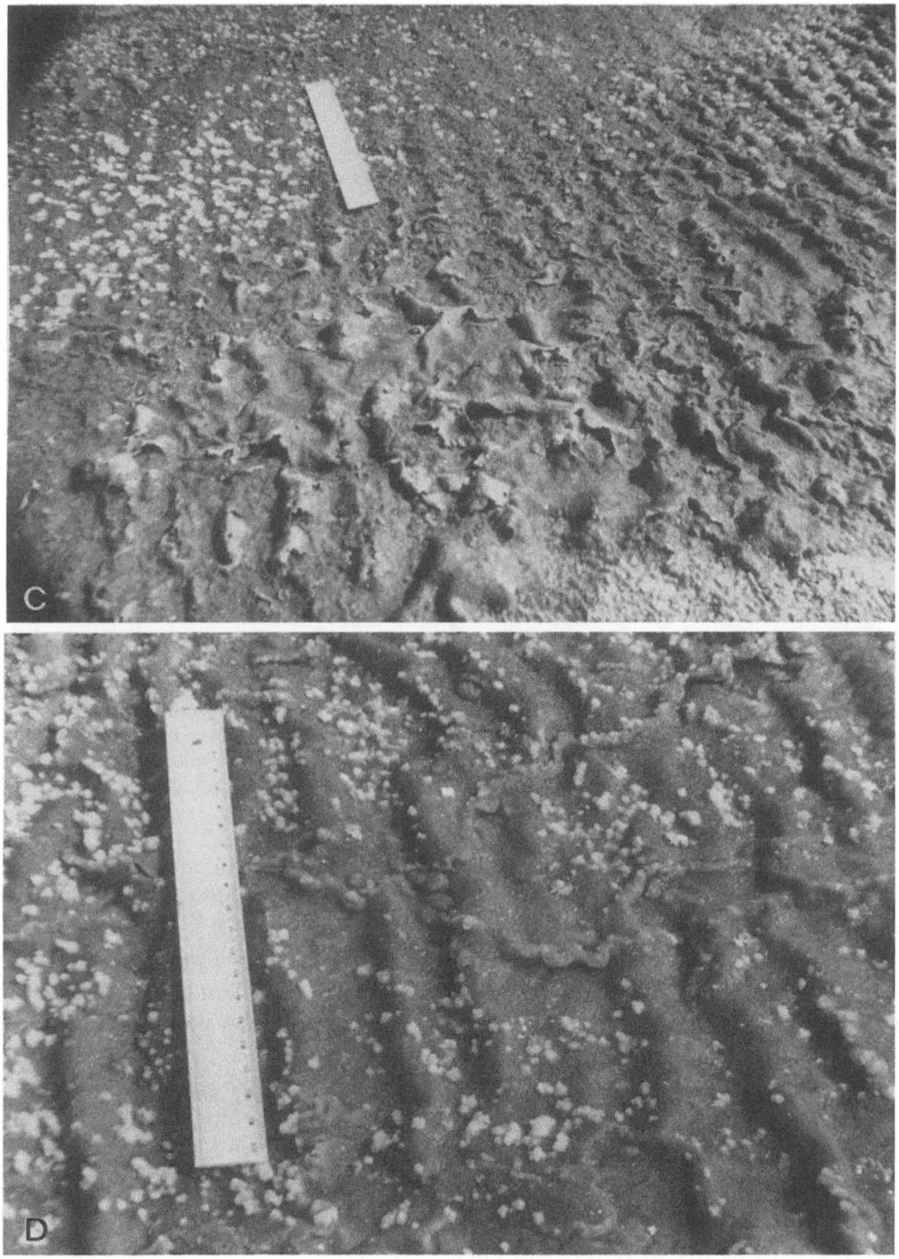

Fig. 9.7 C–D

Fig. 9.8. A view toward the northeastern part of the Sabkha. The *dark rim* at the upper part indicates the edge of the carbonate cemented sand plate (Zone 2). Most of the flow channels rise from below this plate. The open water rim (Zone 6), the white salt crusts (Zone 7) and the dark salt and gypsum crusts (Zone 8) are visible in the *middle and the lower part* of the photograph

6. Evaporative pumping contributes water to the central high parts of the Gavish Sabkha, which are visible as clear white and dark zones in the center of Figs. 9.2 and 9.3, and as white and dark lower parts in Fig. 9.8.

7. Reflux of heavy brine has to be expected as one other principle ruling the hydrodynamics of the Gavish Sabkha inasmuch as the sediment cores contain far too much gypsum and carbonates and not enough halite to be consistent with 2,000 to 3,000 years of evaporation in a closed system.

8. The heavy northerly and southerly winds which blow through the wind channel of the rift valley several times in the year possibly also contribute to the final shaping of the total system; the southern parts of the Gavish Sabkha and southwestern rim are filled faster with windblown sediments than the northern parts and the northeastern rim. This leads to the characteristic north-south zonation of the Gavish Sabkha in its central parts. The northernmost part is frequently flooded by high tides and rain floods, the darker part adjacent to the clear white zone (Figs. 9.2 and 9.3) indicates sporadic water cover. The southwest with its central salt crust is rarely, if ever, water covered at present. This situation is in part due to an elevated fossil reef complex in the southwest of the Sabkha which is clearly visible as plateaus with erosional channels in Figs. 9.2 and 9.3.

9. Water movement parallel to the coast during very high tides and storms as well as during heavy rains and desert sheetfloods may also contribute to the small

tributary systems south of the Gavish Sabkha. These factors also contribute to the accumulation of the bulk of the sediments in the southern and southwestern part, and especially in the southeastern part of the round Sabkha. If one compares Figs. 9.2 and 9.3, this becomes quite evident. The same has actually been observed at the Solar Lake, where sediment was also contributed by water flows parallel to the coast from a wadi south of the Solar Lake which was cut off from the sea by a coastal bar. Therefore, high storm tides and desert floods contribute to the greatest thickness of fine silty sediments into the southeastern area of the Solar Lake with decreasing thicknesses toward the north (Krumbein and Y. Cohen 1974).

10. Last but not least, human activities are also influencing the sedimentology and water regime of the Gavish Sabkha. Bedouins have always used the Sabkha as a salt-mining site and have therefore created depressions in the central parts of it (Fig. 9.8), and also channels and water borderlines by coming down to specific sites with camels and by foot. Car visits recently have superimposed and enhanced some effects caused by Bedouins (Chaps. 2 and 11).

These various influences on the geomorphology and hydrology of the Gavish Sabkha have produced a more or less clearly discernible zonation which we present here as a basis for the somewhat different ecological zonations of microorganisms and macroorganisms, further elaborated in Chapters 11 and 15.

9.4 Mineralogical and Hydrological Zones

The individual zones are marked in Fig. 9.9.

Zone 1: The sand bar and sand dunes, surrounding the Gavish Sabkha. Very little water is supplied by sporadic rains. The sands carry no flora, or very little flora and fauna, practically no microflora, and are not cemented by any evaporative processes. A very thin salt crust often occurs due to windblown marine splash-water.

Zone 2: The carbonate cemented zones, which are locally uncovered by gravel or sand.

Zone 3: The gypsum-halite crusts with Petee structures. This zone is only sporadically wet by heavy desert rains, sheetfloods, and by extremely high tide evaporative pumping. Cores 1 and 8 were taken in this zone (Figs. 9.10 and 9.11).

Zone 4: The periodically-to-permanently sandy sediments, frequently evolving halite incrustations which are redissolved as well.

Zone 5: The permanently water-soaked sediments in the intertidal level. Sometimes trickling films of water flow over and between the sand grains in this zone. Evaporative deposition of minerals is almost not observed. Core 2 originates from this zone (Fig. 9.12).

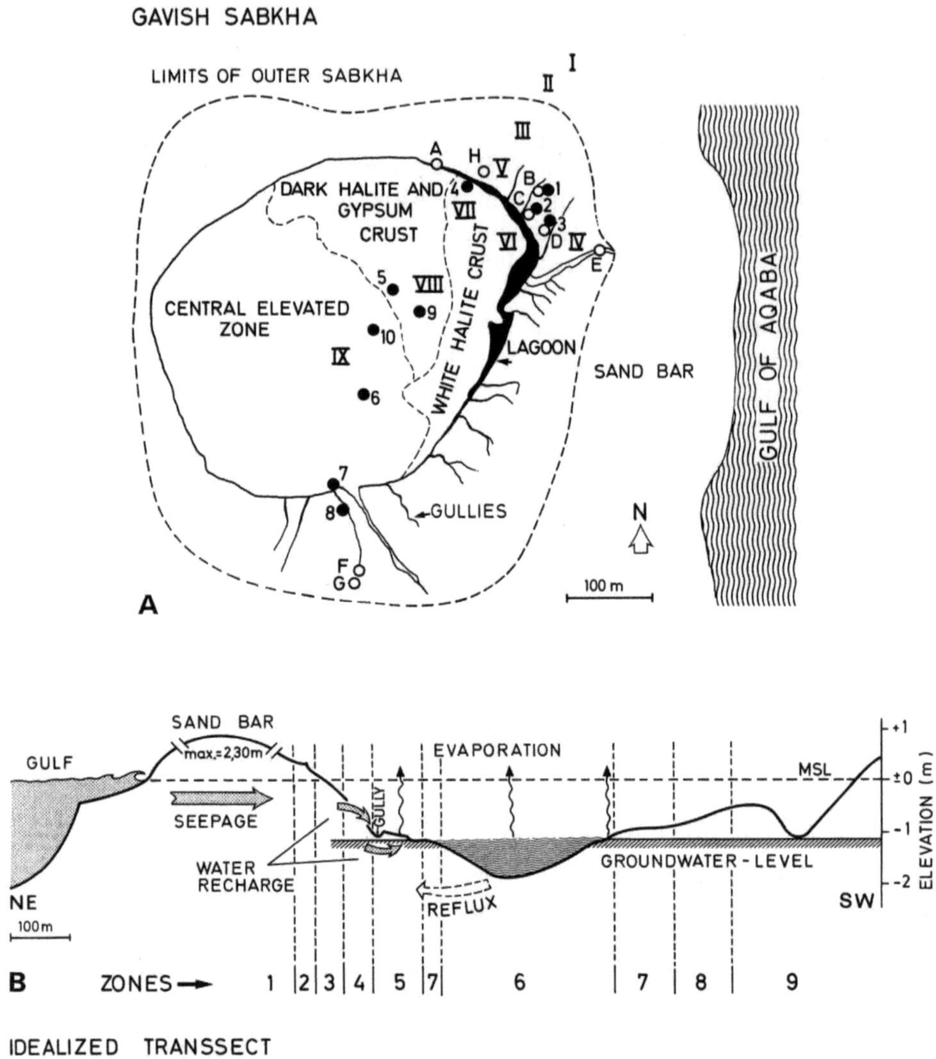

Fig. 9.9. A Generalized map of the study area with locations of sampling. Sites *1–10* (o): Cores and water sampled; Sites *A–H* (*o*): Only water sampled. **B** Generalized transect through Sabkha zones: *1* Sand bar; *2* Carbonate cementation zone; *3* Gypsum-halite crusts; *4* Periodically-to-permanently moist sandy sediments; *5* Permanently water-soaked sediments in the intertidal level (gully = flow channel); *6* Open water rim (lagoon); *7* White salt crusts; *8* Dark salt and gypsum crusts; *9* Central elevated zone. *Arrows* show the hydrodynamic mechanisms operating in the Sabkha

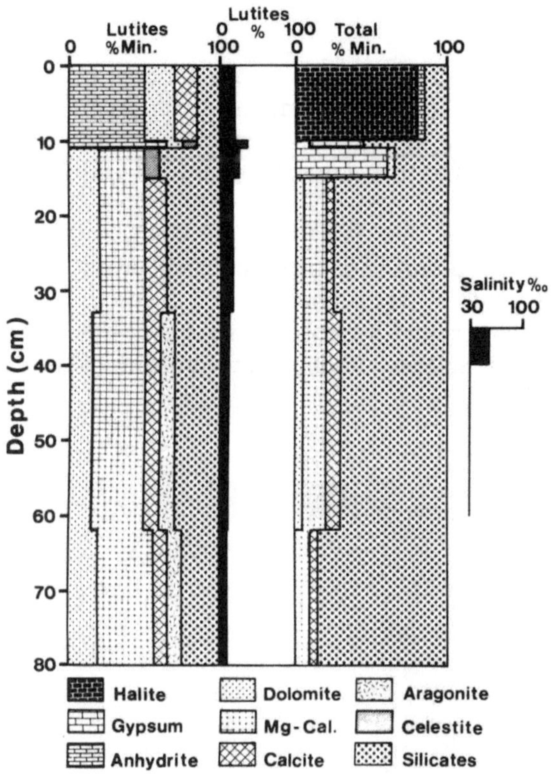

Fig. 9.10

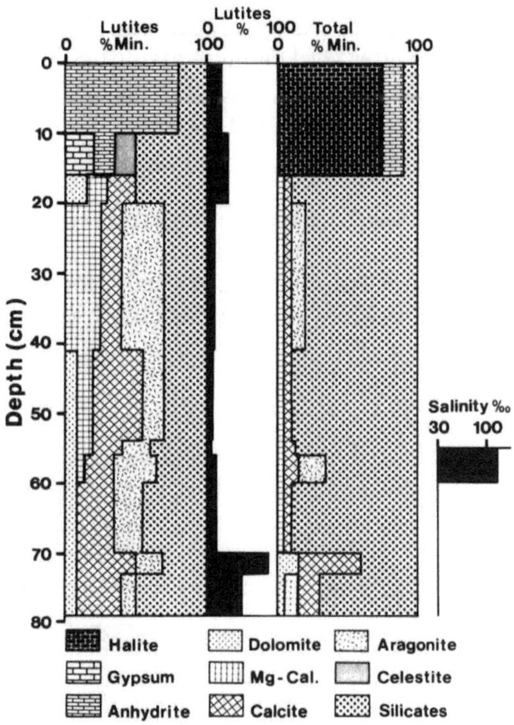

Fig. 9.11

Fig. 9.10. Mineralogy, content of lutites and groundwater salinity in Core 1, Zone 3 (Gypsum-halite crusts, northeastern part). Minerals with 5% abundance are not presented

Fig. 9.11. Mineralogy, content of lutites and groundwater salinity in Core 8, Zone 3 (southern part)

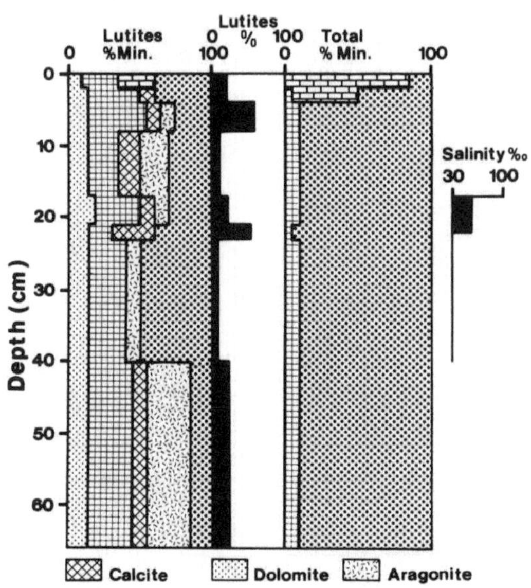

Fig. 9.12. Mineralogy, content of lutites and groundwater salinity in Core 2, Zone 5 (permanently water-soaked sediments in the intertidal level)

Zone 6: The open water rim. The extent of this zone variates with water depth. The water level changes are difficult to estimate, but a fair estimation is an average annual level change from summer to winter of 2–8 cm. In summer halite crusts develop in the upper parts, which mostly redissolve in winter. Cores 3 and 7 (Figs. 9.13 and 9.14) originate from this zone.

Zone 7: The white salt crusts with very little admixed terrigenous clastic wind blown or desert sheetflood sediments. The lack of terrigenous clastics is due to the fact that the zone is partly flooded in winter, and by storms throughout the year, washing out the fine clastic material. The salts accumulate in the bottom of

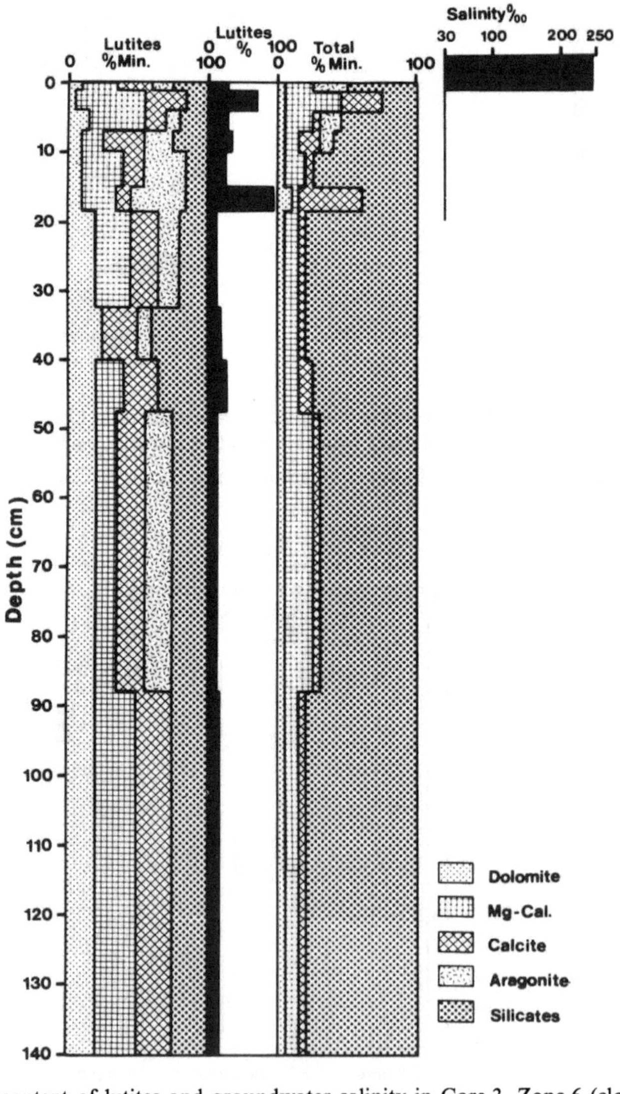

Fig. 9.13. Mineralogy, content of lutites and groundwater salinity in Core 3, Zone 6 (close to open water rim, northeastern part)

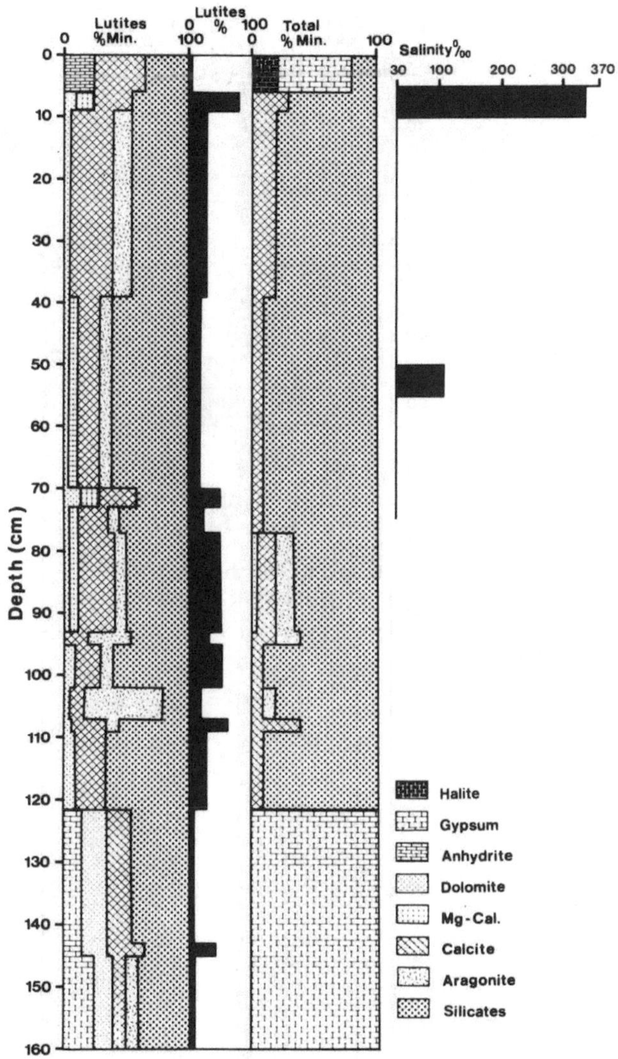

Fig. 9.14. Mineralogy, content of lutites and groundwater salinity in Core 7, Zone 6 (southern part)

the small open water rim recrystallizing all year round to form new clear white deposits. Core 4 was sampled at the borderline of this zone, showing that from the uppermost 70 cm clastic material is washed down into the rim of the lagoon (Fig. 9.15). Halite precipitating in summer is washed out and dissolved in winter.

Zone 8: The dark salt and gypsum crusts. From the water regime this zone is very similar to the white crusts of Zone 7. The zone is slightly higher and silt and clay admixed by wind or desert sheetfloods are not completely washed into the open rim during winter seasons. As visible in Fig. 9.8, the Bedouins knew as well

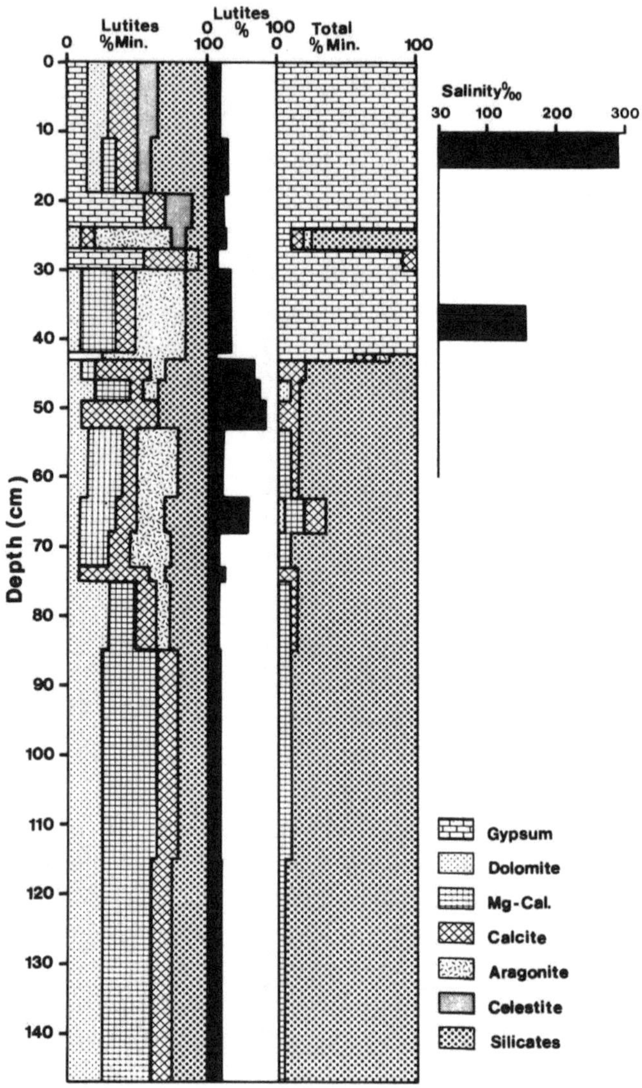

Fig. 9.15. Mineralogy, content of lutites and groundwater salinity in Core 4, Zone 7 (white gypsum crusts)

that the zones do not differ very much and mined both, the dark and the bright halite zone for halite and potash, making round minute evaporation salt pans. Samples 5 and 9 originate from this area (Figs. 9.16 and 9.17).

Zone 9: The central elevated zone. This zone is practically never water covered and only sporadically wet by evaporative pumping and/or floods. It is nearly always emergent and was dry even when the whole Sabkha was flooded after the

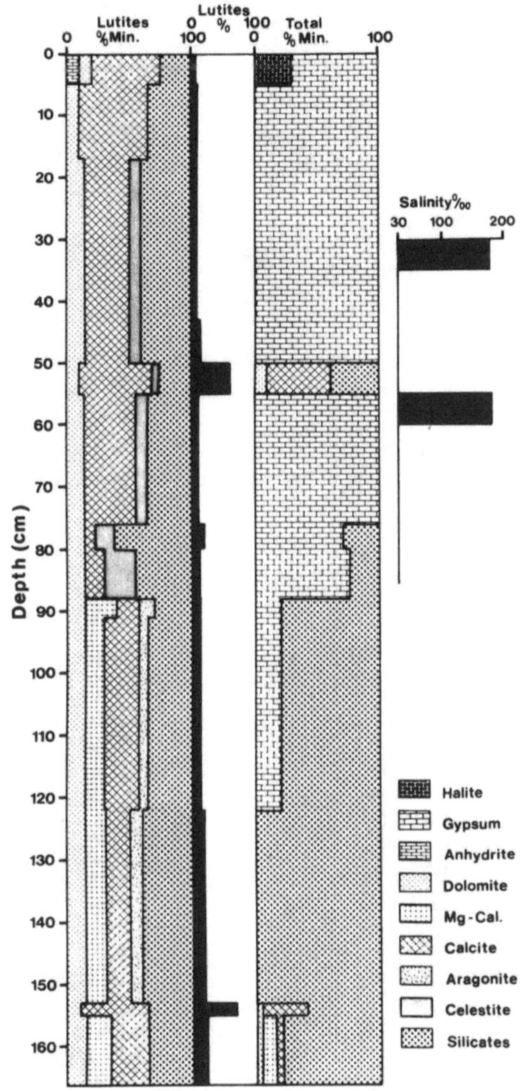

Fig. 9.16. Mineralogy, content of lutites and groundwater salinity in Core 5, Zone 8 (dark salt and gypsum crusts)

1979 desert sheetflood. Large desiccation polygons are coming up to the surface, indicating water cover in former times and during very extreme floods. These polygons are very different from the wrinkled surface structure of Zone 3 and were mentioned before in the context of Tepee and Petee structures (Figs. 9.2, 9.3, 9.6 and 9.7). The surface halite crusts with gypsum below in cores 6 and 10 indicate clearly the evaporation effect. Total dryness is indicated by anhydrite occurrence at the surface (Figs. 9.18 and 9.19).

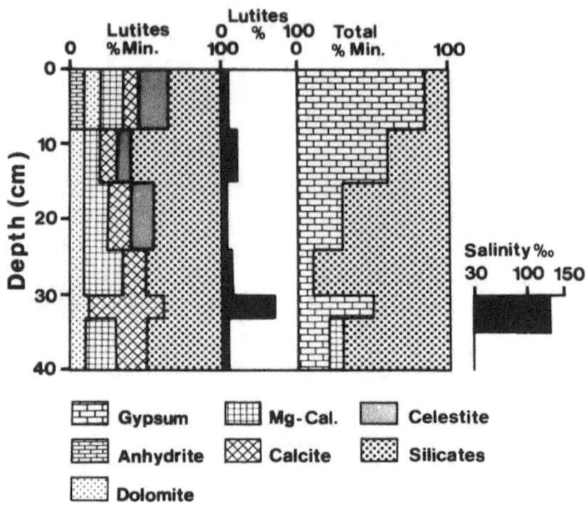

Fig. 9.17. Mineralogy, content of lutites and groundwater salinity in Core 9, Zone 8

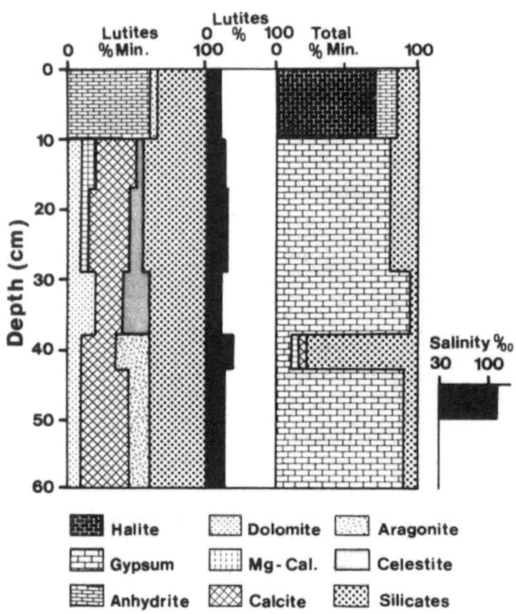

Fig. 9.18

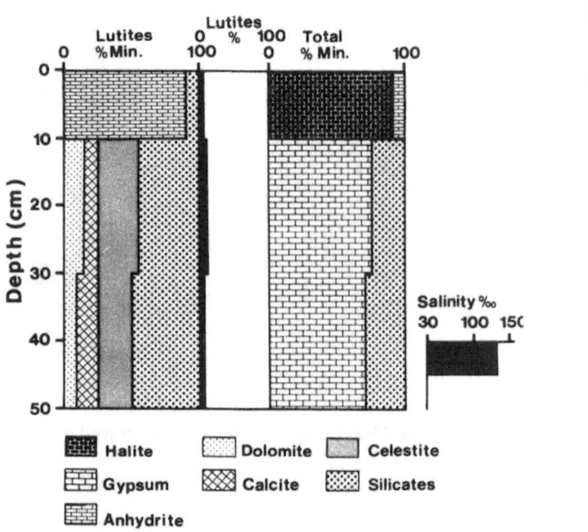

Fig. 9.19

Fig. 9.18. Mineralogy, content of lutites and groundwater salinity in Core 6, Zone 9 (central elevated zone)

Fig. 9.19. Mineralogy, content of lutites and groundwater salinity in Core 10, Zone 9

9.5 Water Geochemistry

Water samples were taken in the immediate vicinity of each core as well as from additional critical places within the Sabkha (samples A, B, C, D and H in Fig. 9.9). The results of the chemical analyses for the major ions are shown in Table 9.1. A number of samples were also taken near the shore of the Gulf of Elat (Aqaba) for comparison with the Sabkha brines. Since the variation in the chemistry of these Gulf samples is small, the results of only one typical sample are given in Table 9.1. Analyses is given of a sample of floodwater that reached the Sabkha with occasional sheetfloods could not be obtained.

Samples E, F and G (Fig. 9.9) were taken from the channels through which seawater is recharged into the Sabkha. Samples F and G were taken during the winter from a southern channel which had considerable surface flow at that time (1972). After the floods of 1979 and 1980, the major water flow came through the big channel immediately to the east, thus initiating complete recolonization. Sample 3 was taken during the summer from an eastern channel (Fig. 9.8), which was dry but kept flowing about 10 cm below the surface. The sampling and analysis show that the water keeps flowing from at least some of the springs all year round and its chemistry does not change much. Aerial photographs since 1956 show exactly the same location of the channels, until 1979/1980. Their size and the strong lithification around the springs indicate that water has been flowing in them for a long time. The location of these springs and the chemistry of the water in them leave little doubt that these are saline Gulf waters recharging the Sabkha through a system of underground conduits, as proposed by Murray (1969) for the Pekelmeer of South Bonaire (Pekelmeer comes from the Dutch/German word for pickle, brine).

The recharge is not only through the above-mentioned subaerial springs but also through springs underneath the Sabkha surface, as it does in the Solar Lake (Y. Cohen et al. 1977a–c). Sample D was probably taken near such a spring underneath the groundwater level in Zone 5. Though the salinity of the surface water in that zone is generally high, the salinity of sample D, which was taken only a few meters away from it, is less than a third (70‰). This is explained by seepage mixing and/or fast evaporation (see Chap. 11).

The aridity of this area excludes signifcant addition of rain water, making this Sabkha an evaporating pan of marine water, only occasionally modified by the heavy metal and trace metal-rich desert sheetflood waters, which are, however, poor in sodium and calcium (Chaps. 16 and 22; Renfro 1974). Figure 9.20 is a generalized north-south cross-section of the Sabkha, showing graphically the chemical properties of the groundwater table. Around the Sabkha rim, where the groundwater table comes to the surface (samples D, 3, 4 and 7 in Fig. 9.9), its salinity is extremely high and the temperatures range between 30° and 37 °C in summer. In the central zones, where the groundwater table is covered by evaporites, and flood and storm sediments (samples 5, 6, 9 and 10), the groundwater salinity is considerably lower with temperatures ranging between 25° and 30 °C. Outside the Sabkha rim, the salinity of the groundwater seepage flow is relatively low and about equal to that in the springs. In north-south transects the ground-

Table 9.1. Chemical composition of groundwater at the surface and at depth in the sabkha of Nabq

Sample location	Depth below groundwater in cm	Ion concentration in ppm							Total salinity ‰	Molar Mg^{2+}/Ca^{2+}	Molar $Cl^-/SO_2^=$
		Ca^{2+}	MG^{2+}	K^+	Na^+	Cl^-	$SO_4^=$	$HCO_3^=$			
A	0	920	3,745	1,115	33,166	56,800	7,416	244	103,406	6.6	10.3
	30	490	1,819	573	16,640	30,743	3,840	213	54,562	6.1	10.6
1	0	680	2,201	780	16,951	29,181	4,608	170	54,571	5.3	8.6
2	0	520	1,975	682	17,250	30,650	4,176	244	55,497	6.2	10.0
3	0	950	9,096	2,652	80,500	141,645	13,373	244	248,460	15.7	14.3
4	0	760	10,944	3,315	89,700	163,726	16,896	366	285,707	23.7	13.9
	40	1,000	5,837	1,638	48,475	90,205	9,024	183	156,762	9.6	13.5
5	0	1,270	6,955	1,950	56,419	100,820	10,224	110	177,748	9.1	13.3
	30	1,220	7,551	2,106	55,200	101,707	10,416	134	178,334	10.2	14.1
6	0	1,250	4,256	1,170	35,880	63,261	7,488	91	113,396	5.6	10.9
7	0	780	13,753	3,939	103,500	194,682	17,472	311	334,437	29.0	15.1
	50	1,200	4,207	1,248	33,488	59,888	8,064	183	108,278	5.8	10.0
8	0	1,180	3,648	975	25,300	47,002	7,488	122	85,715	5.1	8.5
B	0	550	1,949	624	18,170	32,078	3,840	244	57,455	5.8	11.2
C	10	1,050	3,830	1,131	28,451	51,013	7,512	152	93,139	6.0	9.2
	45	580	1,824	585	14,651	25,169	3,619	134	46,562	5.2	9.4
D	0	640	2,626	858	22,425	38,979	4,944	189	70,661	6.8	10.6
	15	830	2,772	955	23,000	39,405	4,488	110	71,560	5.5	11.9
	50	520	1,897	546	13,639	25,915	3,312	85	45,914	6.0	10.6
E	0	520	1,897	604	15,150	26,997	3,552	305	49,025	6.0	10.2
9	0	1,100	4,499	1,462	40,250	73,289	8,640	104	129,344	6.7	11.4
10	0	1,124	5,000	1,657	41,860	75,615	9,600	104	134,960	7.3	10.6
F	0	574	2,220	628	18,055	32,300	4,272	165	58,214	6.4	10.2
G	0	574	2,128	628	16,732	30,405	3,600	165	54,232	6.1	11.4
H	0	900	3,003	920	23,000	43,036	6,864	183	77,906	5.5	8.5
	40	540	1,800	515	12,742	24,246	3,326	134	43,303	5.5	9.9
	80	540	1,800	491	12,316	24,058	3,302	152	42,659	5.5	9.8
Gulf of Elat		500	1,592	487	13,018	23,430	3,216	152	42,395	5.4	9.8

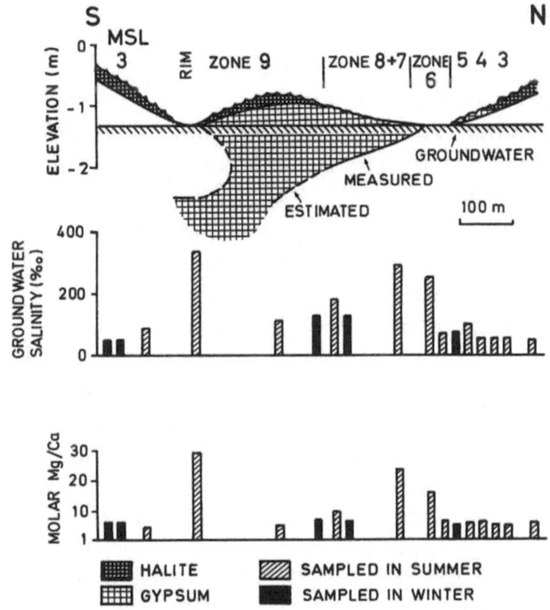

Fig. 9.20. Generalized cross-section of the Sabkha in north-south direction, showing the observed and inferred distribution of evaporites. The geochemical pore water values given are only of the uppermost layer of the groundwater table

water table seems to be horizontal. In east-west transects, however, it occurs at different levels, as influenced by (a) tidal pressure, (b) surface outcrops and (c) evaporative pumping (Figs. 9.20 and 9.24).

The ionic ratios of Na^+, K^+, Mg^{2+} and Cl^- in all the water samples are almost constant and similar to those in the Gulf waters (Table 9.1), which further indicates that their origin was primarily marine. The Mg^{2+}, though it participates in the diagenetic processes of the Sabkha, shows little change in its ratio to Cl^-. The Ca^{2+} and SO_4^{2-} values show considerable variability in their ratios with respect to other ions, as is shown by the Mg^{2+}/Ca^{2+} and Cl^-/SO_4^{2-} ratios in Table 9.1 and Fig. 9.20. These changes, especially associated with sulfate reduction at the open water rim, are explained to more detail in Chap. 11.

An interesting phenomenon is the sharp decrease of the groundwater salinity from the surface downward. This was also observed by Lucia (1968) in one core of the Pekelmeer at Bonaire. The salinity gradient shows a large decrease in the upper 50 cm of the groundwater table (samples A, C, D, and 4) and below that, salinity does not change much (sample H). Most probably, all samples with salinities higher than that of marine and tidal origin indicate water sampled above the groundwater table level with salinities thus increased by evaporation.

9.6 Evaporites – Evidence of Evaporative Pumping and Reflux

Within the Sabkha halite is an accumulative component which is relatively much more abundant than in the Persian Gulf (Illing et al. 1965, Kinsman 1969 a) and in the Pekelmeer of Bonaire (Lucia 1968). In places within the Gavish Sabkha halite forms a crust which is only 10–20 cm thick over the much thicker gypsum (Figs. 9.18 and 9.19). But outside the Sabkha rim, where the groundwater table is deeper, halite forms an even thicker crust and its abundance is far greater than that of the sulfates underneath (Figs. 9.10 and 9.11). Since these sediments are not flooded by marine water, the halite must have accumulated in situ from the groundwater migrating upward and being evaporated near the surface. Because the texture of the sediments is too coarse to support capillary movement (Shinn et al. 1965), the mechanism of the upward brine movement must have been that of "evaporative pumping," as proposed by G. M. Friedman and Sanders (1967) and Hsü and Siegenthaler (1969). This mechanism clearly operates in the outer Sabkha (Zone 3), where the volume of halite, relative to that of gypsum underneath, may indicate total evaporation of marine water. Within the Sabkha rim evaporative pumping is less intensive but still existent, although modified by capillary movement and by organisms which admix silt and clay.

Anhydrite occurs together with halite only in the upper dry crusts, where it is concentrated mainly in the lutitic fraction (Figs. 9.10, 9.11, 9.18 and 9.19). The origin of this anhydrite is not clear, since it was hard to determine its crystallographic form, and no pseudomorphs after gypsum were observed. The association of anhydrite with halite may indicate conditions of gypsum dehydration. But its concentration in the lutitic fraction, while the gypsum crystals are always much coarser, may also indicate that anhydrite as well as gypsum are primary precipitates from the brines migrating upward. The hot and dry climate of this area makes both processes possible. Recrystallization of fine-grained gypsum into coarse grains has, however, been observed as well. Within the central parts of the Sabkha gypsum is the most abundant authigenic mineral with layers reaching an average thickness of about 1 m, which reaches well below the groundwater table surface (Fig. 9.20). The crystal shapes of the gypsum are elongate to discoidal, resembling those in the Persian Gulf, and range in size from 0.1 mm to about 0.5 mm. Generally, there is a clear crystal orientation and bedding within the gypsum-rich sediments close to the surface, getting less distinctive with increasing depth. Such sedimentary features are also very commonly destroyed during coring (Fig. 9.21). Deeper samples cored by Boon, Holtkamp, Gerdes and Krumbein contain gypsum crystals of up to 1 cm and, in the Solar Lake, Krumbein cored gypsum of more than 4 cm crystal size with laminated mats embedded, thus clearly proving recrystallization processes during early diagenesis (Fig. 9.22).

Growth of gypsum in the Gavish Sabkha progressed within the sediment above and below the groundwater table surface. The crystals precipitating interstitially displaced the host grains, causing a "swelling" of the sediment. Gypsum growth progressed not only at the surface but also interstitially, as can be seen in the white area (Zone 7). Such growth must be faster than the accumulation of the sporadically deposited clastic sediments, since the bulk volume of gypsum is

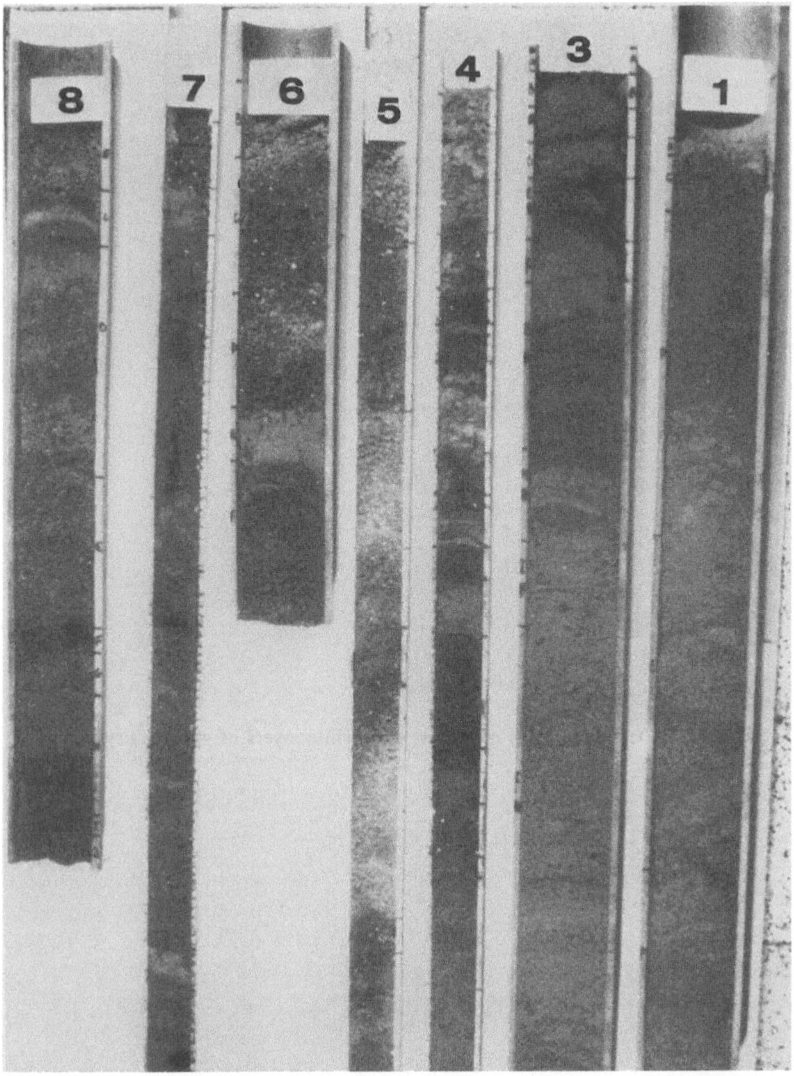

Fig. 9.21. Some of the cores taken in the Sabkha. *Dark zones* indicate fine and laminated sediments

over 90% (core 4). Hence, today, gypsum is being intensively precipitated at the surface and below it in Zones 7 and 8, where the shallow groundwater table is by far supersaturated with respect to gypsum. In Zone 9, where the groundwater table is deeper, the water is only slightly oversaturated with respect to gypsum, the accumulation of gypsum has almost stopped, and it continues mainly above the groundwater through the process of evaporative pumping. The large volume of gypsum found in this zone must have precipitated, therefore, at an earlier stage, when the conditions resembled those in Zones 7 and 8.

Fig. 9.22. Deeper sections of a core, with white layers of gypsum crystals

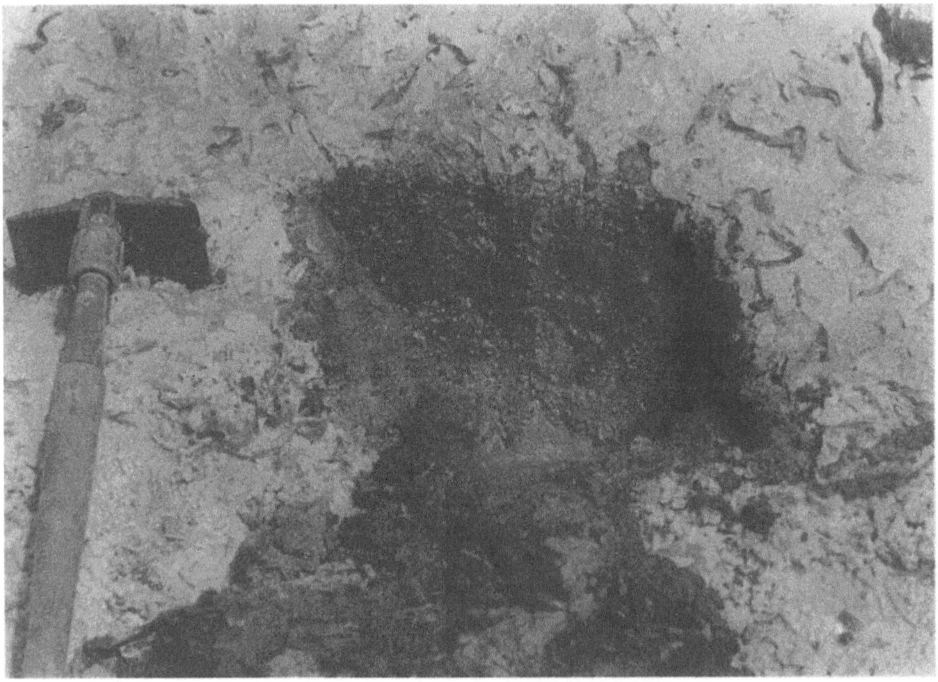

Fig. 9.23. Dark-colored, reduced sediments devoid of gypsum near Core 3 in Zone 6

Within Zone 6 (core 3; Fig. 9.13) no significant accumulation of gypsum is found at and immediately below the groundwater table level, although the salinity of the groundwater is the highest and by far supersaturated with respect to gypsum. The main reason for the lack of gypsum accumulation is probably the destruction of the gypsum by sulfate-reducing bacteria. Y. Cohen et al. (1977 b), Jorgensen and Y. Cohen (1977) and Krumbein et al. (1977, 1979 a) have shown that this process frequently operates in hypersaline environments. Such bacterial activity is indeed very high in this zone, as indicated by black iron sulfide deposits (Fig. 9.23) and the smell of H_2S. Figure 9.21 shows some of the cores which, though partially dry, still show the dark colors of reduction, especially cores 3 and 4. Sulfate reduction was probably more intensive above layers with relatively fine texture (flood-derived silt sediments), as can be observed from the dark color and absence of gypsum in the fine layers of cores 4 and 5. In the Gavish Sabkha the bacteria attack both the gypsum mineral and the sulfate ion in solution. The concentration of SO_4^{2-} in the brines, however, is still high (samples 3 and 7 in Table 9.1) and the smell of H_2S emanates only when the sediment is dug open, because fresh brine is constantly flowing in from the sea and reoxidizing some of the hydrogen sulfide.

In core 7 (Fig. 9.14) gypsum occurs in large amounts only at a depth of 120 cm and below. This is surprising because sulfate-reducing bacteria are active in these sediments, and especially because the salinity of the brines in this core drops sharply with depth and the brines are slightly undersaturated with respect to gypsum even at a depth as shallow as 50 cm below the surface. Without gypsum precipitation the salinity of the interstitial brines in this core should not keep dropping with depth, but stay high or even slightly increase to the point of supersaturation with respect to gypsum. This is probably not an isolated phenomenon in these coarse sediments, where the permeability is good. Hence, this relatively deep occurrence of gypsum precipitation could be taken at least as a suggestion of evidence for a reflux downward of heavy brines, as proposed by Adams and Rhodes (1960) and other supporters of the seepage reflux mechanism, or as demonstrated in the Solar Lake (Y. Cohen et al. 1977 a). This must be a dynamic system, where groundwater near the surface, becoming hypersaline and dense due to evaporation, moves constantly downward, and new seawater supplied by seepage moves constantly upward to replace the loss by evaporation (upward) and reflux (downward). Thus, in the given conditions both mechanisms, evaporative pumping and reflux, may operate simultaneously in the same sediment as shown graphically in Fig. 9.24. In the relatively deep horizons below the groundwater table surface only reflux will be operating, while in the sediments above the groundwater table surface only evaporative pumping proceeds. Reflux of brines from the sediments above the groundwater table surface is improbable in arid conditions, as was already shown by Gavish (1974 b).

The balance between the two mechanisms is largely determined by the depth of the groundwater table below the surface. In places where groundwater is close to the surface (Zones 6 and 7), the dominant mechanism must be reflux of the heavy brines downward with precipitation of gypsum below the sediment surface. In places where the groundwater table is not too deep (Zone 9), reflux is negligible, because the brines are not extremely saline and evaporative pumping with

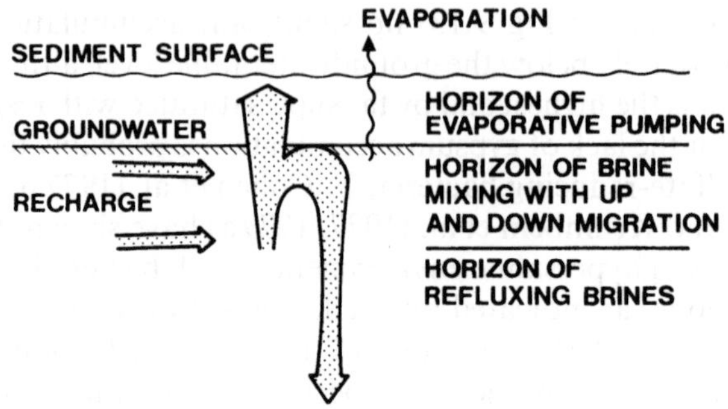

Fig. 9.24. Schematic presentation of possible simultaneous upward and downward migration of brines near the surface of the Sabkha sediments. The *width of the arrow* represents roughly the brine salinity

precipitation of halite at the surface is the dominant mechanism (Figs. 9.18 and 9.19).

The possibility of mixing of incoming and refluxing brines in the upper horizon of the groundwater table (Fig. 9.24) raises doubts about an increase in brine salinity downward. The refluxing brine is definitely being diluted in the mixing zone, and in the case of core 7 to about one-third of its salinity near the surface. But apparently enough of the refluxing brines seep down through separate pores, allowing a constant or even slightly rising salinity gradient below the depth of mixing of the incoming water sheet with the sinking reflux brine.

9.7 Carbonate Diagenesis in the Sabkha

Most calcium and magnesium carbonate minerals, as aragonite, calcite, monohydro-calcite, Mg-calcite, protodolomite and dolomite are found in the Sabkha and in its surroundings. Therefore it is difficult to decide if they are products of in situ generation of minerals (biogenic or abiogenic), of in situ diagenesis, or of clastic origin. Still there is enough direct and indirect evidence to suggest that the carbonates have undergone at least partial diagenesis within this system. Also SEM screening of the Sabkha sediments revealed little if any debris of reef detritus (fossil and recent). Only *Pirenella* debris native to the sabkha has been found frequently.

Aragonite is an abundant carbonate component of the clastic sediments outside the Sabkha. But within the Sabkha, aragonite is mostly a minor (biogenic) component or completely missing. Within all the cores a clear antipathetic relationship exists between aragonite and celestite other than in sediments from outside the Sabkha. Thus the abstite must be an in situ diagenetic product, with the Sr^{2+} contributed by the decomposing aragonite. Celestite crystals were briefly mentioned in the studies of the Persian Gulf (Butler 1973, Butler et al. 1982, Illing et al. 1965, Kinsman 1969b) and in Coorong district (Skinner 1963), where they were also considered as products of aragonite replacement, but their relationship to the total sediment was not elaborated. The celestite in the Gavish Sabkha is in places so abundant (Figs. 9.18 and 9.19, cores 6 and 10) that a reasonable amount of aragonite replaced in its horizons could not have supplied Sr^{2+} content. Thus the celestite must form as an accumulative phase with its Sr^{2+} being contributed by decomposition of aragonite from a much larger sediment volume than that in which the celestite is found, as well as from the concentrated brines near the surface. That migration of ions is possible in the Gavish Sabkha with its relatively intensive upward and downward migration and mixing of brines.

The aragonite must, therefore, dissolve in the hypersaline conditions of the Gavish Sabkha. Liebermann (1967) has already shown that the solubility of $CaCO_3$ increases with rising water salinity and has a sudden jump at the point of gypsum precipitation. Indeed, in places where the brine salinities have not exceeded the gypsum saturation point, aragonite is still, at least partly, preserved, but where they exceeded it, aragonite is excluded and celestite occurs instead. The similarities in the solubilities of celestite and gypsum (G. Müller 1962) allow the precipitation of celestite soon after that of gypsum, accompanied by the dissolution of aragonite, as can be seen from the close association of the two minerals in the cores.

Calcite and Mg-calcite are the major components of the lutite fraction in the Sabkha sediments. Often it is hard to make a clear distinction between the two, because the calcite is relatively rich in Mg^{2+}, with 2–6 mol percent $MgCO_3$, while the Mg-calcite has 10–14 mol percent $MgCO_3$. Both minerals are also present in the clastic material around the Sabkha, but in smaller amounts. It is possible, therefore, that these minerals, and especially calcite, participated, at least partly, in the diagenetic processes that affected the aragonite. Calcite has been considered for some time a product of freshwater diagenesis, but is has been suggested that it may form also under marine conditions (Milliman 1966, Beall and Fischer 1969, A. Katz et al. 1972, Friedman 1972, Krumbein and Y. Cohen 1974, Krumbein 1975, Krumbein et al. 1977).

In order to have an indication of the origin of the calcite, a number of carbon and oxygen isotope determinations were made on lutite samples. Samples with abundant dolomite, aragonite or Mg-calcite had [δ-^{13}C] values between 1.9‰ and 3.8‰ and [δ-^{18}O] values from $+0.6$‰ to $+0.3$‰, which are well within the range of recent marine waters (Berner 1971 b). Samples composed mainly of calcite showed [δ^{13}C] values from -0.43 to -0.5, and [δ-^{18}O] values of -3.3 to -3.9, which are still at the borderline of marine water values, mainly on the basis of the carbon isotope. Had that calcite been completely of terrestrial origin, even of Pleistocene age, its carbon isotope values would have been considerably lighter ([δ-^{13}C]~ -6‰), as shown by Gavish and Friedman (1969) for Pleistocene carbonates along the Israeli coast. Hence a part of the lutite fraction calcite in the Gavish Sabkha seems to be of marine origin (see also Chap. 19).

Within the Gavish Sabkha the calcite of the lutite fraction seems to be relatively concentrated in the upper sediments (Figs. 9.14–9.16, 9.18 and 9.19: cores 4, 5, 6, 7 and 10), where it has been in contact with the most saline brines with the highest MG/Ca ratios (Fig. 9.20). Possibly under these highly saline conditions "the Mg^{2+} ion is less strongly hydrated and does not inhibit the growth of calcite" (Bathurst 1971, p. 538), perhaps the case in the Gavish Sabkha. The release of Ca^{2+} due to the decomposition of gypsum by reducing bacteria could, according to G. M. Friedman (1972), also contribute to the growth of calcite (see also Krumbein et al. 1977, Krumbein and Y. Cohen 1977, Jorgensen and Y. Cohen 1977).

Dolomite could certainly also form in these highly saline brines which have a high Mg/Ca ratio (Alderman and Skinner 1957, Alderman 1965, Deffeyes et al. 1965, von der Borch 1965, Shinn 1968, Berner 1971a, Bathurst 1971). However, Mg-calcite is more abundant than dolomite. This could be related to the larger abundance of Mg-calcite in the clastic sediments, but may also result from the faster growth rate of Mg-calcite if compared to dolomite in these brines (Glover and Sippel 1967, Bathurst 1971). Still, generally more dolomite is found in the Sabkha than in the clastic sediments, indicating the existence of a process of dolomitization. This dolomite is not very well ordered, slightly calcic, and could perhaps be termed protodolomite. The distribution of the dolomite in the total sediments (taking into account the ratios of lutite to total samples) shows no clear preferential concentrations in a particular sediment horizon. Dolomitization can logically be attributed to a biological mechanism and to episodical Mg-binding and release during the production of organic compounds and their degradation within the microbial mats (Chap. 11). In the Sabkha, carbonate diagenesis has been progressing, therefore, under hypersaline brine conditions according to the two-step process suggested by Liebermann (1967). In the first step there was dissolution of calcium carbonates, such as aragonite and possibly some calcite or Mg-calcite, and in the second step coprecipitation of calcite, Mg-calcite and dolomite, all of which may be influenced biologically by the microbial laminated mat systems, frequently observed in the sediment cores.

9.8 The Hydrological System and Its Evolution

The location and isolated condition of the Gavish Sabkha enable the reconstruction of its history and hydrological processes. The main processes operating in the Sabkha today and possibly since its separation from the sea are shown in Figs. 9.9 and 9.24. At present the sabkha seems to have reached a steady state, where the groundwater level and extent of gypsum do not change much. Therefore, if terrestial recharge is disregarded, the hydrological system of this Sabkha can be expressed in the following equation:

$$\text{seepage} = \text{evaporation} + \text{reflux}.$$

If the history of this Sabkha started about 2,000–2,500 years ago, as suggested before, and the average thickness of its gypsum layer is about 1.0 m (Fig. 9.20), then the rate of gypsum accumulation in the Sabkha has been about 3.6 mm sediment per year (see also Krumbein et al. 1977). If the sediment has 50% porosity, the accumulation is 1.8 mm per year of compact gypsum. This accumulation fits very well with the yearly sulfate cycles of 1.63 mm, calculated by Udden (1924) in the Permian of West Texas. It also compares well with the rate of sulfate sedimentation in the Pekelmeer of Bonaire (Deffeyes et al. 1965), where it is about one-sixth (0.6 mm/yr) for an area with about one-fifth (90 cm) of the evaporation recorded for the area of the Gavish Sabkha.

Had this round Sabkha been covered completely by seawater at a certain time at the present rate of evaporation (400 cm/yr), about 3.2 millimol of gypsum would have been precipitated per year per cm^2. That would form a yearly layer of about 2.2 mm of gypsum or about 4.4 mm of gypsum sediment of 50% porosity. The calculated (3.6 mm/yr) and hypothetical (4.4 mm/yr) rates of gypsum sediment accumulation can be even closer, if we consider that not all the sulfates from the enriched brines were precipitated in the Sabkha but in part left the system with reflux brines. The good agreement between the rates of gypsum accumulation in this study with those in other studies, as well as between the calculated and hypothetical rates within the Gavish Sabkha may indicate the following: (a) the assumed date of the start of the Gavish Sabkha (2,000–2,500 years B. P.) is reasonable; (b) unlike today, gypsum has been precipitating in the past over most parts of the Sabkha, which is very probable, if one compares other sabkha systems to this one (Krumbein et al. 1977; see also Chap. 11).

Today, much less gypsum is precipitated in the Sabkha as a whole, because most of its surface is covered by a "protective" halite crust which greatly inhibits the evaporation of groundwater. Today, intensive evaporation is mainly from areas where the groundwater table is at or near the surface (Zones 4, 5, 6 and 7), a condition that exists over about one-fourth of the Sabkha surface. Thus, in comparison with the past, one-fourth of the yearly total of gypsum accumulation in the Sabkha would be produced at present. 40,000 m^2 of evaporating area (one-fourth of the Sabkha), at a given evaporation rate, would evaporate about 160,000 m^3 of seawater per year. If the hydrologic system of the Gavish Sabkha is at equilibrium, as mentioned before, then the seepage from the sea must supply that amount of water plus a smaller amount lost by reflux. Calculations of Krumbein et al. (1977) for the Solar Lake indicate that this is very probable.

The history of this Sabkha could be reconstructed as follows: At the beginning the Sabkha depression had about the same dimensions as today, but the sediment surface, especially in its central parts, was slightly lower with shallow water covering most, or all, of this area. Gypsum precipitated intensively in the arid climate, primarily in the central and southern parts of the Sabkha (Zones 8 and 9). This precipitation had to be fairly continuous and rapid, since there are few interruptions in the gypsum horizons of this zone. Accordingly, the water supply by seepage, and possibly also by overflow from the south, had to be much larger than today. With time, the area of intensive gypsum precipitation moved upward to the water table level and above it, as it is observed today in Zone 3, thus decreasing the loss by evaporation. Simultaneously, or somewhat before, the amount of recharge from the sea decreased; carbonate incrustation of the bar and upheaval of sealing microbial mats could have been the predominant mechanisms (Fig. 9.22). The reduced evaporation loss from a smaller surface area of exposed groundwater was, therefore, balanced by reduced recharge, leading to the present state of hydrological balance. In the areas which were raised above groundwater (Zone 9) gypsum continued to precipitate, but at a much slower rate, while halite started accumulating at the surface. The increased salinity enabled increased production and persistence of microbial mats with the formation of stromatolitic parts practically free of gypsum and leading to the inverted soup-dish effect of less evaporitic sediments accumulating in the rim with its high biological activity and subsequent sulfate reduction.

At present, seepage from the Gulf into the Sabkha must continue all year round at an almost constant rate, since the groundwater level does not change drastically. The amount of seepage is apparently not controlled by the hydraulic head and its pressure is not balanced by the heavier brines, since the Sabkha is situated too low relatively to the sea level to justify such a balance. Thus the seepage must be along conduits with independently limited flow. Many of these conduits, which are probably cracks in the bedrock (reef), are shallow enough to loose water on the way by evaporation. As mentioned before, seasonal changes are to a lesser degree due to compensation of stronger winterly tidal pressure by stronger evaporation pumping in summer.

Major reflux from the Sabkha cannot be limited to a particular time of the year, as it is the case in the Pekelmeer of Bonaire (Murray 1969), although generally the model proposed there may hold true for the Gavish Sabkha as well. The reflux probably continues through most of the year and, as in the Pekelmeer, it may operate along the same conduit system, but through its deeper levels (see also Y. Cohen et al. 1977a). The occurrence of gypsum in the lower horizon of core 7 is especially significant, because of the location of this core. This core is at the point where the major seepage conduits run from the south into the Sabkha (Figs. 9.2 and 9.9). Through the same conduit system the main reflux may proceed from the sabkha southward, explaining the lopsided distribution of gypsum, as shown in Fig. 9.20.

9.9 Conclusions

1. The Gavish Sabkha was isolated from the sea around the beginning of the Christian era and since then has been recharged by seepage through subsurface conduits forming authigenic evaporites, and later stromatolitic microbial mats and desert sheetflood sediment layers.
2. In parts of the Sabkha where groundwater is relatively far below the surface (40 cm), halite is accumulating due to the mechanism of evaporative pumping.
3. In places where the groundwater table is close to the surface (10 cm), extensive gypsum is precipitated below and above it. In these places hypersaline brines must reflux down, as is also shown by the presence of gypsum in sediments at depths greater than 120 cm after an interval with little or no gypsum.
4. It is possible that both mechanisms, evaporative pumping and reflux, operate simultaneously in the same sediment system.
5. In the most saline parts of the Sabkha with very wet surfaces or open standing water, gypsum is being decomposed completely by sulfate-reducing bacteria (see Chap. 11).
6. The closer the groundwater table is to the surface, the more hypersaline is the water, with the salinity dropping sharply with depth. This salinity gradient in

the groundwater does not prevent the reflux of heavy brines downward and back to the sea far below the seepage water table.

7. Diagenesis of carbonate minerals proceeds also in the most hypersaline environments of the Sabkha, without intervention of meteoric water, except during occasional desert rain floods (approximately every 20 to 30 years). In the upper horizons where brines are most saline (supersaturated with respect to gypsum), aragonite is replaced by celestite, other carbonates, and possibly also by low Mg-calcite.
8. Dolomite is forming in the Sabkha possibly by three mechanisms: reflux, evaporative pumping, and by biological fixation and subsequent release of magnesium.
9. The relatively constant amount of recharge by seepage is not necessarily controlled by the groundwater level of the Sabkha and its brine salinity, but by an apparently coincidental seasonal equilibration, i.e., low tides in summer combined with high evaporative pumping, and high tidal pressure in winter combined with low evaporative pumping, respectively.

10. The Ras Muhammad Pool: Implications for the Gavish Sabkha

Gerald M. Friedman, Amihai Sneh, and Roy W. Owen

Both thick and widespread, stromatolites in the rock record testify to the existence of ancient peritidal flat terraines that were far more extensive than today. Not only are stromatolitic strata widespread, they are also economically important because of their association with resources of hydrocarbons and minerals (especially lead and zinc). Stromatolites formed in peritidal environments where microbial mats flourished and led to the formation of laminated carbonates (often dolomitized); evaporites commonly formed and sometimes survived as interlayers. The broad shallow epicontinental seas of the geologic past provided vast areas of such hypersaline peritidal environments; today only in a few restricted areas do similar environments exist. Studies of these modern hypersaline microbial mat environments lead to a better understanding of the vast body of ancient stromatolitic strata. For this purpose, previous studies of modern peritidal environments in arid climates were carried out in the Persian Gulf (Kinsman 1964, 1966, Kendall and Skipwith 1969, Purser and Evans 1973, Purser and Seibold 1973), in Shark Bay, Australia (Logan and Cebulski 1970, Davies 1970a, b) and the Solar Lake Sinai (G. M. Friedman et al. 1973a, Krumbein et al. 1977).

This paper reports on a continuing program to study the hypersaline pools of the northern Gulfs of the Red Sea, in particular the pool at Ras Muhammad, the southernmost tip of the Sinai Peninsula (Fig. 10.1). Three other pools exist, all of which are along the west (Sinai) coast of the Gulf of Elat. They are (1) the so-called sun pool 195 km to the north near Elat (Por 1968, Mazor 1969, Eckstein 1970, Klein et al. 1970, Amiel et al. 1971, 1972, G.M. Friedman et al. 1973a, Krumbein and Y. Cohen 1974, Y. Cohen 1975, Zalcman and Por 1975, Aharon et al. 1977, Y. Cohen et al. 1977a–c, Krumbein et al. 1977), (2) the Dahab pool (70 km to the north), and (3) 40 km north, the pool at Nabq that is part of what is now called Gavish Sabkha (Gavish 1974a, 1975a).

Earlier reports referring to the Ras Muhammad pool include Karsten (1846), Walther (1912), G.M. Friedman et al. (1968), and G.M. Friedman et al. (1973b).

The overall geologic/tectonic setting and climate of the sea is discussed in detail in G. M. Friedman (1968) and elsewhere in this volume (see Chap. 3). In particular, Ras Muhammad is located at the triple junction of the Red Sea, Gulf of Elat (Aqaba), and Gulf of Suez and extends into the Red Sea (Ras means cape). The area is very active tectonically; the climate hot and very dry.

The hypersaline pool at Ras Muhammad is completely and at all times separated from direct contact with the Red Sea; its rapidly evaporating waters (loss 3 m per year, Rosenan 1963, Rosenan and Mane 1970) are replenished only by

Fig. 10.1. Oblique satellite photograph showing location of Ras Muhammad at southernmost tip of Sinai Peninsula

underground percolation through the 200 m of highly permeable and jointed reef rock that separates it from a narrow inlet of the Red Sea (Fig. 10.2). The inlet extends from the Red Sea into the cape and occupies a shallow elongate depression (300 × 250 m) which is a northwest-southeast trending graben. Only at extremely low spring tide is the inlet itself cut off from the Red Sea, briefly becoming an 800 × 200 m isolated pool.

The Ras Muhammad pool is small, 250 × 150 m (Fig. 10.3), and is surrounded by terraces built of Pleistocene and Miocene carbonate rocks (Fig. 10.2). A small alluvial fan, 60 m long, spreads out over the northwest side of the pool. Surface water rarely occurs and can have minimal effect on the water regime of the pool, as annual precipitation runs almost to zero in this area, and the catchment area for the alluvial fan is very limited (about 75 acres). The pool is very shallow – about 50 cm at most – and the water depth is seasonally controlled (Fig. 10.4); its bottom is paved with microbial mats (G. M. Friedman et al. 1968).

A supratidal sabkha surrounds the pool. Its sediments are sand-sized and are cemented by evaporite minerals, the growth of which has raised the sabkha surface, creating irregular hummocks and pits. Microbial mats do not form in channels crossing the sabkha; cerithid gastropods live there and would eat the mats (Garrett 1970). The plant *Halocnenum strobilaceum* is abundant in places on the sabkha.

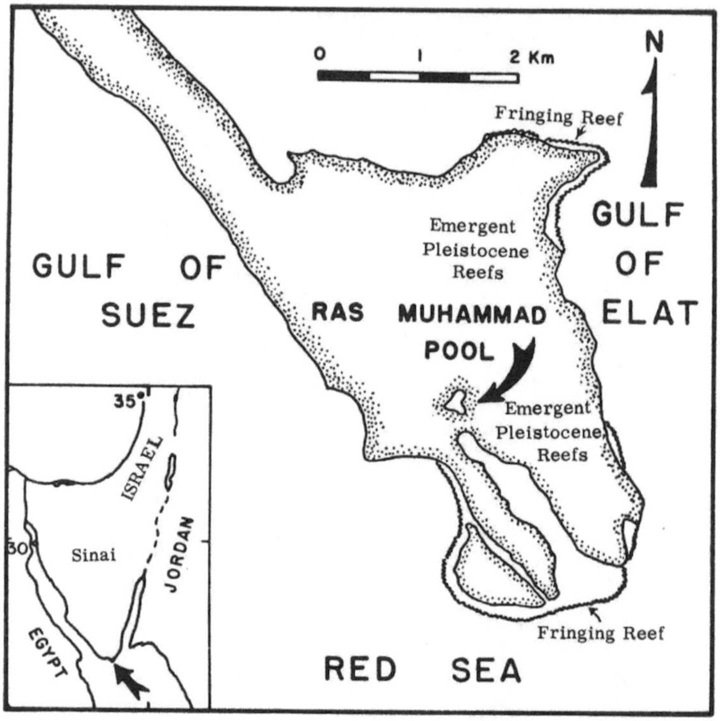

Fig. 10.2. Map of Ras Muhammad, a peninsula projecting into the Red Sea. Sea-marginal hypersaline pool and elongated embayment from the Red Sea occupy a fault-bounded depression within emergent Pleistocene reefs

In a previous study of the more northerly sun pool near Elat (G. M. Friedman et al. 1973 a) the shelf of the pool was seen to be carpeted with finely laminated microbial mats containing carbonate pellets, ooids, oncolites, grapestones, flakes and laminae. Calcite there is enriched with magnesium (up to 40%) and the organic material is magnesium-rich as well, but dolomite is not found in the carbonate of the mats except some questionable peaks recorded by Gavish (Krumbein and Y. Cohen 1974, Krumbein et al. 1977). Evaporites are also missing from the microbial mat deposits on the shelf of the sun pool. The slope and bottom of the pool, however, were found to be floored with gypsum, and dolomite is present in the evaporites.

The results suggested that the dolomite in ancient microbial laminated deposits is postdepositional, resulting from concentration of magnesium in high-magnesian calcite and organic matter in the depositional environment; authigenic or syndepositional dolomite formation was not indicated.

Preliminary work at Ras Muhammad indicated that the pool becomes more hypersaline than sun pool and that seasonal changes are important. The present study was undertaken to compare and contrast the geochemistry of the waters and the nature of the sediments being deposited at the Ras Muhammad pool with those of the previously studied sun pool. Because the pool of Gavish Sabkha is just 40 km away and is similar in many ways to the Ras Muhammad pool, one

Fig. 10.3. Oblique aerial photograph, taken from helicopter, of sea-marginal hypersaline pool and vicinity. View is to the south. Bedrock within which pool is located consists of emergent Pleistocene reef rock; note, for instance, scarp face (*south or upper right* of pool) where best exposures of reef rock may be viewed. This photograph, taken in August, shows that water cover of pool has almost vanished. *Dark area* in pool is underlain by microbial mats beneath approx. 30 cm of water. *White area* is floored by halite which has spread across the microbial mats. Four *black spots*, denoting vegetation, demarkate seepages of water from Red Sea, two on each side of remaining water cover, one on margin of pool (*extreme left*), and one near margin (*lower left*). Area surrounding pool is floored by sabkha. Note sparse vegetation along margin of pool. *Upper left area* of photograph shows elongated embayment connecting with the Red Sea. Pool is about 250 m long and 150 m wide

can anticipate that lessons learned in this study might be applied at Gavish Sabkha. The studies of all the hypersaline pools of the northern Red Sea gulfs are leading to the hoped for better understanding of the sedimentology of ancient stromatolites.

10.1 Procedure

Aerial photography of the pool taken during each of the four seasons (though not during the same year) was used to map both the shoreline configurations during the waxing and waning of the water cover and areas where replenishing marine waters from the Red Sea emerge from the intervening rock and seep into the

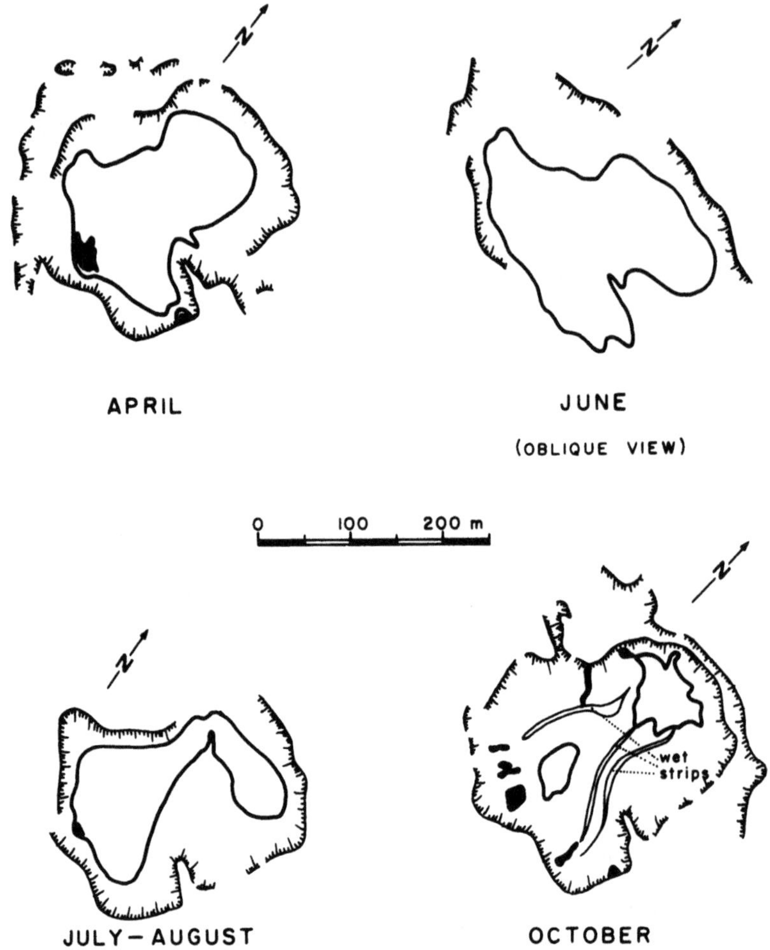

Fig. 10.4. Maps, prepared from aerial photographs, showing the seasonal shifting in shoreline of pool and the waxing and waning of the water cover

pool. Surface waters were sampled for geochemical analysis during the fall and winter of 1970–1971 (November, January), spring and summer of 1971 (May, August), and again during the winter of 1977–1978 (December, February), summer and fall of 1978 (July, September), and spring of 1979 (April). Subbottom sediment cores were taken from the pool. The cores were X-rayed for internal structure, cut open and described. The petrography of the sediments was examined using binocular and petrographic microscopes. X-ray diffraction was employed to determine the composition of the carbonates. The organic matter of the microbial mat sediments was analyzed at several depths below the bottom of the pool. The carbon and oxygen isotopes were also analyzed in samples at various depths below the bottom (radiocarbon dates were determined). The porosity and permeability of the reef rock underlying the pool and separating it from the inlet were also determined.

10.2 Results

10.2.1 Seasonal Changes in Water Level and Shoreline

Shoreline configurations of the pool during the four seasons are shown in Fig. 10.4. The pool generally is fully flooded in April. By July and August it has partially dried up; the floor of the pool, including its photosynthetic microbial carpet, is subaerially exposed, and cyanobacterial mats are broken into desiccation polygons (Fig. 10.5). In October, the water cover has almost vanished; two small isolated pools and several wet strips remain. At that time, microbial mats are extensively desiccated (Fig. 10.5), and in places a cover of halite, including hopper crystals, spreads across the mats (Fig. 10.6). Water percolating from the inlet through the intervening reef rock replenishes the pool at seepage areas, which are visible on aerial photographs and were mapped (Fig. 10.4).

10.2.2 Seasonal Changes in Geochemistry of Water

Geochemical changes in the composition of the water reflect the annual cycle of changes in water level. Salinity increases from late winter to late fall, i.e., the total concentration of dissolved solids, both cations and anions increases (Fig. 10.7; Table 10.1). In very late fall and winter the rise of water level is accompanied by a decrease in salinity.

During 1970–1971, between the months of June and August, when the total salinity increased from approximately 130 to 315×10^3 mg/l, the sulfate level de-

Fig. 10.5. Microbial mats of floor of pool exposed in summer after water cover has evaporated. Mats are polygonally cracked. Halite has not spread across these mats

Fig. 10.6. View of part of sea-marginal hypersaline pool in which water cover has completely dried up. Area of pool in view is floored by halite which spreads across microbial mats. Note irregular hummocky surface of sabkha *in foreground*

creased sharply from 9,700 to 4,600 mg/l (Fig. 10.7; Table 10.1). The calcium-ion concentration (which was increasing at a rate similar to that of the increasing salinity) increased even more sharply from 1,480 mg/l in May to 5,300 mg/l in August, when the sulfate-ion concentration decreased; this is 3.6 × increase in calcium-ion concentration compared to a 2.5 × increase in salinity. No samples were taken later in the fall, so it is not known if the salinity continued to increase, nor if the sulfate concentration returned to a high level (trends observed previously in sun pool).

During the period from 1977 to 1979 sampling took place in winter of 1978–1979, summer and fall of 1978, and spring of 1979 (Fig. 10.7; Table 10.1). Between the months of February and July 1978 the salinity increase from 80 to 260×10^3 mg/l was accompanied by a similar sulfate concentration increase from 6,400 to 11,000 mg/l. Continued increase in salinity was measured in September (to 304×10^3 mg/l) and, although this is similar to the level of salinity that in 1971 was accompanied by a sharp decrease in sulfate concentration, in 1978 the sulfate concentration level in September continued to increase with the salinity (to 17,000 mg/l). No samples were taken between July and September, so it is not known if the sulfate concentration level dropped sharply during that time period and subsequently rose to the level measured in September (again, these trends were observed in sun pool, see Sect. 10.3).

During the 1970–1971 seasons the Mg/salinity, Ca/salinity and Mg/Ca ratios all remained fairly constant even when the sulfate concentration level dropped sharply in August 1971. By way of contrast during the 1977 to 1979 period only the Mg/salinity ratio remained somewhat constant. The Ca/salinity ratio de-

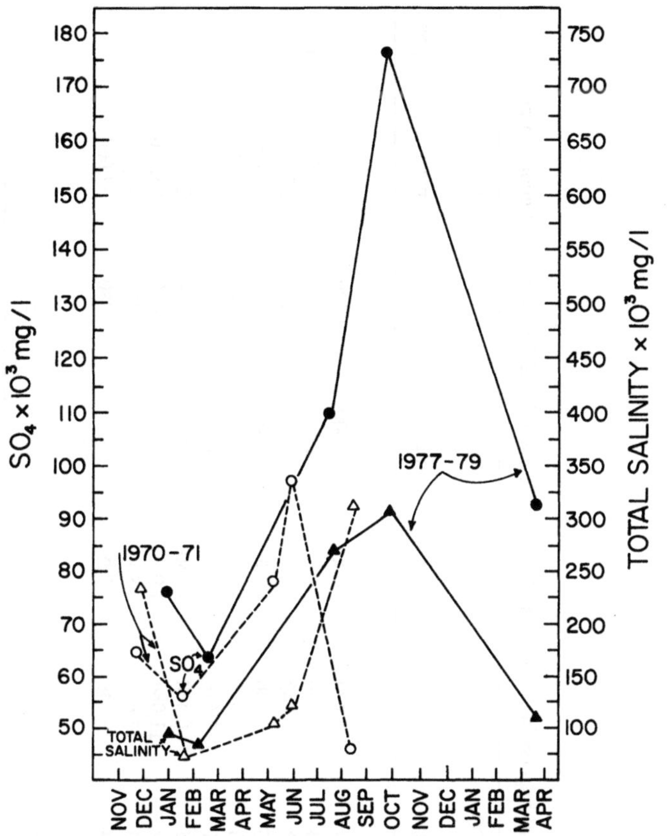

Fig. 10.7. Seasonal changes in salinity and sulfate concentration in water of Ras Muhammad sea-marginal pool. Notice sharp drop in sulfate concentration in August 1971. Further explanation in text. *Open circles* and *triangles* are data for 1970–1971; *filled circles* and *triangles* are for 1977–1979 (September 1978 and April 1979 data from G. M. Friedman and Foner 1982)

creased markedly over the period from February to July to September; the calcium concentration actually decreased as the salinity (and magnesium concentration) increased. The Mg/Ca ratio increased by nearly an order of magnitude from 2.4 (2.9/1.2), to 9.4 (10.2/1.1), to 19.8 (11.9/0.6) over the period (see data in Table 10.1). In April of 1978 the salinity had decreased to 121×10^3 mg/l and the Mg/Ca ratio had returned to 2.6 (4.4/1.7).

Figure 10.8, taken from G. M. Friedman et al. (1973a), illustrates seasonal changes on the geochemistry of waters of the hypersaline pool near Elat (sun pool). Comparison of these data will be made in the Sect. 10.3.

Table 10.1. Chemical composition of surface waters, sea-marginal hypersaline pool, Ras Muhammad, and of Red-Sea water contiguous to pool

Date	Salinity $\times 10^3$ mg/l	Cl mg/l	Br mg/l	HCO_3 mg/l	SO_4 mg/l	Ca mg/l	Mg mg/l	K mg/l	Na mg/l	Sr ppm
Hypersaline Pool										
November 28, 1970	230.7	135,800	445	354	6,420	3,300	7,000	2,400	75,000	
January 16, 1971	72.1	39,657	131	134	5,596	870	2,657	680	22,400	15
May 5, 1971	103.3	56,297	189	152	7,843	1,204	3,827	955	32,800	21
May 21, 1971	126.8	69,900		195	9,700	1,480	4,650	1,290	39,000	
August 8, 1971	314.3	189,070	424	573	4,569	5,300	11,953	3,740	98,701	
December 23, 1977	92.	50,765		181	7,576	1,375	3,326	890	28,100	
February 6, 1978	79.3	43,914		168	6,438	1,180	2,887	780	24,000	
July 15, 1978	263.2	83,050		422	11,084	1,088	10,185	3,000	83,050	
September 28, 1978[a]	304.2	172,548	530	249	17,653	800	11,868	3,500	97,299	
April 23, 1979[a]	121.0	66,713	221	184	9,449	1,708	4,400	1,280	37,000	
Red Sea Contiguous to Hypersaline Pool										
November 27, 1970	43.6	24,260	81	134	3,360	580	1,678	510	13,000	

Chemical Analyses by the Geological Survey of Israel
[a] Data from G. M. Friedman and H. A. Foner (1982)

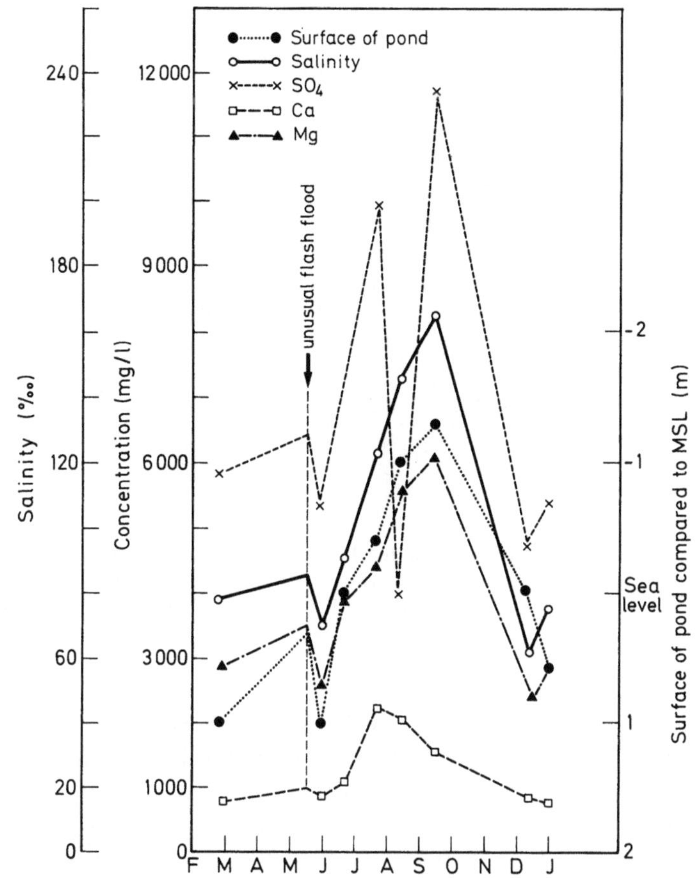

Fig. 10.8. Seasonal changes in level of "sun pool" near Elat and in chemical composition of surface water (salinity, sulfate, calcium, and magnesium concentration). Notice sharp drop in sulfate concentration in August followed by sharp increase in September. Data on level of pool and chemical composition in February, July, and September from Eckstein (1970) (Fig. 10.4 from G. M. Friedman et al. 1973 a)

10.2.3 Bottom and Subbottom Sediments of the Pool

Cores of subbottom sediments (Fig. 10.9) of the pool reveal alternating light and dark laminae; the light laminae are mostly composed of a meshwork of gypsum crystals, whereas the dark consist of the organic matter of cyanobacterial mats (Fig. 10.10). Also present, associated with the cyanobacterial laminae, are particles, crystals, and laminae of aragonite, high-magnesian calcite, and dolomite.

Gypsum. Light-colored laminae of gypsum ranging in thickness from 5 to 20 mm alternate with cyanobacterial laminae in the subbottom sediments. Gypsum crystals are lath-shaped and range in length from 1.5 mm to 15 mm. The surfaces of the crystals typically are slightly pitted, probably indicating partial dissolution during lower-salinity episodes. Crystals normally possess random orienta-

Fig. 10.9. Hypersaline pool *(foreground)* floored by microbial mats and covered by approximately 30 cm of water. Cores are being extracted for study of subbottom sediments (Fig. 10.10). Scarp and terraces *(background)* expose highly porous emergent Pleistocene reefrock. Sabkha occupies a belt between pool and reef

tion, but some intervals containing radially arranged crystal growths were noted (Fig. 10.11). Randomly orientated crystals exhibit a milky appearance, but those with radial fabrics are much clearer and their surfaces are less pitted. The euhedral crystal shape and clarity of the radial crystals may reflect more recent growth. The size of the gypsum crystals increases with depth below the floor of the pool; growth is optically continuous.

Microbial Mats. The dark laminae in the cores alternating with the gypsum layers are the cyanobacterial mats which floor the bottom of the pool (Fig. 10.5). The layers of microbial laminae range in thickness from only 1 to 65 mm, and are dark-brown to almost black. Carbonate laminae, ooids and grapestones occur within the mats. Detrital particles and authigenic crystals also are associated with the cyanobacterial laminae. Microbial mats deteriorate with progressive burial below the floor of the pool. This visual observation of the destruction of microbial mats below the sediment bottom is confirmed by analysis for organic matter (Table 10.2).

Examination of the mats under the binocular microscope during disaggregation of samples reveals that cyanobacterial filaments or sheaths are present to depths of nearly 0.5 m below the floor of the pool. The cyanobacterial fibers consist only of the sheaths of filamentous cyanobacteria. No cells were found within the sheaths; even 3 mm below the bottom of the pool cells have disintegrated and vanished (Fritch 1945). Hence, decomposition of the cyanobacterial cell material, excluding the sheaths, is quite rapid. Attached to the filaments, at all depths, are numerous deci- to centimicron-size carbonate skeletal particles that were trapped and bound by bacterial mucilage.

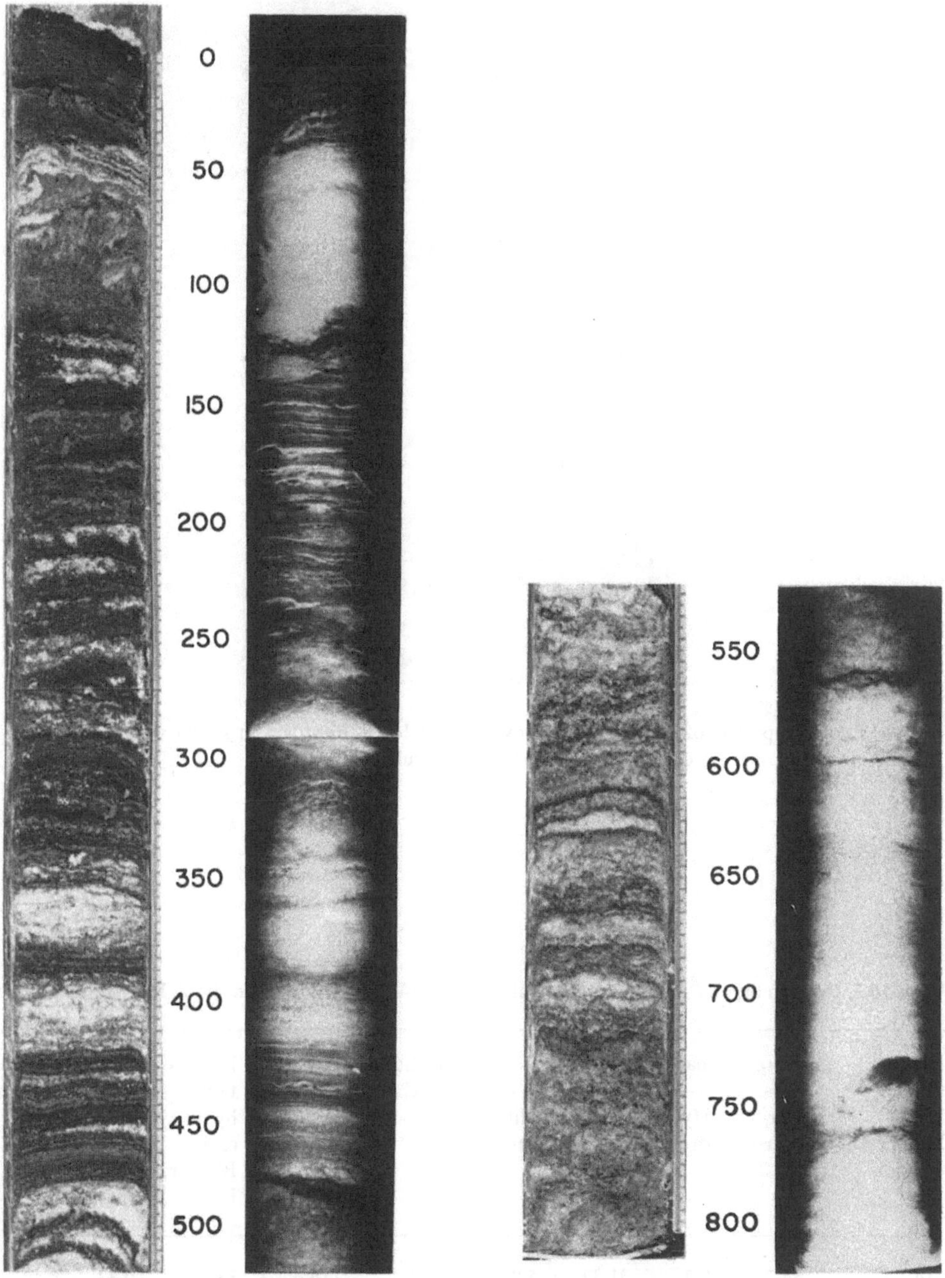

Fig. 10.10. Core of subbottom microbial mats and sediment from hypersaline pool. Two photographs are shown: a radiograph *(on right)* and a photograph of the same core after it was split open. Fine laminae (note especially in radiograph) are intercalated microbial mats *(black)* and gypsum *(gray to white)*. Core exhibits progressive destruction of microbial mats below floor of pond. Indicated *numbers* are depth in millimeter below floor of pool

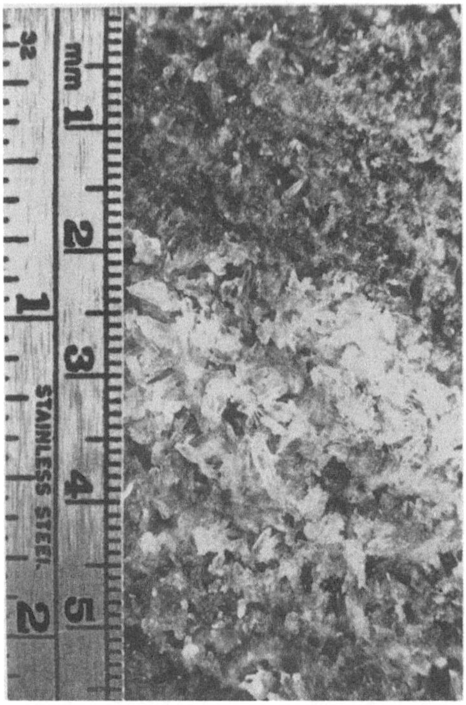

Fig. 10.11. Close-up view of detail of core shown in Fig. 10.10. Photograph shows mush of gypsum crystals. White crystals show subradial pattern

Table 10.2. Data on organic matter of microbial mats, sea-marginal hypersaline pool, Ras Muhammad

Depth (mm) below floor of pool	Element analysis, percent				Ratio H/C	Extractable organic Bbl/acre Ft.	Extractable hydrocarbon Bbl/acre Ft.	Extractable hydrocarbon / Extractable organic matter
	Carbon	Hydrogen	Oxygen	Nitrogen				
14– 35	64.7	8.2	21.3	5.8	1.51	114.6	38.6	0.33
40– 50	62.9	8.2	21.8	7.1	1.57	75.0	11.3	0.15
75– 90	64.4	8.3	20.8	6.5	1.54	65.5	2.0	0.03
340–350	65.6	8.4	19.5	6.5	1.54	96.1	21.3	0.22
575–585	66.6	8.5	18.8	6.1	1.54	38.5	1.2	0.03

Analyses by Research Center of Amoco Production Company

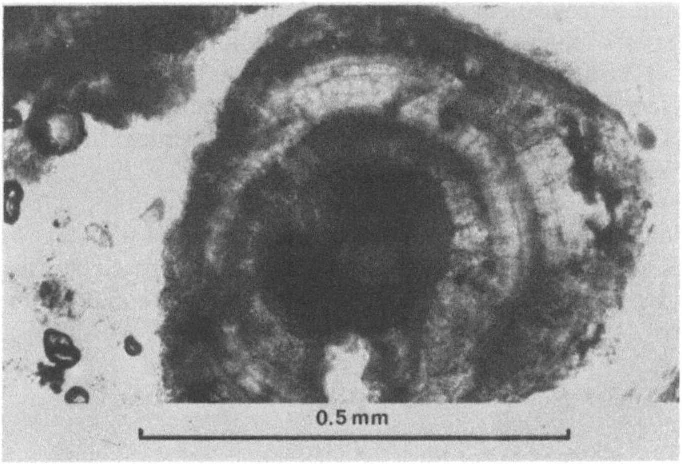

Fig. 10.12. Photomicrograph of thin section of ooid with cryptocrystalline nucleus and radial rim (from microbial mats, Fig. 10.10)

Carbonate Laminae, Ooids and Grapestones. Of particular interest are the carbonate laminae and ooids within the microbial mats. Laminae and ooids are composed of microcrystalline aragonite and, less commonly, high-magnesian calcite. Laminae range in thickness from about 0.1 to 1.5 mm. Thinner laminae are characteristically light-tan, whereas thicker ones are light-gray colored. Most laminae are continuous across cores, but some are discontinuous.

The ooids within the microbial mats are pearly white to tannish white in color, and range from 0.25 to 1 mm in diameter (Fig. 10.12). They are round to slightly ellipsoidal in cross-section and may form aggregates which tend to be flattened parallel to the bedding. Ooids are composed of dense, opaque, microcrystalline centers and transparent to opaque concentric and radial rims (Fig. 10.11). These ooids display apparently the pattern of calcified coccoid cells which Cyanobacteria or chemoorganotrophic bacteria degrading the cyanobacterial organic matter precipitate (Krumbein 1979 b, 1983 a). Ooids which are cemented together resemble the "lumps" of Illing (1954) and "grapestones" of Purdy (1963). These ooids are very similar to those from sun pool of the Gulf of Elat (G. M. Friedman et al. 1973 a, especially Fig. 13, p. 551, G. M. Friedman and Sanders 1978, Figs. 2–30, p. 54).

Ooids of the kind described in the preceding paragraph are the oncolites of some authors, a product precipitated by the microenvironment of cyanobacterial mats (Pia 1927, Johnson 1954, Endo 1961, G. M. Friedman et al. 1973 a, Kalkowsky 1908, Nadson 1903). In the studied pool ooids are always found within microbial mats. However, in the lower parts of the cores, where the mats have disintegrated, ooids occur within the mush of gypsum crystals. The presence of ooids and of broken-up chips of carbonate laminae in these lower, dark laminae of gypsum confirms the former presence of microbial mats, now destroyed.

Dolomite. Dolomite is present among gypsum laminae and microbial mats. The dolomite crystals are tiny, hence indistinguishable from other carbonate material under the petrographic microscope. Among the laminae, in the samples studied, the ratio of high-magnesian calcite to dolomite, based on height of the hkl 104 peak of X-ray diffractograms, is estimated at approximately 80:20. Thus, although present, dolomite is not an abundant constituent. Nadson (1903) reported already microbially formed dolomite upon disintegration of cyanobacterial mats (W.E. Krumbein, pers. comm.).

Halite. Halite is precipitated and deposited during the summer. At that time halite spreads across the cyanobacterial mats, and may be 40 cm thick. During the winter, however, refreshening of the water takes place and halite goes back into solution.

10.2.4 Isotopic Changes Below the Floor of the Pool

Changes in isotopic composition of carbonate sediments below the floor of the pool roughly correlate with increasing depth (Fig. 10.13; Table 10.3). The isotopes of carbon and, to a lesser extent, of oxygen, become progressively heavier with depth (see Chap. 19).

10.2.5 Sabkha Bordering the Pool

The sabkha surrounding the pool is underlain by sediment composed of sand-size particles cemented by halite, gypsum, anhydrite, and dolomite. Force of crystallization during precipitation of these minerals has raised the floor of the sabkha and created irregular hummocks and pits. In places a plant, *Halocnemum strobilaceum*, grows in abundance on the floor of the sabkha. The gastropod *Pirenella conica* is abundant in small channels crossing the sabkha. As mentioned above, because these gastropods eat cyanobacterial mats, the latter do not occur in the channels.

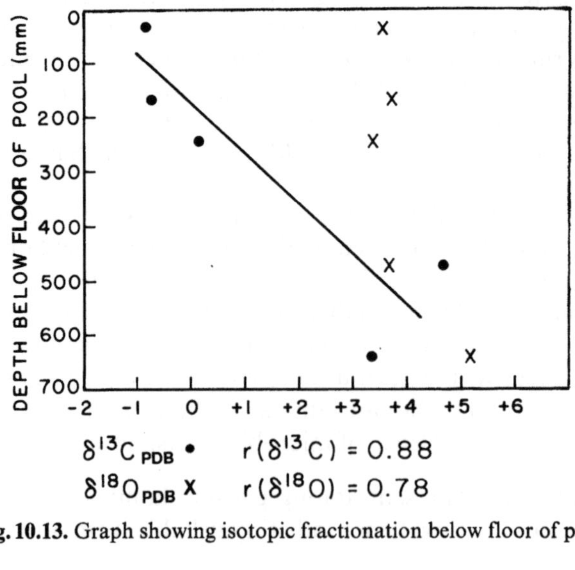

Fig. 10.13. Graph showing isotopic fractionation below floor of pool

Table 10.3. Isotopic fractionation of carbon and oxygen below floor of sea-marginal pool

Depth mm	Carbonate CO_2, % yield	$\delta[^{13}C]_{PDB}$	$\delta[^{18}O]_{PDB}$	Radiocarbon dates (BP[a])
22	67	−0.85	+3.53	26±327 years
170	63	−0.76	+3.72	
245	74	+0.16	+3.37	1,050±160 years
472	36	+4.68	+3.70	
640	62	+3.30	+5.20	
For comparison: sea-marginal pool of Gulf of Elat (Sun Lake), 195 km north of Ras Muhammad				
65	88	+2.52	+7.31	

[a] BP = before present

Stable isotope data by laboratory of I.R. Kaplan (University of California at Los Angeles). Radiocarbon dates by Mobil Research and Development Corporation (courtesy H. F. Nelson) and Krueger Enterprises

10.2.6 Emergent Pleistocene Reefs

The emergent Pleistocene reefs that underlie the pool are highly porous and permeable. Freshwater leaching of aragonite produced widespread moldic porosity. Effective porosity ranges from 50% to 64%, and permeability from 140 to over 10,000 md (Gvirtzman and G. M. Friedman 1977).

10.3 Discussion of Results

10.3.1 Water Level and Shoreline

Aerial photography and field visits have documented that the water level and configuration of the shoreline change with the seasons. These phenomena reflect the changing balance between evaporative loss of water from the pool surface and the amount of water seeping into the pool from the inlet via the intervening highly permeable reef rock.

10.3.2 Geochemistry of Surface Waters

The geochemistry of the pool waters also reflects the above-mentioned balance. When the pool becomes increasingly smaller (water level lower) evaporation is predominant and the salinity and concentrations of individual ions in the water increases.

At variance with this general trend are the sulfate-ion and calcium-ion concentrations in 1971 and the calcium-ion concentrations in 1978. In 1971 both the salinity and sulfate-ion concentrations, which increased from January to May, dropped sharply in August, while the salinity increased 2.5 times, and calcium-ion concentration 3.6 times. In 1978 the calcium-ion concentration decreased from

February to July and again from July to September, while the salinity increased nearly fourfold. Sulfate-ion concentrations apparently increased progressively over the period in 1978.

Study of the sun pool near Elat revealed somewhat similar anomalous trends (Fig. 10.8). Salinity increased progressively in samples taken in June, July, August and September; sulfate-ion concentration also increased from June to July, but dropped sharply in August, only to reach its highest level of the year in September. Calcium-ion concentration increased with salinity during June and July, but began to decrease in August and continued to decrease from August to September. The plunge in sulfate-ion concentration in August was explained (G. M. Friedman et al. 1973 a, p. 544) by noting that experimental work on seawater shows that gypsum precipitates from hypersaline water when its salinity exceeds 3.35 times that of normal seawater. Precipitation of gypsum was believed to have occurred, resulting in the reduction of sulfate-ion concentration in the water. The subsequent (September) increase to the highest sulfate-ion concentration of the year at even higher salinities, apparently without gypsum precipitation occurring to bring the ion concentration down, suggested that the salinity level of 3.35 times seawater was only one of the necessary factors for gypsum precipitation (G. M. Friedman et al. 1973 a, pp. 545, 546). The slight calcium-ion concentration decrease in August was ascribed to gypsum formation as the sulfate-ion concentration plunged. Subsequent calcium-ion concentration decrease was considered to have resulted from carbonate formation.

Now let us consider the observed variations in ion concentrations in the Ras Muhammad pool in the light of those previously described from sun pool near Elat. In the summer of 1971 the sharp decrease in sulfate-ion concentration while the pool water was becoming increasingly hypersaline is very similar to the phenomenon observed in sun pool. No sample was taken in September, so we do not know if the sulfate-ion concentration rebounded and if salinity continued to increase. In the Ras Muhammad pool, however, the calcium-ion concentration increased sharply, outpacing the salinity increase when the sulfate-ion concentration plunged. Gypsum precipitation alone cannot fully account for these data. An explosion in the activity of sulfate-reducing bacteria might be invoked to account for the sharp decrease in sulfate-ion concentration. The bacteria might have removed the sulfate ions directly from solution, as suggested by G. M. Friedman (1972) according to the equation (G. M. Friedman and Sanders 1978, p. 130):

$$\underset{}{SO_4^{2-}} + \underset{\substack{\text{organic}\\\text{material}}}{2CH_2O} = \underset{\text{gas}}{H_2S} + 2HCO_3^- . \qquad (1)$$

Organic matter is present and no one who has worked in these pools can avoid noticing the presence of H_2S gas. When gypsum, either newly precipitated or in the bottom sediments, is first attacked by sulfate-reducing bacteria, calcium and sulfate ions are produced according to the equation:

$$CaSO_4 = Ca^{2+} + SO_4^{2-} . \qquad (2)$$

Subsequent reduction of the sulfate by the bacteria would leave the calcium ions either to form carbonates or, if for some reason conditions were temporarily unsuitable for carbonate formation, to increase the calcium-ion concentration of the pond waters.

It is tentatively suggested that in August of 1971 sulfate-reducing bacteria removed sulfate ions directly from the pool waters and attacked gypsum bottom sediments as well. Gypsum precipitation may or may not have occurred but, if it did, previously deposited gypsum must have been attacked by the bacteria as well to account for a net increase in calcium-ion concentration in the pool water. Carbonate formation was inhibited for some unknown reason, allowing the calcium-ion concentration in the pool waters to rise faster than the salinity. In 1978 both salinity and sulfate-ion concentration increased from February to July and from July to September. No rapid sulfate-ion concentration decrease was detected. In the light of the experience at sun pool, however, we know that it is possible for the sulfate-ion concentration in these hypersaline pools to drop sharply in one month (e.g., July to August), and rebound to an even higher level after a second month (August to Sepember, Fig. 10.8). We cannot rule out the possibility that the sulfate-ion concentration dropped sharply in the summer of 1978 but was not detected by the sampling. If it did, it quickly rebounded to its highest recorded level in September. The decreasing calcium-ion concentration from February to July to August 1978 is unlike previously measured trends; neither the Ras Muhammad pool in 1971 nor the sun pool near Elat showed similar trends. In the normal course of events as the salinity of the pool increases, the calcium-ion concentration would also increase. The calcium ions must have been progressively removed during the 1978 season, presumably by mineral formation. Gypsum is not a likely candidate because of progressively increasing sulfate-ion concentration.

We must believe that carbonates were precipitating. Microorganisms in the mats could precipitate carbonates from the hypersaline waters (Oppenheimer and Master 1965, Krumbein 1979a). Work by G. M. Friedman and Foner (1982) using cyanobacterial mats from the Ras Muhammad pool indicates that the pH fluctuation mechanism of carbonate precipitation proposed by Oppenheimer and Master (1965) does not occur in this case; the CO_2 generation by photosynthesis, which in Oppenheimer and Master's experiment caused the pH shift, is probably offset by H_2S production in the pool. Krumbein (1979a) has shown that the organotrophic bacteria, which live on the remainders of cyanobacteria do precipitate ooids and carbonate needles (both calcite and aragonite). By whatever mechanism mentioned above the geochemistry of the waters suggests that carbonate precipitation took place. The increasing Mg/Ca ratio over the course of the summer and fall implies that the carbonates produced were progressively more magnesium-rich, as the ratio reached nearly 20:1 (12:1 mol/l).

10.3.3 Bottom and Subbottom Sediments

Just as the water level and the geochemistry of the pool water change seasonally, so does the sedimentation on the bottom of the pool. Growth of the cyanobacteria predominates in winter and spring when the pool waters are less hypersaline. At that time the microorganisms operate at two levels, (1) binding and trapping of detrital particles and authigenic crystals and (2) precipitation of carbonate laminae and ooids. Buried in the subbottom sediments the microbial layers are dark-colored and undergo progressive deterioration with depth of burial. Disintegration of cells within the filaments occurs very quickly after burial as evidenced by their disappearance only 3 mm below the surface. Krumbein and Swart (1983) have made a model comparison of this process. Filament or sheath disintegration is slower; they survive to nearly 0.5 m below the floor of the pool. Finally all trace of the actual cyanobacterial material is gone with only broken carbonate laminae and ooids between the gypsum layers left to attest to the former presence of the microbial mats. Kazmierczak and Krumbein (1983 a, b) made a comparison of the modern material with ancient stromatolites.

During periods of highest hypersalinity during late summer and fall the microbial mats are relatively inactive, microbial growth is essentially reduced and restricted to the layers below the gypsum crusts. At this time a layer of gypsum (light-colored crystals), 5 to 20 mm thick with whatever carbonates have accumulated, may bury the mats. Alternatively, or in another part of the pool, the mats may be covered with halite and/or be subaerially exposed and desiccated. In the following winter and spring the water level rises and the salinity drops, dissolving whatever halite was deposited, and microbial growth resumes.

In the subbottom sediments gypsum crystals increase in size with depth of burial. This increase in crystal size suggests continuing postburial precipitation in which the original crystals serve as nuclei for further crystal growth. Continued precipitation and growth are in crystallographic continuity with the original gypsum crystals.

It bears repeating that the type of ooids that are associated with the microbial mats in the Ras Muhammad pool are the oncolites of some authors, a product precipitated in the environment of cyanobacteria (Pia 1927, Johnson 1954, Endo 1961, G. M. Friedman et al. 1973 a, Nadson 1903, Kalkowsky 1908, Krumbein 1979 a, b). As explained above, the mechanism by which the carbonates are precipitated is probably not by pH fluctuation associated with CO_2 generation of microorganisms during photosynthesis. We think, however, that the laboratory experiments of Nadson (1903) and Krumbein (1979 a, b) give sufficient proof of the bacterial origin.

Dolomite, which did not occur associated with the microbial mats in Solar Lake, but did occur with gypsum there, here occurs with the interlayered cyanobacterial mats, carbonate and gypsum sediments, giving further evidence to Nadson's findings. It is presumably syngenetic with its formation enhanced by periodic high Mg/Ca-ion concentration ratios in the pond waters. Although only a small percentage of the total carbonate present, this dolomite could be the first step in the dolomitization of the microbial laminated sediments when they become stromatolitic rock strata. Isotopic change appears to be one of several processes that occur actively below the floor of the pool and accompany gypsum precipitation. Although degradation of organic matter below the floor of the pool commonly leads to a decrease in δ-^{13}C concentration, here δ-^{13}C concentration increases with depth. This increase may be explained as follows: CO_2 formed by the decomposition of organic matter produces a residuum of bicarbonate which is enriched with δ-^{13}C. Such bicarbonate produces carbonate enriched with the heavier isotope (Nissenbaum et al. 1972, see also Chap. 19). With progressive burial, as $CaCO_3$ continues to be precipitated from interstitial waters, isotopically heavier carbonate is added to original carbonate present, resulting in enrichment of δ-^{13}C in the total carbonate fraction. Thus, with increasing distance below the floor of the pool, δ-^{13}C increases. Also the closed system effect described by Schidlowski et al. (Chap. 19) increases the δ-^{13}C of the organic matter and carbonates.

The explanation for the changes in δ-^{18}O below the water/sediment interface is not definitive and although a linear relationship between depth below the floor of the pool and δ-^{18}O may be demonstrated, no satisfactory explanation can be offered for the observed isotopic changes.

10.4 Conclusions

1. Seasonal changes in water level and size of the pool occur which are reflected in the geochemistry of the water and in the sediments accumulating on the bottom. In winter and early spring the water is high and is relatively low in salinity, supporting abundant microbial mat growth and carbonate accumulation. In spring through late fall the water level falls, salinity increases, and microbial growth is reduced and carbonate formation gives way to the formation of evaporites (gypsum and halite).
2. The fluctuation of sulfate-ion concentrations in August 1971 is probably due to a combination of gypsum precipitation and the activity of sulfate-reducing bacteria on sulfate-ions in solution and in both newly deposited and older gypsum on the floor of the pool. The increase of calcium-ion concentration suggests that carbonate formation was inhibited at that time.
3. Decrease in calcium-ion concentration throughout summer and fall of 1978 suggests that carbonates were forming throughout the period.
4. Upon burial, cells in microbial mats disintegrate almost immediately followed by disintegration of the cyanobacterial filaments (after burial beyond 0.5 m). Only broken carbonate laminae and ooids in lamellae between the gypsum attest to the former presence of the mats. Gypsum crystals increase in size with increasing depth of burial.
5. Progressive isotope fractionation of carbon and oxygen isotopes with depth of burial below the floor of the pool (both become enriched with the heavier isotopes) implies an active ion exchange between the interstitial waters and the subbottom sediments.
6. A small percentage of the carbonate in the microbial laminae and gypsum layers accumulating is dolomite. This dolomite, presumably syngenetic, would be the first step in the dolomitization of these microbial laminated sediments when they become stromatolitic rock strata.

11. Salinity and Water Activity Related Zonation of Microbial Communities and Potential Stromatolites of the Gavish Sabkha

GISELA GERDES, WOLFGANG E. KRUMBEIN, and ELISABETH HOLTKAMP

11.1 Introduction

Laminated microbial mats are found in a variety of extreme environments, including hypersaline ecosystems, thermal hot springs, deserts, and even permanently ice-covered lakes in the Antarctic (Bauld 1981 a, b, Borowitzka 1981, Brock 1978, Castenholz 1973, Y. Cohen et al. 1977 b, c, Horodyski 1977 a, Javor 1979, Krumbein et al. 1977, Krumbein and Lange-Giele 1979, Parker et al. 1981, Wright and Burton 1981).

Hypersaline lagoons and lakes in arid areas are subject to a complex interplay of physical and chemical factors. The concentration of sodium chloride, other salts, heavy metals and nutrients often reach unusually high levels (Dexter-Dyer et al. 1984, Renfro 1974). Temperature and pH fluctuate erratically within wide ranges. Extremely low redox potentials, oxygen concentrations and water activity are frequently encountered. The light intensity is also exceedingly high. Some of these systems are subject to long periods of total dryness.

In spite of these harsh conditions, or perhaps because of them, extremely thick laminated microbial mats can develop. Biological processes leading to the diagenetic transformation into stromatolites can be observed in such mats (Brock 1976, Eugster and Hardie 1975, Handford et al. 1982, Purser 1973).

A definition of stromatolites and living microbial systems which may become stromatolites has been proposed by Krumbein (1983 a, b):

"Stromatolites are laminated rocks, the origin of which can clearly be related to the activity of microbial communities, which by their morphology, physiology and arrangement in time and space interact with the physical environment to produce a laminated pattern, which is retained in the final rock structure."

Unconsolidated laminated systems, clearly related to the activity of microbial communities, often called "recent stromatolites" or "living stromatolites", are termed potential stromatolites.

The Gavish Sabkha is a hypersaline environment with salinities of up to 340‰. It harbors an extremely diversified ecosystem centered on various cyanobacterial communities (Krumbein et al. 1979 a, Potts 1980, Potts and Krumbein 1978). High vertical variation in microbial distribution and productivity was found in particularly thick laminated microbial mats. In these mats, H_2S and O_2 were found in each other's presence (Krumbein 1978 b, Krumbein et al. 1979 a). The Gavish Sabkha was found to be an excellent model of a biologically controlled hypersaline ecosystem with unusual mineral assemblages (Gavish 1975a, 1980, Gavish et al. 1978; see also Chap. 9). Studying such hypersaline microbial ecosystems could help solve questions of how heavy metals, phosphate and hydrocarbons are spatially accumulated (G. M. Friedman 1980, Gill 1977, Dexter-Dyer et al. 1984, Krajewski 1981, Krumbein 1982, Krumbein et al. 1977, Renfro 1974). Such environments are also a commercially important source of salt (J. S. Davis 1978, Schneider 1979, Schneider and Herrmann 1980). Chapter 19 deals in some detail with the applied and economic questions of sabkhas.

In this chapter we describe the variation and distribution of microbial communities (potential stromatolites) in the Gavish Sabkha. We related our findings to salinity and water availability. We discuss our findings from an ecological and paleoecological viewpoint.

11.2 Materials and Methods

We made field observations and collected samples in the Gavish Sabkha from 1977 to 1982. In this chapter we present data collected before the big sheetfloods during the winters of 1979–80 and 1980–81.

We divided the Gavish Sabkha into ten visibly distinguishable zones. These zones had remained relatively stable for the 15-year period following the sheetfloods of 1964–66. We observed and categorized the micro- and macrobiota of these zones during the various seasons.

11.2.1 Sampling Procedures

We collected samples of surface and interstitial water for nutrient analysis.

We collected samples of sediment both by scraping the surface with a spoon and taking cores with a corer 50 cm long and 5 cm in diameter.

For micromorphological identification, isolation of cyanobacteria and scanning electron microscopy we collected samples of the mat material with a corer 7 cm long and 10 mm in diameter. Immediately after collection, we fixed mat material for scanning electron microscopy with 4% (aq.) glutaraldehyde, diluted with water from the sampling site.

11.2.2 Measurements

Special light conditions and light penetration of specific sections of the visible spectrum were measured in both the field and the laboratory with a quantaspectrometer (Umea Instruments, Sweden). Schott interference filters were mounted above dissected mats, using either daylight or movie camera halogen lamps.

Temperature and salinity were measured with classical methods, using both mercury and resistance thermometers, and chlorinity titration as well as refractometry (Amer. Optical Inc. Corp.).

Productivity was measured with the oxygen exchange method as well as with the ^{14}C-incorporation method (Krumbein 1979a). Productivity and oxygen evolution were also measured with the bell jar method described in Krumbein et al. (1979a).

Phosphorus, nitrate, and ammonia were measured by the methods of Strickland and Parsons (1968), Goltermann (1969), and Sorokin and Kadota (1972), respectively and presented in Table 11.1.

pH and redox potential were measured with a field pH meter (Knick-Portamess 902), Ingold electrodes and a special platinum multiple microelectrode kindly provided by B. B. Jorgensen. Measurements were taken every 15 min during two 24-h periods.

11.2.3 Identification of Microorganisms

Microorganisms were identified according to the generic assignment suggested by Rippka et al. (1979), and Krumbein et al. (1979b). The taxonomic keys of Geitler (1932) and Desikachary (1959), and descriptions given by Potts (1980) and Ehrlich and Dor (Chap. 14), were also used. Classical botanical names of the cyanobacteria are given in parentheses.

We could correlate all the cyanobacteria we isolated with their morphology in fresh microscopic preparations from mat samples, even though there were considerable morphological differences between laboratory-cultured and naturally occurring hypersaline organisms due to salinity, light and nutrient levels. In the natural environment, considerable differences in morphology were seen. These differences are not completely described by the "hyperspace" theory of Golubic (1979), since the different morphologies of one single species in laboratory culture would represent at least three previously described genera.

The genus names *Entophysalis* or *Eoentophysalis* are often found in literature on stromatolites and potential stromatolites (Elenkin and Danilov 1916, Golubic 1976a, b, Golubic 1980, Horodyski et al. 1977, Knoll and Golubic 1979, Mendelsohn and Schopf 1982, Potts 1980). We do not use these genus names because they do not correspond to a clearly distinguishable taxonomic form. Furthermore, pleurocapsalean and chroococcacean cyanobacteria can easily be mistaken for *Entophysalis,* since the borderlines between these groups and *Entophysalis* cannot be easily seen in field samples, even less so in fossil stromatolites (see Chap. 21). Therefore we adopted the group name pleurocapsalean or the genus name *Gloeocapsa* instead.

11.3 Results

11.3.1 Biological Zones of the Gavish Sabkha

The Gavish Sabkha morphology has been described in Chap. 9. This circular depression (Fig. 11.1) is divided into a central elevated mound and a surrounding circular channel. Permanent water cover is restricted to a crescent-shaped lagoon in the northwestern part of the channel. The water depth of the lagoon is up to

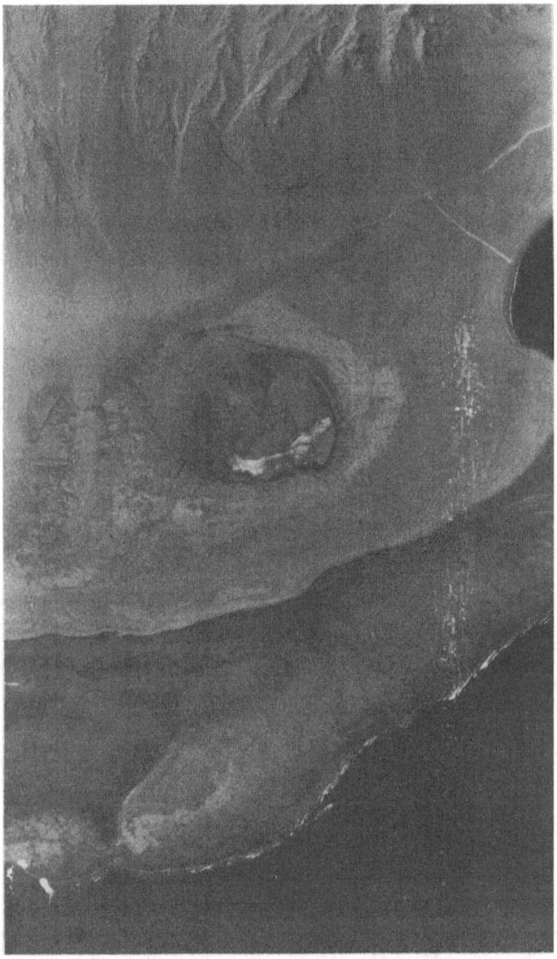

Fig. 11.1. Location of the Gavish Sabkha on a coastal plain between the shore of the Gulf of Aqaba (*bottom*, left) and the Sinai mountains (magmatites and metamorphites). The irregularly shaped morphology south of the Sabkha is built by ancient coral reefs. Recent reefs with surf line are visible at the *bottom*

70 cm. Seawater springs feed the lagoon at the seaward side of the depression. Several large springs have created gullies up to 40 m long and 2 m wide. In these gullies, water flows permanently from fall to spring and occasionally during high tides in the summer.

The tidal water table fluctuates and wets the sediment surface in the eastern parts of the channel. In times of extremely high tides, groundwater wets the sediment throughout the entire channel.

Daily variations in light, temperature and oxygen concentration in the lagoon are shown in Fig. 11.2. Salinity in the open waters of the lagoon ranges from 120‰ to 300‰. Evaporitic crusts are found along the inner and outer shores and on the floor of the lagoon.

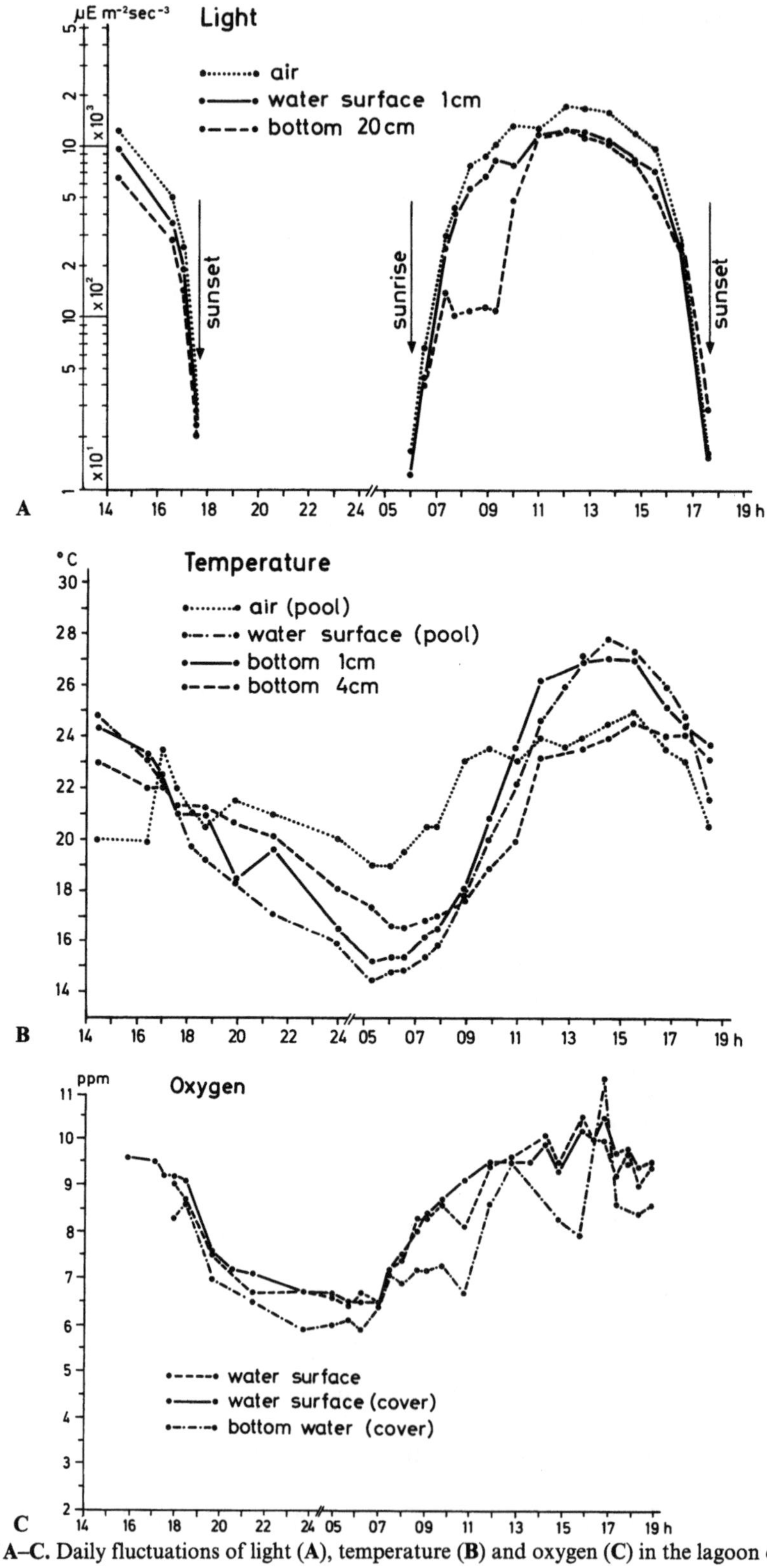

Fig. 11.2 A–C. Daily fluctuations of light (**A**), temperature (**B**) and oxygen (**C**) in the lagoon of the GS

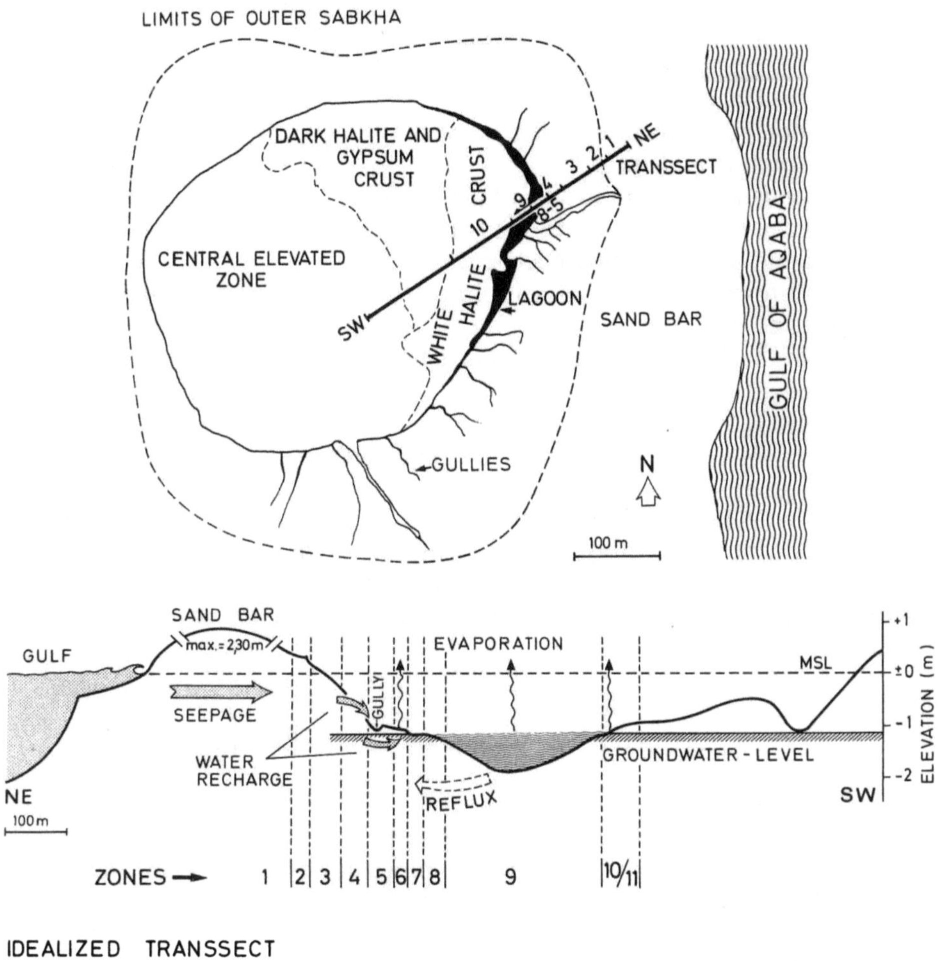

Fig. 11.3. Schematic horizontal transsect through the Gavish Sabkha with 11 zones. *1* Sand bar; *2* Carbonate cementation zone; *3* Gypsum-halite crusts; *4 Bledius*-zone; *5 Pirenella-Enteromorpha*-zone; *6 Pirenella*-zone; *7* Nodule-zone; *8* Cyanobacterial mats I (seasonally air exposed); *9* Cyanobacterial mats II (permanently water-covered); *10* Salt-crust embedded cyanobacterial mats; *11* Bedouin salt pans

The evaporation rate is 4.0 to 5.9 m/yr. This influences the hydrodynamic system and results in a concentration of nutrients and high productivity. This leads to the formation of laminated microbial ecosystems, or potential stromatolites. These systems accumulate organic and inorganic compounds and metals which would otherwise be washed out with the residual brine.

The Gavish Sabkha can be divided into ten different zones (Fig. 11.3). This zonation differs slightly from the zonations presented in the chapters on geomorphology and geochemistry (Chap. 9) and zoology (Chap. 15).

Our zonation reflects the influences of several diverse forces:

1. geomorphological and tectonic setting,
2. sediment composition and sedimentation processes,
3. seawater seepage,
4. evaporation of ground and surface water,
5. solar irradiation and energy,
6. hydrochemistry of ground and surface water,
7. seasonal change in water cover and water activity,
8. biogenic and abiogenic physicochemical gradients,
9. biological forces such as competition, mutual exclusion or grazing stress by animals, and
10. activity of the Bedouins, in particular their salt and potash industry.

11.3.1.1 Zone 1: The Sandbar

The sandbar consists of terrigenous and marine sediments. Cyanobacteria colonize the sediment surface only briefly during periods of rain or storm flooding. A cemented layer is found between 30 and 180 cm below the sediment surface. It is formed by the precipitation of carbonates and gypsum from seawater seeping into the Gavish Sabkha.

11.3.1.2 Zone 2: The Carbonate Cementation Plate

On the upper slope of the depression, cemented carbonate banks from plates at sea level. Seawater seeps under these plates, emerges through springs and flows down gullies into the lagoon. In contrast to biogenic beach rocks and endolithic crusts, these plates are not of biological origin. They are not readily colonized by epilithic or endolithic microbial communities because they are only partially covered by sand. Rainfall and humidity are also too low throughout the year.

11.3.1.3 Zone 3: The Dry Gypsum/Halite Crusts with Petee Structures

This zone is covered by uneven and hard gypsum/halite crusts which show many cracks and cavities. Petee structures, described and defined in Chap. 9, are found throughout this zone. The surface becomes wet from below during extremely high tides, and from above during storms and floods. A microbial mat then develops in the sediment. The short-lived mat community creates a typical surface pattern by consolidating the sediment and reducing wind erosion. We have occasionally seen how wind shifts and folds these mats like tablecloths and forms ripple-like structures on the surface. Similar structures are produced where such mats glide downslope. Petee structures are further modified by the formation of gas bubbles and gas domes underneath the mat. When these mats dry, hardly any sign is left of them, although the surface is smoother where they had been. These biogenic, wavy, upward convex and sometimes pyramidal structures are called Petee structures, in contrast to the abiogenic Tepee structures. Since periods of moisture are relatively rare, this zone is usually considered to be devoid of life, except for occasional spiders and the few remaining mat remnants.

11.3.1.4 Zone 4: The *Bledius* Zone

Percolating seawater reaches the surface and forms dark wet patches. These patches are covered with numerous sandy pellets, colonized by coccoid and filamentous cyanobacteria (Fig. 11.4). The little granulated sediment pellets are produced by staphilinid beetles (*Bledius capra* and *Bledius angustus*), which are highly abundant where the sediment becomes moister. Cyanobacteria and diatoms form a dense endobenthic mat between 2 and 4 mm below the surface. The cyanobacteria belong to the genera *Gloeocapsa, Chroococcodiopsis, Synechococcus, Spirulina, Phormidium* and *Microcoleus*. The dominant diatom genera are *Rhopalodia, Nitzschia* and *Mastogloia*. During the summer, parts of this zone are covered with halite crusts. The cyanobacteria migrate upward and form dense colonies below the crusts. The species composition does not vary seasonally. Coccoid forms predominate under moist conditions, whereas filamentous forms predominate under drier conditions. *Microcoleus* is particularly resistant to drying, since its trichomes are well protected by thick multilayered sheaths. *Spirulina* seems to be less tolerant of lower water potential. Microbial productivity increases toward the gullies, where water is more available.

A sulfate-reducing layer is found below the compact photosynthetic mat. It is interrupted in places by *Bledius* burrows, through which oxygen penetrates (Fig. 11.5). The aeration chimneys of *Bledius angustus* are colonized by cyanobacteria, diatoms and aerobic chemoorganotrophic bacteria (Fig. 11.6).

Fig. 11.4. Irregular surface crust (Zone 4). Excavation pellets of *Bledius* form the granular surface on top of the sediment, solidified by cyanobacterial endobenthic mats

Fig. 11.5. Exposition of deeper layers below the surface, shown in Fig. 11.4. *From left to right*: Original surface, cyanobacterial layer, and sulfate-reducing zone with oxidation patterns around *Bledius* burrows

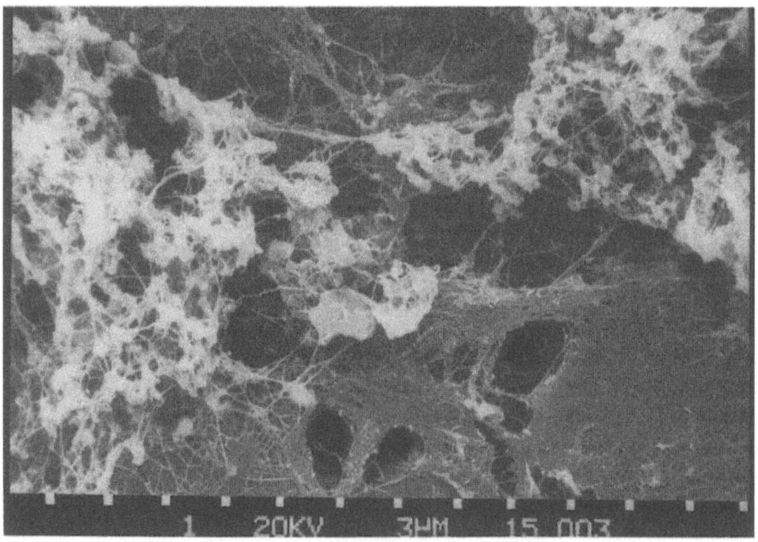

Fig. 11.6. Colonization of an aeration chimney of *Bledius* by microorganisms (SEM photomicrograph); distance between white bars = 3 µm)

11.3.1.5 Zone 5: The *Pirenella-Enteromorpha* Zone

Green algae (*Enteromorpha* and other genera) are dominant near the seawater springs on the eastern slope of the Gavish Sabkha and in small water outlets which form in the gullies during winter (Fig. 11.11). In this zone, salinity ranges from 50‰ to 60‰. *Enteromorpha* and other green algae apparently cannot actively grow when salinity is higher than 70‰. Cyanobacteria occur almost entirely as epiphytes (*Johannesbaptistia, Spirulina* and other genera). Diatoms are also common (*Nitzschia, Mastogloia, Achnanthes*). Gastropods (*Pirenella conica*) colonize the *Enteromorpha* mats in high numbers. They feed on the epiphytic microorganisms and not on the green algae themselves.

11.3.1.6 Zone 6: The *Pirenella* Zone Without *Enteromorpha*

As the salinity increases to over 70‰, *Enteromorpha* is gradually excluded. The microbial community, however, becomes more diverse. We have observed *Mastogloia* (dominant), *Nitzschia,* LPP forms of cyanobacteria (*Phormidium* and *Microcoleus*), *Oscillatoria, Spirulina, Synechococcus, Pleurocapsa, Gloeothece, Chromatium, Thiocapsa* and *Beggiatoa*. These organisms form an irregularly stratified system. *Pirenella* plows through these aggregates and grazes on them (Fig. 11.7). Analysis of the fecal pellets revealed that many microbes pass through the nail's gut undigested (Gerdes and Krumbein 1984). *Thiocapsa* is apparently toxic and is excreted in slimy strings before it is digested.

The zone receives seawater only from seepage. The highly saline brine of the lagoon rarely if ever reaches the zone, since it is held back by a row of mounds of laminated microbial mats. These mounds are seasonally exposed to the air and give the impression of "microbial reefs". Salinity ranges from 50‰ to 120‰, and is influenced by evaporation of the shallow water. As the salinity increases, *Pirenella* gradually becomes excluded, and empty shells become more common.

Fig. 11.7. *Pirenella* zone (Zone 6). Grazing trails of *Pirenella* increase toward decaying *Enteromorpha at the top*. Accumulation of shells *at the bottom*. The three dark leafs are sea weeds, blown into the Gavish Sabkha from the litoral area of the Gulf

11.3.1.7 Zone 7: The Nodule Zone

In this zone, salinity ranges from 65‰ to 150‰, and is higher in summer than in winter. Through the year the zone is fed by a trickling film of seawater from the seawater springs. In winter, highly saline water from the lagoon reaches this zone. However, since seawater seepage is more efficient in winter, there is a complex equilibrium of salt-crust formation and desintegration. Because of this, salinity and nutrient concentrations are optimal for cyanobacterial growth in the fall, winter and spring. In the summer, evaporation lowers the level of the lagoon by 2 to 3 cm, which leads to the formation of a 2 to 3-cm-thick salt crust. This crust contains laminated tabular microbial systems.

As salinity and water depth further increase, the sediment surface begins to show increasing numbers of rigid, granular aggregates of cyanobacteria (Figs. 11.8 and 11.12). The cyanobacterial aggregates form cauliflower-shaped structures up to 2 cm in diameter which resemble fossilized oncoids or oncoidal stromatolites (Krumbein and Cohen 1974, Friedman 1978).

The bulk of the nodules is made up of *Pleurocapsa* (the dominant species), *Myxosarcina* and *Chroococcodiopsis*. *Gloeocapsa, Synechococcus (Aphanocapsa),*

Fig. 11.8. Nodule zone (Zone 7). The oncoid-shaped, dark cyanobacterial nodules form a characteristic rim between Zones 6 and 8. The size, density and cover of the nodules increase toward higher salinity (indicated *at the right* by the formation of salt crusts). The nodular salt crusts *in the center* and close to the *cigarette box* indicate the mode of formation of the nodules

Synechocystis (Aphanothece), *Spirulina* and LPP forms *(Lyngbya, Microcoleus, Phormidium* and *Plectonema)* are also present. Obligate anoxygenic photosynthetic bacteria (Chlorobiaceae, Chloroflexaceae, *Chromatium* and *Thiocapsa*) can be found near the center of the nodules. Spirillae, *Beggiatoa* and budding bacteria can also be seen. Some of the nodules (geologically and after diagenesis one would talk of oncoids) are also formed around the numerous empty houses of *Pirenella conica*.

The size, density and numbers of these nodules increase with salinity and water depth (at right in Fig. 11.8). They are smallest closer to the *Pirenella-Enteromorpha* zone. The larger nodules gradually give way to continuous tabular mats toward areas covered by brine for most of the year. Higher salinity tends to favor filamentous forms. As salinity further increases, the initial stages of horizontally extended laminated systems begin to take place instead of nodule formation. In these intermediate mats, coccoid cyanobacteria are found near the surface, while filamentous forms are found in the lower layers.

How nodules forms is not yet clearly known. It is possible that they form around the grazing remainders and feces of *Pirenella* after the snail dies off in the summer because of increasing salinity. With grazing stress removed, cyanobacteria would develop in concentric layers around the detrital grains. Another possibility is that when the snails die off in summer, the sediment surface is no longer held together by their slimy feces and is subject to mechanical disruption by wind or wave action. Nodules could then form from particles of the disrupted surface growth. A third possibility is that nodules form around fragments of thick summer salt crusts which disintegrate because of lower salinity in winter. Cyanobacteria would grow around the fragments and anoxygenic phototrophs would gradually replace the salt inside the forming nodule. In this way, concentrically laminated microbial nodules would be formed and may become embedded in the salt crusts formed during the next summer before they are totally degraded.

Formation of nodules or similar spherical aggregations of microbes has been observed not only at the Gavish Sabkha, but also at the Ras Muhammed Peninsula, at the mangal of the Sinai, at Mormona Bay, in hot springs in Oregon and Yellowstone Park, and at Mare's Egg Lake in Oregon. Nodule formation seems always to be related to "schizoecological" situations in which abiotic and biotic conditions are continually shifting.

11.3.1.8 Zone 8: The Seasonally Air-exposed Cyanobacterial Mats

Dome-shaped laminated microbial mats form a small but distinct bioherm between the nodule zone and the brine-filled lagoon (Figs. 11.9 and 11.13). In summer, the mats are 5 to 8 mm thick. In winter, the mats, covered with water, are actively growing and increase in thickness to around 10 to 13 mm. Salinity in this zone can be as high as 250‰ in summer, or as low as 180‰ in winter.

The top layer consists mainly of *Synechocystis, Synechococcus (Aphanocapsa)*, interspered with pleurocapsalean colonies. This layer produces a thick covering of slime which dries out and becomes tough and leathery in summer. The cyanobacteria produce large amounts of carotenoids which protect the mat from

Fig. 11.9. Partially air-exposed laminated microbial mats (Zone 8) form domal-shaped mounds. *Arrow* indicates the sampling site of 1-cm diameter cores for [^{14}C] incorporation productivity experiments

solar radiation and give the mat a bright red color (Fig. 11.13). Chlorophyll *a* is degraded and as a result, the concentration of pheophytin *a* can be 20 times that of chlorophyll *a*. When covered with water or shaded in field experiments, the coccoid cyanobacteria, which are still alive, quickly synthesize chlorophylls and phycobiliproteins. In winter, and where the lower slopes of the domes dip into the water of the lagoon, less carotenoids are formed, and the mat is colored green.

The second layer is made up of *Pleurocapsa* and *Gloeothece*. The third layer consists mainly of LPP forms (*Microcoleus, Hydrocoleus* and *Hydrocoleum*). The fourth layer is bright pink and is formed by a large phototrophic anoxygenic, though at least microaerophilic, bacterium (Fig. 11.10) tentatively identified as a member of the *Thiocapsa* group. The fifth layer is made up of very thin LPP forms (*Schizothrix* and *Phormidium hendersonii*). This layer produces large amounts of oxygen. Photosynthetic flexibacteria are also found within the mat. When the salinity is low, a few diatoms (*Nitschia* and *Amphora*) can be found as well. A layer of purple bacteria and green photosynthetic bacteria usually lies immediately above the black sulfate-reducing zone.

This zone tapers into the Zone 9. The borderline is rather sharp, however, due to the slope and the surf, with waves up to 5 cm high. The slope is formed by the accumulation of sediment washed down toward the mounds from the nodule zone.

11.3.1.9 Zone 9: The Permanently Water-covered Cyanobacterial Mats

The crescent-shaped lagoon is up to 70 cm deep. Salinity ranges from 120‰ to 300‰, and is higher in summer than in winter. Salinity is lower near the outer shore, where seawater seeps in, than near the inner shore, where a mound of eva-

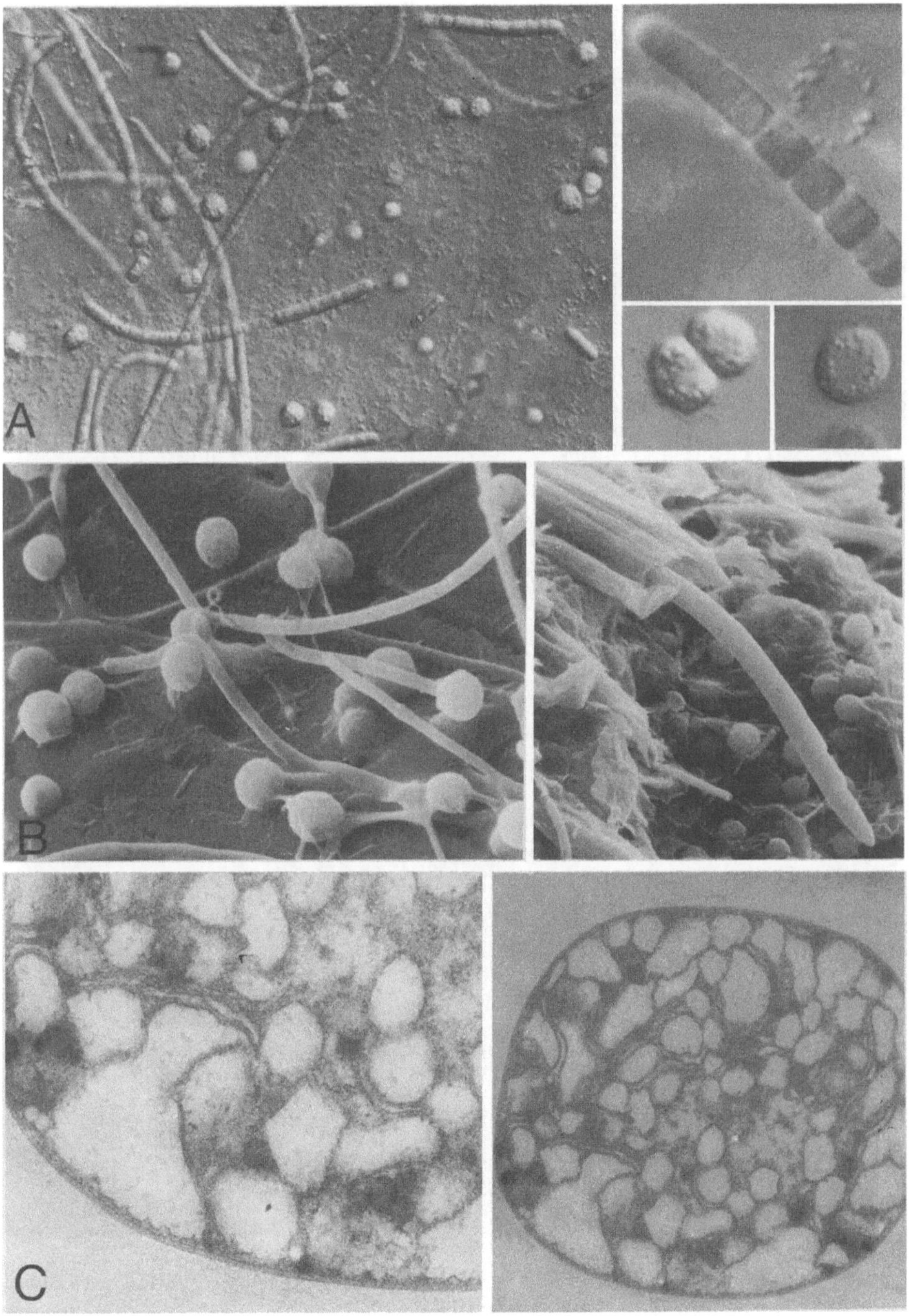

Fig. 11.10 A–C. Microorganisms of Zones 4, 5 and 6 of Fig. 11.14. Two filamentous oxygenic cyanobacteria species of the LPP-group (*Microcoleus* sp. and *Schizothrix*) inclose the giant *Thiocapsa* sp. **A** Two different micrographs of *Microcoleus* and *Thiocapsa* in differential interference contrast. **B** SEM photomicrographs of *Thiocapsa* with *Schizothrix* and *Microcoleus*. **C** TEM photomicrographs of *Thiocapsa* sp. Diameters: *Microcoleus* 3 μm, *Schizothrix* 1 μm, *Thiocapsa* 4 to 7 μm

12

13

Fig. 11.12. Intermediate zone of salt crust-related nodule formation and leathery air-exposed cynobacterial mats

Fig. 11.13. A mound of seasonally air-exposed laminated microbial mats (*left*). The *orange-reddish colors* indicate (a) diatoms, (b) carotenoids and (c) possibly protective glycocalyx embedded pigments. 1-mm-diameter *black spots* in the water-covered mat *to the right* indicate oxygen bubble channels reaching the surface from depths down to 18 mm (see Fig. 11.14). Insects hide in these tracks but do not produce them

14

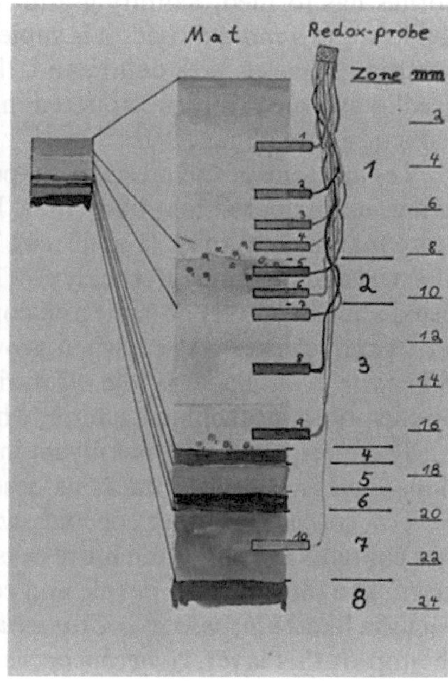

11

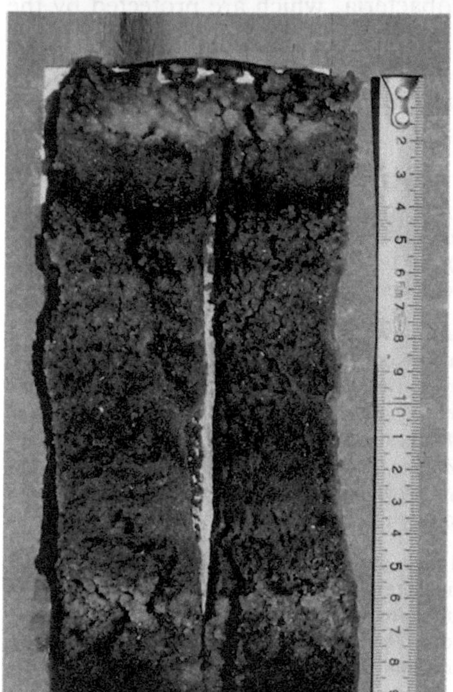

15

16

Legends see p. 254

porites has formed. Salinity gradually increases with depth. Since the lagoon is shallow and wind-exposed, it is subject to frequent mixing. Normally, water temperature increases with depth, and, therefore, with salinity. Aragonite and calcite needles are selectively precipitated instead of gypsum because of the high sulfate-reduction rate (Chap. 19).

The bottom of the crescent-shaped lagoon is covered by an extremely thick, living, multilayered microbial mat. Prior to the sheetflood in 1979 it measured more than 22 mm thick (Fig. 11.14). The mat organisms produce a great amount of extracellular slime (glycocalyx). The uppermost layer of this mat, in fact, consisted almost entirely of slime and contained only a few cells. Coccoid cyanobacteria produce excess slime when growth is inhibited by extreme salinity or light. When the cyanobacteria die off during the winter, the cells are rapidly degraded by chemoorganotrophic bacteria, whereas the slime persists for some time.

There are seven distinct layers in this mat. The first layer consists or orange slime, and is relatively thick, as described above. The dominant organisms are *Synechococcus* and other coccoid cyanobacteria. The second layer is light-green and contains less slime and more cells. *Synechococcus* is the dominant form in this layer. The third layer is purple and consists mostly of anoxygenic photosynthetic bacteria like *Thiocapsa*. No *Cromatium* cells could be found in this layer. Toward the top of this layer, *Synechococcus* Type 2 is often found. Near the bottom of this layer pleurocapsalean cyanobacteria occur. The forth layer contains filamentous cyanobacteria of the LPP group and other phototrophic and chemoorganotrophic bacteria. The fifth layer is bright pink and is made up of *Thiocapsa*. The sixth layer resembles the fourth. The seventh and lowest layer of phototrophic organisms consists of *Cromatium* and cyanobacteria, which are protected by the overlying layers.

As a whole, this mat contains a large number of cyanobacterial genera. We observed *Microcoleus, Hydrocoleum, Schizothrix, Phormidium, Pseudanabaena,*

Fig. 11.11. View from seawater springs (*right* and *bottom*, 50‰ S) toward high-salinity salt crusts (*top left*, around 250‰ S). *Colors* from bottom to top and right to left indicate: Diatoms (*brown*), *Enteromorpha* (*green* to *yellow*), cyanobacteria (*dark green*), few *dark* nodules, *white* salt crusts and *orange-colored* pools with halobacteria. The *white spots* around the ruler indicate grazing *Pirenella*

Fig. 11.14. Schematic drawing of the permanently water-covered microbial mats (see also Fig. 11.19). The figure illustrates a most unusual sequence of microbiota in a potential stromatolite. The *upper orange zone* contains few cells, mostly mucus and protective pigments. *Zone 2* indicates a maximum of coccoid cyanobacteria with oxygen bubbles embedded. *Zone 3* consists of decreasing numbers of cyanobacteria, decay products and *Thiocapsa*. *Zone 4* indicates a dense layer of *Microcoleus* producing large amounts of oxygen. *Zone 5* is domianted by *Thiocapsa*. *Zone 6* consists of a thin filamentous oxygenic cyanobacterium. *Zone 7* is made up mainly by *Chromatium*. *Zone 8* is a sulfate-reducing zone. A microredox probe has been placed in varying positions within this mat

Fig. 11.15. Example of the most expanded microbial mats of the Gavish Sabkha. The typical yellow to orange top colors, intermediate dark green cyanobacteria colors and purple to violet colors of anoxygenic phototrophs are followed by the black sulfate-reducing zone only at 3 cm depth. This zonation can expand to 6 to 8 cm under natural conditions. The expansion is caused by salt deposits acting as a light transmitter

Fig. 11.16. Sinai Bedouin collecting salt for various purposes at the Gavish Sabkha

Lyngbya, *Oscillatoria*, and *Calothrix*, three types of *Spirulina*, and other cyanobacteria. The obligately anoxygenic phototrophic *Chloroflexus* sp. occurs together with *Microcoleus* and chemoorganotrophic bacteria and two different forms of *Beggiatoa*. During the summer, a bloom of *Dunaliella* often occurs and its remainders and cysts are deposited in the mats when they die off.

11.3.1.10 Zone 10: Cyanobacterial Mats Embedded in Salt Crusts

The inner shore of the lagoon is covered by water for only short periods during the winter. During the summer, thick mineral crusts form, which are colored white, yellow or orange by embedded colonies of cyanobacteria or dead *Dunaliella*. Toward the center of this zone, halobacteria increase in numbers and give the salt crusts a typical pink color. This pink color may also be caused by *Thiocapsa* and *Ectothiorhodospira* which migrate toward the salt crusts. Cyanobac-

Fig. 11.17. Bedouin salinas. The Bedouins mine salt following the same rules as the Venetians at the Yugoslav coast or the French at the Atlantic coast, except that they use their hands instead of primitive instruments (see finger trails in the *center*). Thus the evaporation pans are as small as the range of their arms. They also do not destroy the microbial mat below the salt to guarantee the clear white quality of the salt. Thus small piles, in which the iron is reduced by bacteria originating from the "petola" (Venetian term for microbial mat), are collected besides each "mini-salina"

teria and anoxygenic photosynthetic bacteria can be found on the edges of the crusts. These organisms are vertically distributed like those of Zones 8 and 9. The laminated microbial systems embedded in the salt crusts are up to 60 mm thick. Apparently, the salt crusts act as a light-transferring system which favors the formation of extremely thick mats (Fig. 11.15).

The salinity of interstitial water in core samples is 300‰ to 340‰. The concentrations of sodium and magnesium ions are very high. The ratio of chloride to sulfate is 15:1 (calculated by E. Gavish, see Chap. 9).

In this zone, the Bedouins dig their salt pans (Figs. 11.16 and 11.17). During the summer months these salt pans are covered by a hard layer of salt. The salinity in the salt pans ranges from 300‰ to 360‰. Halobacteria and square bacteria cover the bottoms of the salt crusts. Many of these bacteria are vacuolated (Chap. 12). Some filamentous and coccoid cyanobacteria can be found below the salt crusts. They form laminated layers similar to those seen in Zones 8 and 9. The Bedouins do not destroy these slimy layers of cyanobacteria, because they impregnate the salt crusts and reduce ferric ion (Chap. 9). This keeps the salt white by preventing the formation of red, brown or orange precipitates.

11.3.2 Field Analyses and Experiments

11.3.2.1 Light and Temperature

In tropical regions light intensities increase and decrease very rapidly (Fig. 11.2a). In the Gavish Sabkha, this phenomenon is increased by the proximity of high mountain ranges to the west. The rippling of the water surface by wind is also a factor in the light activity.

The diurnal variations of air and water temperatures characteristic for the springtime are shown in Fig. 11.2b. The graph illustrates the temperature buffering capacity of the sediments in which the microorganisms are often embedded.

11.3.2.2 Oxygen, Hydrogen Sulfide and Productivity

The oxygen values presented in Fig. 11.2c were obtained from the water surface of the lagoon close to the mats of Zone 9. In bell jar productivity experiments, we measured diurnal changes in the concentration of oxygen. We compared these results with those obtained using defined incubation periods with radioactive-labeled bicarbonate (Fig. 11.18 and Table 11.2). The increase of oxygen in bell jar 3 (dark experiment) probably is due to an error during reference Winkler determinations or by light penetrating through the aluminium foil wrapping. Oxygen was also measured by H_2S-insensitive oxygen electrodes and by using a difference micro-Winkler method (Ingvorsen and Jorgensen 1979) on individual layers of mats suspended in Winkler bottles. Known amounts of mat material from individual depths within the mat were suspended in Winkler bottles filled with water of known oxygen concentration. We estimated the amount of oxygen present in individual mat layers during day and night. The frequent occur-

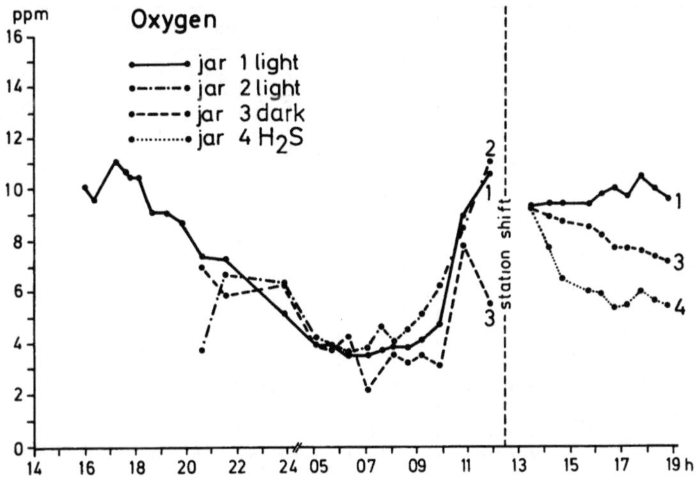

Fig. 11.18. Bell jar experiments for primary productivity determination in situ. Jars *1* and *2* are parallel oxygen production experiments in the light by recording oxygen concentration in the water column above defined mat areas. Jar *3* is the *dark reference* indicating respirative oxygen consumption. Jars *1* to *3* represent microbial mats dissected from the lagoon bottom, sewed on mm paper and placed on a plate in order to prevent upfloating during the experiment. Jar *4* was placed directly on the sediment. Hereby the influence of H_2S on photosynthetic oxygen cycling could be demonstrated

rence of oxygen bubbles (see Figs. 11.14 and 11.19) may have affected some of the measurements. The combination of the oxygen probe and the Winkler method was used to obtain relatively safe data on the diurnal oxygen cycle in individual layers of the mats of Zone 9. A representative set of data is given in Figs. 11.20 and 11.21.

11.3.2.3 Oxidation-reduction Potential

A multiple electrode (by courtesy of B. B. Jorgensen) was used to measure oxidation-reduction potentials in mats of different thickness. Results taken over the course of one day are shown in Fig. 11.22. One characteristic daytime and one characteristic night situation are shown in Figs. 11.20 and 11.21. The effect of hydrogen sulfide on the oxygen concentration is shown in Fig. 11.18. The positioning of the individual platinum threads in relation to the microbial communities is shown in Fig. 11.19.

The irregularities in the oxidation-reduction potentials in the mat of Zone 9 are due to the alternation of oxygenic photosynthetic layers with anoxygenic photosynthetic layers.

11.3.2.4 Nutrients

Most of the data we collected on nutrient levels in the Gavish Sabkha were lost in the fire which recently destroyed our laboratory. Only one set of data from

Colour zones and redox-probes	Probes	MM	Dominant form	Diatoms	Colourless sulfur bacteria	Desulfovibrio	Filamentous bacteria	Budding bacteria	Synechococcus	Gloeotece	Gloeocapsa	Pleurocapsa	Spirulina	LPP group A	LPP group B	Chloroflexus	Thiocapsa	Chromatium	Chlorobium	Non defined photos. bacteria
1 ORANGE – YELLOW – PALE GREEN	1, 2, 3, 4, 5	5	SYNECHOCOCCUS (TYPE 1)	+	++	++	+	+	+++	+			+	++	+		+			
2 GREEN	6, 7	10	SYNECHOCOCCUS		+	+			++++	++	+	++	+		++					
3 WHITE – PINKISH – YELLOW	8, 9	15	SYNECHOCOCCUS (TYPE 2), THIOCAPSA, PLEUROCAPSA		++	++	+++	++	+++		+	+++	+		++		++	+		+
4 GREEN (BLUE)			LPP – B			+			+	+		+++	++		++++		++			
5 PINK			THIOCAPSA			++	+		+			+	+		++		++++		++	+++
6 GREEN (OLIVE)			LPP – B			++			+						++++	+	++			
7 VIOLET	10	20	CHROMATIUM, CHLOROBIACEAE			++	++	+							++	+	++	++++	+++	++++
8 BLACK			DESULFOVIBRIO			++++														

Fig. 11.19. Distribution of microorganisms in the permanently water-covered microbial mat. *Arabic numbers to the left* refer to the microbial zonation, the *right ones* indicate the number of thin platinum wire of the multiple redox-probe. + present; + + few; + + + many; + + + + abundant (see Fig. 11.14)

an early stage in our investigations survived without substantial loss (Table 11.1). To the best of our knowledge, the data presented in Table 11.1 were in accordance with the lost data. It appears that the biological system of the Gavish Sabkha is trapping nutrients coming in from the open sea. This is typical of lagoon ecosystems in general (see Chap. 22; Krumbein 1982).

Fig. 11.20. Concentrations of oxygen and hydrogen sulfide and oxydation-reduction potential within the mat of Fig. 11.13 at daytime. The most characteristic results of these measurements are coexistence of large amounts of oxygen and hydrogen sulfide, and several oxygen peaks at the site of individual cyanobacterial populations. Thus, a sandwiching of O_2 and H_2S is produced

Fig. 11.21. The same parameters as in Fig. 11.20, measured at night. Although molecular diffusion of H_2S should immediately reduce oxygen accumulated in the mats during day, some oxygen is still recorded. This is explained by the 1,000- to 10,000-fold lower diffusivity of gases within the extracellular slimes of the cyanobacterial community

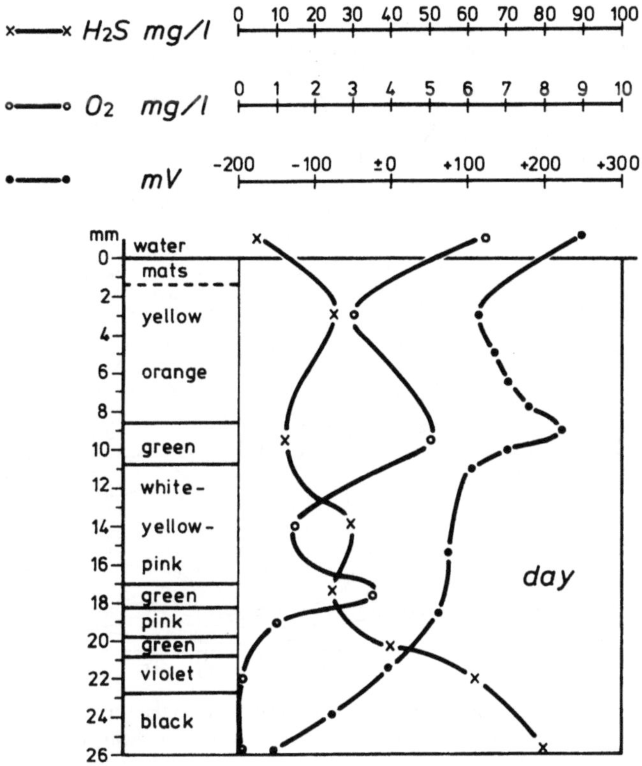

Fig. 11.20

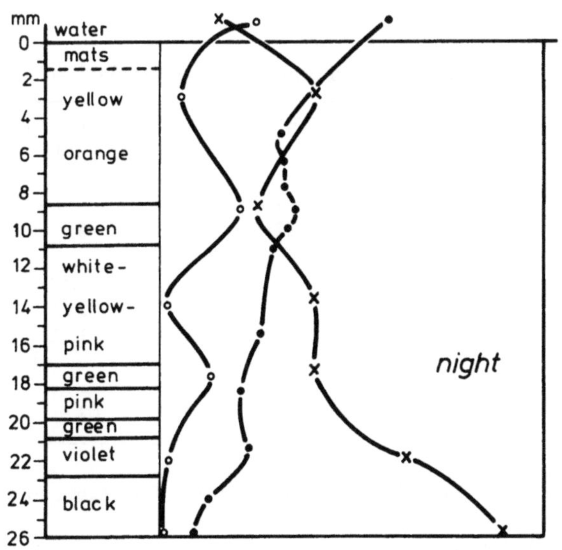

Fig. 11.21

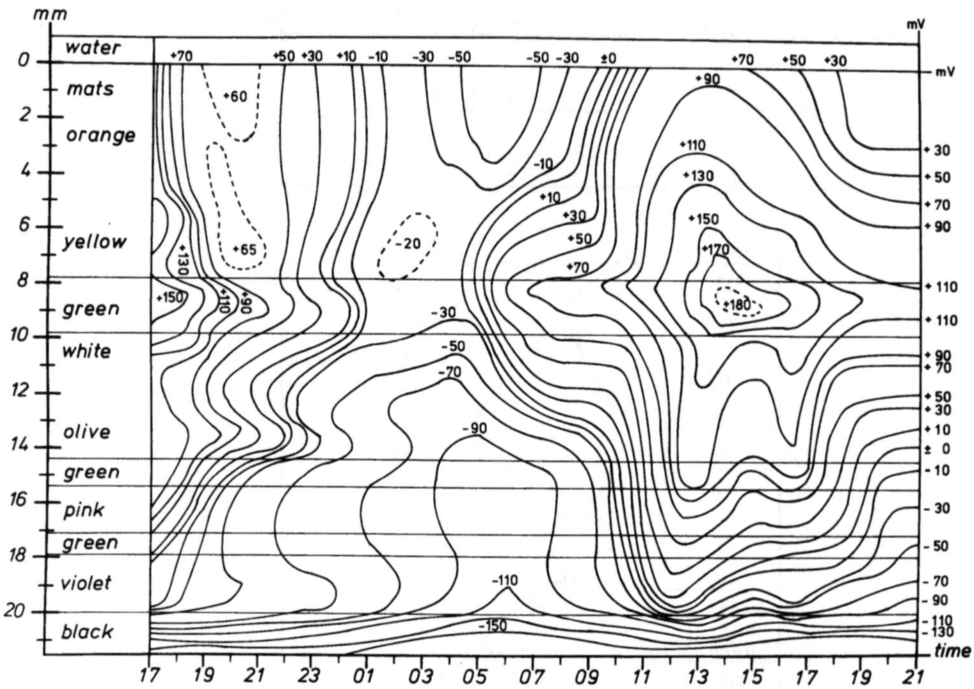

Fig. 11.22. Oxidation-reduction potentials (corrected for hydrogen electrode) in the permanently water-covered mats during a 24-h cycle. The continuous recordings of the microprobe indicated in Figs. 11.13 and 11.19 show the complex relations of aerobic and anaerobic conditions within individual layers of the mat. It is noteworthy that the oxygen maxima are always far below the surface. The surface layer consists mainly of mucus and decaying cells, while actively photosynthesizing cyanobacteria reach the first maximum only at 8 to 10-mm depth

Table 11.1. Chemical data on nutrients and pigments of "Gavish Sabkha" samples

Zone	Salinity ‰	Chl. a μg cm		BChl. a μg cm^{-3}		PO_4^{--} μM l^{-1}	$NO_3^- + NO_2^-$ μM l^{-1}	SiO_2 μM l^{-1}
		x	s	x	s			
(5)	55– 70[a]	24.6	12.7	1.97	3.42	1.8	5.9	19.8
(6)	60– 70	–	–	–	–	–	–	–
(7)	65– 70	45.6	3.14	1.81	3.14	2.1	1.58	22.4
(8)	100–180	55.1	3.28	7.54	1.89	8.2	2.15	69.4
(9)	180–200	–	–	–	–	8.9	1.72	71.2
(10)	200–250	–	–	–	–	–	–	–
(11)	320–330	–	–	–	–	–	–	–
RFSW	41.9	–	–	–	–	1.2	3.6	11.7

[a] Salinity range in spring

Chl. a = chlorophyll a; BChl. a = bacteriochlorophyll a; x = mean of 10 extractions; s = standard deviation; – = no data analyzed; RFSW = reef flat seawater

Chlorophyll data refer to the mat material itself, while nutrient data and salinities refer to the water above the mats

Table 11.2. Primary productivity of the mats measured by the oxygen method and ^{14}C method in the field and laboratory, using different methods and sections of the mats. Field data measured from 11–13.00. Laboratory data using 10,000 lux m^{-2}

Sample	Oxygen method ($gCM^{-2}h^{-1}$)	^{14}C method ($gCm^{-2}h^{-1}$)
Dome inserted into mat		
light	0.15	0.75
dark	−0.10	0.08
Active mat separated from sediment and washed		
light	0.47	0.62
dark	−0.12	0.05
Layer 1–6		
light (lab.)	0.52	0.64
dark (lab.)	−0.08	0.07
Layer 1–4		
light (lab.)	0.46	0.55
dark (lab.)	−0.1	0.05

11.4 Discussion

11.4.1 Distribution of Cyanobacteria as Related to Salinity

A large number of filamentous and coccoid cyanobacteria were found where the salinity was higher than 200‰. This surpasses by far the uppermost limit of 100‰ reported by Brock (1979). Walsby (1982) extended this limit to 230‰ on the basis of isolates from the Solar Lake (Y. Cohen et al. 1977b, Krumbein et al. 1977, Krumbein and Y. Cohen 1977). Our observations of the different mat types at the Gavish Sabkha revealed that increasing salinity tends to favor filamentous cyanobacteria. However, some coccoid forms were observed even in the Bedouin salt pans, where salinity often exceeded 350‰.

The salinity in the mats of Zone 9 is almost twice as high as the salinity of the mats of the Solar Lake (Revsbech and Jorgensen 1983). The overall physiological processes are slower. Both the production and decay rates of organic material are low, while the accumulation rate is high. This may explain why the mat at Zone 9 is so thick.

We have isolated about 30 strains of halotolerant cyanobacteria from the Gavish Sabkha. Many of them continue to grow in ASN III medium enriched with sodium chloride (up to 230‰). Some of the coccoid forms exhibited peculiar morphologies in the field and in culture, apparently due to variations of salinity.

Adding sodium chloride or magnesium chloride to cultures from the Gavish Sabkha had different effects on cell morphologies. Increasing salinity generally resulted in an increase in cell thickness or length.

11.4.2 Vertical Extension of Actively Photosynthesizing Laminated Mats

The Gavish Sabkha contains a large number of different mat types. This is caused by flooding, seepage, salinity grades, evaporation, sedimentation, and changes between water cover and exposure to the atmosphere. In response to these factors, different microbial communities establish themselves and adapt within limits to changing environmental conditions. The physioecological re-

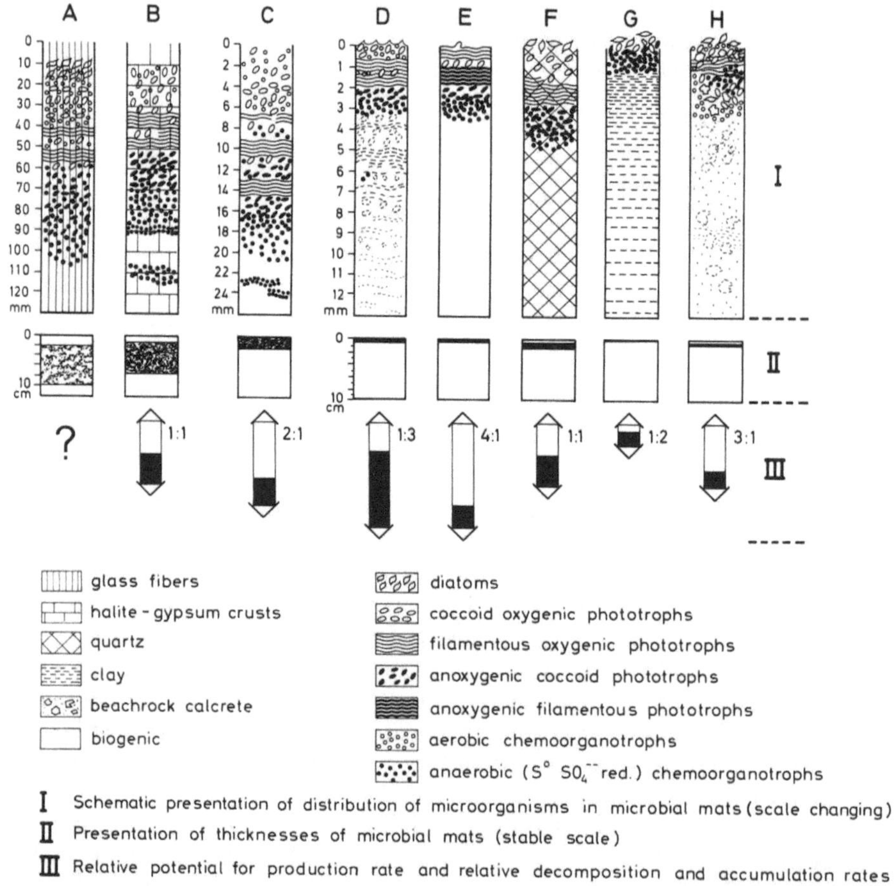

Fig. 11.23. Schematic summary of various possibilities of expansion of photosynthetic microbial mats in the Gavish Sabkha and similar environments as related to sedimentary environments and physical-chemical conditions. The mats can be epibenthic or endobenthic. They may contain eukaryotes or not. Light penetration and salinity regulate the thickness of the active mat as well as the degradation-preservation potential during early diagenesis. *I, II* representation of different mats. *A* Microbial mat developed within glass fibers; *B* Salt-encrusted microbial mat; *C* Extracellular slime-regulated, multilaminated microbial mat; *D* Pinnacle mat of Solar Lake, slime-regulated to a lesser extent; *E* Classical, compact, completely organic microbial mat; *F* Siliciclastic potential stromatolite; sand-embedded microbial mat; *G* Silt-clay dominated microbial mat; *H* Stromatolitic beach rock, calcrete or rock varnish. *III* The *arrows* indicate *in the white range* the measured or observed annual production of organic matter; the *black range* describes the relative amount of degradation. The *first number* indicates production, the *second number* degradation, thus indicating the ratios

sponse of the community to the environmental conditions also influences mat structure and thickness. Thus, the constantly changing Sabkha is an excellent location for the study of the evolution of various mat systems and the processes leading to the buildup of potential stromatolites.

The thickest actively photosynthesizing microbial mat ever recorded to our knowledge was found embedded in the salt crusts of Zone 10. This mat was extremely thick, possibly because of the relatively high amount of gypsum needles. Vertically oriented salt needles play an important role in light transmission in these mats. This phenomenon is comparable to the action of glass fiber light sources or etiolated plant tissues (Mandoli and Briggs 1982). The various patterns of light transmission in different types of photosynthetic microbial mats are shown in Fig. 11.23. In mats into which glass fibers were inserted, the mat was not substantially thicker along the fibers (Fig. 11.23 a). Data we have collected from other mat environments are also shown in Fig. 11.23. The effects of sediments and sedimentation, annual productivity cycles, species composition and degradation patterns of buried mats are also shown in Fig. 11.23.

11.4.3 Distribution of Oxygen and Hydrogen Sulfide

The unusual sandwiched stratification of various photosynthetic layers in Zone 9 remains a puzzle. This alternation of oxygenated and anoxic layers within one productive mat may explain the mineral variability in stromatolitic evaporitic rocks (Krumbein et al. 1979 a).

Until now, it has been generally assumed that only relatively minute amounts of oxygen and hydrogen sulfide can occur together in stratified bodies of water (Sorokin 1970, Jorgensen et al. 1979). Krumbein et al. (1979 a), when studying the unusually thick mats in the Gavish Sabkha, showed for the first time that the two gases could coexist in high concentration for extended periods of time. Revsbach et al. (in press) have subsequently shown that this also occurs in thinner mats from the Solar Lake. We measured the concentrations of oxygen and hydrogen sulfide and oxidation-reduction potentials in the mats of Zone 9 (Figs. 11.20–11.22). We found that oxygen remains in deeper layers of mats which had been kept in darkness for more than 36 h. This is probably caused by the low diffusivity of gases through cyanobacterial slimes (see Fig. 11.6 in Krumbein et al. 1979 a).

11.4.4 Productivity

Both the Gavish Sabkha and the Solar Lake mats have extremely high productivities (Cohen et al. 1977 b, Krumbein et al. 1977, Krumbein et al. 1979 a). This is in contrast to some data reported by Larsen (1980) and Brock (1979). The Solar Lake is protected from the wind by high surrounding mountain ridges. This allows stable stratification of the water body during the fall, winter and spring. Salinity in the Solar Lake is lower than in the Dead Sea and the Great Salt Lake. This may also influence productivity.

In the Gavish Sabkha the productivity is as high as in the Solar Lake, although the salinity is almost twice as high. This is influenced by the following factors: (1) High temperatures and high irradiation rates throughout the year are balanced and buffered by constant supply of seawater and evaporation cooling; (2) The mats produce large amounts of protective slime; (3) The high salinity and low oxygen content of the water may reduce the dangers of photooxidative conditions; (4) Seawater constantly seeping into the sabkha brings in nutrients, in particular phosphorus (Table 11.1). Caumette (1978) has reported similar observations of the moderately hypersaline lagoons of Prevost (France). Krumbein (1982) has reported that these factors play a significant role in the ecology of hypersaline lagoons in general.

Within the saturated brine of the Gavish Sabkha almost all productivity is restricted to the benthic and endobenthic systems. The open water column rarely if ever shows the typical red colors of *Dunaliella* or halobacteria blooms. At the surface of the brine pools these organisms color the salt surface bright red. The stinking red waters described in the biblical book of Exodus and also by Ehrenberg (1830) can be found only rarely in the Gavish Sabkha. The occurence of this phenomenon in the Dead Sea and in the Great Salt Lake is due to the greater depth and different chemical composition of the water in these systems. We also never observed the characteristic milkiness due to high elemental sulfur content (Labourg 1975). This milky appearance, however, may also be caused by blooms of vacuolated cyanobacteria, as it is the case in the Solar Lake. The Gavish Sabkas waters usually were clear and clean-smelling.

11.4.5 Extracellular Polysaccharides

It was demonstrated (Krumbein et al. 1979a) that benthic cyanobacteria, especially of the coccoid genera, produce large amounts of extracellular polysaccharide slimes. These may efficiently buffer the otherwise steep gradients and diurnal changes of oxygen and hydrogen sulfide by trapping oxygen bubbles produced by oxygenic photosynthesis. Gases diffuse within concentrated sugar solutions up to 10,000 times slower than in water. We could safely assume that this applies even more to extracellular slimes since they are extremely viscous polysaccharide solutions. This has several important effects on the microbial mat system. Oxygen produced during photosynthesis is trapped by the slime. Carbon dioxide excreted during nighttime respiration is trapped as well, thus enhancing the closed system effect and leading to the reported unusually heavy organic carbon isotope fractionation data (Chap. 19). The aerobic organisms are protected against the poisonous effects of hydrogen sulfide produced underneath the mats and within decay microenvironments within the mats. The large amounts of slime also permit light to penetrate deeper into the mat.

These slimes are also important for the preservation of cell structures in deeper layers of the mats and possibly for their fossilization (Kazmierczak and Krumbein 1983a, b, Knoll and Golubic 1979, Krumbein and Swart 1983). It is difficult for chemoorganotrophic bacteria to degrade the slimes. Bacteria, however, destroy most of the cells, leaving hollow structures which become fossilized. Kazmierczak and Krumbein have shown that the structures remaining are practically the same from the very early stages of diagenesis to the oldest fossils. Bacterial structures of the decay environment have been shown to be incorporated into the extracellular slimes (see also Krumbein and Swart 1983).

The relations between salinity and slime production by individual species of cyanobacteria can only be studied on isolates from highly hypersaline brines, and also the taxonomy of the species occurring is still open to discussion, because not only the amount of slime produced by one and the same species under different salinities is altered but also the morphology of the cells. Golubic (1979) makes some very critical comments on the species identification of cyanobacteria observed in the Solar Lake and other hypersaline environments. Yopp et al. (1978) and Walsby (1982) have worked on strains which were isolated from salinities between 120‰ and 80‰, while our isolates are originating from salinities of up to 340‰.

Preliminary results on some strains of coccoid cyanobacteria from the Gavish Sabkha indicate that slime production increases and cell morphology varies with increasing salinity. One single strain of *Synechococcus* sp. may increase its diameter from 6 µm to 15 µm and may change into filamentous growth forms strikingly similar to *Dactylococcopsis*. We have to wait for further results on the physioecology of these strains until some of the protection patterns are understood in some more detail. Up to now we can only state that many of our slime embedded isolates are halotolerant rather than halophilic and often grow better and with less slime production under lower salinities. It appears that the capability of cell expansion and slime production does not necessitate the evolutionary adaptation to a narrow salinity range as it is the case with true halophiles. Also, we must keep in mind that in contrast to the Dead Sea and the Great Salt Lake the Gavish Sabkha is a schizohaline environment which favors those forms that can tolerate large variations in salinity. In such environments the typical adaptational pattern to increased salinity stress may be size variation and slime production rather than complex biochemical adaptations, such as the production of complex organic compounds or the evolution of a light-driven proton pump (see Chap. 13). Different Mg, Ca concentrations and requirements additionally complicate the picture.

11.4.6 Nodular (Oncoidal) and Tepee-shaped Cyanobacterial Aggregations

The nodules which form at the rim of the lagoon resemble evaporative laminated microbial structures and biscuits observed in many Precambrian rocks (Lowe 1980, Buick et al. 1981). The microbial population of the nodules is extremely rich and diversified.

Many oncoids in coastal series may not be formed by sand agglutination and water movement, as often described (e.g., Peryt 1981), but rather by the formation and dissolution of bizarrely shaped salt crusts.

The Tepee and Petee structure problem has been mentioned already in Chapter 9 and has been explained earlier in this chapter. The biogenic Petee structures resemble Conophyton-type stromatolites. This calls for caution in the interpretation of biogenic and abiogenic structures, as well as of stromatolites and stromatoloids.

11.4.7 Faunal Elements in Hypersaline Potential Stromatolites

In the Gavish Sabkha several zones are relatively rich in animals (Chapter 15; Gerdes and Krumbein 1984). Only the snail *Pirenella conica* completely destroys and bioturbates microbial mats. Most of the other species (e.g., crustacean meiofauna and insects) live in coexistence with and feed on the microbial mats without destroying them. Staphilinid beetles *(Bledius capra)* form burrows in the mats which resemble fossil worm burrows.

Coexistence of faunal elements and potential stromatolites is not very often reported. In most cases, faunal elements are reported to destroy microbial mats. Microbial mats are scarce in moderate marine and freshwater environments, where the faunal communities are rich and very active. Therefore, it has been generally assumed that productive microbial mats and a thriving faunal community are mutually exclusive. However, in the Gavish Sabkha there is an active faunal community, in spite of the harsher conditions.

We may state that the appearance of faunal elements is not the only cause of the decline of stromatolites in the late Proterozoic. From the relationship between the animals and the mats in the Gavish Sabkha we can postulate that the fauna can stabilize the mats and enhance productivity.

11.4.8 New Bacteria from the Gavish Sabkha

The largest *Thiocapsa* species (probably nov. spec.) as yet described was found in the mat at Zone 9 (Fig. 11.10). Apparently, many microorganisms in hypersaline environments are larger than their equivalents in less saline environments. This *Thiocapsa,* like *Ectothiorhodospira* from the Wadi Natrun (Trüper and Imhoff 1981, Imhoff et al. 1979), can grow at the surface of hypersaline brines, because the solubility of oxygen in hypersaline water is low (1.7 mg O_2/l). We do not have cultures of the *Thiocapsa* and therefore cannot take far-reaching conclusions. Within the mats of Zones 8 and 9, this organism can make up the bulk of organic material preserved within the sediments. Since this species contains substances not produced by cyanobacteria, we may be able to use it as an additional aid in characterizing ancient stromatolitic aerobic and anaerobic ecosystems (see Chaps. 17 and 18).

The square and box-shaped halophilic bacteria (Chaps. 12 and 13) were first detected in the Gavish Sabkha. The physical stress of the formation of halite crystals may be responsible for their peculiar morphology, although the pseudocrystal X-ray detectable pattern of many membrane proteins might indicate also evolutionary relationships between complex inorganic crystal lattices, which may have served as templates for organic molecules. This would lead us too far beyond the scope of this chapter.

In conclusion we would like to state that we have been and still are fascinated by the tremendous adaptational flexibility of the Gavish Sabkha microbial communities. We have heard that the State of Egypt intends to place both environments – the Solar Lake and the Gavish Sabkha – under strict Nature Reserve conditions. We sincerely hope that this will be achieved.

12. Structure and Physiology of Square-shaped and Other Halophilic Bacteria from the Gavish Sabkha

MARTIN KESSEL, YEHUDA COHEN, and ANTHONY E. WALSBY

The extremely halophilic bacteria described in the following two chapters occur in the zones of highest salinity in the Gavish Sabkha (see Chap. 9). When one of us (A. E. Walsby) visited the Gavish Sabkha (together with W. E. Krumbein in April 1979) he was mainly in search of gas-vacuolate halophilic bacteria. Samples of the pink salt and brine slush of the pools of the elevated central parts of the Gavish Sabkha, when investigated microscopically, revealed an exciting new group of bacteria. Floating ethereally among the microorganisms were diaphanous squares with glistening gas vacuoles attached to them. Further inspection revealed that many of the squares had just discernible thickness and were made up of subsidiary squares linked like postage stamps in a sheet. It transpired that each square was a bacterium of unprecedented form. After the first description of these unique bacteria from the Gavish Sabkha (Walsby 1980), W. Stoeckenius (see Chap. 11) and Y. Cohen isolated other bacteria of square shapes which turned out to contain bacteriorhodopsin and to be able to change from the square shape into the normal form of bacteria.

The Gavish Sabkha contains also other highly pleomorphic bacteria (see Chap. 11), cyanobacteria which are able to change from small to large and from filamentous to coccoid, as well as from tapered to ovoid. These organisms are able to cope with astonishingly different and widely fluctuating.

Depending on the season of the year various niches occur, each representing different conditions of salinity. Our descriptions of the microbial population will be confined to those areas of the Sabkha where conditions of highest salinity prevail. This means we will be mainly examining the population of microorganisms in closest association with the salt crusts that are found in various locations in the Sabkha. Our observations are also confined to the population of microorganisms that is found at the hottest time of the year when maximum evaporation occurs from the water body. Examination of samples collected under the above conditions reveals a bewildering array of microorganisms of varying shape and size. To isolate and identify all the various components of the population would be an almost insurmountable task. The most fascinating group of organisms observed are those displaying square or rectangular shape, either with or without gas vacuoles. The square organisms have already been the subject of a number of publications (Stoeckenius 1981, Javor et al. 1982, Kessel and Y. Cohen 1982), all stemming from Walsby's communication to *Nature* 1980 mentioned above.

The material presented in this chapter will attempt to consolidate our current knowledge of these organisms from a structural and physiological point of view. A separate chapter (13) will deal with the light-harvesting apparatus of these organisms. It should be pointed out that since Walsby's recognition of these organisms at the Gavish Sabkha they have been reported from a number of locations elsewhere in the world, where similar high-salt conditions prevail (Stoeckenius 1981, Javor et al. 1982). A section of this chapter will also discuss the importance of gas vacuoles for halophilic microorganisms.

12.1 Collection of Samples

The samples were collected mainly from areas where small pools of water are interspersed with salt crusts. Such locations are usually characterized by a dominant pink color of the salt crusts due to the carotenoid pigments of the bacteria. The upper layer of the crust was sampled together with the surrounding water, and after initial agitation in an Erlenmeyer flask was left stationary. Due to the large number of gas vacuoles present in the cells, the square bacteria floated up and became concentrated at the water surface and could thus be collected with relatively high efficiency for viewing in the light microscope, and for electron microscopy preparation.

12.2 Isolation and Growth of Halophilic Bacteria

In order to obtain information on physiological aspects of growth of the bacteria from the Gavish Sabkha a large number of isolates were obtained, using the methods outlined below.

It should be emphasized at this point that the use of the term square bacteria refers to cells with a principally flat rectangular or square shape, and may sometimes appear in various pleomorphically angular forms. The only types that have been isolated so far are those lacking gas vacuoles. The gas-vacuolate forms may require different culture conditions.

Media: Standard medium used for growth of the halophilic bacteria were used (Oesterhelt and Stoeckenius 1973).

Growth media were modified in the various experiments, as indicated in 10.2.2–10.2.4. All cultures were grown at 37 °C under continuous light provided by a bank of cool white fluorescent lamps.

Bacteria were isolated on the same medium in agar plates without enrichments.

Growth was determined by optical density measurements at 750 nm, by protein assay according to Lowry et al. (1951), and by triplicate viable counts of colonies from dilution series prepared in salt solution. The salt solution used was identical to that used for growth without the addition of peptone.

Determination of bacteriorhodopsin: Cells were collected by centrifugation at 14,000 g for 15 min. Pellets were suspended in 5 ml distilled water and were sonicated for 30 min at 4 °C. Lysates were centrifuged at 1,000 g to remove particles, and samples were incubated for 1 h at 37 °C with the addition of 10 $\mu g\ ml^{-1}$ of deoxyribonuclease (Sigma). Samples then were centrifuged at 35,000 g for

30 min at 4 °C. Pellets were resuspended in 5 ml 50 mM Tris buffer pH 8.0. Optical density was determined as 580 nm against bleached sample (0.05 mM cetyltrimethyl-ammoniumbromide) in a Zeiss spectrophotometer.

12.2.1 Viable Count

Three samples of saturated salt solution were collected at the western rim of the Gavish Sabkha permanent water body. Samples were diluted in sterile Sabkha water and the dilutions were incubated on agar plates according to Javor et al. (1982). From each of the three samples a total of 2×10^8, 1.6×10^8 and 2.7×10^8 colonies developed, 90% of which were pigmented. In all, 103 isolates were identified from these samples, 59 of which were square bacteria, 22 were pleomorphis rods, and 22 were cocci. One of the square bacteria isolated (81830-2) was examined for its growth requirement.

12.2.2 NaCl Requirements

Cells were grown in growth medium containing 250 g l^{-1} NaCl (Javor et al. 1982). Cells from exponentially growing cultures were inoculated in the same media containing 200, 150, 100 and 75 g l^{-1} NaCl (Fig. 12.1). Growth rate and yield dropped considerably in media containing 150 and 200 g l^{-1} NaCl compared to cells grown with 250 g l^{-1} NaCl. No morphological changes were observed when cells grew at lower NaCl concentration (Kelier and Henis 1970).

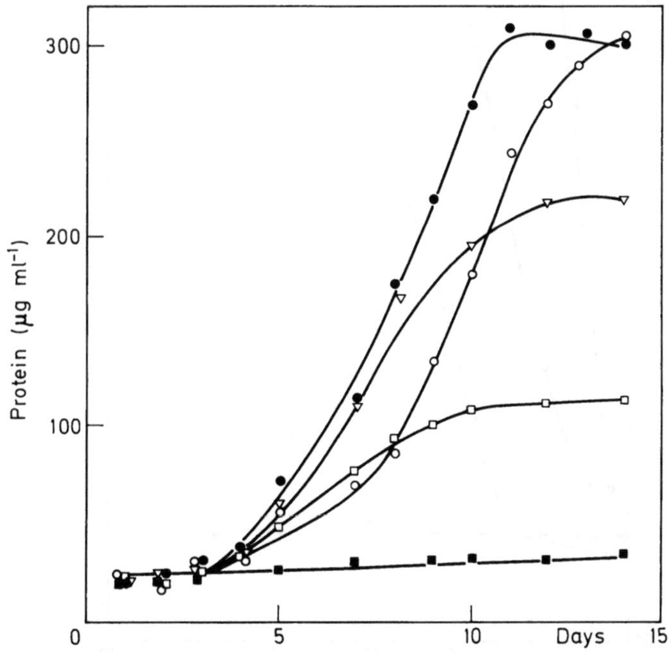

Fig. 12.1. Growth rate of the square bacterium 81830-2 as a function of NaCl concentration and Ca^{2+}. —●— 250 g l^{-1} NaCl and 0.2 g l^{-1} $CaCl_2$, —○— 250 g l^{-1} NaCl without Ca^{2+}, —▽— 200 g l^{-1} NaCl, —□— 150 g l^{-1} NaCl, —■— 100 g l^{-1} and 75 g l^{-1} NaCl

12.2.3 Ca^{2+} Requirements

When cells grown on complex medium containing 0.2 g l^{-1} $CaCl_2$ were transferred to the same medium without calcium cloride a prolonged lag period was observed. Yet after the lag period cells grew at the same growth rate as the control and the growth yield was identical to that of the control (Fig. 12.1). Examination under the phase contrast microscope revealed that cells were rounded in the lag and log phase and only on the 10th day was the typical square morphology observed. At stationary phase all cells were morphologically identical to that of the typical square shape.

12.2.4 Mg^{2+} Requirements

Optimal growth was observed in cells grown in complex medium containing 80 mM $MgSO_4 \cdot 7H_2O$ (Fig. 12.2). No growth was observed in cells incubated in

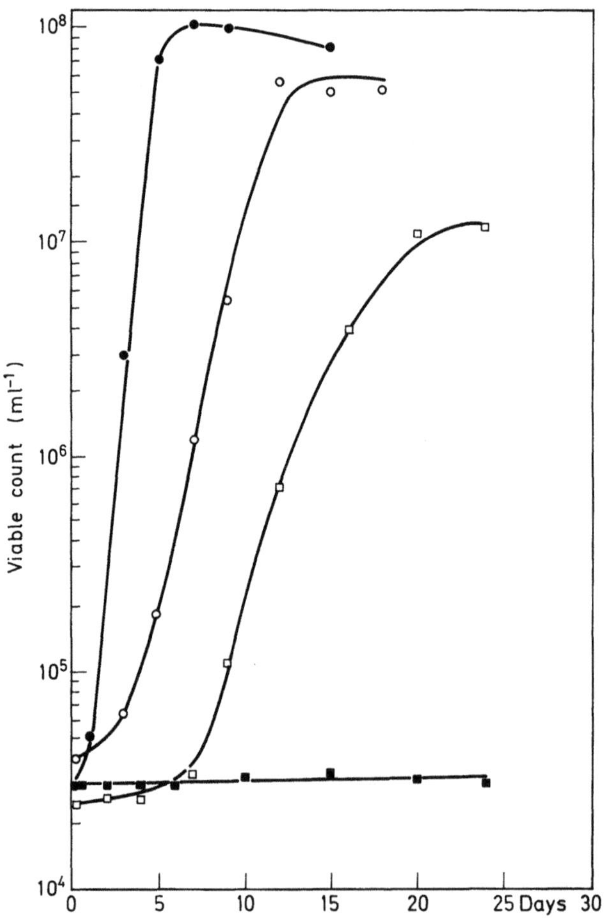

Fig. 12.2. Viable count of 81830-2 as a function of Mg^{2+} concentration. —●— 80 mM, —○— 30 mM, —□— 10 mM, —■— 5, 2 and 1 mM

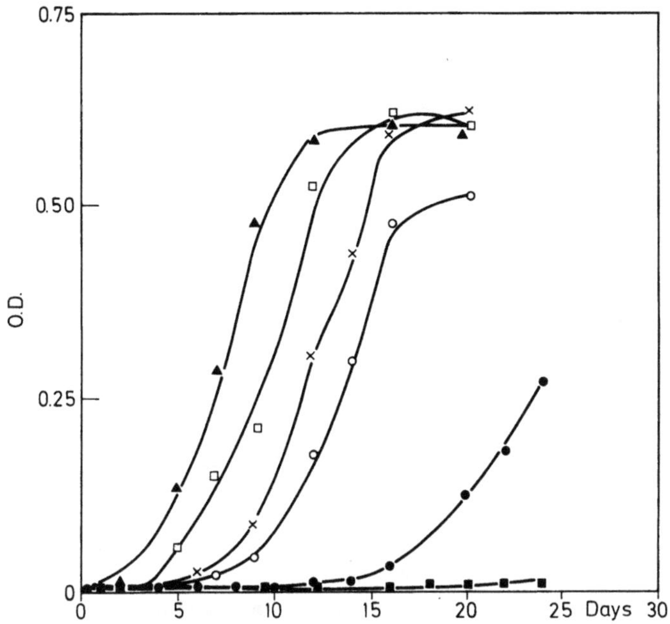

Fig. 12.3. Production of bacteriorhodopsin, measured as optical density at 580 nm as described under methods, as a function of Mg^{2+} concentration. —▲— 80 mM, —□— 40 mM, —×— 30 mM, —○— 20 mM, —●— 10 mM, —■— 5, 2 and 1 mM

media containing 5, 2 and 1 mM Mg^{2+} after 24 days of incubation at 37 °C in the light. The cell yield dropped to 55% of the control when cells were grown at 30 mM Mg^{2+}. When Mg^{2+} concentration was 10 mM, the cell yield was only 2% of the control. A drop in growth rate and prolongation of the lag phase was observed in cultures grown at 30 and 10 mM Mg^{2+}. The Mg^{2+} requirement for bacteriorhodopsin production is plotted in Fig. 12.3. The yield of bacteriorhodopsin did not change when the Mg^{2+} concentration was lowered from 80 mM to 30 mM, yet at lower Mg^{2+} concentration of 20 and 10 mM an apparent drop in yield of bacteriorhodopsin was observed. The rate of production of bacteriorhodopsin and the duration of the lag phase before production commenced is Mg^{2+} -dependent (Fig. 12.4). Both parameters changed only slightly when Mg^{2+} concentration was lowered from 80 to 40 mM, but at lower Mg^{2+} concentration a logarithmic drop of both the production rate of bacteriorhodopsin and the duration of the lag phase was observed.

The morphology of the cells was dependent on magnesium in a similar way to that observed with calcium. When cells were grown at 20 or 10 mM Mg^{2+}, they appeared rounded until the stationary phase of growth when the typical square structure reappeared.

12.2.5 Carbon Source Requirements

Cells grown in complex medium were incubated in media containing various single carbon sources (Fig. 12.5). Little growth was detected when glucose served

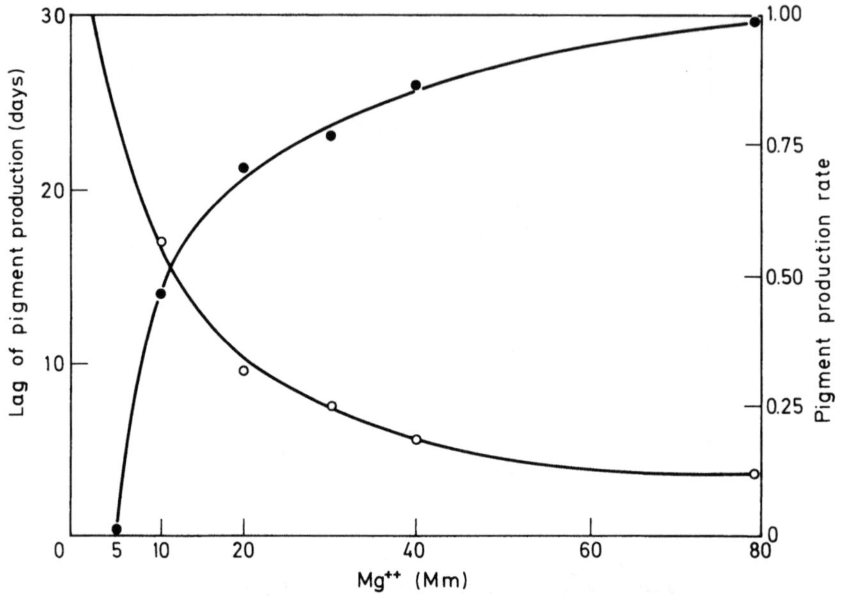

Fig. 12.4. Relative rate of bacteriorhodopsin production (—●—) and duration of lag of pigment production (—○—) as a function of Mg^{2+} concentration, in the square bacterium isolate 81830-2

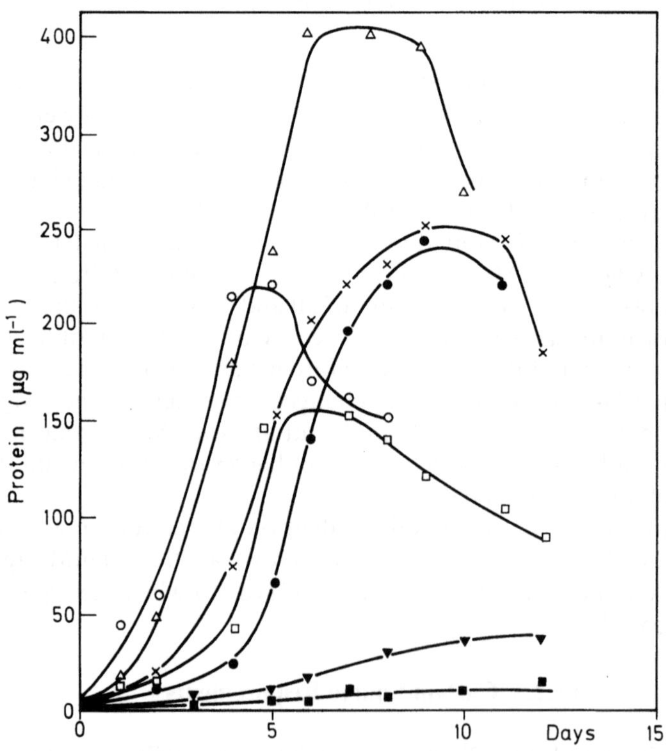

Fig. 12.5. Growth of the square bacterium 81830-2 on different single carbon sources supplied at 1% concentration in salt solution. —■— glucose, sucrose, formate, —▼— glycerol, —□— glycolate, —○— malate, —●— acetate, —×— lactate, —△— caseine hydrolysate

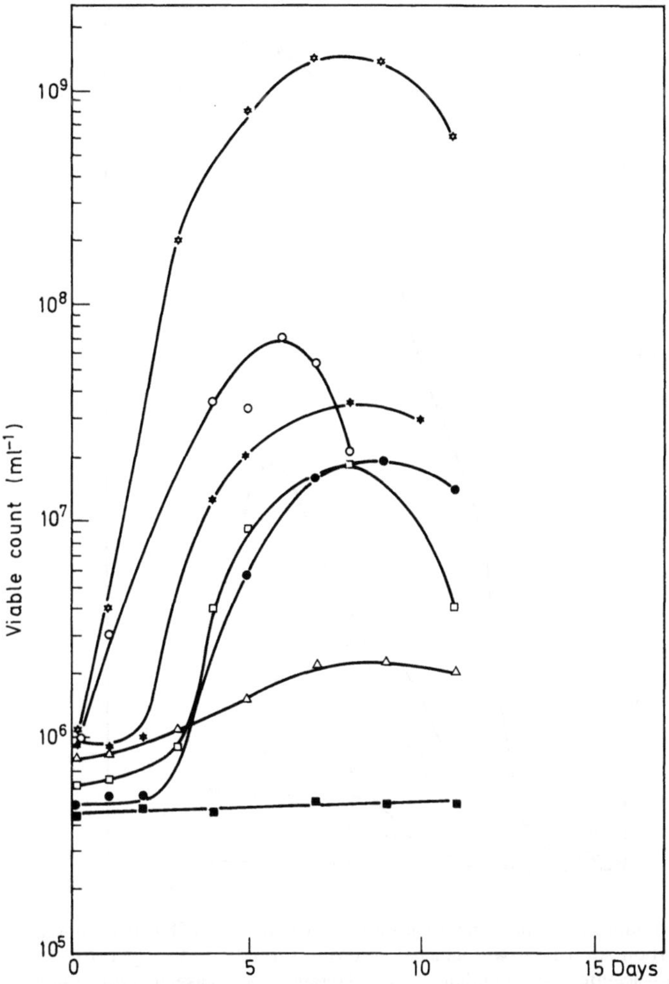

Fig. 12.6. Viable count of the square bacterium 81830-2 grown on different single carbon sources $10\,g\,l^{-1}$. —■— glucose, sucrose, formate, —△— glycerol, —□— glycolate, —●— acetate, —★— lactate, —○— malate, —☆— caseine hydrolysate

as the only carbon source. No growth was observed when sucrose, xylose or starch were used as the sole carbon source, yet some growth was detected in the presence of 2% sorbitol. It has previously been reported that halobacteria, with the exception of *H. marismortui,* do not metabolize glucose or other carbohydrates (Larsen 1967). Yet Gochnauer and Kushner (1969) report growth of several species of halobacteria on glucose, fructose and galactose.

Glycerol supported little growth, though glycolate served as a much better carbon source. The square bacteria were found at the Gavish Sabkha when salinity increased above 22%. Preceding the bloom of the square bacteria was a bloom of the halophilic alga *Dunaliella* sp. These green algae grew well at salinities of up to 17% but lysed at high salt concentrations releasing two major compounds, glycerol and glycolate, which were then available for growth of the square bacteria.

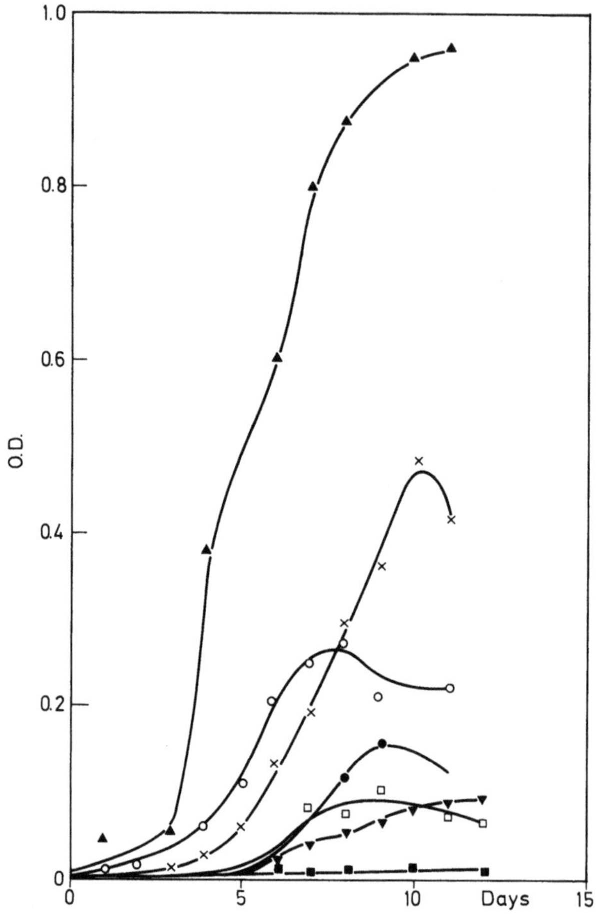

Fig. 12.7. Bacteriorhodopsin production measured as optical density at 580 nm as function of growth at different single carbon sources. —■— glucose, sucrose, formate, —▼— glycerol, —□— glycolate, —●—acetate, —○— malate, —×— lactate, —▲— caseine hydrolysate

Acetate, malate and lactate also supported growth of the square bacteria. As expected, growth on lactate was better than on malate, while growth rate and cell yield on acetate was lower, as indicated by protein production (Fig. 12.5), viable count (Fig. 12.6) and bacteriorhodopsin production (Fig. 12.7). Caseine hydrolysate supported even more vigorous growth (Figs. 12.5 and 12.6).

Table 12.1 indicates all of the carbon sources examined for their ability to support growth of the square bacteria. They showed considerable versatility with regard to carbon source. The production of bacteriorhodopsin in all cultures was a function of growth rate and not dependent on the specific carbon source.

The comparison of growth rate (Figs. 12.5 and 12.6) and bacteriorhodopsin production (Fig. 12.7) indicates that bacteriorhodopsin production per cell was similar in cells grown in all the different carbon sources. Thus the pigment production was dependent on availability of divalent cations, namely Ca^{2+} and Mg^{2+}, but not on the carbon source for growth.

12.3 Light Microscopy

It was found that the best way to examine the population of cells in brine samples was to place a drop of the material to be examined on a glass slide covered with agar. This had the immediate effect of immobilizing the cells, preventing Brownian or swimming motion. The preparations were examined with phase contrast or Nomarski interference optics (Figs. 12.8 and 12.9).

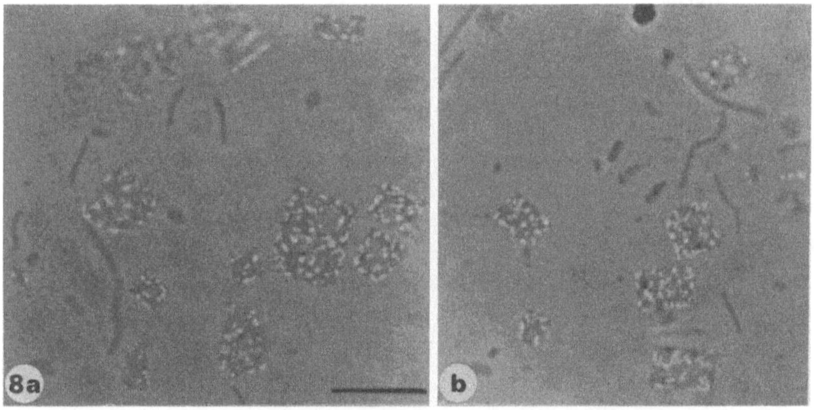

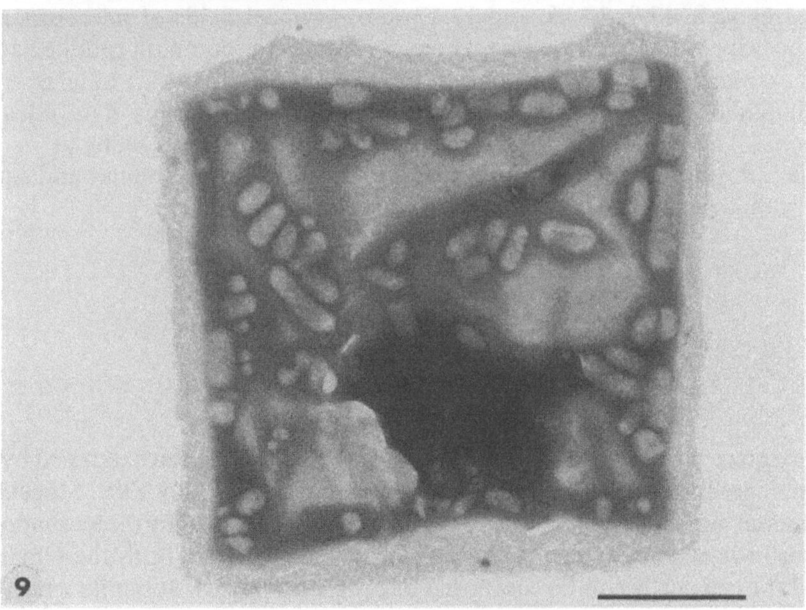

Fig. 12.8 a, b. Phase contrast micrographs of samples taken directly from brine water at Gavish Sabkha. Square-shaped organisms display large numbers of highly refractile intracellular inclusions. Smaller square-shaped bacteria show no such inclusions. Bar = 10 m

Fig. 12.9. Unstained square bacterium showing large numbers of gas vesicles. Bar = 1 μm

Table 12.1. Qualitative growth and bacteriorhodopsin production of the square bacterium 81830-2 grown on different single carbon sources

Carbon source	Growth	Pigmentation
Glucose	+−	−
Sucrose	−	−
Xylose	−	−
Starch	−	−
Sorbitol	++	++
Acetate	++	++
Malate	++	++
Formate	−	−
Lactate	++	++
Succinate	++	++
Glycolate	++	++
Glycerol	+	+
Citrate	++	+−
Ascorbate	−	−
Benzoate	−	−
Oxalate	++	−
Alanine	++	++
Caseine hydrolysate	+++	+++
Glutamine	+++	+++

Figures 12.8, 12.9, 12.16, and 12.17 shows typical fields of microorganisms when the brine water was examined directly. Besides the common rodlike bacteria and an extremely long rod form, attention is immediately drawn to large square or rectangular objects which are characterized by numerous light-refracting inclusions. These are the organisms described by Walsby (1980) as "square bacteria". Besides these organisms there are also smaller rectangular and square-shaped organisms, but without the refracting inclusions.

12.4 Electron Microscopy

12.4.1 Square Bacteria

The square bacteria occurring in the natural brine have been examined by conventional methods of electron microscopy (Parkes and Walsby 1981, Stoeckenius 1981, Kessel and Y. Cohen 1982). First results were obtained by direct shadowing of the cell surface or by the preparation of surface replicas. Both these methods showed the cell surface to be composed of ordered arrays of subunits which were predominantly hexagonally arranged (Fig. 12.10 a, b) but which occasionally also exhibited a square arrangement of subunits (Kessel and Y. Cohen 1982). Optical diffraction of micrographs of such preparations clearly revealed the periodic arrangement of subunits. Evidence of periodic subunits could also be obtained from freeze fracture replicas, although this method showed only small areas of the sur-

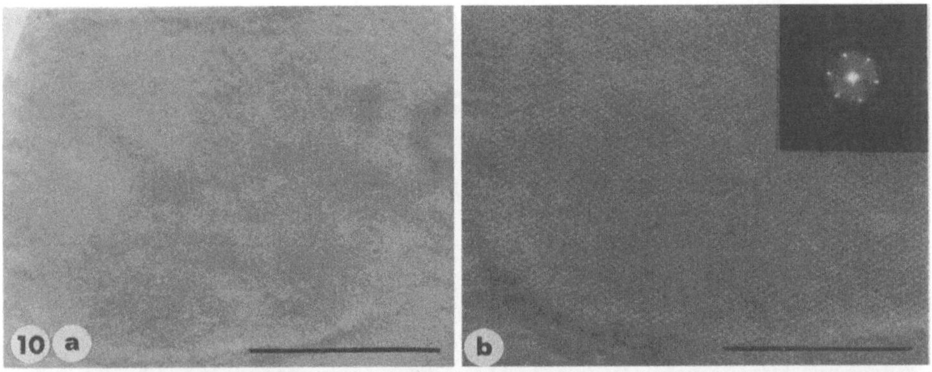

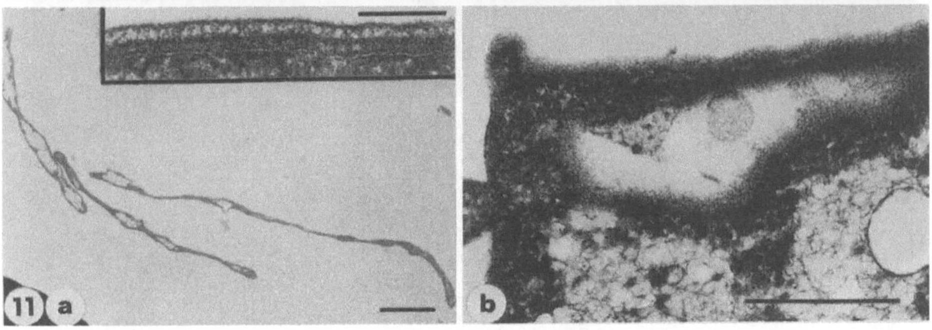

Fig. 12.10. a Platinum-carbon shadowed surface replica of a square bacterium. Bar = 1 μm. **b** Higher manification showing the periodic arrangement of surface subunits. Bar = 0.5 μm. *Inset* shows an optical diffractogram of this micrograph, showing the clear hexagonal symmetry present

Fig. 12.11. a Low-power micrograph showing the extremely thin profile of the square bacteria when sectioned normal to the flat surface. Bar = 1 μm. *Inset* shows the detail of the cell wall at high magnification with the outermost component exhibiting periodicity. Bar = 0.1 μm. **b** Grazing section of the cell wall showing periodically arranged surface subunits. Bar = 0.5 μm

face. The center-to-center spacing between subunits was typically of the order of 160 Å. Shadowed replicas of whole cells revealed bulges randomly distributed within the cells. These are most likely the storage bodies shown by freeze fracture to behave like β-polyhydroxybutyrate, and like the empty swollen regions, as seen by thin sectioning (Stoeckenius 1981).

Conventional thin sectioning of cells from the natural population showed the variety of types present. All of the bacteria examined, however, showed the extremely thin profile with or without inclusions (Fig. 12.11 a).

The external layers of the bacteria were sometimes of variable appearance and cells with and without an extracellular sheath were observed (Kessel and Y. Cohen 1982, Parkes and Walsby 1981). Detail of the cell wall, as seen in thin section in Fig. 12.11, gives a hint of the repeating structure of the wall. The cell wall showed a typical halobacterial profile, with the notable absence of a peptidoglycan layer. The outer wall had a multilayered appearance and the subunit periodicity was evident in favorable planes of section, such as grazing sections (Fig. 12.11 b).

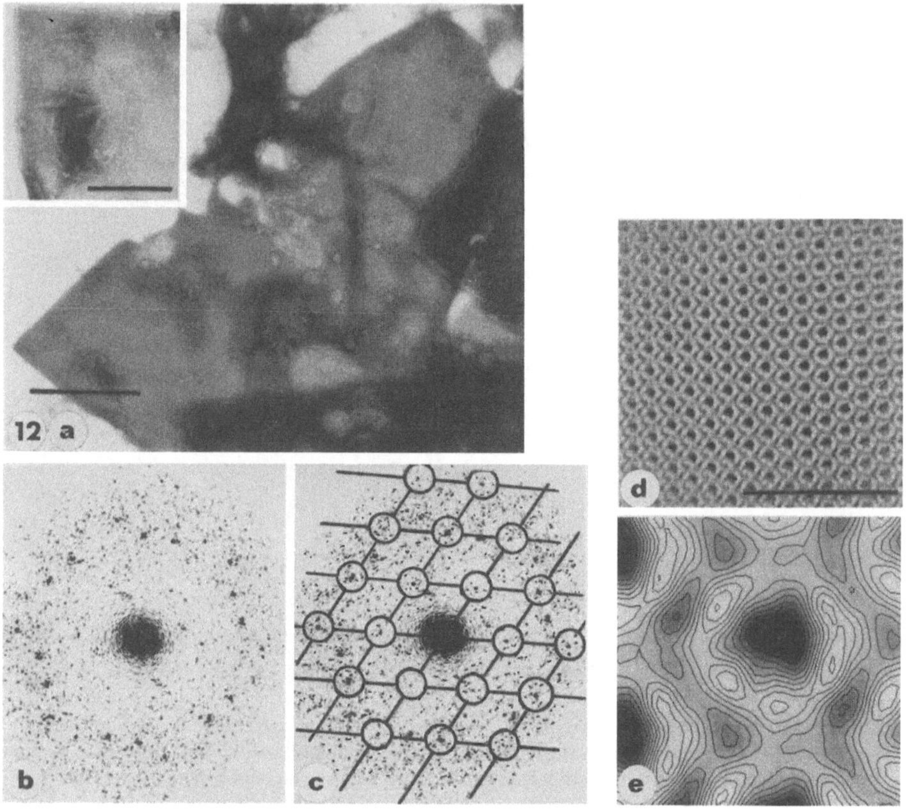

Fig. 12.12. **a** Lysed envelope of a square bacterium negatively stained with 1% uranyl acetate. Bar = 1 μm. *Inset* shows the periodic appearance at high magnification. Bar = 0.5 μm. **b** Optical diffraction from the *inset area* in a showing a clear double set of reflections. **c** The diffraction pattern from **b** with one set of the reflections indexed on a hexagonal lattice. **d** Fourier filtered image of a single cell wall. Bar = 0.1 μm. **e** A single periodic element from **d** after interpolation and superposition of a contour plot

Negative staining of intact cells showed only the outline of the cell and the positions occupied by inclusion bodies and gas vacuoles. In nonstained cells, however, the gas vacuoles were more easily seen (Fig. 12.9). In the natural brine organisms were found which had lysed either as a result of natural cell death or due to bacteriophage infection. In these lysed cells the periodic wall structure was very well maintained and was clearly seen in cells devoid of cellular contents, providing an opportunity of examining the cell-wall structure of these organisms in the natural state without any laboratory manipulation.

When the lysed cells found in the natural brine water were negatively stained with 1% aqueous uranyl acetate (pH 4.0), the structure of the cell wall was clearly revealed as an extended periodic array (Fig. 12.12). However, due to the fact that the negative stain attaches to both sides of the flat bacterial envelope, the resultant image is a superposition usually appearing as a moiré pattern. This situation is very confusing to the human eye and its interpretation requires image-processing techniques.

12.4.1.1 Image Processing

The relatively new methodology of image processing of electron micrographs (Frank et al. 1981) is outlined in the block diagram (Fig. 12.13). The first step is to examine the micrographs for evidence of periodic structure, using an optical diffractometer. This apparatus permits observation of the diffraction plane when the electron micrograph negative is used as the diffracting object. If a periodic arrangement is present in the micrograph the diffracted rays will give rise to a series of discrete reflections which themselves are arranged in the order corresponding to the image from which they were derived. Thus a hexagonal arrangement of subunits gives rise to a hexagonal arrangement of diffraction spots. When dealing with a two-sided object, such as the envelope of the square bacterium, each side will give rise to its own diffraction pattern. The two sides will almost always be rotated more or less with respect to each other and consequently a double set of reflections appears in the optical diffraction pattern. It is now possible to construct either an optical or digital filter, using only one set of the periodic compo-

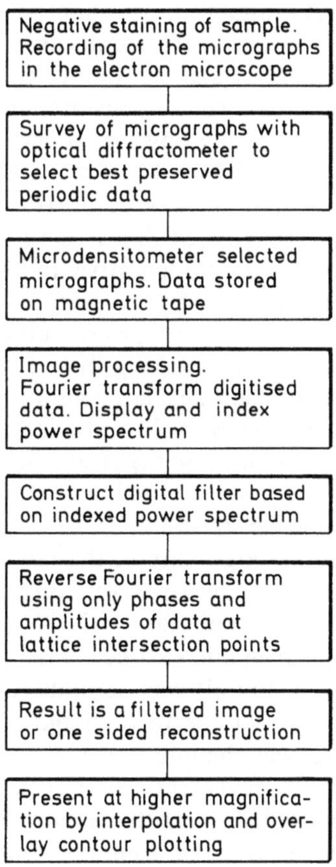

Fig. 12.13. Flow diagram of the steps in image processing of electron micrographs of periodic assemblies

nents of the image (the diffraction spots) and observe a reconstruction of one side of the bacterium. This is the only way that such information can be retrieved.

All image processing described in this chapter was carried out using the SPIDER system as developed by Joachim Frank at the Center for Laboratories and Research, New York Department of Health, Albany, N. Y. The results of the processing presented were obtained by using the computer facilities in Albany and Jerusalem, the latter using the Hadassah University Hospital Vax 11/780.

An example of a reconstruction made by the method described above is shown in Fig. 12.12. Figure 12.12a is an overall view of a negatively stained square bacterium which appears to be lysed. The optical diffraction pattern obtained from the area inset in Fig. 12.12a is shown in Fig. 12.12b. The diffraction pattern clearly has two sets of reflections, each arising from one of the sides of the bacterium. In Fig. 12.12c one of the sets of the reflections has been indexed as lying on a hexagonal lattice. Based on the optical diffraction analysis, the area marked on the micrograph and shown in the inset was chosen for image processing. Microdensitometer measurements were made of the area and the digital equivalent of the optical diffraction pattern known as the power spectrum was computed from the Fourier transform. In the power spectrum the two sets of reflections are clearly evident, and a digital filter is constructed corresponding to the lattice drawn, based on only one set of reflections as in Fig. 12.12c. The resulting filtered image is shown in Fig. 12.12d and is seen to be comprised of a hexagonal array of ringlike structures with a center-to-center spacing of 140Å. A clearer understanding of the image is obtained by superimposing a contour plot on the image which highlights areas of high and low density (Fig. 12.12e). The reconstruction presented shows the subunit to be comprised of a circular ridge of high density enclosing a stain-filled cavity. Such a structure contains the elements of a pore, but any such conclusion is only tentative and must await a three-dimensional reconstruction. In our studies, so far based on the one-sided reconstructions, we have evidence for at least two types of subunit structure: (1) the pore type described above and (2) a closer packing shared hexamer arrangement.

We have been successful in obtaining a large number of isolates of the bacteria from the Gavish Sabkha which grow well in the laboratory, and we have subsequently focused our attention on those organisms which have a square or close to square shape, and have used these for further characterization of their growth physiology, biochemistry and fine structure.

Cultures were grown to late log phase in the laboratory in the medium described by Javor et al. (1982). Cell envelopes were prepared by freezing and thawing in liquid nitrogen followed by incubation with DNAse. A drop of the envelope suspension was applied to electron microscope grids and examined either by negative staining or by shadow casting. It was possible to obtain negatively stained images of sufficiently high quality to perform image processing as described previously. In no case, however, was the image as clear as that due to natural lysis, but image processing was able to reveal the essential elements of the subunit structure of the wall. In all cases studied so far the reconstructions appear to be of the pore type. It should be mentioned that a dominant feature of the laboratory cultures is that they do not have gas vascuoles and seem therefore to correspond to the nonvacuolate type seen in the natural brine.

12.4.2 Stalked Bacteria

An additional class of bacteria discovered in the Gavish Sabkha pool are the stalked bacteria. We have tentatively identified these as a *Pedomicrobium* type, but a more positive classification awaits examination of more representatives of this class of organisms.

Nevertheless, negatively stained images of these organisms proved to be highly amenable to image processing. Figure 12.14a shows a low magnification micrograph of the stalked bacterium illustrating the extremely long hyphal tube emanating from the cell body. At higher magnification the surface of the cell body is seen to be comprised of a periodic structure, which is also evident on the sleeve-like extension of the cell body (Fig. 12.14b). The long hyphal tube itself also has strong periodic features (Fig. 12.14c). These are all superposition images whose structure can only be elucidated by the methods of image processing described before. Figure 12.14d is the power spectrum derived from Fig. 12.14c and following the scheme in the flow diagram. Figures 12.14e and f show the result of image processing of the hyphal tube, which reveals the subunits to be arranged in an exquisite trimeric array. Each element of the trimer appears to be a pore, as evi-

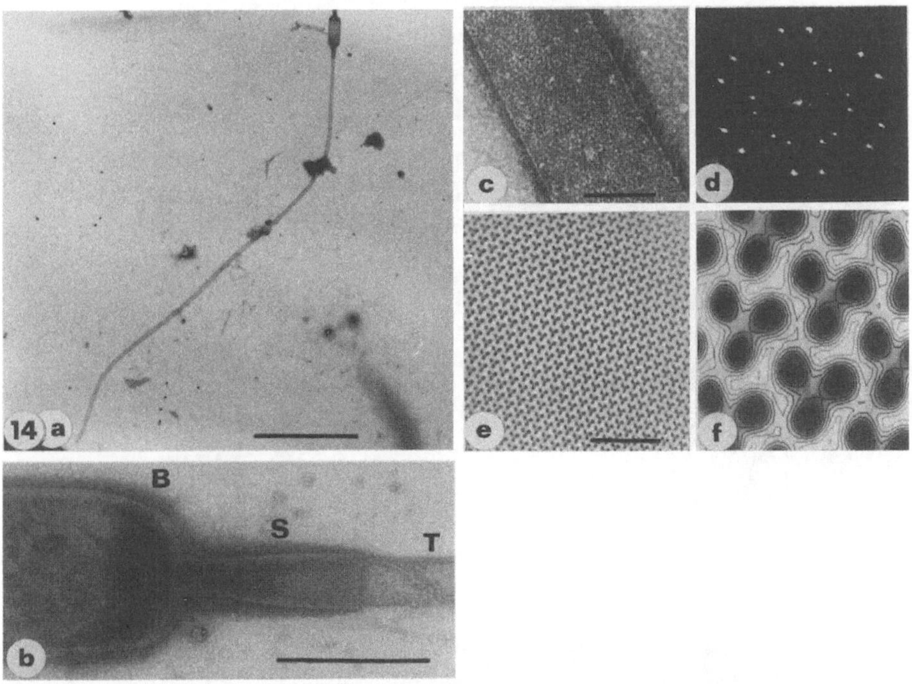

Fig. 12.14. a Low-power micrograph of a stalked bacterium showing the extremely long stalk. Bar = 5 μm. **b** Higher power of the cell body from which the tube emanates. *B* cell body, *S* sleeve, *T* tube. Bar = 0.5 μm. **c** High power of a small area of the stalk. Bar = 0.1 μm. **d** Power spectrum of the Fourier transform of **c** showing two sets of reflections rotated relative to each other by 23°. **e** Fourier filtered image based on the indexing of one of the sets of reflections. Bar = 0.1 μm. **f** Higher magnification of the periodic elements after interpolation and contour overlay showing the elements to be trimers. Each component of the trimer appears as a strain-filled depression

denced by a strongly staining central region. Again, this conclusion is tentative and must await further processing. Evidence for a similar trimeric structure has been obtained by Steven et al. (1977) and by Engel, Rosenbusch and their colleagues (pers. comm.) in their study of the *E. coli* outer membrane porin. It is tempting to speculate that the structure described here could be implicated in a function of nutrient assimilation as has been shown for porin and other outer membrane proteins.

12.5 Cell Motility

During our observations of the isolated strains from the Sabkha, with the cells observed in the phase contrast microscope, in a hanging drop preparation to pre-

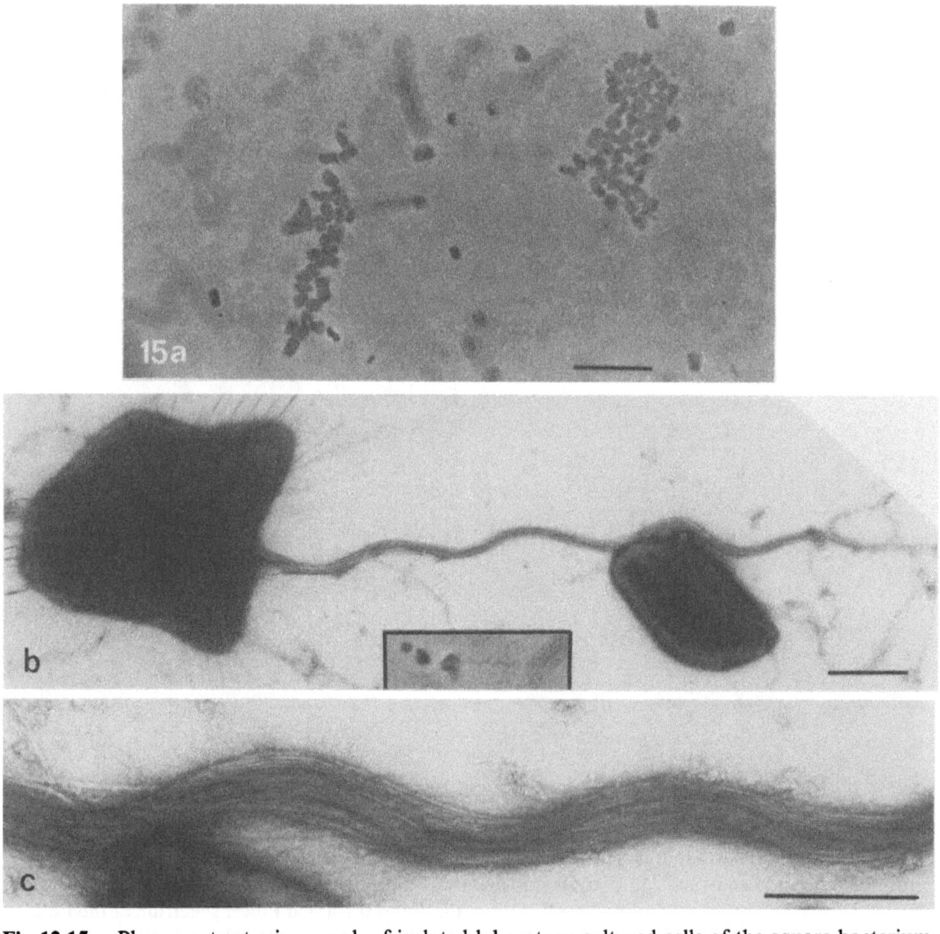

Fig. 12.15. a Phase contrast micrograph of isolated laboratory-cultured cells of the square bacterium type. Cells cluster along an imaginary line. **b** Negatively stained cell from the above population showing square shape and trailing a distinct flagellar bundle. *Inset* is a phase contrast micrograph showing the same. **c** Higher-powered view of the flagellar bundle. Bar = 1 μm

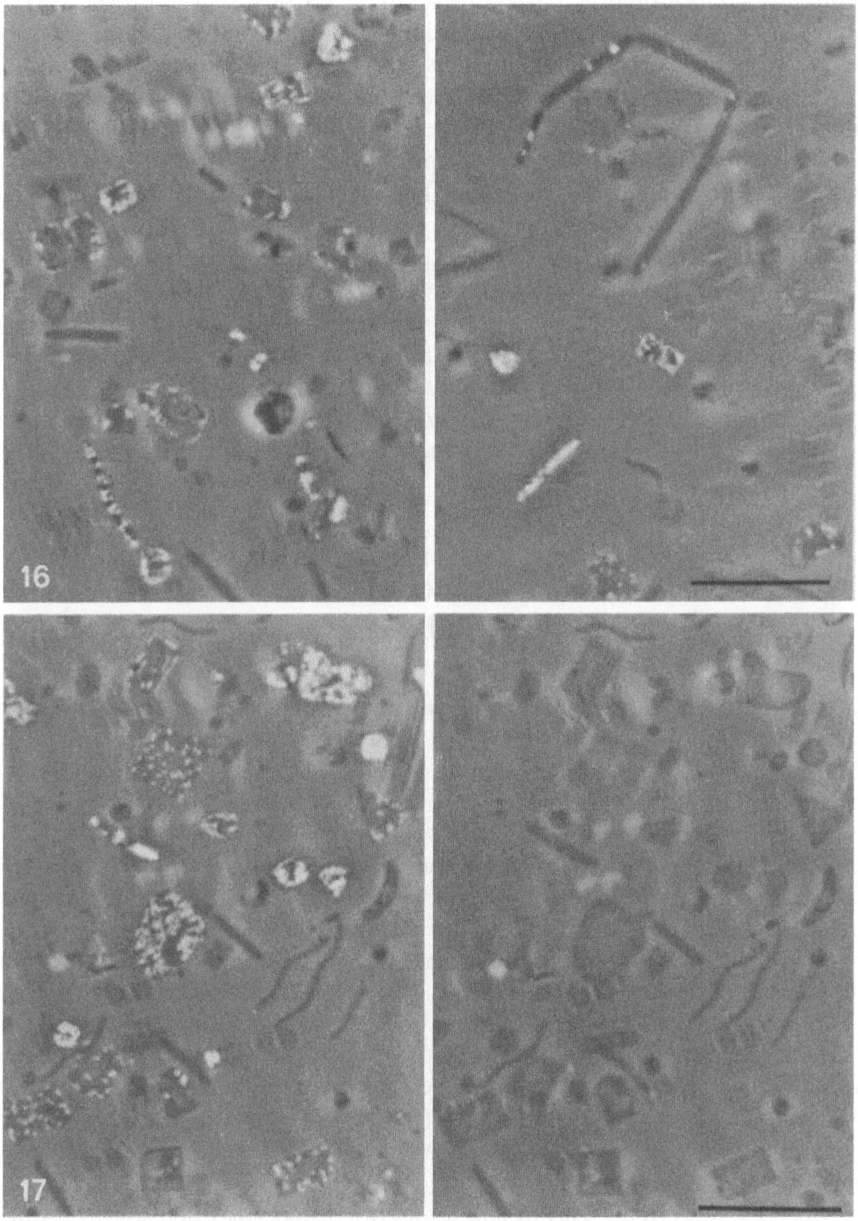

Fig. 12.16. Phase contrast light micrographs showing different gas-vacuolated organisms from Gavish Sabkha. Bar = 10 μm

Fig. 12.17. Phase contrast light micrographs of a field of square bacteria seen before *(on the left)* and after *(right)* application of 5 bar pressure which causes the gas vacuoles to disappear. Bar = 10 μm (Walsby 1980)

vent directional motion due to pressure from the cover slip, we noticed a peculiar tumbling behavior of the cells. In addition, in older cultures, a number of cells appeared to be tied together like clothes-pegs on a line (Fig. 12.15a), providing a most striking observation that the flat square-shaped cells were seen rotating in either a clockwise or counterclockwise fashion with one of the four cell corners acting as the pivot of a swivel. Our interpretation of the observation was that the cells must be flagellated and that the flagella of several cells had become entangled. This observation of the rotatory motion of the immobilized bacteria is strongly reminiscent of the tethering experiments conducted by Silverman and Simon (1974) with *E. coli* in order to provide evidence that the bacterial flagellum rotates rather than performs like a whip. Confirmation of the cell flagellation was obtained by examination of negatively stained cells from a culture in the logarithmic phase of growth. A number of flagella were seen emanating from a single cell (Fig. 12.15b) and these invariably became twisted into helical flagellar bundles. The flagella are nonsheathed and arranged in a peritrichous fashion.

12.6 The Importance of Gas Vacuoles for Halophilic Bacteria

Initial microscopic observations of natural samples from the Gavish Sabkha revealed an abundance of square-shaped structures containing numerous refractile bodies (Fig. 12.16). It was the demonstration that these refractile bodies were gas vacuoles that led to the conclusion that the observed square structures were gas-vacuolate bacteria (Fig. 12.17).

Gas vacuoles are found in a number of different bacteria in the Gavish Sabkha. The gas-vacuolate organisms show a range of morphological form. There are simple rods, coccoids, filaments and the square bacteria found by Walsby (1980). There is evidence that gas vacuoles cause these different bacteria to float, as in the case of cyanobacteria and other bacteria from freshwater and marine habitats (Walsby 1972, 1974, 1981a, b). The advantages accruing from flotation are different in the various groups and habitats. In halobacteria the buoyancy provided by gas vacuoles may take the cells to the brine surface where they may benefit from the higher oxygen tension (Petter 1931) and higher light intensity (Walsby 1981b). These ideas, which have yet to be tested experimentally, are examined in further detail below.

It was Klebahn (1895) who first demonstrated that gas-filled spaces occurred in certain cyanobacteria and he coined the term gas vacuole to describe them. He subsequently carried out many experiments on gas vacuoles in an attempt to find out what the nature of the gas was (Klebahn 1922, 1925, 1929), though the solution to this problem eluded him. One of Klebahn's other great interests was the halobacteria and he wrote a detailed review of these organisms (Klebahn 1919). Remarkably, it contains no mention that halobacteria too possessed gas vacuoles. This discovery was made by H. F. M. Petter (1931, 1932). She recognized that the gas vacuoles enabled halobacteria to float, and pointed out that in saturated brine pools, where the solubility of oxygen is very low, these obligate aerobes would accumulate at the brine surface, ensuring their access to air.

The gas vacuoles seen under the light microscope as bright refractile areas of the cell are made up of clusters of minute hollow structures termed gas vesicles. They were first demonstrated by Houwink (1956) in *Halobacterium halobium*. Larsen et al. (1967) found similar structures in *Halobacterium* sp. strain 5 and showed that the gas vesicles possessed a membrane with a regular substructure, like those found in cyanobacteria by Bowen and Jensen (1965).

The gas vesicles in halobacteria are usually lemonshaped and formed from 4.5 nm-wide ribs that are oriented normal to the long axis (see micrographs in Larsen et al. 1967, Stoeckenius and Kunau 1968, Walsby 1972). Less frequently the vesicles may be of a slightly different shape, cylinders closed by conical caps at each and like those in cyanobacteria but considerably wider.

The gas vesicles of halobacteria have the same physical properties and chemical constitution as those of Cyanobacteria, which have been studied more extensively. They are rigid structures, highly permeable to gases (Walsby 1969, 1971). Liquid water is excluded from the inside of the structure by the hydrophobicity of the inner surface. Although the gas vesicle shows little elastic compressibility, it is brittle and will collapse when subjected to sufficient hydrostatic pressure. The gas vesicles of halobacteria are much weaker than those of other bacteria; they have average critical collapse pressures of only 1 bar or less, compared with 6 bar or more in cyanobacteria. However, they are perfectly adequate to resist the pressures they are likely to encounter: the halobacteria have little or no cell turgor pressure (Walsby 1971), and the hydrostatic pressure in the shallow brine pools they inhabit will be only 0.1 bar for each meter depth. There has, therefore, been no selective advantage in these organisms' developing strong gas vesicles. It appears that the same mechanical considerations apply to gas vesicles as to the engineer's pressure vessel: its strength is related to the ratio of wall thickness to diameter. The weaker halobacterial gas vesicle is wider than the cyanobacterial one (Walsby 1972), though both have the same wall thickness (cf. Blaurock and Wober 1976, Blaurock and Walsby 1976). Consequently, the ratio of gas space to protein is substantially higher in the halobacterial vesicle and it is therefore more efficient at providing buoyancy. This is a nice instance of natural selection working to produce a structure of maximum efficiency (Walsby 1972). Simon (1981) has demonstrated that in a mutant of *Halobacterium salinarum*, the majority of gas vesicles are of the cylindrical type and narrower than the lemon-shaped vesicles that predominate in the wild type. In keeping with the trends described above, these mutant gas vesicles have a mean critical collapse pressure nearly twice that of the wild type (A. E. Walsby and R. D. Simon, unpubl.). This observation indicates that the *Halobacterium* would be capable of making stronger gas vesicles if required by constraints of pressure. Such stronger gas vesicles might be expected, for example, in halobacteria from the Dead Sea.

In all cases so far investigated gas vesicles are made solely of protein, and almost certainly of only one type of protein, called gas vesicle protein–GVP (see Armstrong and Walsby 1981, for a detailed review). The first 65 of the estimated 195 amino acid residues of GVP from the cyanobacterium *Anabaena flos-aquae* have now been sequenced (Walker and Walsby 1983). The amino-terminal sequences of other cyanobacteria are now being investigated and they are highly homologous. Of particular interest, though, is the finding that 50% of the first 26 amino-terminal residues in halobacterial GVP are identical to those in the sequence of the cyanobacterial GVP and many of the other residues show weak homology (Walker, Hayes and Walsby, in prep.). This is a much higher degree of homology than that found between other proteins of halobacteria and cyanobacteria, and is suggests that the GVP structural gene may have been transferred laterally between these two unrelated groups of organisms.

There is some circumstantial evidence that the GVP structural gene might be carried on a plasmid in halobacteria. Mutants defective in gas vesicle production arise with very high frequency (Larsen et al. 1967, Stoeckenius and Kunau 1968). Simon (1978) showed that the gas vesicle defectives of Larsen's *Halobacterium* sp. strain 5 had lost one of the three plasmids present in the wild type, though in subsequent studies it was found that the defective was capable of gas vesicle production at a very low rate (Simon 1978, 1981). These results might indicate that either there are two structural GVP genes (one on the deleted plasmid), or that the plasmid carries a gene that affects the expression of the structural gene. Weidinger, Klotz and Goebel (1979) found that loss of gas vesicle production in a mutant of *Halobacterium halobium* was accompanied by a change in the 100 Mdalton plasmid. Restriction mapping indicated that there was an insertion, presumably of DNA from the chromosome, in the plasmid of the mutant (Pfeifer et al. 1981). Here again the evidence that the plasmid carries the GVP structural gene is not conclusive. The insertion might have occured within the plasmid GVP gene, rendering it nonfunctional, but it is also possible that the inserted sequence came from part of the main chromosome carrying the structural gene. Identification of the coding DNA base sequence would be invaluable for locating the gene. It would also enable us to determine the relatedness of GVP in different organisms. There is a possibility that this gene may spread as a cross-infection between organisms sharing the same brine pool and this idea could be tested by comparing the sequence data.

Although information is accumulating on the physiology and chemistry of gas vesicles in halobacteria, we still need experimental confirmation of their ecological significance in these organisms. Walsby (1976) demonstrated with another obligate aerobe, *Prosthecomicrobium pneumaticum,* that gas vacuoles could bring the sort of benefits suggested by Petter (1931). The gas-vacuolate wild type and a gas vacuole-deficient mutant were found to grow at about the same rate in shaken culture but in static culture the wild type, which floated to the culture surface, outcompeted the mutant, which sank out. In a vessel constructed to allow sinking cells contact with a gas-permeable membrane in contact with air, while floating cells rose to a membrane in contact with nitrogen, the mutant won the competition.

Parallel experiments are needed with the halobacteria. It has been demonstrated that in the absence of oxygen halobacteria can still make ATP, using light absorbed by bacteriorhodopsin (Oesterhelt and Stoeckenius 1973); the buoyancy provided by gas vesicles should therefore also bring benefits if the halobacteria floated up into layers receiving a higher light intensity. These advantages could also be tested in competition experiments.

Laboratory experiments of the type discussed will at best only test whether it is *feasible* for the gas vacuole to provide benefits through buoyancy. The actual benefits can only be assessed by making observations and performing experiments in the field, as has been done in the case of gas-vacuolate cyanobacteria (Walsby and Klemer 1974, Konopka et al. 1978, Walsby et al. 1983a, b). The Gavish Sabkha would make the ideal site to test these ideas on gas-vacuolate bacteria and should remain accessible for microbiological and other scientific exploration.

13. Photoactive Pigments in Halobacteria from the Gavish Sabkha

Walther Stoeckenius, Don Bivin, and Kathleen McGinnis

The Gavish Dabkha in late summer and early fall typically shows a large area covered with a salt crust and a residual small and shallow, crescent-shaped brine pool with a strong orange-red color. The color is due to the presence of halobacteria at a cell density of $\sim 10^7$ cells ml. The population of the unicellular algae (*Dunaliella*), more abundant earlier in the year, has almost disappeared at this stage and in addition to the halobacteria only a few, small unidentified flagellates are seen upon microscopic examination of the brine. The color of the brine is due to a high concentration of carotenoid pigments present in the cell membranes of these bacteria, mainly the C_{50} carotenoid bacterioruberin. The distribution of microorganisms in the Gavish Sabkha is described in Chapters 11, 12 and 14. The physiological aspects of research on halobacteria are dealt with in Chapter 12. Ecological considerations are presented in Chapter 11. In this chapter, we deal mainly with the relations of halobacteria to light.

Recently, halobacteria have attracted considerable attention because they belong to the newly recognized kingdom of archaebacteria (Fox et al. 1980) and because they contain photoactive pigments structurally similar to the visual pigments of animals. These pigments not only function as light sensors for the phototactic responses of the organisms, but also serve als light-energy transducers, enabling the cells to use light as their energy source for metabolic work [Stoeckenius et al. (1979), Stoeckenius and Bogomolni (1982), for a definition of energy vs signal transducers see Oesterhelt and Stoeckenius (1971)]. The retinal-based photosynthetic mechanism operating in these concentrated brines is much simpler than the only other known photosynthetic system, which is based on chlorophyll.

Three such rhodopsin-like pigments have been found in the cell membrane of *H. halobium,* the most extensively studied halobacterium. Bacteriorhodopsin (bR) functions as a light-driven proton pump, ejecting protons from the cell (Stoeckenius et al. 1979). Halorhodopsin (hR) acts as a light-driven pump for the uptake of chloride ions (Schobert and Lanyi 1982). The recently discovered third pigment (sR) apparently mediates a phototactic response of the organism (Bogomolni and Spudich 1982). All three pigments are chemically and structurally very similar; the maxima of their visible absorption bands are located between 560 and 590 nm. Bacteriorhodopsin is the best characterized of the three. Its chromophore consists of retinal linked to the 26,000 MW polypeptide chain via a protonated Schiff base with the ε-aminogroup of a lysine residue, Additional, strong, noncovalent interactions, probably with charged groups in the protein, determine the position of the visible absorption maximum at 570 nm. The other two pigments apparently have very similar chromophores. For a recent review on the photoactive pigments of *H. halobium* see (Stoeckenius and Bogomolni 1982).

Conventional absorption spectra do not afford a sensitive assay for these pigments because of the high bacterioruberin concentration and light scattering of cells or cell membrane suspensions. Light adaptation, a 10 nm red shift and 13% increase in absorbance of the illuminated pigment, may be used to determine bR, but the other two pigments do not share this phenomenon (Oesterhelt and Stoeckenius 1971, Casadio et al. 1980, Casadio and Stoeckenius 1980, Bogomolni et al. 1980). However, bR is much more sensitively detected through transient absorbance changes accompanying its fast cyclic photoreaction. A photocycle intermediate, M_{412}, which absorbs maximally at 412 nm, is formed in 40 μs and decays with a half time of ~3 ms; the transient absorbance change near 412 nm can therefore be used to measure pigment concentration. The two other pigments, hR and sR, undergo similar photoreaction cycles and their photocycle intermediates have sufficiently different absorbance maxima and kinetics so that the pigments can be distinguished on this basis. The presence of bR and hR, but not sR, can also be detected by their effects on H^+ translocation in the presence or absence of uncouplers, e.g., carbonylcyanide 3-chlorophenyl hydrazone (CCCP), which render the cell membrane highly permeable to protons.

So far these photoactive retinal pigments have only been found in *H. halobium*. Using flash spectroscopy, we have detected the presence of similar or identical photoactive pigments in cell concentrates obtained by centrifugation of the brine from the Gavish Sabkha (Stoeckenius 1981). We describe here the pigment-related properties of some halobacteria strains isolated from the brine and solid salt and briefly discuss their physiological roles.

13.1 Techniques

All strains were isolated from samples collected between the times of late August and early November in 1980, 1981 and 1982. The techniques used for isolation and characterization, including electron microscopy, flash and conventional absorption spectroscopy, and the recording of light-dependent proton movements have been described elsewhere (Stoeckenius 1981, Javor et al. 1982).

13.2 Results

So far we have collected preliminary data for eight isolates from single red-colored colonies on agar plates; some of these were selected from the large collection of isolates obtained by Yehuda Cohen from the same source under essentially the same conditions.

The morphology of these strains as seen in the light microscope varies widely from long slender rods to short plump ones to coccoid shapes. Strains showing flat, square or triangular cells or with similar somewhat more irregular shapes are isolated frequently. We have termed these "haloarculae" and described them in more detail elsewhere (Javor et al. 1982). It may be noted that very similar shapes have recently been recognized in another group of archaebacteria, the methanogens (Wildgruber et al. 1982). All the isolates examined show a regular cell-wall structure in the electron microscope and many are motile.

The absorption spectra of cell suspensions or isolated membranes from all our isolates are dominated by absorption bands near 470, 500, 540 nm, which are indistinguishable from the bacterioruberin spectra of *H. halobium* or *H. cutirubrum* or of cell concentrates from the Gavish Sabkha brine (see Fig. 1 in Stoeckenius 1981). Flash spectroscopy of the same preparations clearly revealed the presence of photoactive pigments. All difference spectra looked similar. At 1 ms after the flash we see a marked decrease in absorbance near 570 nm and a somewhat smaller increase near 400 nm. Slightly later, at 6 ms, a more or less pronounced absorbance increase occurs near 640 nm. Our analysis of light-induced absorbance changes in *H. halobium* (Stoeckenius and Bogomolni 1982) indicates that the main features in the time-resolved spectra are caused by the presence of a bR-like pigment and may be modified by varying but generally smaller amounts of sR- and hR-like pigments. We have selected two isolates for further investigation: H.YC-5 has cells with a short rod or lanceolate shape; *H.s.* is a typical *Haloarcula* indistinguishable from *Haloarcula sinaiiensis* briefly described in Javor et al. (1982); both isolates have a regular cell-wall structure (Fig. 13.1), are flagellated, motile and show phototactic responses.

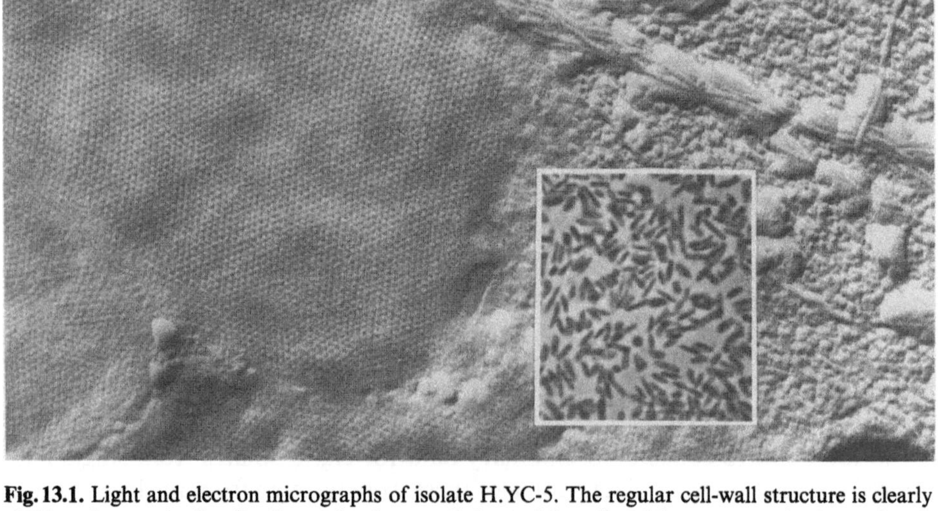

Fig. 13.1. Light and electron micrographs of isolate H.YC-5. The regular cell-wall structure is clearly visible and so are the flagella. Except for the overall shape of the cell and the presence of only one flagellum *H. sinaiiensis* has a very similar appearance. The cells were prepared for electron microscopy by drying a drop of the cell suspension on a Nuclepore filter, which rested on a pad of filter paper, so that the brine rapidly drains through the filter. This minimizes salt crystal formation. After shadowing with Pt-C and backing with a thicker layer of carbon the replica was floated off and cleaned with Clorox. Magnification: × 77,000. Inset: Phase contrast. Magnification: × 2,000

Figure 13.2 shows the difference spectra of membrane suspensions from H.YC-5 and *H.s.* before and after illumination and compares them with the to a hig concentration of carotenoid pigments present in the cell membranes of physiological aspects of research on halobacteria are dealt with in Chapter 12.

It is difficult to demonstrate convincingly hR and sR by flash spectroscopy in the presence of large amounts of bR. However, in the 1-ms difference spectra a rise of absorbance at 500 nm, where bR shows a decrease, would indicate the presence of hR, and a blue shift of the short wavelength peak from the 412 nm position of bR preparations suggests the presence of sR. Comparing the difference spectra for H.YC-5 and *H.s.* (Fig. 13.3) we see that they are dominated by features of a bR-like photoreaction cycle and that contributions from any hR- and/or sR-like pigments are slight.

The third retinal pigment of *H. halobium*, sR, can also be recognized by its much slower photoreaction cycle (Bogomolni and Spudich 1982). One observes that after a light flash the absorbance of the sample returns to its original value much later than can be accounted for by the photoreaction cycles of hR and bR,

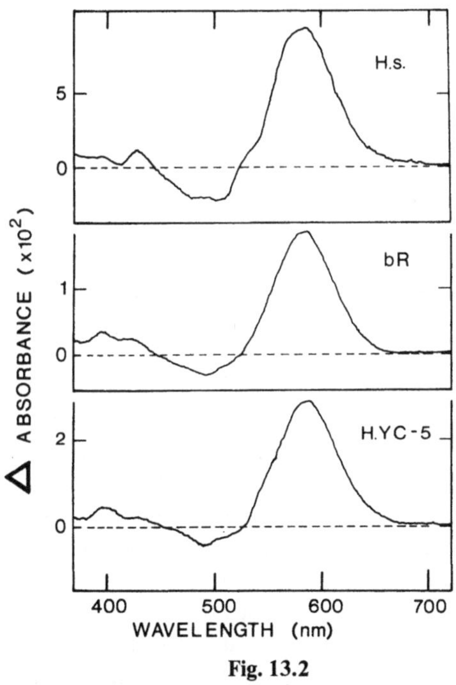

Fig. 13.2

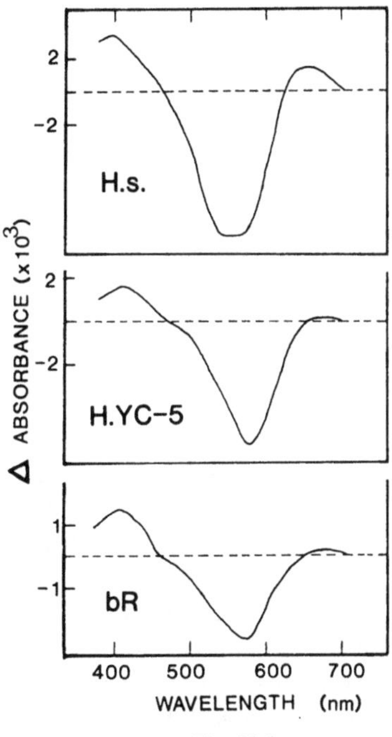

Fig. 13.3

Fig. 13.2. Difference spectra of light-adapted minus dark-adapted H.YC-5 and *H.S.* membranes compared to the corresponding spectrum for authentic bR. The membrane preparations were kept in the dark at room temperature overnight and then exposed to orange light (3×10^3 erg $cm^{-2} s^{-1}$) for 3 min. The similarity of all three spectra indicates the presence of bR-like pigments in both new isolates

Fig. 13.3. Flash-induced absorbance difference spectra of H.YC-5 and *H.s.* membranes 1 ms after excitation with a 5-ns laser flash. The corresponding difference spectrum for a *H. halobium* membrane preparation with a high bR content is shown for comparison. The decrease in absorbance near 570 nm and the increase near 400 nm are due to the formation of the short wavelength intermediates. Very similar spectra have been obtained from all red-pigmented isolates so far

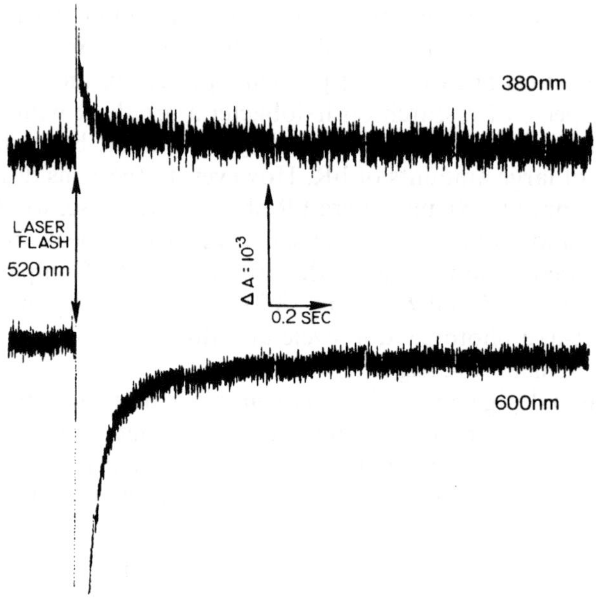

Fig. 13.4. Time-resolved absorbance changes at 380 and 600 nm in a H.YC-5 membrane preparation exposed to a 520-nm laser flash. For explanation see text

which last only a few milliseconds. Both isolates show this phenomenon, but it is more clearly seen in H.YC-5. Figure 13.4 shows the time-resolved absorbance changes at 600 nm and 380 nnm after excitation by a 520 nm laser flash. The rapid phase of the absorbance return is due to hR- and bR-like pigments, the slow phase – incomplete at the time-scale used – may be attributed to the presence of an sR-like pigment. Surprisingly, it is easier to demonstrate phototactic responses in *H.s.* than in H.YC-5.

Additional evidence for the presence of bR- and hR-like pigments can be obtained by analyzing the light-driven ion translocations in intact cells or cell-envelope vesicles and by other effects of light on cells, e.g., ATP synthesis. While the data and interpretations for H.YC-5 and *H.s.* are preliminary, they tend to confirm the spectroscopic results. Illumination of bR causes an outside-positive membrane potential and acidification of the medium through proton ejection and this effect is abolished by proton ionophores. Halorhodopsin also generates an outside positive membrane potential by translocating Cl^- into the cell. However, illumination will drive protons into hR-containing cells or vesicles and this effect will be enhanced by the addition of protonophores and inhibited by an increase in the permeability for other ions. The absorption spectrum of bR is so similar to that of hR that any light that effectively excites one will also excite the other, and the usually higher concentration of bR will tend to obscure the function of hR when light-driven proton movements are montiored. However, the effect of protonophores and other permeability-modifying agents is the opposite for hR- and bR-driven proton fluxes. This makes it possible to detect an hR-like pigment even in the presence of relatively large amounts of bR. Figure 13.5 shows the results of such an assay for our two isolates; it corroborates the tentative interpre-

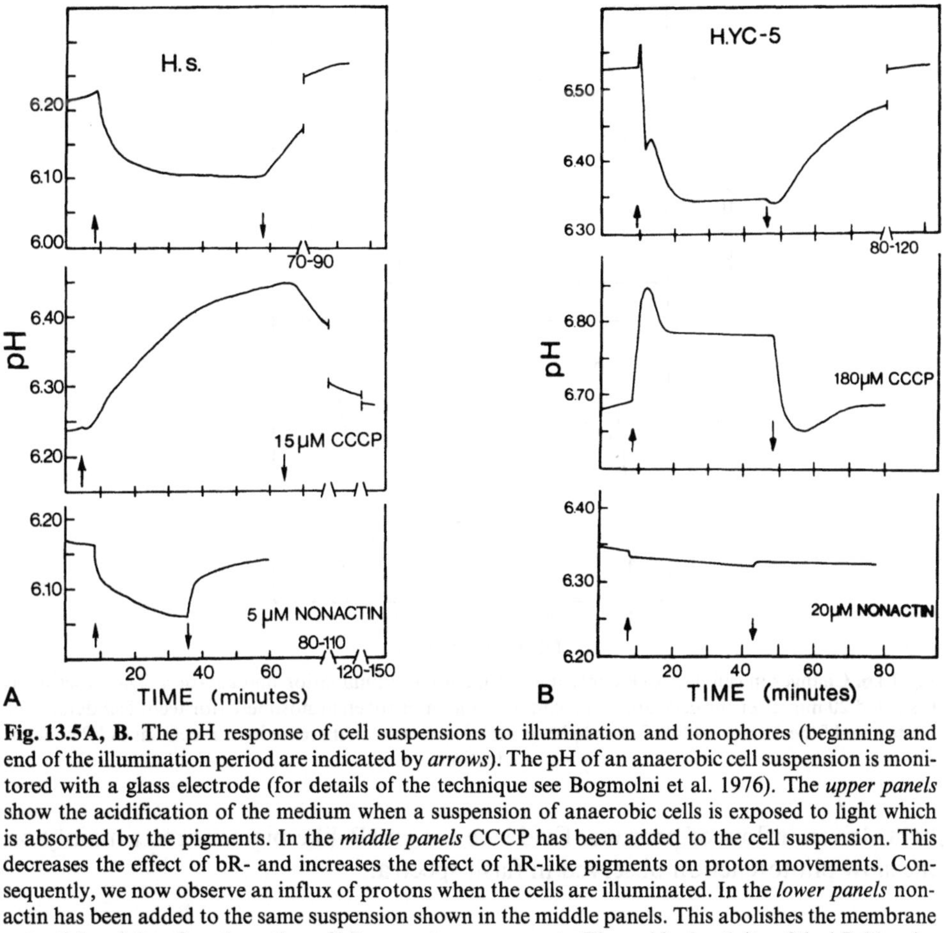

Fig. 13.5A, B. The pH response of cell suspensions to illumination and ionophores (beginning and end of the illumination period are indicated by *arrows*). The pH of an anaerobic cell suspension is monitored with a glass electrode (for details of the technique see Bogmolni et al. 1976). The *upper panels* show the acidification of the medium when a suspension of anaerobic cells is exposed to light which is absorbed by the pigments. In the *middle panels* CCCP has been added to the cell suspension. This decreases the effect of bR- and increases the effect of hR-like pigments on proton movements. Consequently, we now observe an influx of protons when the cells are illuminated. In the *lower panels* nonactin has been added to the same suspension shown in the middle panels. This abolishes the membrane potential and therefore the action of hR on proton movements. The residual activity of the bR-like pigment is revealed in the small acidification. The over- and undershoots of the pH response seen especially in the *upper* and *middle panels* in **B** are an expression of the regulatory functions in the cells. (For a detailed interpretation see Bogomolni et al. 1976)

tation of the flash spectra that in addition to a bR-like pigment an hR-like pigment is also present.

Illumination of H.YC-5 and *H.s.* cells reversibly increases the intracellular ATP concentration (Fig. 13.6). The same effect in *H. halobium* must be attributed to the bR content (Danon and Stoeckenius 1974, Hartmann and Oesterhelt 1977); we therefore assume that the effect in our new isolate is also due to the bR-like pigment, confirming the conclusion reached on the basis of light adaption, flash spectra and light-driven proton efflux. Contrary to earlier reports, hR apparently cannot mediate light-driven ATP synthesis under physiologic conditions (Stoeckenius and Bogomolni 1982). A contribution from sR or an sR-like pigment is unlikely because we have so far failed to find any evidence for light-generated ion gradients in cells lacking bR and hR but containing sR. The strongest evidence for an hR-like pigment in these isolates is therefore the CCCP-facilitated

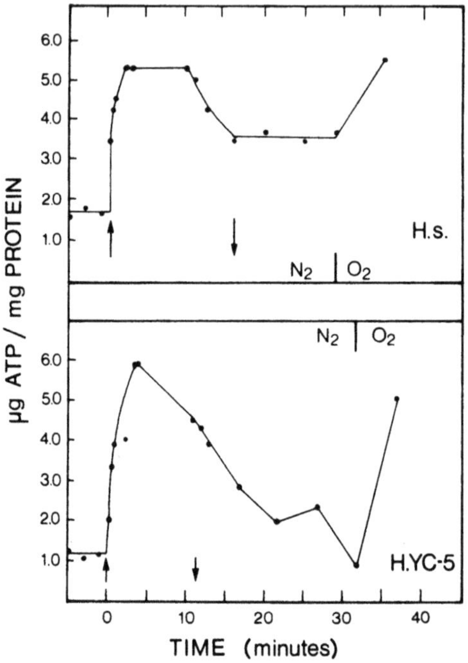

Fig. 13.6. Changes in cellular ATP content as a function of illumination and aeration. The experiment is started 20 min after the cells are in a chamber under a nitrogen atmosphere (for technical details see Danon and Stoeckenius 1974). Either light or aeration causes an increase in intracellular ATP concentration

light-driven inflow of protons. Evidence for an sR-like pigment rests mainly on the slow photoreaction observed in flash spectra.

13.3 Discussion

Our results indicate that retinal pigments similar to or identical with the bR, hR and sR of *H. halobium* are rather common in halobacteria. This is further borne out by similar findings in other halobacterium strains isolated from geographically distant locations (Javor et al. 1982). We are apparently dealing with a family of retinal pigments which mediate both photosensory and photosynthetic functions and we will consider briefly their physiological role.

In all strains isolated bR is the dominant photoactive pigment, but only rarely does it appear in the crystalline patches known as purple membrane. Whether this is simply a question of pigment concentration or has a functional significance is not known. Of the two strains described here H.YC-5 shows purple membrane-like patches in freeze-fracture electron micrographs; *H.s.* does not (our unpublished observations).

The main energy source for extreme halophiles is respiration; it generates an outside-positive electrochemical proton gradient which drives ATP synthesis, locomotion, K^+ uptake, Na^+ extrusion and other energy-requiring processes. Bacteriorhodopsin apparently is an ancillary energy transducer which can substitute for respiration when oxygen tension is low and light is available. This should often be the case in stagnant brine pools at high ambient temperature and in the absence of O_2-producing algae. It is known that in *H. halobium* a small part of the total absorbed light energy suffices to saturate photophosphorylation, the remainder is used mainly to increase the NaCl and KCl gradients (Bogomolni et al. 1976; Wagner et al. 1978, Helgerson et al. 1983). These gradients have a large capacity for energy storage, because the salt concentrations are so high. *H. halobium* cells are able to sustain ATP synthesis for extended times at the expense of the salt gradients and an improved survival of anaerobic starving, but illuminated vs dark cultures has been reported (Brock and Petersen 1976). Oren and Shilo (1981) have recently demonstrated the presence of bR in a halobacteria bloom in the Dead Sea and found that light absorbed by bR was responsible for the CO_2-fixation observed (A. Oren 1983 a, b). However, so far all attempts to grow halobacteria with light as the sole energy source have failed, and further investigations into the physiological role of bR are badly needed.

The role of the second light-energy transducer, hR, is still obscure. However, K^+ appears to be the most permeant anion at least in *H. halobium* and, therefore, Cl^- influxes should be accompanied by a K^+ influx, and should increase the internal salt concentration; the expected light-induced volume increase upon illumination has been observed (Schobert and Lanyi 1982). It may be argued that only relatively low concentrations of this pigment are necessary, because the increase in brine concentration due to evaporation is a slow process.

The phototaxis of *H. halobium* comprises a repellent response by blue and near ultraviolet light and an attractant response by green and red light, which may both be mediated by sR (Bogomolni and Spudich 1982). Since the attractant response has an action spectrum with a maximum close to the absorption maxima of bR and hR, it leads to an accumulation of cells in regions where the intensity at the wavelengths absorbed by the two energy transducers is highest. Thus sR should optimize the conditions for light-energy transduction in the natural environment where light scattering and the strong absorbance of bacterioruberin generate gradients of blue and red light.

Halobacteria are archaebacteria, which generally seem to have survived only in a few rather extreme environmental niches (Fox et al. 1980). They may, therefore, have preserved some rather primitive features. Indeed, bR is clearly the simplest biological light-energy transducer and ion pump known. The apparent structural similarity of hR and sR coupled to a very different function is very intriguing. A detailed comparison of the molecular structure and photoreactions of the three pigments should reveal new insights into the mechanisms of fundamental biological processes. The Gavish Sabkha and similar ecosystems elsewhere have thus proven to be a rich source of material uniquely suited for basic investigations in light-energy and signal transduction and they offer some of the best examples of chemiosmotic mechanisms. Implications of these findings for our ideas about the origin of life have been discussed elsewhere (Stoeckenius 1978).

Thus the halophilic microorganisms living in and isolated from the Gavish Sabkha turned out to be some of the best witnesses of the conditions of life on the early Earth and on Earth's earliest biospheres.

14. Photosynthetic Microorganisms of the Gavish Sabkha

Aline Ehrlich and Inka Dor

Many studies during the past 20 years have dealt with the photosynthetic and lithogenetic processes occurring in benthic microbial mats of saline lakes throughout the world (Bauld 1981 a, b). The primary constituents of the majority of mats are cyanobacteria; however, eukaryotic algae, e.g., diatoms and green algae, are often present, and sometimes even abundant, especially in brackish and metahaline waters. At salt concentrations exceeding 100‰ S, reports on living diatoms are rare. The diatom assemblages thriving in the benthic mats of Solar Lake (northeast Sinai) are the most salt-resistant ones hitherto described from a permanently hypersaline (up to 180‰ S) environment (Ehrlich 1978 a, b). Cyanobacteria, however, have been reported from much more saline waters, some of them able to grow in 300‰ S (Borowitzka 1981).

This chapter presents a detailed taxonomic inventory of the phytobenthic organisms recorded in different types of mats and surface sediments throughout the Gavish Sabkha, including cyanobacteria, diatoms and green algae. The composition of various communities and the relative frequency of their constituents in relation to the salinity gradient will be discussed.

14.1 Materials and Methods

14.1.1 Sampling and Preparation

The material of the present study was taken in January and March 1982 at 33 sampling stations of the Gavish Sabkha (Fig. 14.1), corresponding with the zonation described in Chap. 11. Most of the stations were on the eastern slope of the rim (Zone 4, see Chap 11), in the area of the emerging sea water seepages (Zone 5), and at different depths of the shallow lagoon (Zone 6). A few samples were also collected from the elevated central rim in the shallow water zone, which was covered by a halite crust (Zone 7). Altogether, seventy samples were collected at water depths varying from 0 to 50 cm. The material collected was stored in small plastic bottles, with formaldehyde 3% as a preservative.

In the laboratory the samples were first rinsed with distilled water, and a preliminary examination of the raw material was carried out in order to identify the dominant genera. At this stage, the relative contribution of the various groups to the total biomass was semi-quantitatively estimated (Table 14.1). The samples were later processed separately for the different groups.

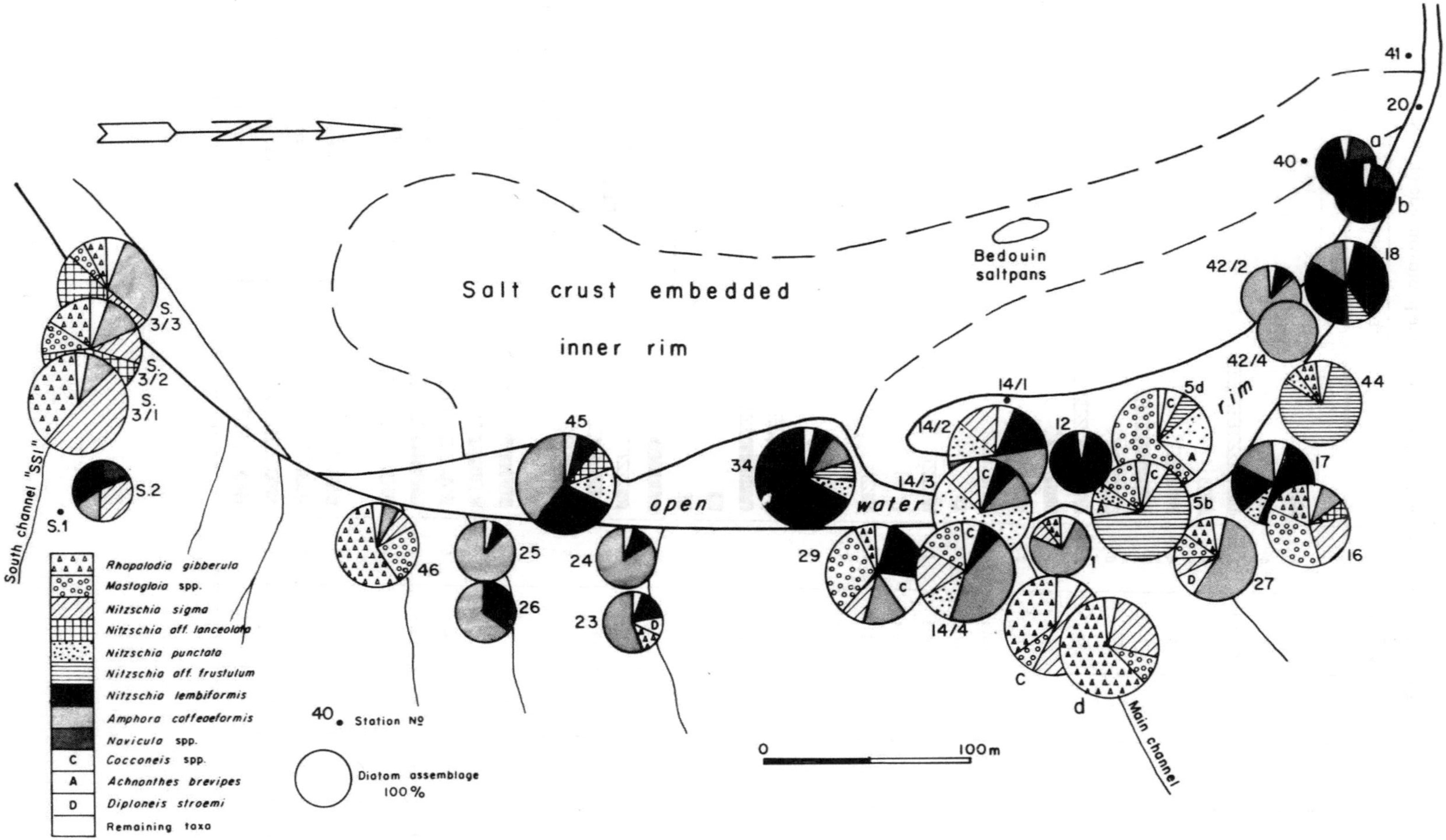

Fig. 14.1. Composition of diatom community at various sampling stations

Table 14.1. Sampling stations and composition of the photosynthetic biomass

Sampling Station No.	Number of Samples	Kind of Samples	Measured Salinity %	Zone* No.
1	1	S	5.8	4
2	1	C	7	4
5	5	M	5.1-6.5	5
12	1	C	19	6
14/1	4	C	15	6
14/2	4	M	8.7-13.7	6
14/3	1	M	7	6
14/4	1	M	7	6
16	1	S	6	4
17	1	C	6	4
18	3	S	20.5	6
20	2	C	17.5	7
23	1	S	6.3	4
24	2	S	5.4	4
25	1	C	6	4
26	2	C	6.5	4
27	1	S	5	4
29	1	M	6	5
34	3	M	17.5	6
40	1	C	32.7	7
41	1	C	>25	7
42	4	M	18.2	6
44	2	S	6.1	5
45	3	M	≈13	6
46	1	M	6	5
SSI				
1	1	C		4
2	2	S	20	4
3/1	3	M	6.5	5
3/2	3	M	6.5	6
3/3	1	M	6.5	6
A	1	S	≈18	6
B	1	S	≈18	6
C	1	M	≈ 6	6
D	2	M	≈ 5.5	5

Estimated biomass of Cyanobacteria, Diatoms and Green Algae

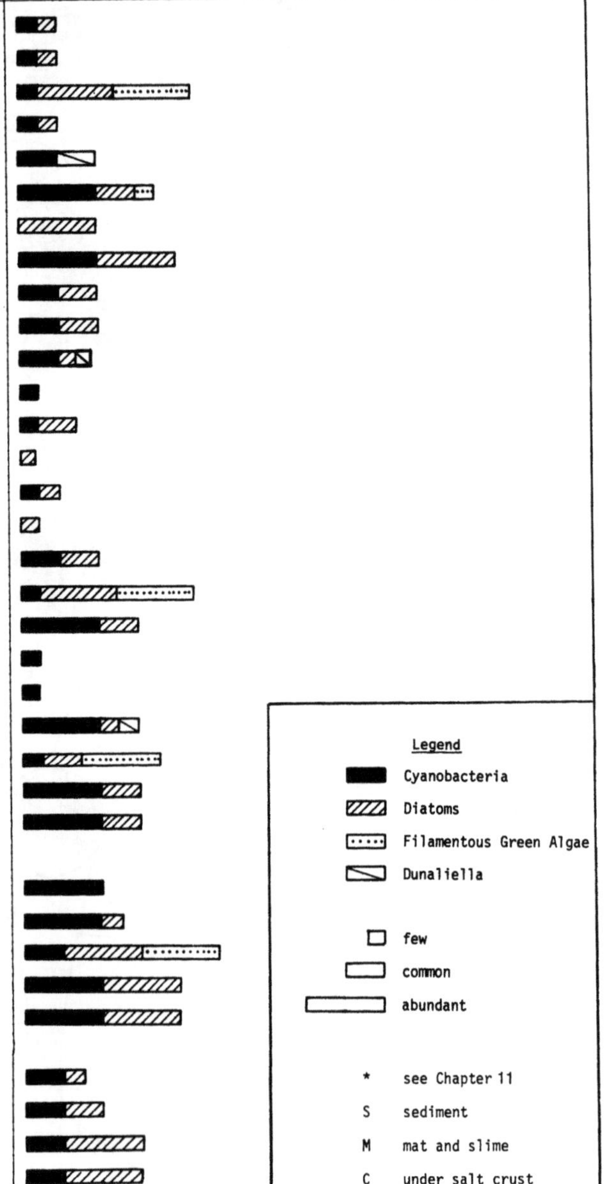

14.1.2 Identification

14.1.2.1 Cyanobacteria (Blue-Green Algae)

For the identification of cyanobacteria, a small subsample was placed on a microscope slide in a drop of warm glycerine-gelatine-phenol embedding mixture (Johansen 1940), a coverslip was added, and the slide allowed to cool. The permanent slides prepared in this way were utilized in further taxonomic work. Altogether, 55 samples were examined for cyanobacteria.

The taxonomy of cyanobacteria has recently been reevaluated and a new approach is in process of being defined (Stanier et al. 1971, Rippka et al. 1979). Until the discriminating properties introduced by the bacteriological approach become more defined, we still have to use the old classical taxonomy based on the botanic code. Accordingly, classification of cyanobacteria was based on Hauck (1885), Gomont (1892), Geitler (1925, 1932, 1942), Frémy (1930, 1934), Frémy and Nasr (1938), Volcani (1940), Desikachary (1959), Bourrelly (1970) and Humm and Wicks (1980). Remarks on distribution in Israel and adjacent area are based on Y. Cohen et al. (1977 b): Solar Lake, 75‰–150‰ S; Dor (1967): Tiberias Hot Springs, 33‰ S; Dor (1974): Lake Kinneret, about 0.2‰ S; Dor (1975) and Dor (1984): Sinai Mangroves, 40‰–60‰ S; Potts (1980): Marine coastol environments of Sinai, 40‰–350‰ S, Rayss (1959 a, b); Rayss and Dor (1963): Red Sea shore; 40‰–41‰ S.

14.1.2.2 Bacillariophyta (Diatoms)

In order to check if diatoms, whenever present, were alive at the time the samples were collected, the material was first examined under the microscope in a drop of water, and cells containing green or brownish plastids were looked for; diatoms devoid of cellular content were regarded as "dead" cells. This, however, did not allow an accurate identification of the taxa, hence for the taxonomic study the samples were further processed as follows.

The fixed material was rinsed with distilled water, cleaned with 5% hydrochloric acid and then heated in a solution of hydrogen peroxide for a few hours (at 80 °C in a boiler) until cleared of the organic matter. After repeated rinsings at time intervals of about 1 h, one drop of the water suspension was deposited on a cover slip, allowed to dry and then mounted on a microscope slide with a resin of high refraction index (Aroclor mounting medium, $n = 1.66$). The permanent slides thus obtained were stored in the collection of the Paleontology Division, Geological Survey, Jerusalem.

The diatoms were identified using a Zeiss light microscope at objective magnification of 40 and 100 (in oil immersion). At least 300 valves (all regarded as "living" at time of collection) were counted for frequency percentages (Fig. 14.1). Photomicrographs were taken under Nomarski differential interference contrast (DIC) with a Reichert Polyvar microscope. For SEM observation, the material was prepared as above and cleaned ultrasonically for 1 min before being dried on a stub and coated by gold. The SEM pictures were taken with a Camscan. Altogether, 63 samples were examined for diatoms.

Diatom identification was based on Hustedt (1930a, b, 1957, 1959, 1961–66), Peragallo and Peragallo (1897–1908), Schmidt's Atlas (Schmidt et al. 1874–1959), O. Müller (1899) and others. Data concerning the geographic distribution of the taxa and their ecology were compiled from the above authors, and also from Brockmann (1950), Simonsen (1962), Cholnoky (1968), Baudrimont (1973) and Gasse (1975). Reports on occurrences in Israel and Sinai were based on A. Ehrlich (1975, 1978a, b), A. Ehrlich and Ortal (1979), and on unpublished data of one of the authors (A. E.). Reference was made to the following locations: brackish springs of the Dead Sea area, including among others Ein Feshkha, Ein Kalia and Ein Rwer on the western shore of the Dead Sea area, 1,5‰–6‰ S; Solar Lake (northeast Sinai), 75‰–150‰ S; Bardawil Lagoon (north Sinai), 40‰–80‰ S; Lake Kinneret, about 0.2‰ S; Jordan River, 0.1‰–0.5‰ S; Jordan R. at Alummot Dam, 5‰–10‰ S.

14.1.2.3 Chlorophyta (Green Algae)

Representatives of this group were examined directly under the microscope, and identified at the generic level only.

14.2 Results

14.2.1 Composition of the Photosynthetic Biomass

Sixty-two samples examined (Table 14.1) yielded varying amounts of cyanobacteria, diatoms and green algae, which contributed in changing proportions to the total biomass. Filamentous green algae were found only in the metahaline waters of Zone 5 and in the *Enteromorpha* belt (both up to 65‰ S). Cells of *Dunaliella* appeared in three stations at salinities of 150‰–200‰. Diatoms were found in most stations up to 205‰ S. Great numbers of them were noticed in some slimes and mats from metahaline waters (50‰–70‰ S), whereas in the higher salinity of the lagoon (located within Zone 6), or in the almost dry sediments (Zone 4), diatoms were less abundant. Cyanobacteria were recorded everywhere in the area studied up to 327‰ S. As a rule, their contribution to the total biomass was preponderant above 100‰ S, and at extremely high salt concentration they became the sole constituent.

14.2.2 List of Taxa

Taxa	Abundance[a]	Salinity (‰ S)	References to plates and figures
CYANOBACTERIA (blue-green algae)			
Order: CHROOCOCCALES			
1. *Aphanocapsa littoralis* Hansg.	+	50– 65	14.1 (2)
2. *Aphanocapsa marina* Hansg.	+++	50–185	(1)
3. *Aphanothece halophytica* Frémy	++(+)	150–330	(3)
4. *Aphanothece saxicola* Näg.	+	60– 65	(4)
5. *Aphanothece stagnina* (Spreng.) A. Br.	+++	50–200	(5)
6. *Chroococcus minor* (Kütz.) Näg.	+++	50–200	(6)
7. *Chroococcus minutus* (Kütz.) Näg.	+	60	(8)
8. *Chroococcus turgidus* (Kütz.) Näg.	+++	50– 70	(7)
9. *Gloeocapsa deusta* (Menegh.) Kütz. (?)	+	60	14.2 (1)
10. *Gloeocapsa polydermatica* Kütz.	+	200	(3)
11. *Gloeothece confluens* Näg.	++	60–200	14.1 (9)
12. *Gomphospheria aponina* Kütz.	+	90–180	(10)
13. *Entophysalis granulosa* Kütz.	+++	50–205	14.2 (2)
14. *Johannesbabtistia pellucida* (Dickie) Tylor et Drouet	+	60– 65	(5)
Order: PLEUROCAPSALES			
15. *Chroococcopsis* sp.	+	175–185	(6)
16. *Pleurocapsa fuliginosa* Hauck	+	50–200	(4)
Order: NOSTOCALES			
17. *Arthrospira miniata* Gom.	++	60	14.5 (8)
18. *Hydrocoleum lyngbyaceum* Kütz.	+	60– 65	14.3 (1)
19. *Lyngbya aestuarii* Liebm.	+	50– 65	14.5 (5)
20. *Lyngbya complectens* Frémy	+	90–135	14.4 (1)
21. *Lyngbya infixa* Frémy	+	70–135	14.5 (6)
22. *Lyngbya* sp.	++(+)	90–135	(7)
23. *Microcoleus chthonoplastes* Thur.	++	50–180	14.3 (2)
24. *Oscillatoria nigro-viridis* Thwaites	+	90–135	14.5 (2)
25. *Oscillatoria salina* Biswas	+	60	14.5 (3)
26. *Oscillatoria tenuis* Ag.	++(+)	60–150	(1)
27. *Phormidium fragile* (Menegh.) Gom.	+	50– 65	14.4 (5)

[a] + rare, ++ common, +++ abundant

Taxa	Abundance[a]	Salinity (‰ S)	References to plates and figures
28. *Schizothrix arenaria* (Berk.) Gom.	+++	50–330	14.3 (3)
29. *Schizothrix calcicola* (C. Ag.) Gom.	+++	50–205	14.4 (2)
30. *Schizothrix nasri* Frémy	++(+)	60–200	14.4 (1)
31. *Spirulina labyrinthiformis* (Menegh.) Gom.	+++	60–175	(4)
32. *Spirulina subsalsa* Oersted	+++	60–200	14.4 (3)
33. *Calothrix parietina* Thur.	+	70	14.5 (4)
BACILLARIOPHYTA (diatoms)			
Order: PENNALES			
1. *Achnanthes brevipes* Ag.	+(++)	55– 75	14.6 (1, 2) 14.8 (4, 5)
2. *Cocconeis placentula* (Ehrenberg) var. *euglypta* (Ehrenberg)	+(++)	60–130	14.6 (3, 4) 14.8 (1–3)
3. *Diploneis stroemi* Hust.	+(++)	50– 70	14.6 (15) 14.8 (9)
4. *Mastogloia brauni* Grunow	++(+++)	50– 70	14.6 (5–7) 14.9 (6, 8, 9)
5. *Mastogloia pumila* (Grun.) Cleve	++(+++)	50– 70	14.6 (9–11) 14.9 (7,10,11)
6. *Mastogloia sirbonensis* Ehrlich	+(++)	65	14.6 (8)
7. *Navicula* sp. cf. *N. cincta* (Ehrenberg) Kütz.	+(++)	50–130	14.7 (15) 14.9 (4,5)
8. *Navicula* sp. cf. *N. complanata* (Grun.) Grun.	+(++)	60–150	14.9 (1–3)
9. *Navicula* sp. cf. *N. ramosissima* Ag.	+	60– 80	14.6 (12)
10. *Navicula* sp. cf. *N. subinflatoides* Hust.	+(++)	50–205	14.6 (12, 16) 14.10 (1–4)
11. *Navicula* sp.	++	70–205	14.10 (2, 3)
12. *Pleurosigma formosum* W. Smith	+(++)	50– 70	14.6 (21,22)
13. *Amphora angusta* (Gregory) Cleve	++	60	14.7 (3, 4) 14.10 (12)
14. *Amphora coffeaeformis* (Ag.) Kütz.	++(+++)	50–205	14.7 (5–8, 11) 14.10 (8, 9)
15. *Amphora* sp. cf. *A. coffeaeformis var. acutiuscula* (Kütz) Hust.	++(+++)	*60–130*	*14.7 (9, 10)* *14.10 (10,11)*
16. *Amphora* sp. 1	++(+++)	*50– 70*	*14.7 (1)* *14.10 (5–7)*
17. *Amphora holsatica* Hust.	++	*50– 70*	*14.7 (2)*
18. *Amphora proteus* Gregory	+	*50– 70*	*14.6 (20)*

Taxa	Abundance[a]	Salinity (‰ S)	References to plates and figures
19. *Rhopalodia gibberula* (Ehrenberg) O. Müller	++	50– 70	
20. *Rhopalodia gibberula* var. *timsahensis* (O. Müller)	++	50– 70	14.6 (16, 17) 14.8 (6–8)
21. *Rhopalodia gibberula* var. 1	+	50	14.6 (18)
22. *Nitzschia frustulum* (Kütz.) Grunow	++	50– 70	14.11 (5, 9)
23. *Nitzschia frustulum* var. *perpusilla* (Rabenhorst) Grun.	++	50–205	14.7 (19–21) 14.11 (3, 4)
24. *Nitzschia lanceolata* W. Smith var. *minor* Van Heurck	++	50–170	14.7 (14)
25. *Nitzschia lembiformis* Meister	++	150–205	14.7 (17, 18) 14.11 (11–14)
26. *Nitzschia punctata* (W. Smith) Grunow	++(+++)	60–150	14.6 (19) 14.11 (1, 2)
27. *Nitzschia sigma* (Kütz.) W. Smith	++(+++)	50–150	14.7 (24, 27) 14.11 (10)
28. *Nitzschia* sp. 1	++	60– 70	14.7 (22, 23) 14.11 (7, 9)

Appendix: Allochthonous freshwater diatom taxa sporadically found in the surface sediment samples

Cyclotella kuetzingiana Thwaites
C. kuetzingiana var. *planetophora* Fricke
Cymbella affinis Kütz.
Fragilaria construens (Ehrenberg) Grunow
F. construens var. *venter* (Ehrenberg) Grunow
F. pinnata Ehrenberg
Aulacosira granulata (Ehrenberg) Simonsen
A. granulata var. *angustissima* (O. Müller) Simonsen
A. italica var. *valida* (Grunow) Simonsen
Stephanodiscus astraea var. *minutula* (Kütz.) Grunow
Synedra ulna (Nitzsch) Ehrenberg

CHLOROPHYTA (green algae)			
Order: VOLVOCALES			
1. *Dunaliella* sp.	+(++)	150–200	
Order: ULVALES			
2. *Enteromorpha* sp.	+++	50– 65	
Order: SIPHONOCLADALES			
3. *Rhizoclonium* sp.	+++	50– 65	
4. *Chaetomorpha* sp.	+++	50– 65	

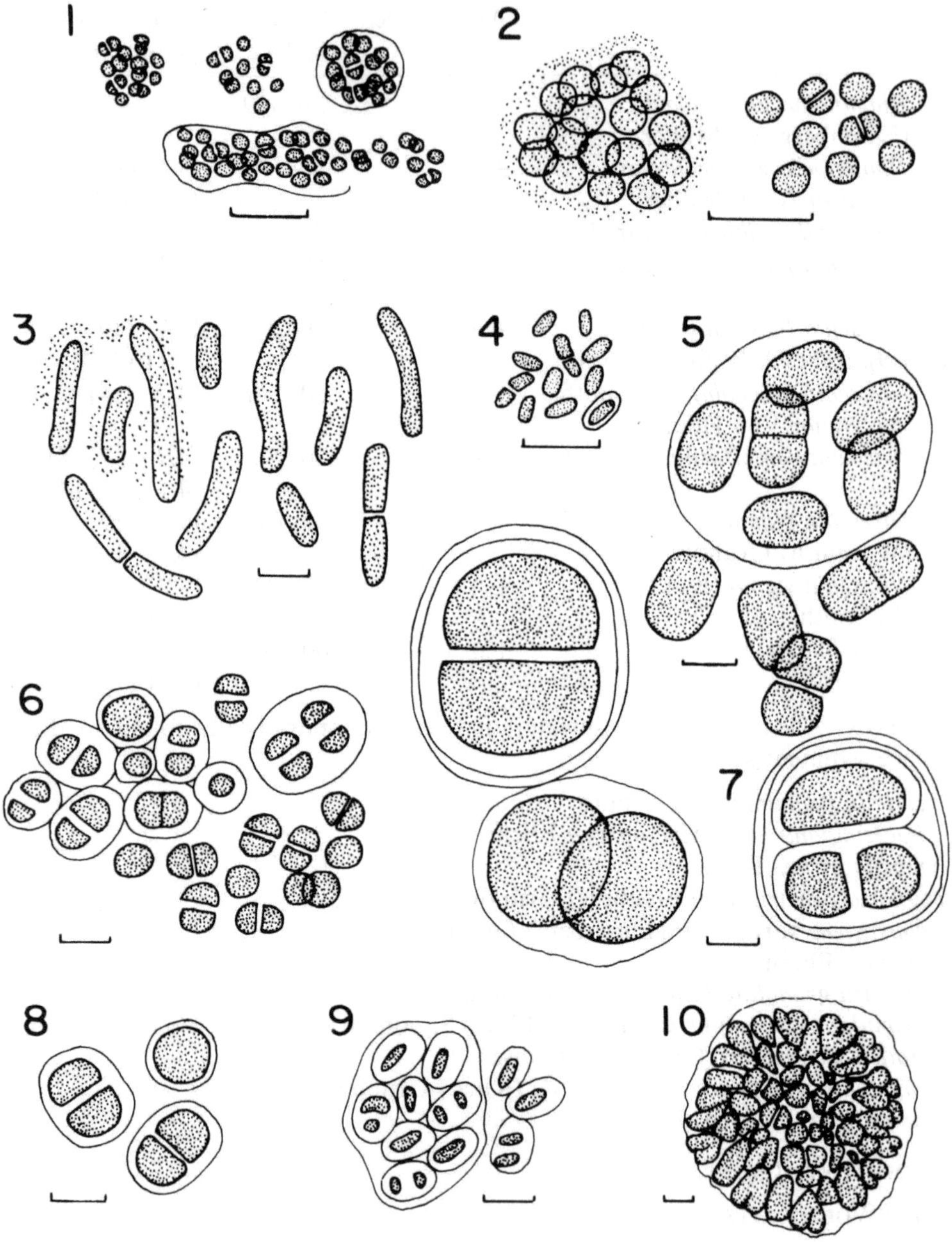

Plate 14.1. Cyanobacteria of the Gavish Sabkha. Scale = 5μm.
1 Aphanocapsa marina; 2 A. littoralis; 3 Aphanothece halophytica; 4 A. saxicola; 5 A. stagnina; 6 Chroococcus minor; 7 Ch. turgidus; 8 Ch. minutus; 9 Gloeothece confluens; 10 Gomphospheria aponina

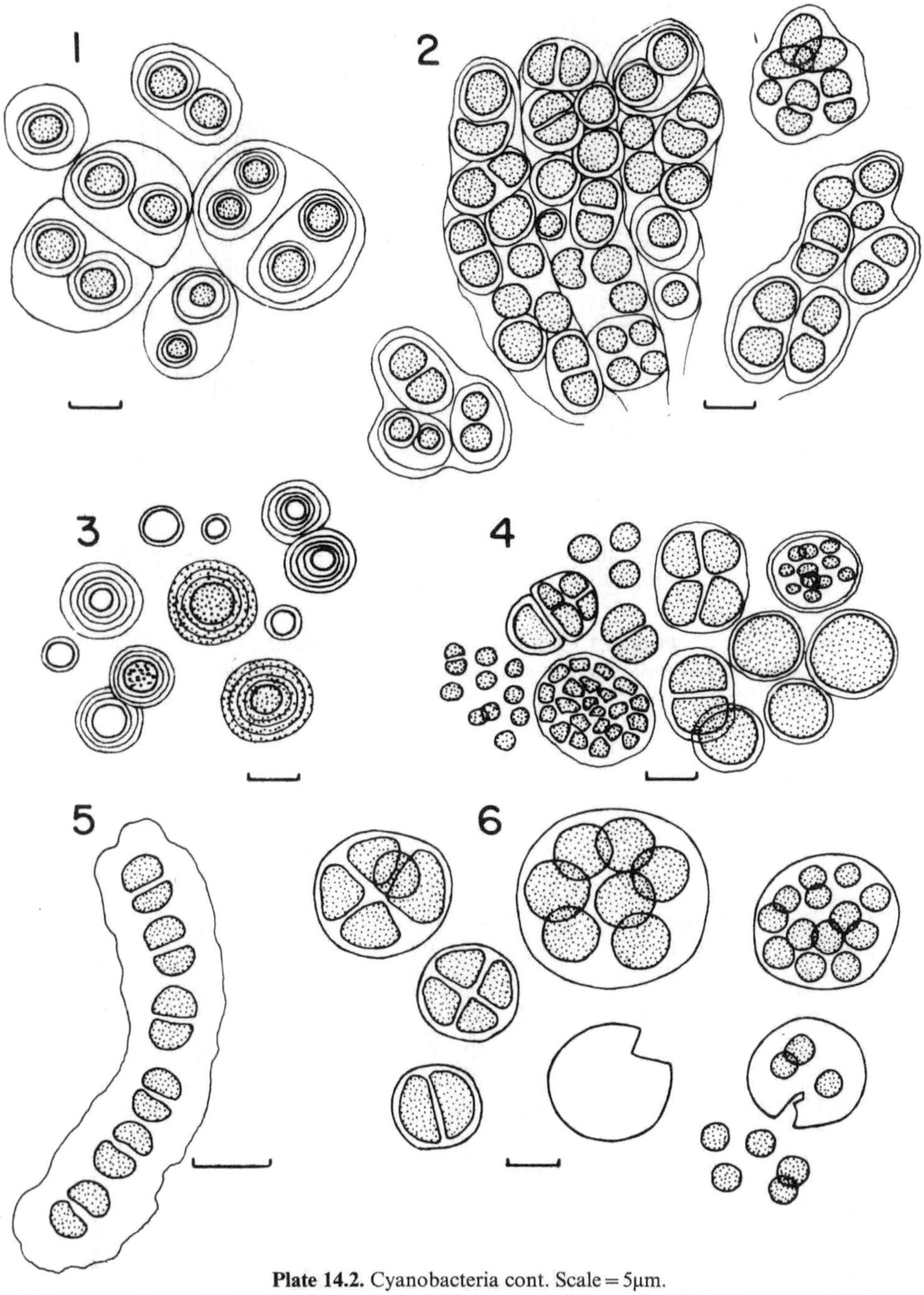

Plate 14.2. Cyanobacteria cont. Scale = 5μm.
1 Gloeocapsa deusta; 2 Entophysalis granulosa; 3 Gloeocapsa polydermatica (?); *4 Pleurocapsa fuliginosa; 5 Johannesbabtistia pellucida; 6 Chroococcopsis* sp.

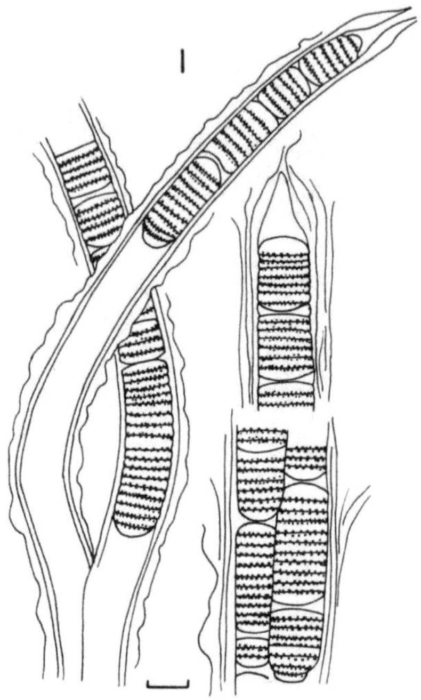

Plate 14.3. Cyanobacteria cont. Scale = 10µm.
1 Hydrocoleum lyngbyaceum; 2 Microcoleus chthonoplastes; 3 Schizothrix arenaria

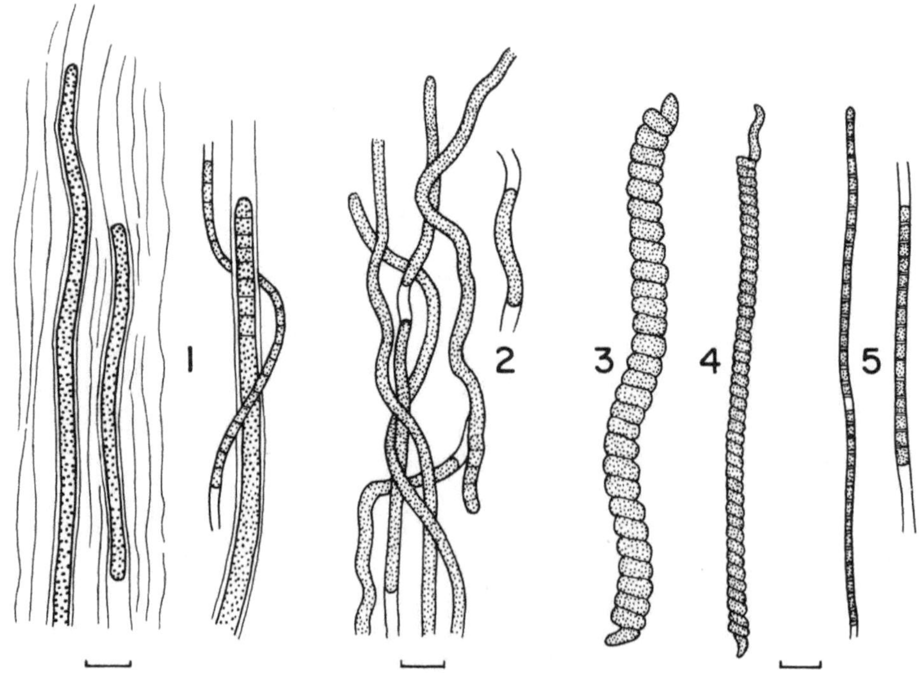

Plate 14.4. Cyanobacteria cont. Scale = 5 µm.
1 Schizothrix nasri and *Lyngbya complectens; 2 Schizothrix calcicola; 3 Spirulina subsalsa; 4 S. labyrinthiformis; 5 Phormidium fragile*

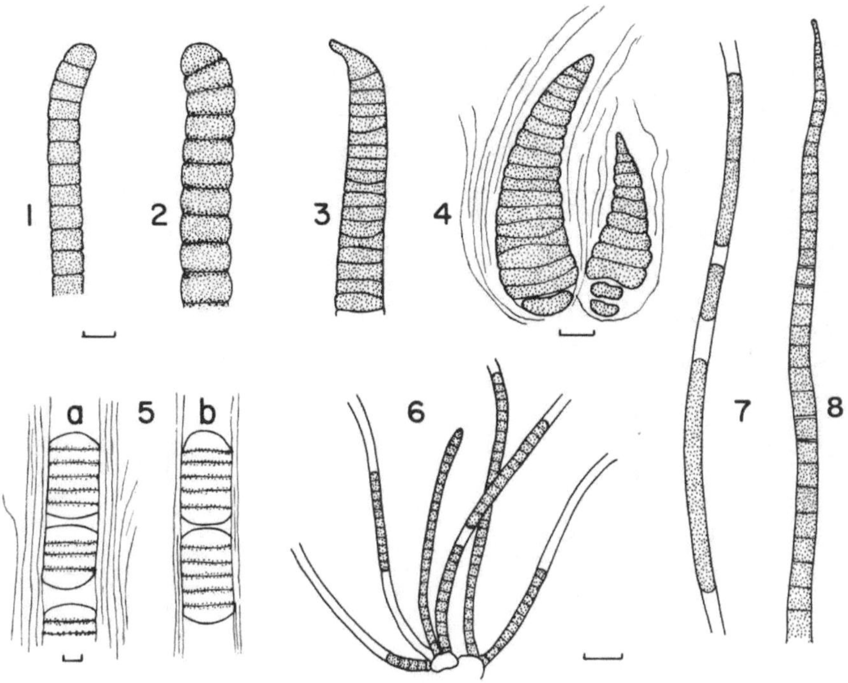

Plate 14.5. Cyanobacteria cont. Scale = 5 μm.
1 Oscillatoria tenuis; 2 O. nigro-viridis; 3 O. salina; 4 Calothrix parietina; 5 Lyngbya aestuarii; 6 L. infixa; 7 Lyngbya sp.; 8 Arthrospira miniata

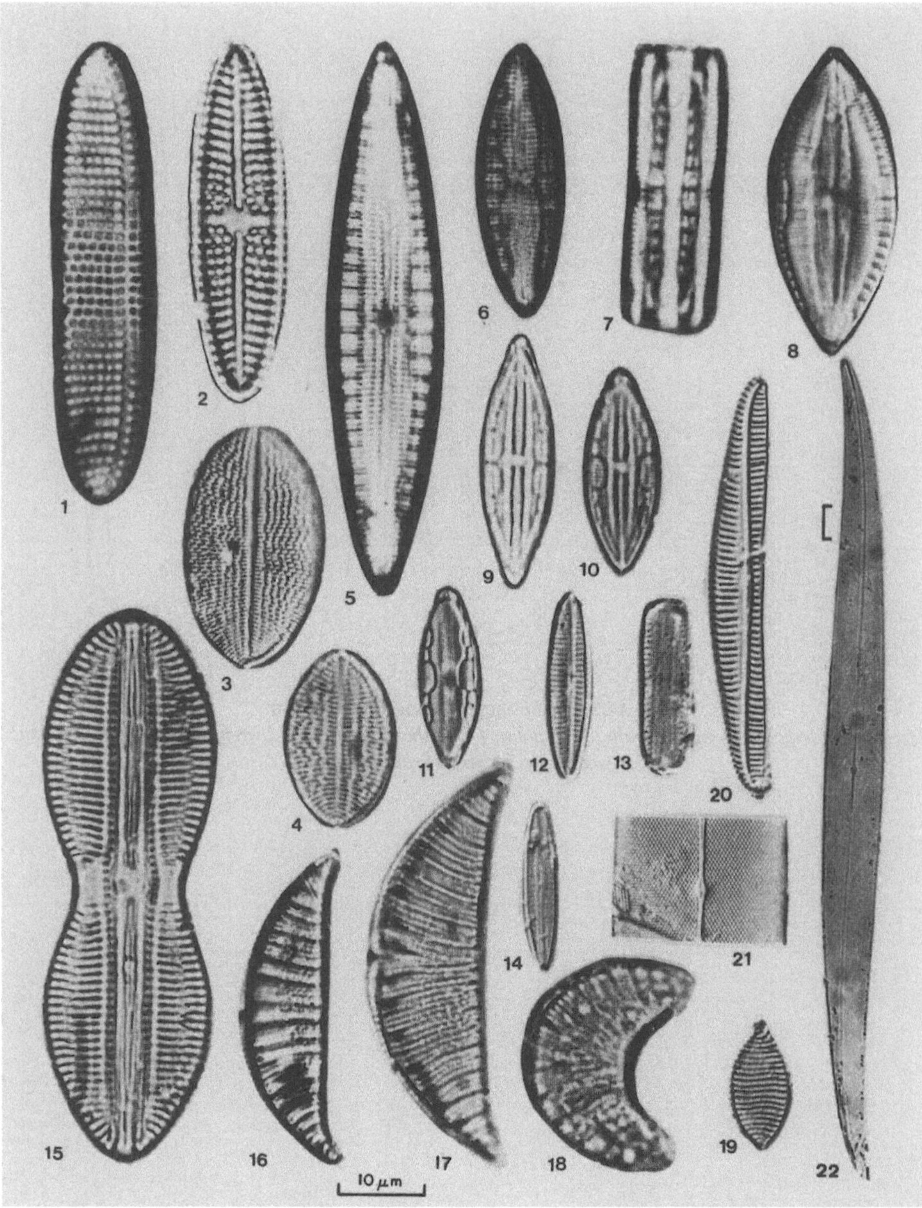

Plate 14.6. Diatoms of the Gavish Sabkha. Scale = 10 μm; all figures <M.
1 Achnanthes brevipes, raphe-less valve; *2 Achnanthes brevipes*, raphe-valve; *3 Cocconeis placentula* var. *euglypta; 4 Cocconeis placentula* var. *euglypta; 5, 6 Mastogloia brauni*, valvar view; *7 Mastogloia brauni*, girdle view; *8 Mastogloia sirbonensis; 9–11 Mastogloia pumila; 12 Navicula* sp. cf. *N. ramosissima; 13, 14 Navicula* sp.; *15 Diploneis stroemi; 16, 17 Rhopalodia gibberula* var. *timsahensis; 18 Rhopalodia gibberula* var.; *19 Nitzschia punctata; 20 Amphora proteus; 21, 22 Pleurosigma formosum; 21* central area of a valve; *22* valve seen at low magnification

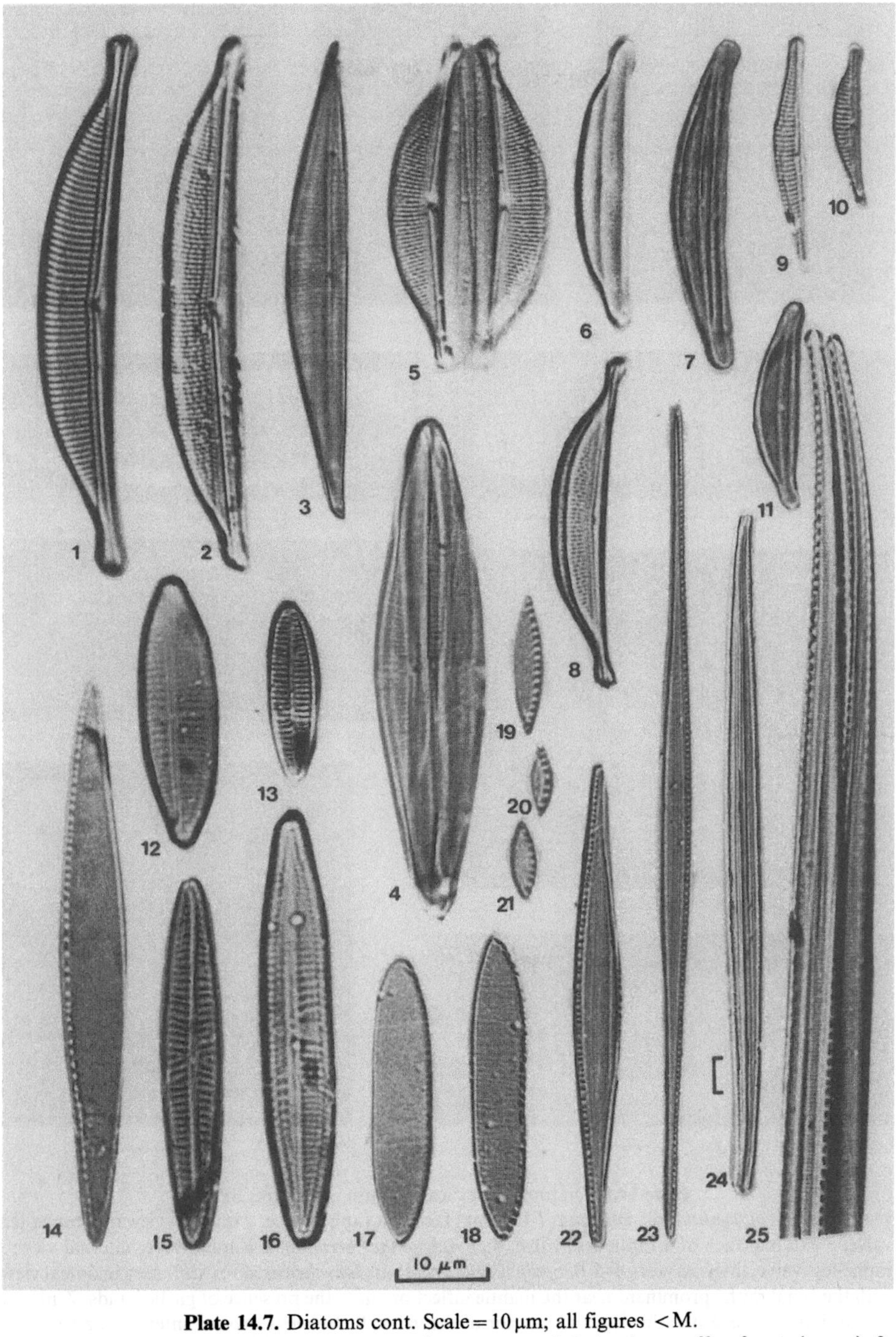

Plate 14.7. Diatoms cont. Scale = 10 μm; all figures <M.
1 Amphora sp. 1; *2 Amphora holsatica; 3, 4 Amphora angusta; 5–8 Amphora coffeaeformis* (group); *9, 10 Amphora* sp. cf. *coffeaeformis* var. *acutiuscula; 11 Amphora coffeaeformis* (group); *12 Navicula subinflatoides; 13 Navicula* sp. cf. *N. ramosissima; 14 Nitzschia lanceolata* var. *minor; 15 Navicula cincta* (?); *16 Navicula subinflatoides; 17, 18 Nitzschia lembiformis; 19–21 Nitzschia frustulum* var. *perpusilla; 22, 23 Nitzschia* sp.; *24, 25 Nitzschia sigma; 24* valves at low magnification; *25* extremities of two valves

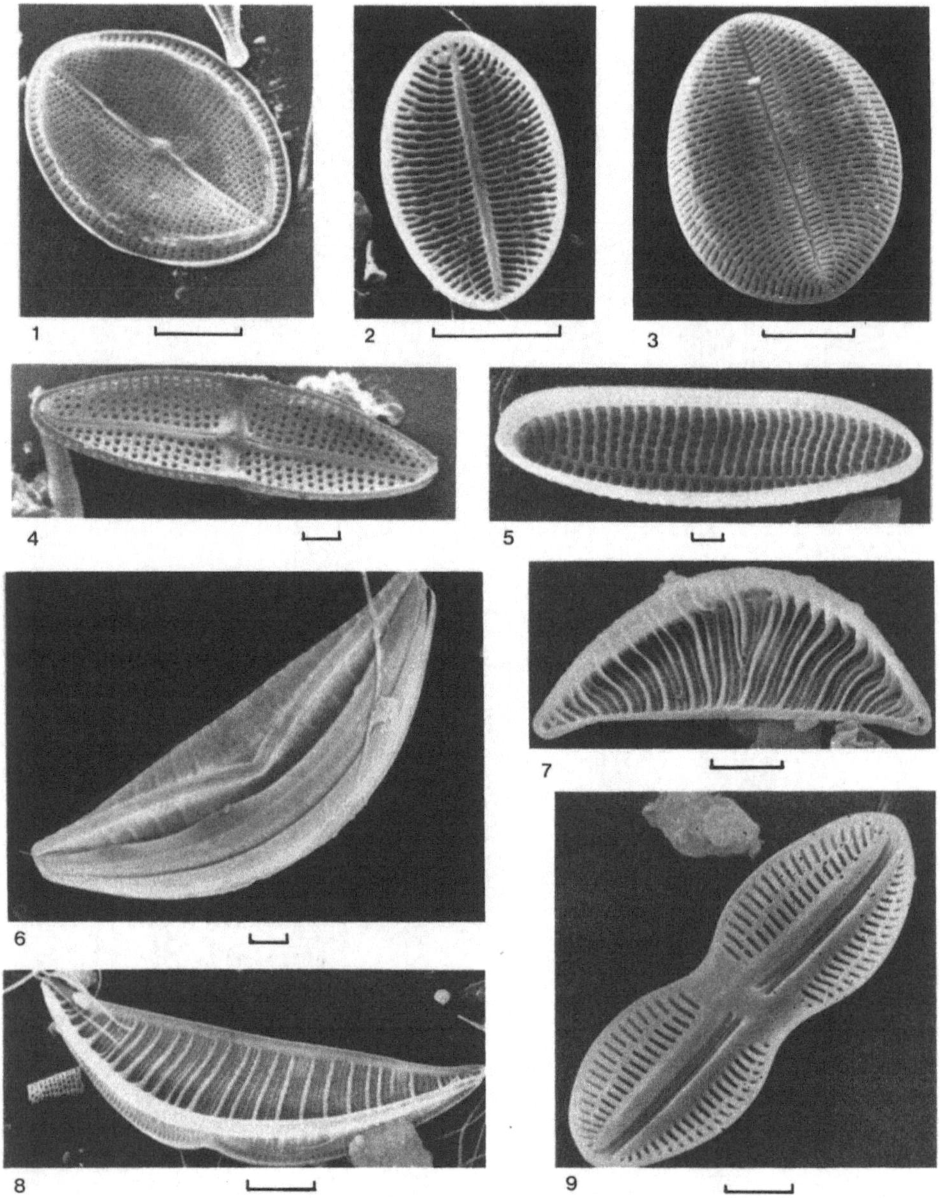

Plate 14.8. Diatoms cont. Scale = 5 µm; all figures SEM.
1–3 Cocconeis placentula var. *euglypta*. *1* External face of a raphe-valve; *2* internal view of a raphe-less valve; *3* external face of a raphe-less valve; *4, 5 Achnanthes brevipes*. *4* Raphe-valve, internal view; *5* raphe-less valve, internal view. *6–8 Rhopalodia gibberula* var. *timsahensis*. *6* Frustule seen in dorsal view, with the canal raphe prominent near the mantle inflection; note the presence of girdle bands; *7* internal face of a valve; *8* mantle view and interior of the valve; *9 Diploneis stroemi*, interior of a valve

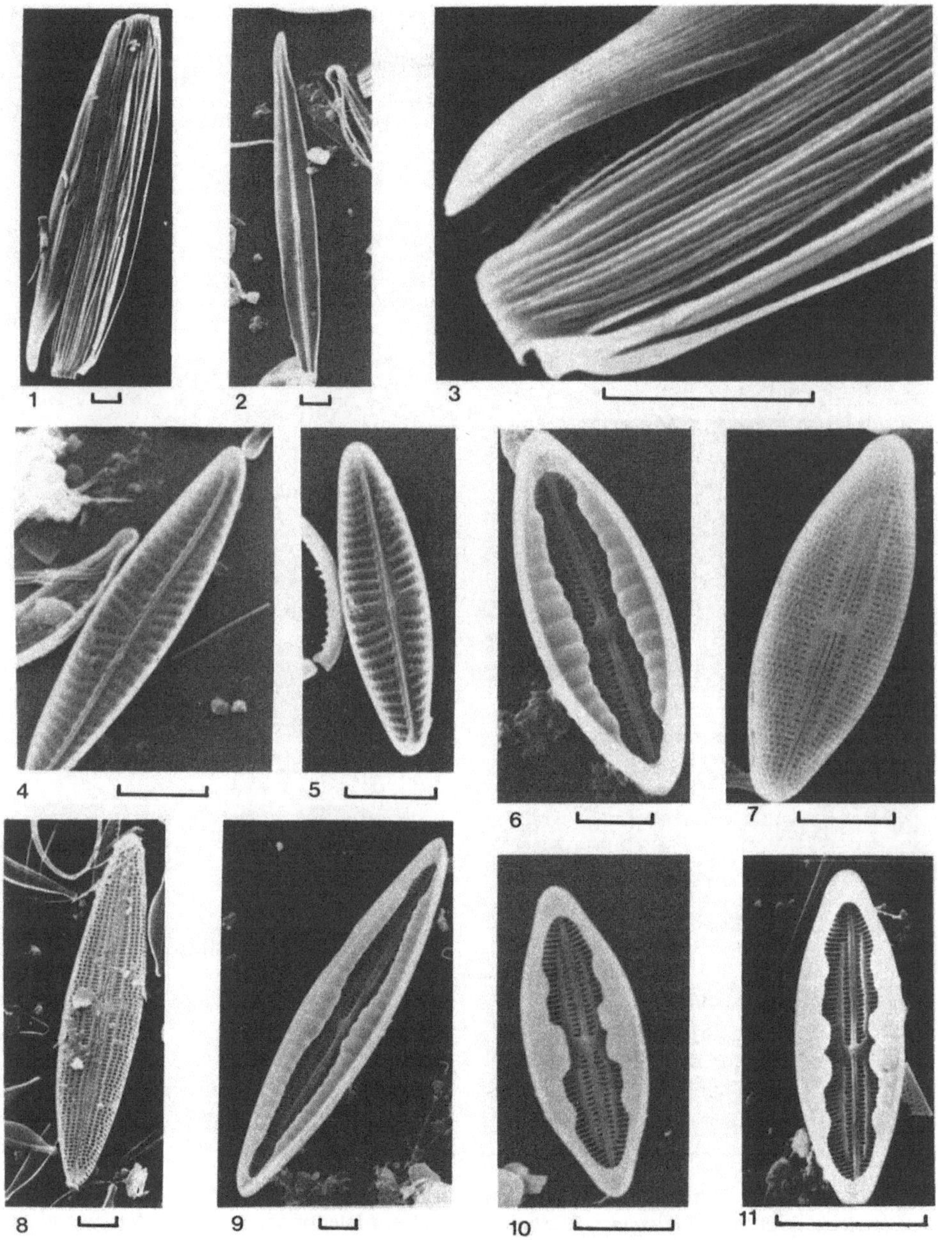

Plate 14.9. Diatoms cont. Scale = 5 μm; all figures SEM.
1–3 Navicula sp. cf. *N. complanata. 1* Open frustule, showing the two valves and the girdle bands; *2* interior of a valve; *3* extremity of a frustule; *4–5 Navicula cincta* (?), internal face of valves; *6, 8, 9 Mastogloia brauni; 6* interior of a valve, showing the lateral loculi; *8* external face of a valve; *9* internal view of a valve; *7, 10, 11 Mastogloia pumila; 7* external face of a valve; *10, 11* internal face with the lateral loculi, the median ones much larger than the others

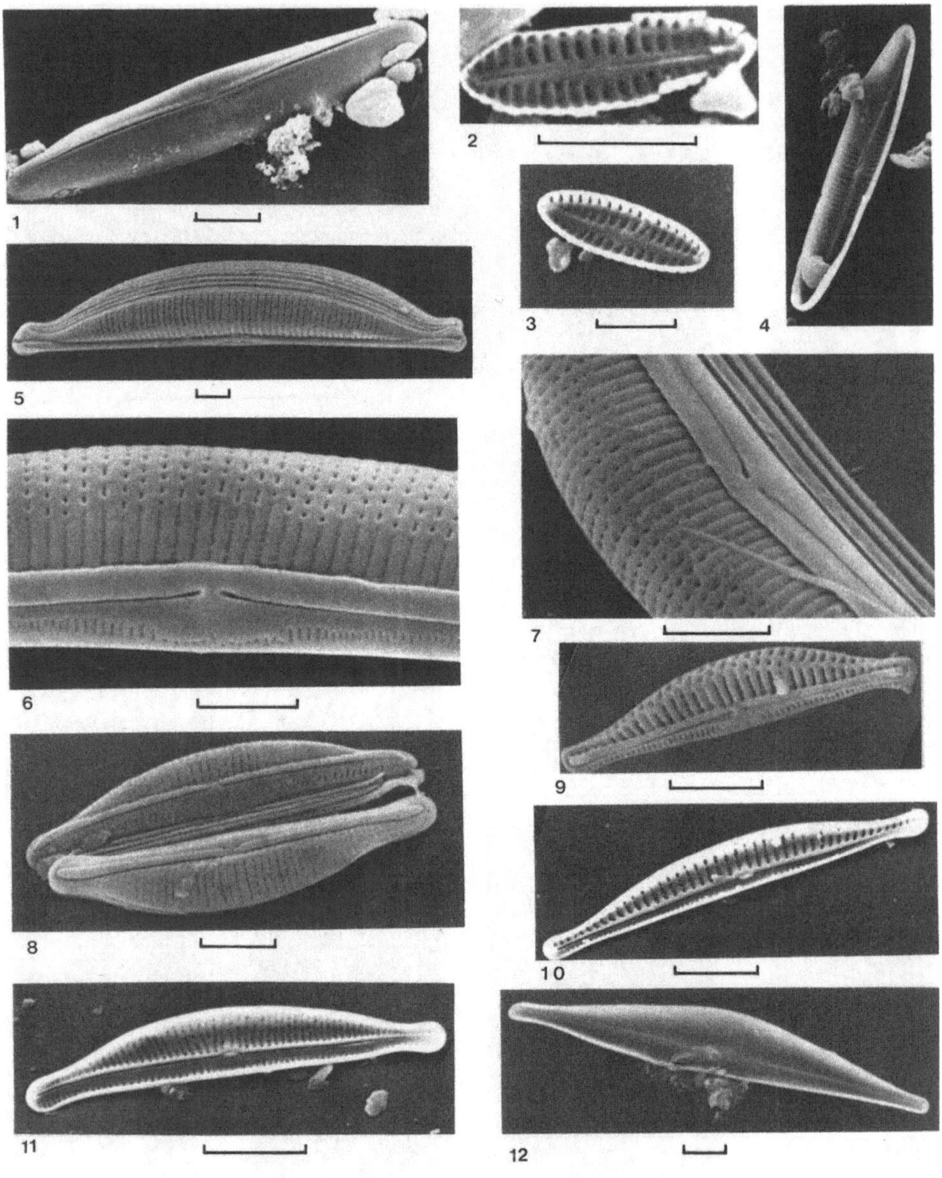

Plate 14.10. Diatoms cont. Scale = 5 μm; all figures SEM.
1, 4 Navicula sp. cf. *N. subinflatoides*. *1* Exterior of a valve; *4* interior of a valve; *2, 3 Navicula* sp., internal faces of two different specimens; *5–7 Amphora* sp. 1; *5* frustule seen in dorso-lateral view; *6* external view of a valve, in the central area; *7* central area of a frustule in ventro-lateral view, showing a silica flap covering the dorsal axial area and the base of the transapical ribs; presence of girdle bands on the ventral side of the frustule; *8, 11 Amphora coffeaeformis*; *8* frustule seen in ventral view, with its two valves open; *11* interior of a valve; *9, 10 Amphora* sp. cf. *A. coffeaeformis* var. *acutiuiscula*. *9* Exterior of a valve, showing a longitudinal rib along the mantle inflection area; *12 Amphora angusta*, interior of a valve, transapical striation very faint

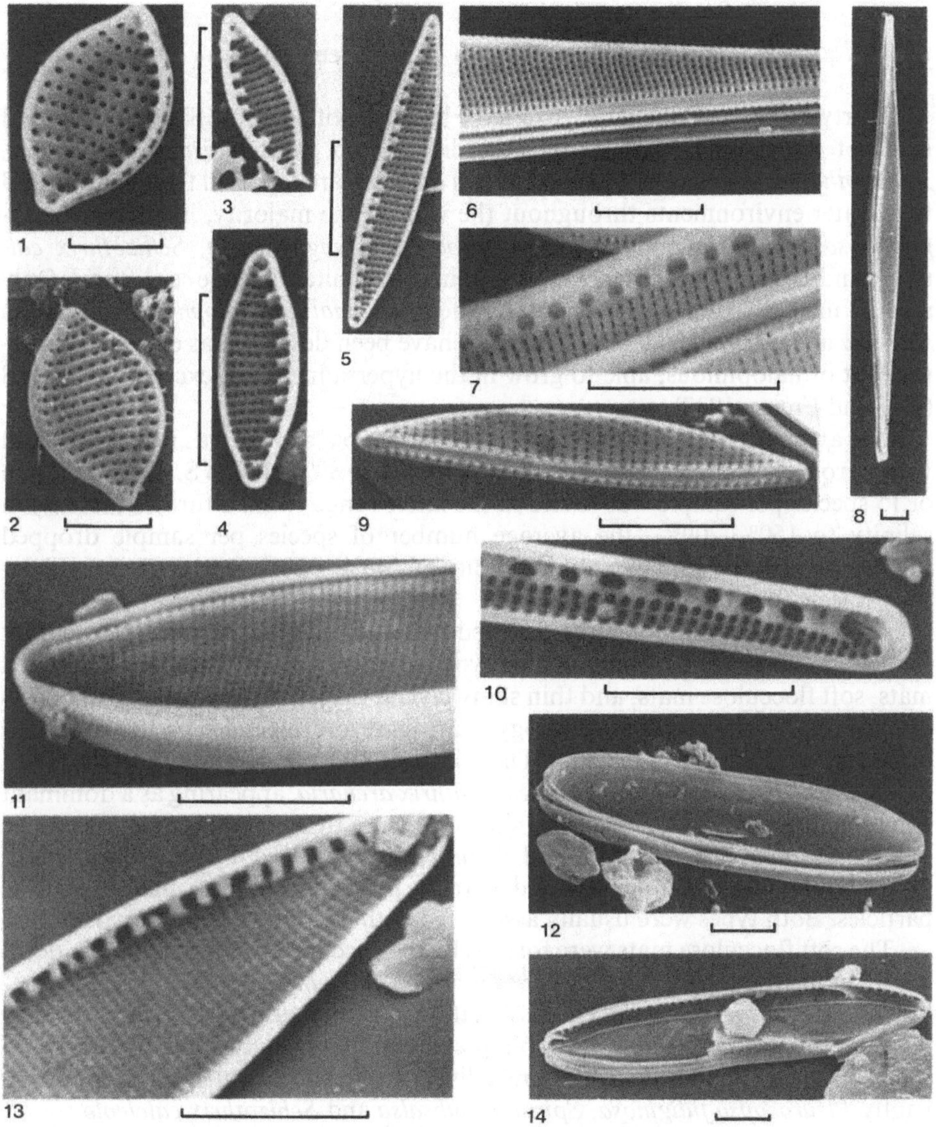

Plate 14.11. Diatoms cont. Scale = 5 μm; all figures SEM.
1, 2 Nitzschia punctata (small specimens). *1* Interior of a valve with the canal raphe in very excentric position (on the *right*); *3, 4 Nitzschia frustulum* var. *perpusilla,* internal face of two specimens; *5, 9 Nitzschia frustulum. 5* Internal face of a valve; *9* a frustule; *6–8 Nitzschia* sp. 1; *6* external face; *7* interior of a frustule (fragment) showing the excentric canal-raphe with the fibulae; *8* a frustule; *10 Nitzschia sigma,* extremity of a valve, interior view showing the raphe fissure within the canal raphe and its distal end; *11–14 Nitzschia lembiformis*; *11* extremity of a frustule; *12* open frustule showing the two valves and the girdle bands; *13* extremity of a valve, internal view showing the excentric canal raphe with narrow fibulae, transapical striation very delicate; *14* frustule (interior)

14.2.3 Salinity-related Distribution and Frequency of the Dominant Taxa

14.2.3.1 Cyanobacteria (Blue-Green Algae)

Thirty-three cyanobacterial taxa have been identified in the Sabkha, recorded in 48 of the samples examined. Some, like *Chroococcus turgidus, C. minutus, Aphanothece stagnina* and *Gomphospheria aponina,* are reported from marine and freshwater environments throughout the world. The majority, like *Entophysalis granulosa, Pleurocapsa fuliginosa, Hydrocoleum lyngbyaceum, Schizothrix calcicola* and *Lyngbya aestuari* are cosmopolitan inhabitants of the sea littoral. Only a small minority of the species,among them *Aphanothece halophytica, Spirulina subsalsa* and *Microcoleus chthonoplastes,* have been described as extremely halotolerant or halophilous, able to grow in the hypersaline water exceeding 150‰ S (Hof and Frémy 1932).

The average number of cyanobacterial species per sample was rather low, and varied from 3 species in 50‰–70‰ S to 9 species in 80‰–130‰ S. The maximum of 15 species per sample was found in the latter range. With a further increase in salinity to 150‰-200‰, the average number of species per sample dropped slightly to 5–7, and in the extreme salinity of 240‰–320‰ S, a single cyanobacterial species per sample was found.

The cyanobacteria constituted a predominant microbial biomass throughout the Sabkha, and formed three major types of aggregations: compact laminated mats, soft flocculose mats, and thin slimy layers. Each type was composed of several species; however, the morphology of an aggregate depended on the dominant organism or group of organisms. The compact mat builders were of two types: (a) *Microcoleus chthonoplastes* and *Schizothrix arenaria,* appearing as a dominant component separately or together, and producing thick multilayered structures, and (b) *Entophysalis granulosa* and *Pleurocapsa fuliginosa,* forming hard gelatinous nodules which disintegrate under strong pressure, yielding smaller ganulose particles. Both types were usually associated with a variety of additional species.

The soft flocculose mats were formed by the coccoid cyanobacteria, the dominant among them being *Aphanocapsa marina, Aphanothece halophytica, A. stagnina* and *Chroococcus minor*. This association appeared also as floating slime supported by entrapped bubbles of photosynthetic oxygen. Within the common slimy matrix of the above community lived many additional species, predominantly *Pleurocapsa fuliginosa, Spirulina subsalsa* and *Schizothrix calcicola.*

The thin slimy layers appeared usually on the top of a cyanobacterial mat or the sediment, and were composed of *Hydrocoleum lyngbyaceum, Lyngbya aestuari* and several species of *Oscillatoria.*

The cyanobacterial microflora of the Gavish Sabkha was composed mainly of euryhaline species, and thrived over an extremely wide salinity range. However, the limits of distribution differed considerably for various taxa, and their frequencies in the samples changed with increasing salinity as summarized in Fig. 14.2.

The 13 taxa listed in Fig. 14.2 appeared in six or more samples, and, because of their wide distribution and quantitative importance, were considered as dominant forms. Two species, *Chroococcus turgidus* and *Hydrocoleum lyngbyaceum,* were limited to 50‰–70‰ S and within this range were found in 30% and 23%

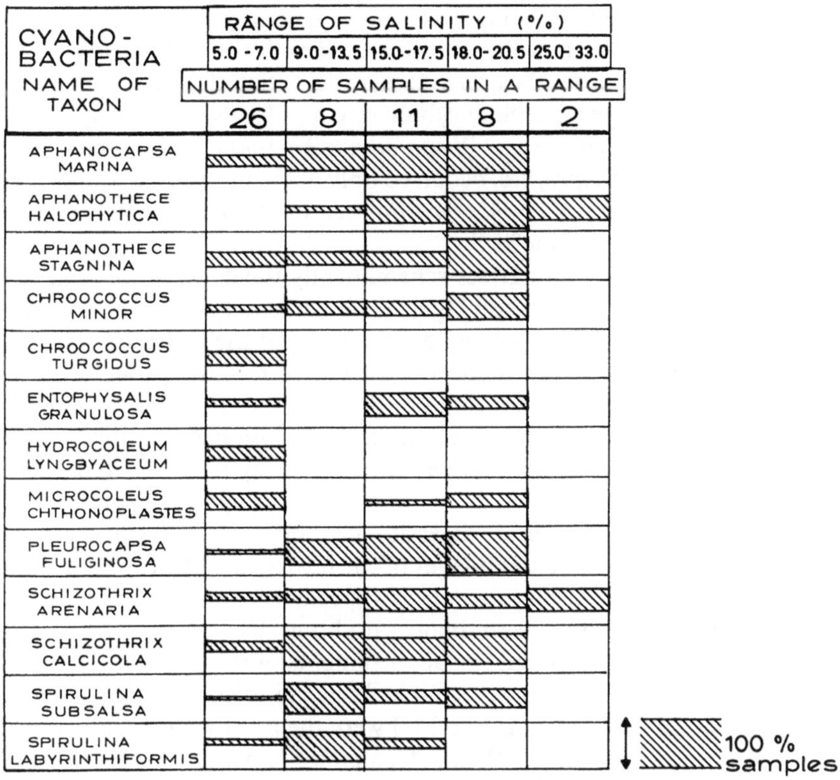

Fig. 14.2. Relative frequency of dominant cyanobacterial species. *Dashed area* represents percentage of the samples populated by a species in a given range of salinity

of the samples, respectively. The frequencies of three species, *Aphanothece stagnina, Chroococcus minor* and *Pleurocapsa fuliginosa,* progressively increased with higher salinity, and in 180‰–205‰ S were present in 75%, 50% und 87% of the samples, respectively. Species of *Spirulina* exhibited maximum frequency in 90‰ –135‰ S, where they were seen in 62% of the samples. Others, such as *Microcoleus chthonoplastes, Schizothrix arenaria* and *Schizothrix calcicola,* seemed to be quite frequent throughout most of the salinity range encountered in the Sabkha.

A unique species, *Aphanothece halophytica,* was absent from sample of lower salinity, began to appear from 90‰–135‰ S on, and attained its maximum frequency of 75% in 180‰–200‰ S. Only two species, the above *A. halophytica* and *Schizothrix arenaria,* appear in the extreme salinities of 250‰–300‰ S. At salinities ranging from 50‰, to 200‰ S, the general number of dominant cyanobacteria did not change, remaining at 9–12 species throughout the Sabkha. The drastic decrease to 2 species was recorded in 250‰–300‰ S, as mentioned above.

14.2.3.2 Bacillariophyta (Diatoms)

Diatoms were recorded in 54 samples. Altogether, 39 taxa were identified, 11 of which were probably allochthonous (see Discussion). The remaining 28 autochthonous taxa, belonging to eight genera, were periphytic euryhaline forms, the most common being *Amphora, Nitzschia* and *Navicula*. Some species were present throughout the entire area, at salinities of 50‰–205‰ S and are therefore considered here as holoeuryhaline, e.g., *Amphora coffeaeformis* and *Nitzschia frustulum*. Others, e.g., *Rhopalodia gibberula, Mastogloia brauni, M. pumila, Achnanthes brevipes,* etc. occurred only in the metahaline waters near the seawater springs, whereas *Nitzschia lembiformis* was found predominantly in the hypersaline lagoon above 150‰ S.

Figure 14.1 summarizes schematically the distribution and specific composition of the diatom assemblages. (The percentages may be somewhat inaccurate, since "dead cells" may have been present in the collected samples together with the "living" ones, but there was no way to recognize them in the studied slides.) Three main assemblages could be distinguished: (1) *Rhopalodia-Mastogloia* in the metahaline zone, (2) *Nitzschia sigma-N. punctata* in the alpha-hypersaline zone up to 150‰ S and (3) *Nitzschia lembiformis* in the beta-hypersaline zone up to 205‰ S.

As a rule, the species diversity was relatively low when compared to that of the marine diatom assemblages of the nearby sea littoral. In the mats of the metahaline zone, the average number of species recorded in one sample was 8, whereas the most diversified assemblage (at about 60‰ S) was composed of 16 different species. In the mats and surface sediments of the hypersaline lagoon, the average was three to four taxa per sample from salinities higher than 150‰ S.

The 16 taxa listed in Fig. 14.3 were the most widespread, appearing in five or more samples. Among them, *Achnanthes brevipes, Amphora holsatica, Diploneis stroemi, Mastogloia brauni, M. pumila, M. sirbonensis, Pleurosigma formosum* and *Rhopalodia gibberula* were restricted to metahaline salinities (50‰–70‰). *Cocconeis placentula* var. *euglypta, Navicula complanata* (sensu lato) and *Nitzschia* aff. *lanceolata* reached their limits of distribution in Gavish Sabkha at 135‰, 150‰ and 175‰ S, respectively. *Nitzschia punctata* and *N. sigma* thrived up to 150‰ S. Valves of these species were found sporadically at higher salinities, but it is possible that they were dead cells transported and redeposited. *Amphora coffeaeformis, Navicula subinflatoides, Navicula* sp. and *Nitzschia frustulum* proved to be extremely euryhaline, and frequented salinities ranging from 50‰–205‰ S. The frequency decrease of these species in the 150‰–175‰ range (Fig. 14.4) was only apparent, as many of these samples were devoid of diatoms. *Nitzschia lembiformis* was very rare, and possibly not even alive, in the metahaline samples; in contrast, its frequency increased dramatically in hypersaline waters, reaching 50% of the total samples in 180‰–205‰ S. No diatoms were recorded for higher salinities.

14.2.3.3 Chlorophyta (Green Algae)

Green algae were represented in the Gavish Sabkha by only four genera, with a rather limited distribution. The filamentous forms *Enteromorpha, Chaetomorpha* and *Rhizoclonium* seemed to be in association, and were found in 5 out of 55 samplex examined, all of them from the seepage area where salinity does not exceed 65‰ S.

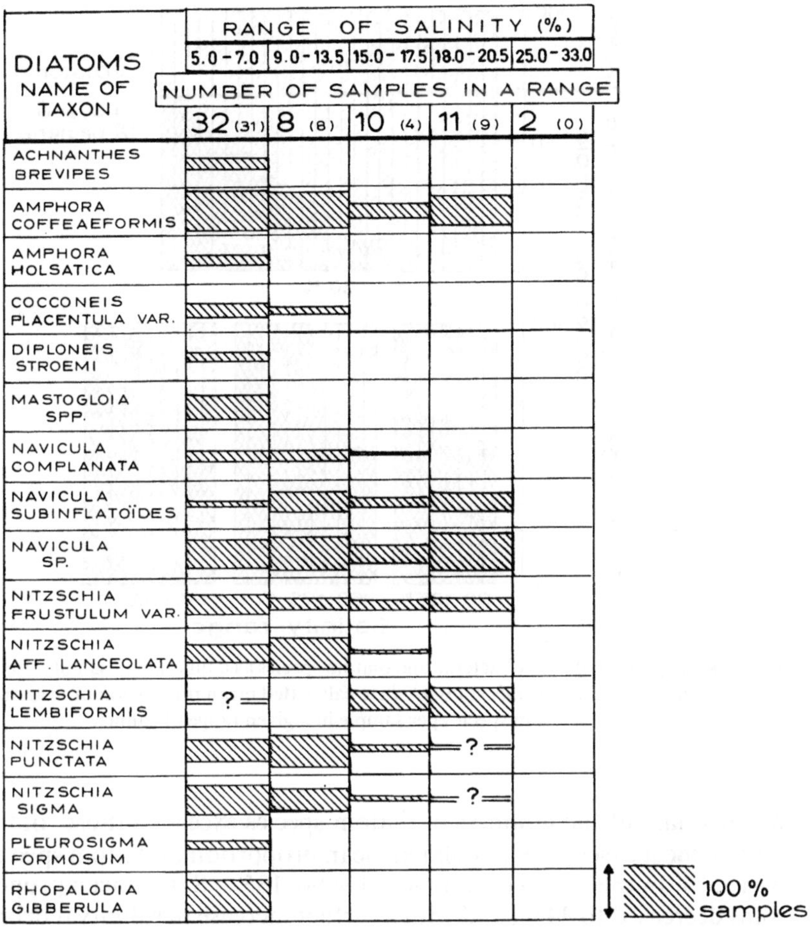

Fig. 14.3. Relative frequency of dominant diatom species. *Dashed area* represents percentage of the samples populated by a species in a given range of salinity. *Number in brackets* indicates the diatom-bearing samples

Rhizoclonium alone was recorded in two samples from 87‰–132‰ S.

Dunaliella appeared in only three samples and was restricted to the higher salinities of 150‰–200‰ S.

14.2.4 Proportion of Cyanobacteria and Diatom Species as a Function of Salinity

Several samples contained either diatoms or cyanobacteria; however, the majority included both in proportions which changed according to sample salinity (Fig. 14.4).

Within the range of metahaline salinities (50‰–70‰ S), 7 of 26 samples analyzed contained only diatoms. The remainder contained both diatoms and cya-

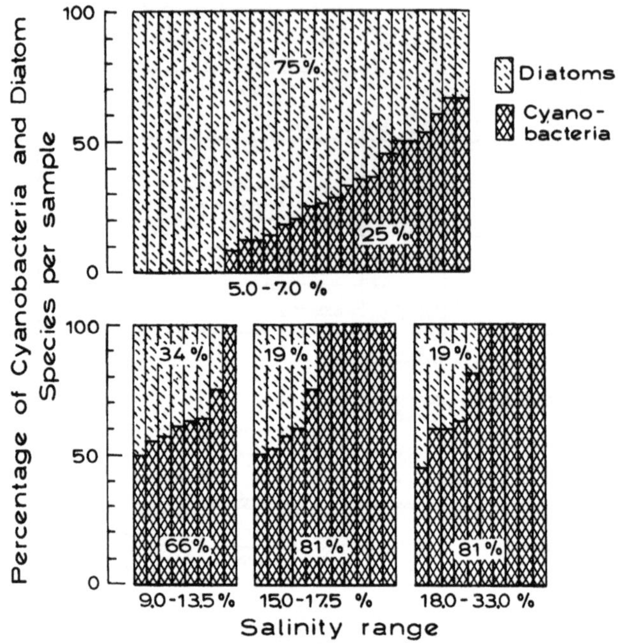

Fig. 14.4. Proportion of cyanobacteria and diatom species as a function of salinity. Each *bar* represents a single sample examined; *numbers inserted* are calculated mean proportion of cyanobacteria and diatom species per sample in a given range of salinity

nobacteria, but the numbers of diatom species exceeded those of the cyanobacteria in most cases. The calculated mean proportion of diatom species per sample was 75%. In the group of samples of 90‰–135‰ S, the mean proportion of diatoms decreased to 34%, and with a further increase in salinity it dropped to 19%. A respective mean proportion of cyanobacterial species per sample increased from 25% in 50‰–70‰ S to 66% in 90‰–135‰ S, and reached 81% in the salinities ranging 150‰–330‰ S.

14.3 Discussion

This study of the Gavish Sabkha has supplied additional evidence that cyanobacteria constitute a widespread and conspicuous component of the biotic community in hypersaline environments. It has also provided some new information on the role of diatoms as contributors to the total biomass at high salinity.

In the habitats undergoing periodic desiccation and air exposure, the cyanobacteria are far more successful than any other halotolerant or halophilous photosynthetic organism. Despite this fact, the cyanobacteria which build benthic mats are often quoted in general terms, such as "filamentous forms", "coccoid blue-greens", or, at best, are identified at generic level (e.g., Kinsman and Park 1976, review by Bauld 1981a, b). This neglect has stemmed from difficulties in identification of an abundance of closely related taxa, and also from uncertain taxonomy following recent revisions (Stanier et al. 1971, Rippka et al. 1979). As a result, both because of vagueness of identification and scarcity of adequate data, boundaries of cyanobacterial communities and species tend to be ill-defined. The Gavish Sabkha, with its extensive salinity gradient, provided an excellent opportunity for analysis of species distribution, as well as of the interplay between green algae, diatoms and cyanobacteria.

The increasing frequency of many cyanobacteria toward higher salinities (Fig. 14.2) may be explained in terms of halotolerancy, and competition between various groups. A well-developed community of filamentous green algae associated with a variety of diatoms, was found flourishing within the metahaline range. Although all major cyanobacteria were present in this zone, their frequency in the samples was low, 7%–30% (Fig. 14.2), probably because of the restricting presence of competing eukaryotic algae. With an increase in salinity to over 70‰ S, the limit of halotolerance of green algae and many diatoms was reached, resulting in their elimination; the frequency of other diatoms considerably diminished (Fig. 14.3). The resulting decrease in competitive pressure was advantageous to more halotolerant cyanobacteria, whose frequency increased, and in salinities of 180‰–202‰ S they appeared in 25% to 87% of the samples. The growing significance of cyanobacteria as a function of increased salinity was demonstrated also by changes in community composition, and is summarized as percentage of diatoms and cyanobacterial species per sample in Fig. 14.4. It must be stressed that a steady increase in the percentage of cyanobacteria in the samples resulted chiefly from the progressive elimination of diatoms, while the total number of cyanobacterial taxa remained similar throughout almost the entire salinity range. However, it must be borne in mind that other environmental factors which were not analyzed in this study, such as desiccation, light, temperature and quality of the bottom substrate, may also have an effect on the competitive ability of various species.

Three cyanobacterial communites: (1) *Microcoleus chthonoplastes-Schizothrix arenaria*, (2) *Entophysalis granulosa – Pleurocapsa fuliginosa* and (3) *Aphanothece stagnina – Aphanothece halophytica – Aphanocapsa marina – Chroococcus minor*, seemed to be particularly well adapted to the sabkha environment, where periods of increasing salinity and desiccation are followed by reflooding. Under such extreme and unstable conditions, the advantageous characteristics of cyanobacteria are fully expressed, the most prominent ones being the common colloidal matrix which preserves water inside the mat, and the yellow coloration of the sheaths which provides a screen against excess solar radiation (Brock 1976).

Frequently, salty crusts formed on the top of a mat as a result of evaporation. These tended to isolate the entire community from the dry air, presumably avoiding or at least slowing down further water loss.

Several diatoms recorded in the Gavish Sabkha from extreme salinities of up to 205‰ S always appeared in association with cyanobacteria (Figs. 14.2 and 14.3). They probably profited from the microniche provided by the slimy matrix of photosynthetic prokaryotes.

Many of the diatom taxa identified in the highly saline Gavish Sabkha are known elsewhere in brackish marine or continental waters. In the literature, some are reported as polyhalobious, others as mesohalobious, and two species are even considered oligohalobious, according to Hustedt's halobian classification (Hustedt 1953). However, since this classification does not take into account the tolerance of the taxa to salt changes, and also does not refer to diatoms of highly saline waters, assignment to the above halobious groups could be misleading when defining the diatom assemblages of the Gavish Sabkha. In fact, all the taxa recorded in the Sabkha were euryhaline, but their halotolerance varied from oligoeuryhaline (species whose osmoregulation potential is rather low) to holoeuryhaline (species with a very high osmoregulation potential). Intermediate euryhaline groups existed between these extremes. Simonsen (1962) proposed a salt tolerance classification for diatoms up to 40‰ S, which was extended toward higher salinities by Ehrlich (1975); however, data on the distribution of euryhaline diatoms, especially at high salinity, are still too scarce and scattered to make this classification useful. Experiments had been carried out in the laboratory on the study of osmotic behavior of some euryhaline species (Fischer 1979), but the results did not necessarily fit the field evidence.

The distribution of the diatoms in the Gavish Sabkha supplied valuable information on the hypertonic resistance of some cosmopolitan and more local euryhaline species. It seems that the metahaline/hypersaline boundary defined at 70‰ S for saline waters (Por 1972, 1980) fits rather well the highest salinity at which diatoms of a "mesoeuryhaline" group can thrive. This group would include species such as *Rhopalodia gibberula, Mastogloia brauni, M. pumila, Achnanthes brevipes* and other taxa wich were found only in the seawater seepage area (Fig. 14.3). These species were also recorded in the Bardawil Lagoon at salinities up to about 80‰, but were missing in the hypersaline Solar Lake. Another boundary, i.e., alpha-hypersaline/beta-hypersaline waters, at about 140‰ S (Por 1980) approximately fits the hypertonic resistance of some species of a "pleioeuryhaline" group, e.g., *Nitzschia sigma, N. punctata* and *Navicula complanata.* The highest salt concentration at which the holoeuryhaline species, such as *Amphora coffeaeformis* (group), can thrive is 205‰ (possibly somewhat more, but definitely less than 250‰).

It is interesting to note that at least one species, *Nitzschia lembiformis,* was found almost exclusively at extremely high salinity, about 150‰. The existence of a "hyperhalobious" diatom, preferring high salt concentrations, has never been reported hitherto. On the contrary, it has been claimed that there are no species thriving preferentially in hypersaline waters (Ehrlich 1975). *Nitzschia lembiformis* was not recorded from Bardawil, nor from Solar Lake, but it occurs sporadically in some hypersaline, artifical ponds in the Dead Sea area. It was first described as a fossil species from the Pleistocene Lisan Formation in the Jordan Rift Valley (Meister 1968, Begin et al. 1974), where it was very abundant at some levels. The highly saline character of the layers yielding *N. lembiformis* in the Lisan Formation is supported by mineralogical evidence, e.g., presence of authigenic aragonite and sometimes gypsum in the sediments.

The sporadic presence of a few allochthonous freshwater diatom species in the surface sediments of the Sabkha was also noteworthy. Most belonged to the order Centrales and are cosmopolitan, planktonic forms widespread in rivers and lakes (e.g., Jordan River and Lake Kinneret). Although oligohalobious, some of them tolerate a slight increase in salinity; in this area, however, they were never found thriving at salinity higher than 1‰. The forms recorded in the Gavish Sabkha were probably not living when the samples were taken. Possibly the valves were transported into the Sabkha by floods during some of the rainy periods which occur in the area every several years (see Chap. 7). These valves were well-preserved and, as a rule, seemed more resistant to dissolution than valves of many euryhaline species whose thin walls tend to corrode rapidly when buried under new sediment. Another alternative can be proposed for the presence of freshwater diatoms in the surface sediments of the hypersaline Sabkha. It has been observed that during rainy periods the water level of the Sabkha increases by a few centimeters, or even decimeters (see Chap. 7). During these periods, freshwater planktonic diatoms may have developed temporarily in the upper water layer, while the Sabkha had become for a short period a stratified lake with a freshwater epilimnion above a hypersaline hypolimnion. After the mixing of the lake waters, and its partial evaporation, the valves of the planktonic diatoms were buried in the sediments and preserved.

15. The Fauna of the Gavish Sabkha and the Solar Lake – a Comparative Study

GISELA GERDES, JONATHAN SPIRA, and CHANAN DIMENTMAN

Episodic or permanent water supply is an important factor which influences the faunal composition and distribution in hypersaline ecosystems. Inland salt lakes may receive water permanently or sporadically from streams, springs or run-off. Playas, qàas, and other ephemeral saline waters are mainly seasonally supplied with water. Coastal lagoons can be in open contact with the sea or have los this contact and receive water by subsurface supply. The ionic composition of the water, and the stability or instability of salinity and temperature gradients influence the structure and spatial distribution of faunal communities in hypersaline ecosystems (Bayly 1967, 1972, Collins 1977, Copland and Jones 1976, Por 1980, see also Chap. 6).

In this chapter, we describe the community structure and distribution of the fauna which lives in the Gavish Sabkha and the Solar Lake. Both environments are supplied with chloride-sulfate waters of marine origin (Por 1972, 1980). Both have lost the open contact with the sea. They have permanent water bodies in which hypersaline conditions prevail, episodic water bodies and surrounding flats where the sediments are permanently wet but not flooded. In both environments, prokaryotic microbial communities occur which form and alter the sediments.

The comparison of the faunal communities, presented here, is based on studies which were carried out separately in the Gavish Sabkha (G. Gerdes) and the Solar Lake (J. Spira). We use the following terms which were defined by Por (1972, 1980):

The term metahaline refers to waters whose salinity is greater than that of seawater and does not exceed 70‰. Hypersaline waters are even more saline and are divided into four categories: (1) Alpha-hypersaline, whose salinity range is 70‰–100‰ S; (2) Beta-hypersaline, whose salinity range is 100‰–140‰ S; (3) Gamma-hypersaline, whose salinity range is 140‰–300‰ S; (4) Delta-hypersaline, whose salinity range is 300‰ S and above. These boundaries refer to biological criteria, such as: presence or absence of marine and/or freshwater-derived species, decrease in faunal richness, presence of predators, and other parameters.

We call groups of organisms as "hyperstenohaline", which tolerate a salinity of up to 140‰, and groups of organisms as "hypereuryhaline", which tolerate salinities of up to 300‰. These terms refer to the distribution of species rather than to their physiology, which has not yet been studied.

15.1 Field and Laboratory Methods

Our investigations were carried out during the following periods: Gavish Sabkha: August 1981; February through March 1982; Solar Lake: February through March 1980, August through September 1980, and March 1981.

15.1.1 Sampling Procedures

Bottom samples were taken from several sites with corers: Gavish Sabkha, corer with 86.5 cm^2 surface area, 12 cm deep; Solar Lake, corer with 280 cm^2 surface area, 5 cm deep. Samples were preserved in the field in formalin 10% or alcohol 70% and delivered to the laboratory for analysis (Krebs 1978, Landin 1976). The fauna was separated from the sediment with the aid stereoscopes, sorted into species and counted. To obtain a comparative picture, individual values were adjusted to 100 cm^2.

Gut contents were examined qualitatively. Faunal behavior was observed at the laboratory and in the field.

15.1.2 Field Measurements

Salinity profiles of groundwater and open water were measured with a refractometer (American Optics, Inc.). Temperature profiles were measured with a telethermometer.

15.1.3 Identification

The distribution of benthic microbiological systems was examined qualitatively and related to detailed microbiological investigations (Cohen et al. 1977a–c, Ehrlich 1978, Hirsch 1978, Krumbein and Cohen 1974, Krumbein et al. 1977, 1979a, Potts 1980, see also Chap. 9). Identification of microorganisms was made according to Geitler (1932), Krumbein et al. (1979b), Rippka et al. (1979), and with the help of Y. Cohen, A. Ehrlich, E. Holtkamp and W. E. Krumbein.

Fauna: The taxonomical identification was based mainly on Bertrand (1972), Cummins (1973), Leach and Chandler (1968) and Pennak (1953) as well as on personal communications with specialists (see acknowledgements).

15.2 Research Areas

The Solar Lake lies 18 km south of Elat and about 150 km north of the Gavish Sabkha (Fig. 15.1). Contrasting with the unprotected site of the Gavish Sabkha which lies in a wide and open coastal plain (Fig. 15.2a), the Solar Lake is pro-

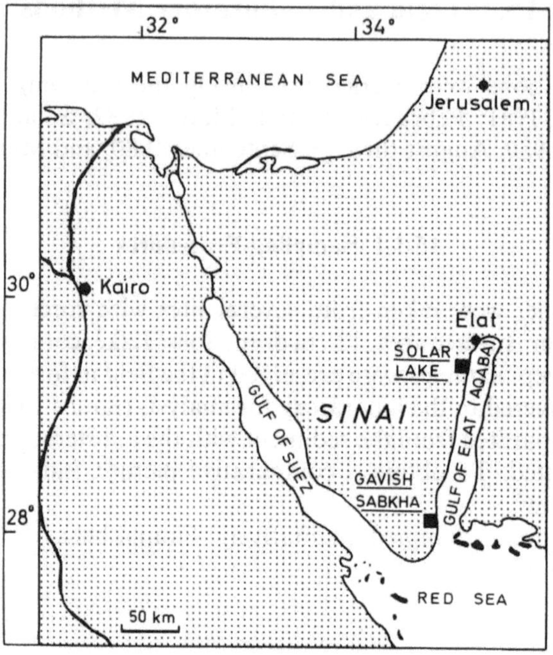

Fig. 15.1. Solar Lake and Gavish Sabkha, two hypersaline ecosystems on the Gulf of Elat

tected by 100-m-high ridges of metamorphic and intrusive rocks which closely surround the lake (Fig. 15.2 b).

We subdivided both areas into zones according to comparable geological and hydrological parameters (see Figs. 15.2 a, b, 15.3 and 15.4):

Zone 1. Coastal bars. The Solar Lake coastal bar is between 45 and 60 m wide, about 3 m above the mean sea level (MSL) and covered with coarse sediments of gravel size. The Gavish Sabkha coastal bar is 400 m wide, about 2.30 m above MSL and covered with sand and shell fragments, supplied from the sea.

Zone 2. Carbonate cemented plates (Gavish Sabkha) and beach rock outcropping with loose igneous and metamorphic pebbles (Solar Lake) border the uppermost rims.

Fig. 15.2 A, B. Aerial views of the Gavish Sabkha (**A**) and the Solar Lake (**B**). The Gavish Sabkha lies open and unprotected on a coastal plain, the Solar Lake is protected by mountain ridges. *Lower parts:* Gulf of Elat (Aqaba) with coral reefs. Both photos show the summer situation where white crusts are formed around the center of the Gavish Sabkha and the littoral zone of the Solar Lake which is exposed by the retreating waterline. The *numbers* denote identical zones: *1* Sand (GS)-(SL) gravel bar; *2* Carbonate cemented plates (only Gavish Sabkha) respectively beach rock plate (only Solar Lake); *3* Gypsum-halite crusts; *4* Wetlands; *5* Littoral; *6* Lagoon; *7* Central elevated hill (only Gavish Sabkha)

Fig. 15.2 A, B

Zone 3. The upper parts of the slopes are covered with hard cemented conglomerate and gypsum crusts.

Zone 4. Wetlands. The lower parts of the slopes and the bottoms of small erosional channels and rills (mainly Gavish Sabkha) consist of permanently wet but not flooded sandy sediments.

Zone 5. Littorals. These are zones of shallow water where the seawater seepage comes to the surface. The littoral zones surround the lower lying and persistently water-filled lagoons.

Zone 6. Lagoons. The Solar Lake lagoon is between 65 and 140 m wide and up to 6 m deep. Seasonal variations in the seawater supply and evaporation, the depth of the basin and the protection by the surrounding ridges allow for a special type of monomixis with solar heating (Cohen et al. 1977, Eckstein 1970, Gerdes et al. 1982, Krumbein and Cohen 1974, Por 1969). During most of the year – September to May – the lake is inversely stratified due to the supply of seepage seawater through the coastal bar and to occasional winter floods. The epilimnion is less saline, lower in temperature and well aerated. The hypolimnion is characterized by high salinities (up to 185‰) and high temperatures (up to 60° C) and is anaerobic. High evaporation rates in summer exceed the water inflow and bring about the seasonal holomixis. During this period, the entire lake is strongly saline (160‰–185‰ S).

The Gavish Sabkha lagoon lies in the northwestern part of the channel which surrounds the central elevated hill (see Chap. 9). The lagoon is between 1 and 30 m wide. The water depth does not exceed 70 cm. The large surface : volume ratio prevents the permanence of thermal stratification in the water body.

Zone 7. The central hill of the Gavish Sabkha, does not exist in the Solar Lake.

15.3 Environmental Parameters, Relevant to Faunal Distribution

15.3.1 Gavish Sabkha

15.3.1.1 Water Supply (Fig. 15.3: Sects. I and II). The sediments of Zones 1–3 and 7 are permanently or periodically dry, and the fauna of the Gavish Sabkha avoids these areas.

Permanently wet but unflooded sediments occur in Zone 4 (the wetlands). This zone is enlarged by various erosional channels (gullies) which emerge below the carbonate cementation plate on the eastern rim of the Gavish Sabkha and run down the slope.

Water, supplied by seepage streams, reaches the surface in the eastern littoral zone (Zone 5). Three different conditions are realized in this zone:

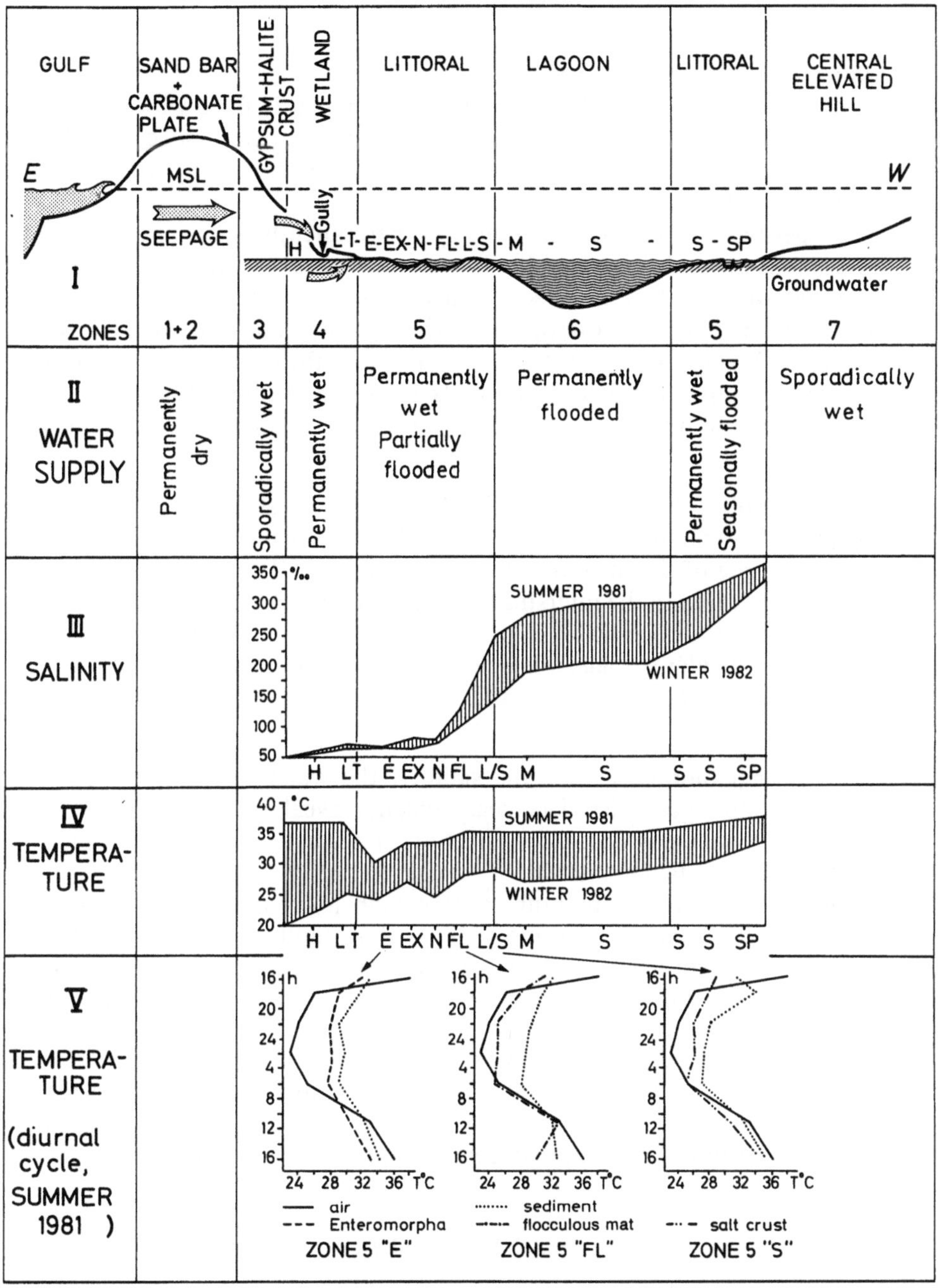

Fig. 15.3. Distribution of abiotic environmental parameters in the Gavish Sabkha. *I* Schematic transsect; *II* Distribution of water along the transsect. *H* Higher-lying areas with groundwater 30–10 cm below surface; *L* Lower-lying areas; *T* Transitions toward the littoral zone (Groundwater 10–1 cm below surface); *E Enteromorpha* mats; *Ex* Exposed sediment without cover of *Enteromorpha* mats; *N* Nodule bays; *FL* Flocculous, spongy or blister-like mats; *L* Leathery mats; *S* Salt-embedded mats; *M* Slime-embedded mats; *SP* Bedouin salt pans. *III* Salinity values from two seasons (February 1982, August 1981). *IV* Temperature values from two seasons (February 1982, August 1981). Mean values from different sampling sites, measured during daytime (13–15 h). *V* Diurnal cycle of temperatures (24-h measurements) at selected sampling sites; August 1981. The values were obtained from *Enteromorpha mats* (*E*), flocculous mats (*Fl*) and below salt crusts (S)

1. Sediments are flooded by a trickling water film.
2. Shallow depressions, 10 to 15 cm deep, are permanently filled with water.
3. Elevated deposits around the depressions are erratically flooded and desiccated, influenced by wind and sun heating. Solid salt crusts are partially developed on the sediment surface.

The deeper basin of the lagoon (Zone 6) is permanently water-covered. Shallow-water areas around the rim of the lagoon are, however, influenced by water table changes, due to tidal and seasonal fluctuations.

The littoral zone on the western side of the lagoon (Zone 5) is covered by the lagoon water. During summer, salts precipitate from the concentrated brine and cover nearly the entire area. Below the crusts, a thin water film is maintained. The water accumulates in small holes which were excavated by Bedouins (Bedouin salt pans).

15.3.1.2 Salinity. A horizontal salinity gradient is established with lowest values (about 50‰) on the eastern side, and with highest values (more than 300‰) on the western side of the littoral zone (Fig. 15.3: Sect. III). Only small seasonal differences were found in the interstitial water of the wetland sediments (Zone 4) and partially in the surface-water film of the littoral (Zone 5). The more the distance from the seawater seepage inlets decreases toward the lagoon (Zone 6), the higher are the values both in winter and summer, as well as the seasonal differences.

15.3.1.3 Temperatures. Unlike the salinity, the water temperature did not show any zonation regularities. The measurements reveal, however, some important aspects about the compensatory effect of the seawater supply by subsurface contact (Fig. 15.3: Sects. IV and V):

Even though the air temperature during high noon in summer often rose above 45° C, water temperatures above 38° C were rarely found. Temperatures of about 40° C were measured only in exposed sediments, for example in the wetland zone. In the flowing seepage seawater, the temperatures were generally lower (Sect. IV).

In the diurnal temperature cycle in summer, the air temperatures fluctuated between 38° C (day) and 24° C (night). The water temperature, mainly inside of the *Enteromorpha* mats, inside the floculous microbial mats and below salt crusts did, however, not reflect these high differences (Sect. V).

15.3.2 Solar Lake

15.3.2.1 Water Supply (Fig. 15.4: Sects. I and II). The conditions in the wetland (Zone 4) are similar to the Gavish Sabkha wetlands. Larger parts of the sediments are, however, solidified by beach rock (calcretes of biogenic origin,

Fig. 15.4. Distribution of abiotic environmental parameters in the Solar Lake. *I* Schematic transsect *II* Distribution of water along the transsect. Abbreviations as in Fig. 15.3. Additional abbreviations: *B* Blister-like mats; *P* Pinnacle mats. *III.* Salinity, temperature and oxygen profiles during winter stratification, intermediate situation, and summer holomixis

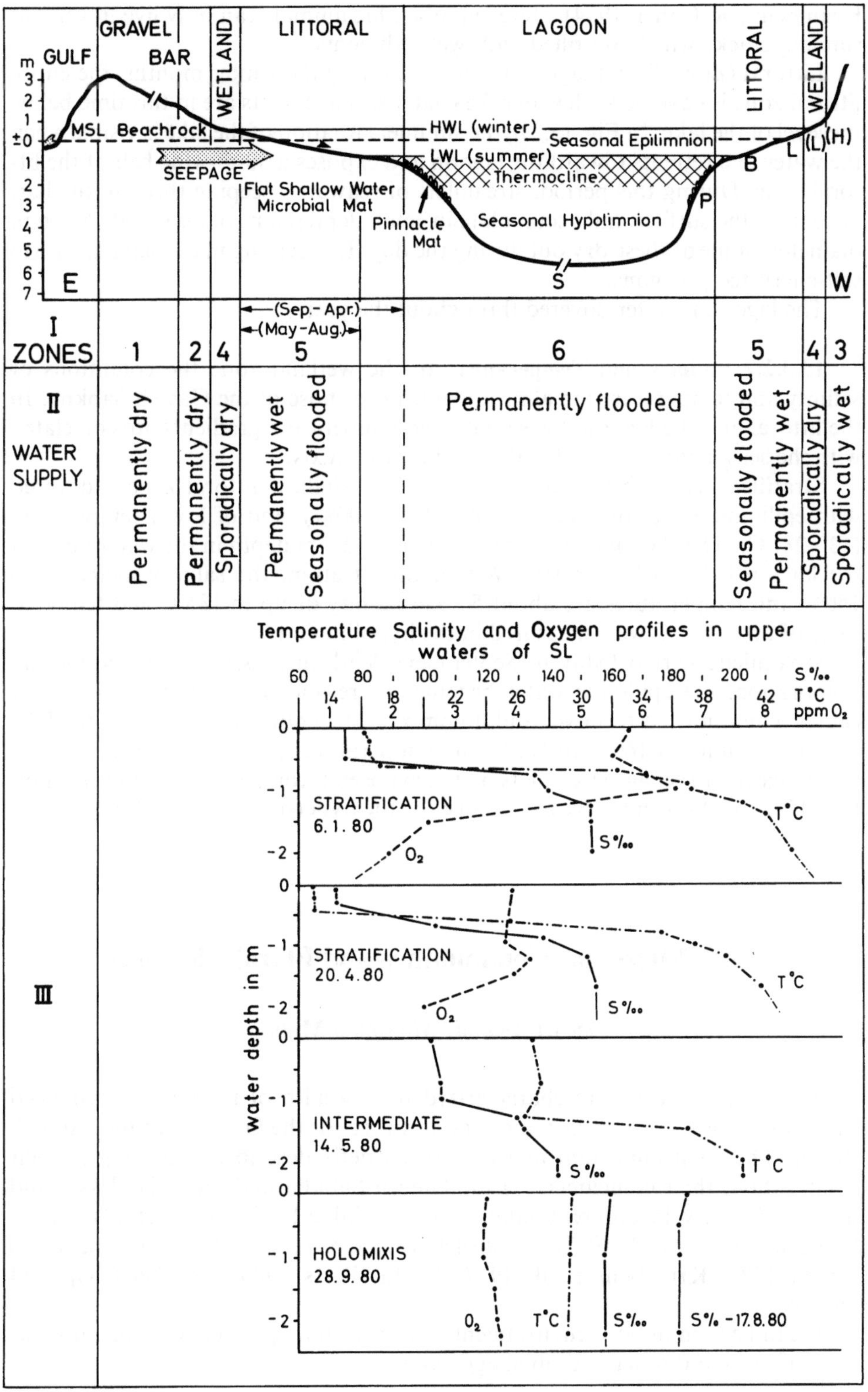

GRAVEL
BAR
WETLAND
LITTORAL
LAGOON
LITTORAL
WETLAND
m GULF
MSL Beachrock
HWL (winter)
Seasonal Epilimnion
LWL (summer)
Thermocline
SEEPAGE
Flat Shallow Water
Microbial Mat
Pinnacle
Mat
Seasonal Hypolimnion
B
L
(L)
(H)
P
S
E
W
(Sep.-Apr.)
(May-Aug.)
I
ZONES 1 2 4 5 6 5 4 3
II
WATER
SUPPLY
Permanently dry
Permanently dry
Sporadically dry
Permanently wet
Seasonally flooded
Permanently flooded
Seasonally flooded
Permanently wet
Sporadically dry
Sporadically wet
III
Temperature Salinity and Oxygen profiles in upper
waters of SL
S ‰
T °C
ppm O2
water depth in m
STRATIFICATION
6. 1. 80
STRATIFICATION
20. 4. 80
INTERMEDIATE
14. 5. 80
HOLOMIXIS
28. 9. 80
O2
T °C
S ‰
S ‰ -17.8.80

Fig. 15.4

Krumbein and Cohen 1974). Here, endobenthic animals can only live in vertical running cracks which are filled with wet sediments.

Littoral (Zone 5) and lagoon (Zone 6): During the winter months, the entire littoral zone is covered with water. Seawater seepage inlets are in this time below the level of the lake. In Fig. 15.2 b, the summer situation of the lake is visible when the water line retreats toward the lagoon and exposes a 30-m-wide belt of the littoral zone. During this period, streamlets of seawater, seeping through the bar, appear on the surface and accumulate in small depressions of about 10–20 cm in diameter. Some of these dry out during the day. Desiccating microbial mats form salt-encrusted polygons.

The lagoon is water-covered throughout the year.

15.3.2.2 Salinity and Temperature. In the wetland zone the conditions of salinity and temperature are almost identical to those of the Gavish Sabkha. In the littoral and the lagoon, the salinity and temperature gradients are correlated with the seasonality of stratification and of holomixis.

Stratified period (September to May): The shallow littoral zone is covered with epilimnic water of lower salinity (70‰–100‰), and of lower temperature (13°–28° C). The thermocline forms at about 1–2.5 m depth in the water body of the lagoon (Fig. 15.4: Sect. III). Within this stratum, the salinity increases to 150‰ and temperatures are about 50° C. Salinity of up to 185‰ and temperatures of up to 62° C were measured in the hypolimnion.

Holomictic period (May to September): With increasing solar heating and evaporation, the epilimnic water becomes more and more saline. This finally causes wind-induced turnover and holomixis. The entire water body of the lake reaches salinities of 160‰ to 185‰ and temperatures of 28° C to 30° C. The exposed seawater seepage streamlets on the exposed littoral zone have lower salinities (45‰ to 70‰) and serve as a refuge for the hyperstenohaline fauna.

15.4 Microbial Communities and Floral Elements

15.4.1 Benthic Microbial Mats

Both environments are characterized by a wealth of microbially produced substrates which serve as the main food resource for the fauna and supply it with shelter. The most important mat types are briefly described according to their composition, their coincidence in the Gavish Sabkha and the Solar Lake, and their availability in different salinity zones (Table 15.1). More details can be found in Cohen et al. 1977 a–c, Ehrlich 1978 a, b, Hirsch 1978, Krumbein and Cohen 1974, Krumbein et al. 1977, 1979 a, Potts 1980 (see also Chaps. 11 and 14).

It can be seen in both environments that the fauna prefers heterogenous surfaces of mats and other microbial aggregates.

Table 15.1. Benthic microbial system in the "Gavish Sabkha" and the "Solar Lake"

Abbreviation	Type	Occurrence	Microbial community
H L T	Sand-embedded microbial mat (1) Single and multilayered	GS and SL Wetland (Zone 4)	Diatoms: Navicula, Amphora Coccoid cyanobacteria: Synechococcus, Synechocystis Filamentous cyanobacteria: Microcoleus, Spirulina Phototrophic bacteria and sulfate-reducing bacteria
 Ex E	Microbial films (2) – epibenthic films – epiphytic films	GS and SL Littoral (Zone 5) Only GS on *Enteromorpha* filaments	Diatoms: Navicula, Amphora, Nitzschia Coccoid cyanobacteria: Gloeocapsa, Gloeothece, Pleurocapsa, Synechococcus, Synechocystis Filamentous cyanobacteria: mainly Spirulina Sulfate-reducing bacteria
N	Nodules (3)	Only GS Littoral: Waterfilled Pools	Diatoms, coccoid and filamentous cyanobacteria: same as 2 Phototrophic and sulfate-reducing bacteria Budding bacteria, Spirillae, Beggiatoa
Fl B	Flocculous, spongy, and blister-like microbial mats (4)	GS and SL Littoral (zone 5) Waterfilled pools (GS) Lower littoral (SL)	Diatoms, coccoid and filamentous cyanobacteria: same as 2 Phototrophic and sulfate reducing bacteria
P	"Pinnacle"-like microbial mats (5)	Only SL Lagoon (zone 6) Upper slope	Diatoms, coccoid and filamentous cyanobacteria: same as 2 Phototrophic and sulfate-reducing bacteria
L M	Flat shallow-water microbial mat (6)	GS and SL Littoral (zone 5) Shallow desiccating parts and lagoon slope (GS) Entire littoral area (SL)	Several layers of coccoid, filamentous cyanobacteria (sandwiching) Diatoms; phototrophic and sulfate-reducing bacteria
S	Salt crust embedded microbial mat (7)	Only GS Littoral (zone 5)	Several layers of coccoid, filamentous cyanobacteria, phototrophic bacteria and sulfate-reducing bacteria (sandwiching) Square bacteria, Halobacteria
S	Gypsum-carbonate or aragonite encrusted microbial mat (8)	GS and SL Lagoon (zone 6)	Coccoid and filamentous cyanobacteria; phototrophic and sulfate-reducing bacteria

Abbreviations:
H, L T: Higher, lower and transitional areas of the wetland
E, Ex: Enteromorpha mats, exposed areas
N, Fl, B: Nodules, flocculous/spongy, blister-like mats
P: Pinnacle mats
L, M: Flat shallow-water mat of leathery or slimy (mucilagenous) consistence
S: Salt crust embedded mats (halite, gypsum-carbonate or aragonite)

15.4.1.1 Gavish Sabkha. Heterogenous surfaces have: (1) The nodules which are described in detail in Chap. 11 (see also Figs. 11.8 and 11.12). Nodules are accumulated on the bottom of the water-filled depressions which lie on the eastern littoral zone. Their cauliflower-shaped irregular surface offers various niches to the meiofauna and smaller aquatic insects. (2) Drifting flocculous aggregates of coccoid cyanobacteria which are also common in several of these depressions. They are well aerated and harbor a rich fauna. The fauna is still abundant in laminated microbial mats with a spongy or blister-like surface, which occur close to the lagoon.

These heterogenous microbial mats and aggregates are covered with water of 70‰ up to 140‰ salinity. Thus, they are available for species which we have denoted as "hyperstenohaline" (see Fig. 15.6).

Mats with homogenous, often leathery, slimy or salt-encrusted surfaces (see Fig. 11.13 in Chap. 11) occur in the Gavish Sabkha chiefly in stronger saline areas (more then 140‰). Though hypereuryhaline species can tolerate these salinities they are not very abundant where the mats are leathery or covered with microbial slimes and salt crusts (Fig. 15.6).

15.4.1.2 Solar Lake. The seasonality of higher and lower water tables allows seepage pools with the typical filling of flocculous microbial aggregates only to form in the summer. Most important are, with respect to heterogenous surfaces in the Solar Lake, the blister-like mats on deeper parts of the littoral zone, and the pinnacle mats on the upper slopes of the lagoon, where the thermocline is already established. Since the surfaces of the pinnacles reach into the epilimnion, the fauna finds well aerated niches.

Both, the blister-like and the pinnacle mats are flooded in the summer holomixis with hypersaline water of more than 140‰. Thus, they are not available in this time for the hyperstenohaline fauna (Fig. 15.7).

Flat laminated microbial mats with homogenous surfaces cover most upper parts of the Solar Lake littoral, which lie exposed during the summer. The microorganisms which build this mat shift with seasonal fluctuations of the water level and changing intensity of illumination (Krumbein et al. 1977). Under a deeper water column in winter, filamentous cyanobacteria dominate the top layers. They give the mat a green appearance. In summer, diatoms and coccoid cyanobacteria form the top layers above the filamentous forms and the mat appears yellow-brown. This shift has an important influence on the distributional behavior of animals which feed preferentially on either filamentous or coccoid cyanobacteria and diatoms (Spira 1981).

15.4.2 Planktonic Microorganisms

15.4.2.1 Dactylococcopsis, sp. This gas-vacuolated cyanobacterium occurs exclusively in the Solar Lake. It is mainly distributed in the hypolimnion. *Artemia* sp. the only planktonic filter feeder in the Solar Lake, dives for short intervals into the inhospitable hypolimnion to feed on *Dactylococcopsis* (Spira and van Rijn 1982).

15.4.2.2 Dunaliella sp. This flagellated chlorophyte occurs exclusively in the Gavish Sabkha where it occasionally develops blooms in the water of the lagoon.

15.4.3 Enteromorpha Mats

Thick scummy mats of *Enteromorpha* sp. and other filamentous green algae (see Chap. 14) occur exclusively in the Gavish Sabkha. They prosper at the level of seawater seepage inlets. High densities of the aquatic fauna were found in these mats (Fig. 1.6). The filaments are covered with epiphytically living diatoms and cyanobacteria. Snails, which are abundant in the *Enteromorpha* mats, chiefly graze on these epiphytic organisms.

15.5 Faunal Composition and Distribution

Thirty-two metazoan species were found in the Gavish Sabkha and 24 in the Solar Lake (Table 15.2). These species belong to ten supraspecific taxa, most of which are represented by one or two species only. Only the coleopterans and dipterans are represented by a wealth of species (Fig. 15.5). Three phyla – Rotifera, Annelida and Mollusca – are part of the faunal assemblage in the Gavish Sabkha but are absent from the Solar Lake. In Fig. 15.11, a comparison of species numbers is given which were found in both environments, or in the one or the other environment exclusively.

According to their spatial distribution, the species were differentiated into three major ecological units (Figs. 15.6 and 15.7, Table 15.2). This differentiation follows Remane (1940), Thorson (1957) and Dörjes and Hertweck (1975).

15.5.1 The Wetland Fauna (Unit I)

In the wetland of both environments the conditions of water availability, salinity and temperature are very similar. The interstitial water keeps the sediments permanently wet. The sediment surface is partially solidified by thin halite or gypsum crusts. Microbial mats are developed endobenthically and consist of a mixture of diatoms, coccoid and filamentous cyanobacteria.

These conditions make the wetland fauna in the Gavish Sabkha and the Solar Lake similar in composition and abundance. Mainly terrestrial arthropods occur, which are adapted to life in the permanently moist sediments. Dominant are beetles of the genus *Bledius*. Two species (*B. angustus* and *B. capra*) occur in the Gavish Sabkha, whereas only the latter was found in the Solar Lake. Both species dig characteristic tunnels into the sediment. The burrows of *B. capra* reach about 12 cm deep. This species is limited by the high groundwater table (Figs. 15.6 and 15.7). Pseudoscorpions (*Halominniza aegyptiaca litoralis*) and spiders (Clubionidae) roam freely above and below the sediment surface and inside the network of tunnels.

Table 15.2. Species list of the Gavish Sabkha and the Solar Lake

		Occurrence in	
		GS	SL
Unit I: Wetland fauna			
Insecta			
Coleoptera	*Bledius angustus*	x	
	Bledius capra	x	x
	Georyssus sp.	x	
	Actidium sp.	x	
	Lophyridia aulica	x	x
Diptera	*Musca crassirostris*	x	
	Orthellia caesarion	x	
Hemiptera	unidentified species	x	
Arachnida			
Pseudoscorpiones	*Halominniza aegyptiaca litoralis*	x	x
Clubionidae	unidentified species	x	x
Crustacea			
Isopoda	*Halophiloscia* sp.	x	x
Nematoda	*Enoplus communis*	x	
	Monhystera sp.		x
Unit II: Hyperstenohaline fauna			
Insecta			
Coleoptera	*Deronectes* sp.	x	x
	Eretes cf. *strictus*		x
	Philihidrus sp.		x
	Enochrus sp.	x	x
	Scarodites sp.		x
Diptera	*Ephydra macellaria*	x	
Crustacea			
Ostracoda	*Cyprideis torosa*	x	x
	Paracyprideinae sp.	x	
Copepoda	*Robertsonia salsa*	x	x
	Nitocra sp.	x	x
Annelida			
Oligochaeta, Enchytraeidae	unidentified species	x	
Polychaeta, Capitellidae	unidentified species	x	
Mollusca			
Gastropoda	*Pirenella conica*	x	
Nematoda	*Oncholaimus fuscus*	x	
	Adoncholaimus oxyuris	x	
	Punctadora sp.		x
Turbellaria	*Macrostomum* sp.	x	x
Rotatoria	*Brachionus plicatilis*	x	
Protozoa			
Ciliata	several species	x	x
Unit III: Hypereuryhaline fauna			
Insecta	*Anacaena* sp.	x	x
Coleoptera	*Ochthebius* cf. *auratus*	x	x
	Paraberosus melanocephalus		x
	Guignotus sp.		x
Diptera	*Hecamede grisescens*	x	
	Ephydra flavipes		x
	Atissini sp.		x
	Atylotus agrestis	x	x
	Bezzia sp.	x	
Crustacea			
Anostraca	*Artemia* sp.		x

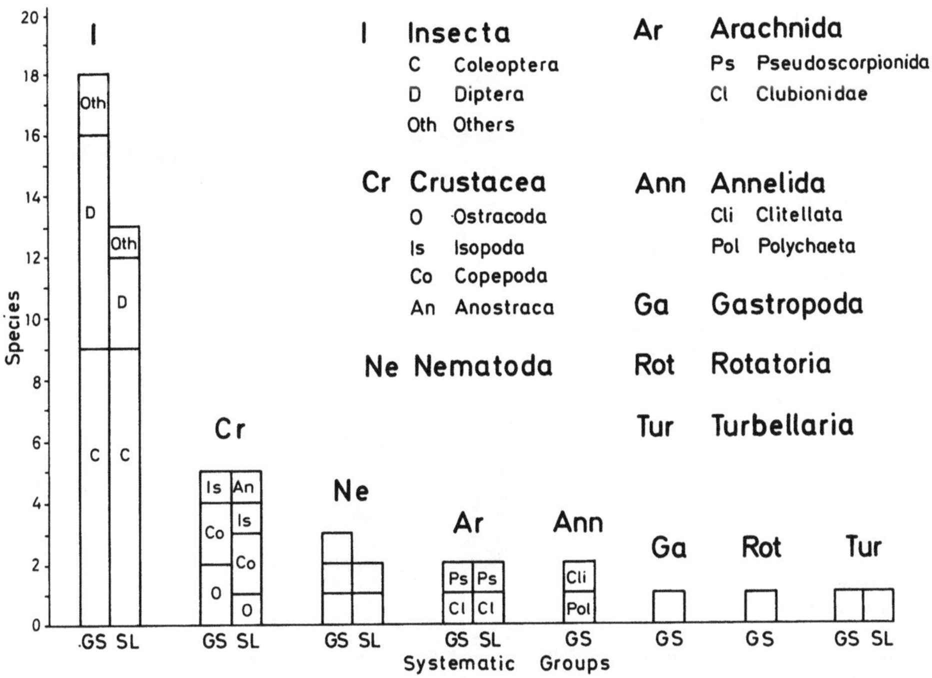

Fig. 15.5. Distribution of species within major taxa (in % of total species number: Gavish Sabkha = 33, Solar Lake = 23)

The wetland fauna migrates along a humidity gradient into the seasonally air-exposed microbial mats of the littoral zones. *B. capra* digs its burrows into the mats (Fig. 15.8.) A population of *Halophiloscia* sp., an amphibious oniscoid isopod, lives in this zone feeding on decaying organic material. The migratory behavior of terrestrial arthropods was mainly observed in the Solar Lake, where the seasonal dryness of sediments in Zone 4 and exposure of microbial mats to the air in Zone 5 is more significant than in the Gavish Sabkha.

15.5.2 The Hyperstenohaline Fauna (Unit II)

This unit is composed mainly of species of limnogenic origin (Por 1980). The fauna consists of small crustaceans (ostracods, copepods), plathyhelminths, oligochaetes, nematodes and coleopterans. They tolerate salinities of up to 140‰ (Figs. 15.6 and 15.7, Table 15.2).

15.5.2.1 Gavish Sabkha. The permanent water supply on the eastern littoral zone with low salinity and temperature, as well as heterogenous microbial systems (nodules, flocculous, spongy and blister-like-mats) and the *Enteromorpha* mats, support high population densities of metazoan species (Fig. 15.6). Salinity conditions are in the metahaline (50‰–70‰) and alpha-hypersaline range (70‰–100‰). With exception of *Pirenella conica* and the oligochaetes, Unit II species

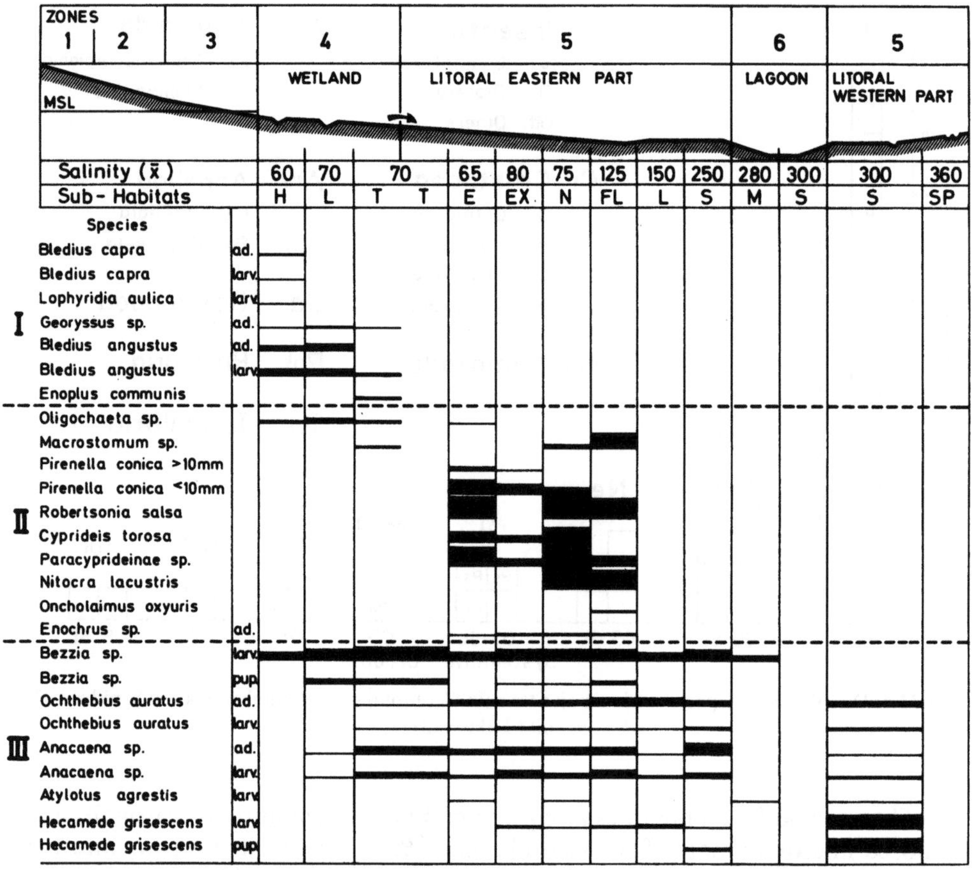

Fig. 15.6 A, B. Gavish Sabkha: Faunal distribution and abundance. **A** Summer (August 1981); **B** Winter (February/March 1982). The distribution is related to spatial gradients of water availability (Fig. 15.3: Sects. *I* and *II*), salinity (Fig. 15.3: Sect. *III*), and to the distribution of microbial mats. *I* Wetland fauna; *II* Salinity limited (hyperstenohaline) aquatic fauna; *III* Salinity tolerant (hypereuryhaline) aquatic fauna

were found up to 140‰ (beta-hypersaline range), though the diversity declines sharply in salinities reaching 140‰. Additionally, the water cover is not permanently maintained in places of this salinity range, the temperature increases, and the former heterogenous mats are replaced by flat laminated microbial mats with a leathery, slimy, or halite-incrusted surface.

P. conica is a euryhaline gastropod which typically lives in the upper littoral zone of the gulf (Taraschewski and Paperna 1981). There, the snails prefer a well saturated and oxygenated sediment, covered with microbial films. These conditions are also maintained in the Gavish Sabkha due to a supply of less saline and oxygenated seawater. Salinities of about 100‰ were observed to be the upper limit for *P. conica* as well as for the oligochaetes (marine enchytraeids). The latter were found in the Gavish Sabkha only in immature stages.

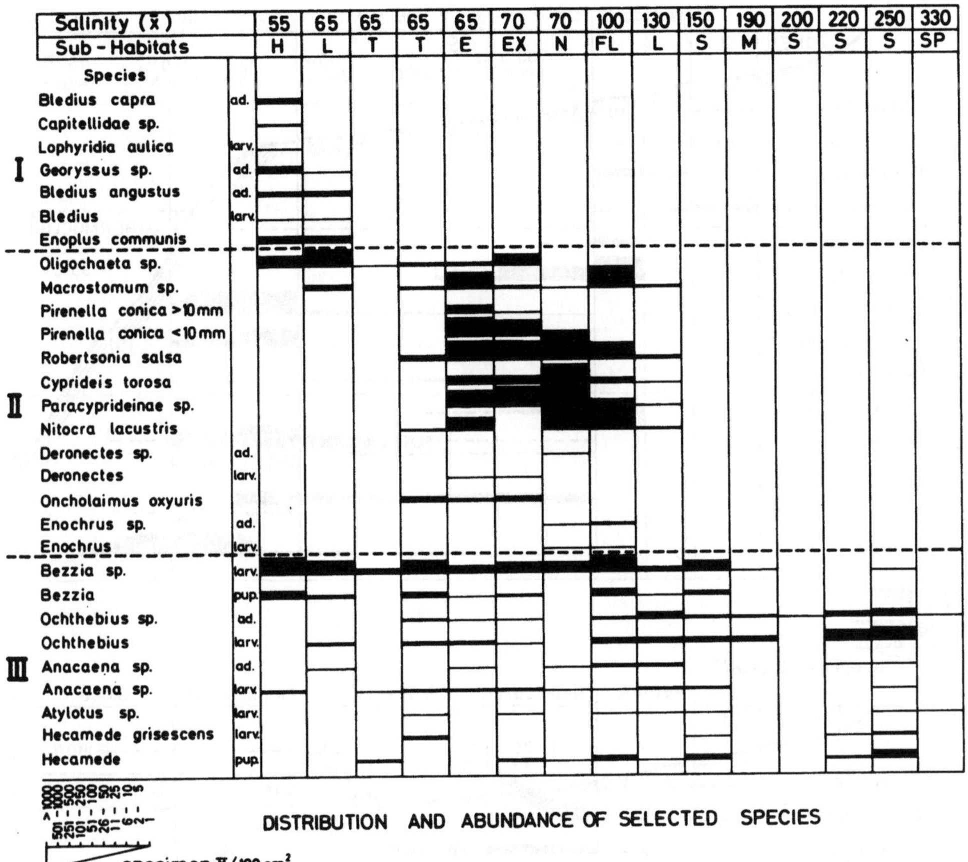

Fig. 15.6B

15.5.2.2 Solar Lake. During the long period of stratification (October to May), the hyperstenohaline fauna concentrates on the flat shallow-water microbial mat of the littoral zone (Zone 5). Here, the well-aerated epilimnic water has alpha- and beta-hypersaline conditions with temperature ranges between 15° and 28° C. In this period, the fauna is diverse and includes benthic elements such as were found in the Gavish Sabkha, with the exception of *P. conica* and oligochaetes. The epilimnion includes a variety of limnetic fauna exclusive to the Solar Lake, most of which are swimming beetles of the dytiscid family (Table 13.2, Unit II).

During the summer, as salinities rise into the gamma-hypersaline range (above 140‰), some species of Unit II fauna migrate into the seewater seepage streamlets and small pools, formed on the air-exposed microbial mat surface. Even though temperature can be high (28°–35° C), salinities are low and in the metahaline range (45‰–70‰). Several species of Unit II (mainly dytiscid imagines) leave the Solar Lake during summer, while others decrease in number and are hard to find (*Nitocra* sp., *Cyprideis torosa*, Fig. 15.7).

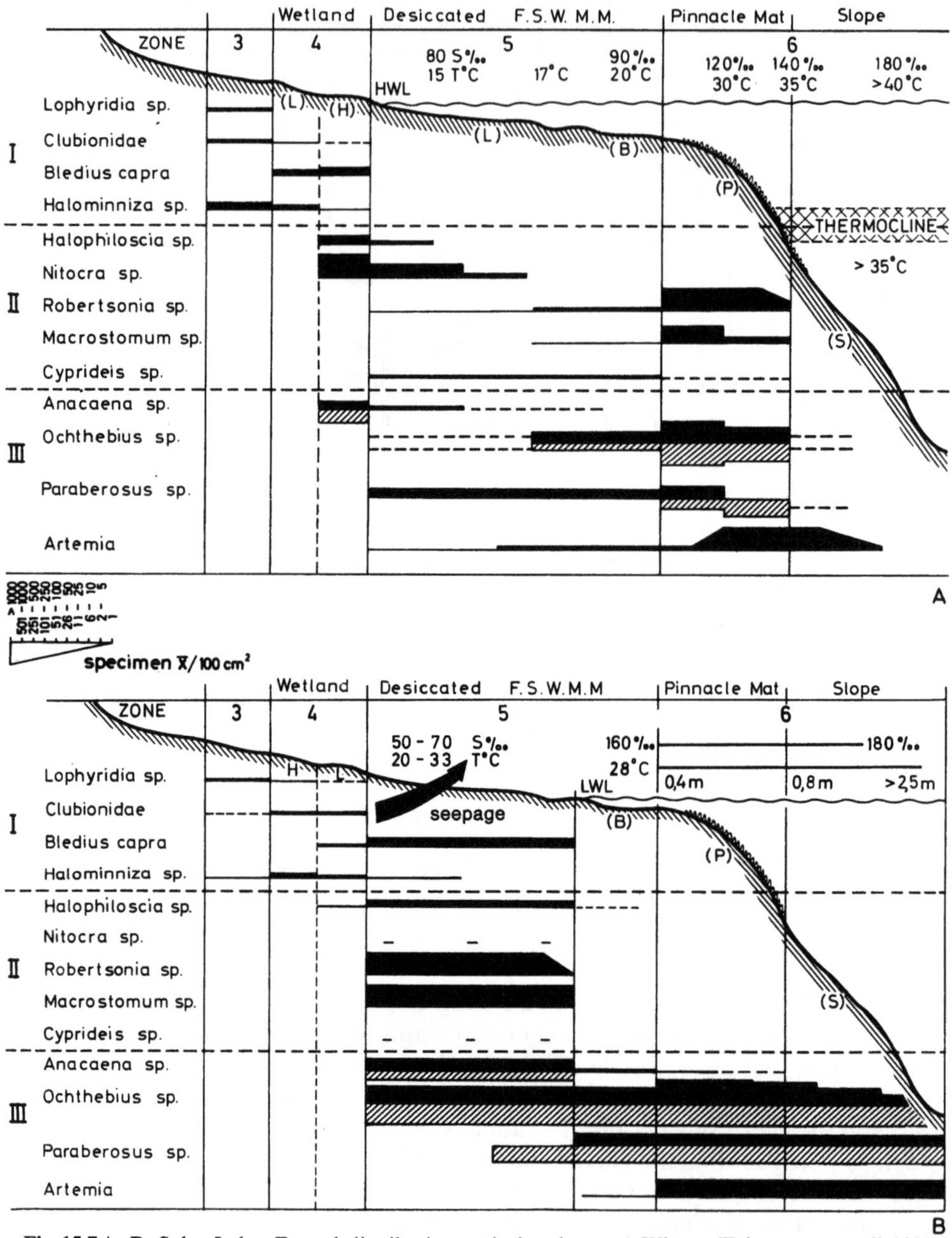

Fig. 15.7 A, B. Solar Lake: Faunal distribution and abundance. **A** Winter (February – April 1980); **B** Summer (August–September 1980). The distribution is related to spatial and seasonal gradients of water availability (Fig. 15.4: Sects. *I and II*), salinity and temperature (Fig. 15.3: Sect *III*), and to the distribution of microbial mats. *I* Wetland fauna; *II* Salinity limited (hyperstenohaline) aquatic fauna; *III* Salinity tolerant (hypereuryhaline) aquatic fauna. *F.S.W.M.M.* Flat shallow-water microbial mat; ■ adults, ▨ larvae

Fig. 15.8. Burrows of *Bledius capra* in shallow-water mats of the Solar Lake. The burrows were produced during summer air exposure of the mats

15.5.3 The Hypereuryhaline Fauna (Unit III)

Species of this unit tolerate a wide range of salinity and temperature. They were found in almost all zones within the range of 50‰–180‰ S (Solar Lake) and up to 300‰ S in the Gavish Sabkha (Figs. 15.6 and 15.7).

Ochthebius cf. *auratus* occurs in salinities up to 250‰. Dense population correspond with the occurrence of spongy, flocculous and pinnacle-like microbial mats. Salt crusts, which occasionally cover mats in the Gavish Sabkha, are not a distributional barrier. This species is almost the only one which retracts into deeper layers of laminated microbial mats with high concentrations of H_2S. The small size allows the beetles to utilize oxygen bubble channels. These channels run toward the surface from deeper layers of filamentous cyanobacteria, which coexist with anoxygenic photosynthetic bacteria (Krumbein et al. 1979a). In the Solar Lake, this species is found in high numbers in the deeper pinnacle mat where temperatures and salinities are high and oxygen is poor during stratification (Fig. 15.4).

Anacaena sp.: This species characterizes the periphery of water bodies in both, Gavish Sabkha and Solar Lake. The inability of larvae and adults to swim restricts them to the shallow-water region. Larvae were found to prefer burrows of *Bledius* spp. where they feed on larvae of the terrestrial beetles and of dipterans. In the littoral zones *Anacaena* sp. lives underneath decaying plant remains and patches of microbial mats as well. Also, small water holes embedded with flocculous microbial aggregates are preferred. In the Gavish Sabkha, the beetles also inhabit the undersurface of salt crusts, coated with spongy microbial mats.

Bezzia sp. was found in the Gavish Sabkha exclusively. The larvae and pupae are distributed over the whole range of zones. The adult mosquitoes prefer open-water holes where they accumulate on the water surface.

Atylotus cf. *agrestis* was found only in its larval stages. The horsefly larvae live in small numbers below the salt crusts and inside the flocculose microbial material that drift in the water.

Larvae and pupae of *Hecamede grisescens* were found mainly on the western littoral zone of the Gavish Sabkha below salt crusts. Adults were observed above open-water holes.

15.6 Trophic Relations

15.6.1 Primary Consumers

The majority of species in both ecosystems are primary consumers (Fig. 15.9). Gut content analyses indicated that these species can be divided according to their feeding behavior into four distinct groups: (1) benthic fauna that feeds primarily on diatoms and coccoid cyanobacteria; (2) benthic fauna that feeds primarily on filamentous cyanobacteria; (3) benthic fauna that prefer decaying organic material; and (4) planktonic filter feeders.

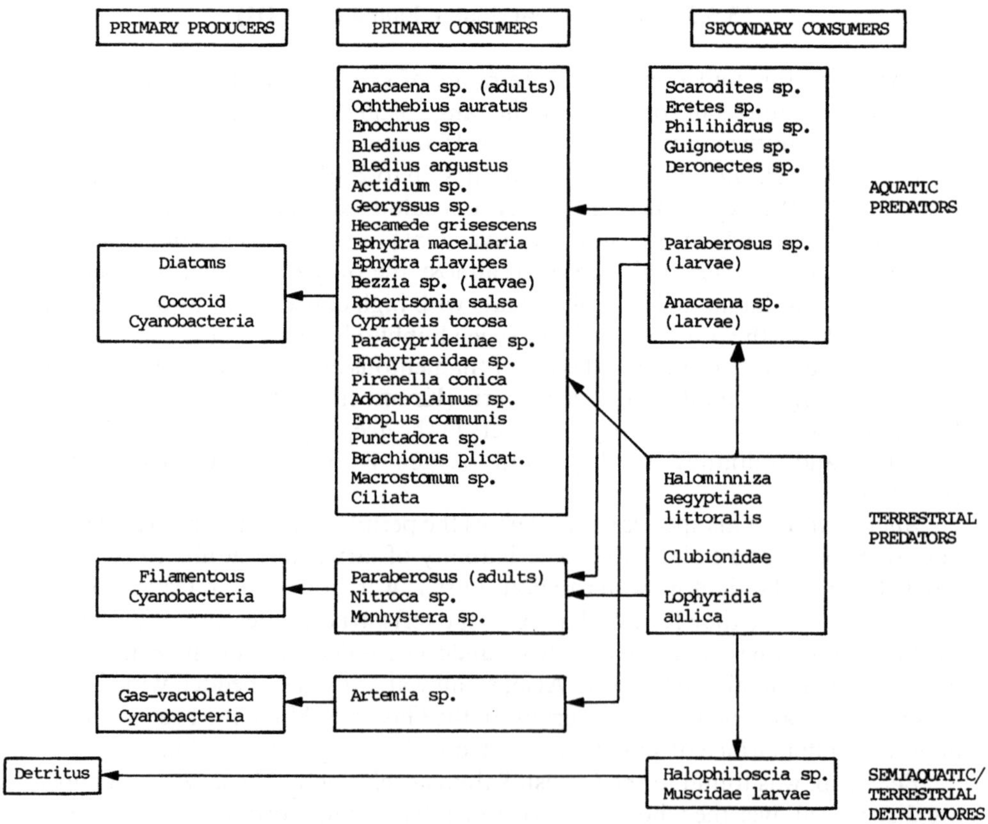

Fig. 15.9. Trophic relations. The species of the Gavish Sabkha and the Solar Lake were combined

Group 1: The water bodies of both ecosystems are characterized by a wealth of species which belong to the first group. This corresponds to the wealth of coccoid cyanobacteria and diatoms in the top layers of the microbial mats and their wide horizontal distribution.

Group 2: Three species of benthic primary consumers feed preferentially on filamentous cyanobacteria (Fig. 13.9). *Nitocra lacustris* is the only one of these species which occurs in both systems. *Paraberosus* sp. and *Monhystera* sp. were found only in the Solar Lake. Here, filamentous cyanobacteria are abundant in top layers during winter months. Prevailing high salinity and shifting of coccoid cyanobacteria and diatoms into the top layers force *Nitocra lacustris* and *Monhystera* sp. to migrate into the small seepage inlets, the *Paraberosus* sp. adults are almost unaffected by increasing salinity and by higher temperatures of up to 35° C. During summer, the adults migrate to deeper water where filamentous cyanobacteria can tolerate the high irradiation. During the stratification period adult beetles preferentially roam the flat shallow water mat where filamentous cyanobacteria are found in the top layer.

Group 3: Three species preferentially feed on decaying organic material. The larvae of Muscidae, belonging to this group, were only found in the Gavish Sabkha. Here, plant litter and camel dung occur along flood marks around the wetland. These deposits indicated occasionally higher water levels and sheet-floods. *Halophiloscia* sp., found around both water bodies, prefer decaying plant litter as well as decaying microbial mats. A seasonal change of biotopes was observed sometimes between *Halophiloscia* sp. and *Bledius capra,* when increasing moisture of the sediments or mats forced the beetles to move into drier habitats, and *Halophiloscia* sp. entered the burrows of the beetles.

Group 4: Artemia sp. is the only example of planktonic filter feeders. It is one of the most important grazers in the Solar Lake, according to its population density. Food organisms are the gas-vacuolated planktonic cyanobacteria *Dactylococcopsis* which accumulate predominantly in the hot (40° C) and unoxygenated hypolimnion. *Artemia* sp. lives in the favorable metalimnion, concentrating just above the thermocline. It dives down into the rich but unhospitable hot brine in order to feed (Spira and van Rijn 1982).

15.6.2 Aquatic Predators (Secondary Consumers)

This trophic level in the aquatic environments of the Gavish Sabkha and the Solar Lake is composed of hydrophilid larvae and some species of dytiscids. *Paraberosus* larvae are good swimmers. They were found only in the Solar Lake and mainly in niches where high concentrations of *Artemia* occur (along the pinnacle-like mat and the thermocline). The larva actively feeds on *Artemia* sp., *Robertsonia salsa, Macrostomum,* sp. and *Ochthebius* larvae.

Anacaena larvae feed on the oniscoid isopod *Halophiloscia* sp., on *Bledius* larvae, dipteran larvae, and on some of the meiofaunal elements. In the Solar Lake, the larva efficiently feeds on *Artemia* resting eggs that accumulate at the shoreline (Dimentman and Spira 1982).

Fig. 15.10. Shore birds in the Gavish Sabkha, feeding on benthic invertebrates

Both larvae belong to the Unit III to which the totality of habitats is accessible.

Dytiscids appear in both environments during the winter. In the Solar Lake, they are represented by a number of species, whereas only one species (*Deronectes* sp.) occurs in the Gavish Sabkha.

15.6.3 Terrestrial Predators

These are mainly terrestrial arthropods (Fig. 15.9). They occur in both environments. Submersed faunal elements are not accessible to them. The very agile adults of *Lophyridia aulica,* the spiders and the pseudoscorpions inhabit, however, the humid sediments of the littoral and the wetland and hunt meiofaunal elements and dipterans.

Shore birds (Fig. 15.10) are common visitors to the Gavish Sabkha, where they feed on benthic invertebrates. Nests were found along the wetland.

15.7 Discussion

At first sight, it seemed surprising to find a high degree of similarity in the faunal assemblages of the Gavish Sabkha and the Solar Lake (Fig. 15.11), because both systems show several differences. The high degree of similarity emphasizes, the prime importance of parameters like the continuous influx of fresh

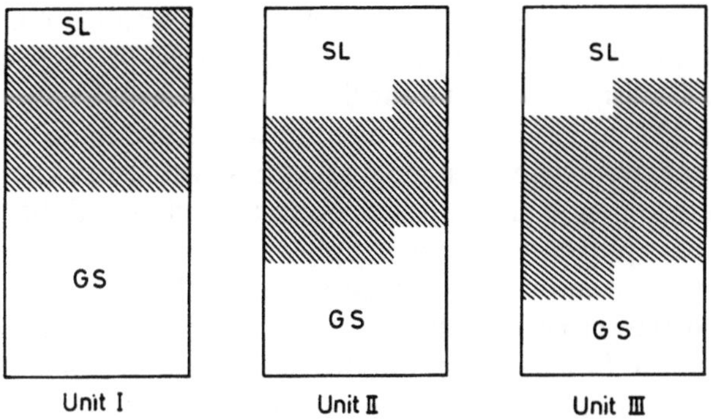

Fig. 15.11. Percentage rate of species which occur in both environments (*etching*), or which occur only in the Gavish Sabkha (*bottom*), or which occur only in the Solar Lake (*top*)

seawater, combined with arid conditions, the separation from the sea and the prevailing hypersalinity.

The continuous seawater recharge in combination with semi-arid conditions has important consequences with respect to the dominant environmental parameters – the water regime, the salinity and the temperature.

The seawater influx prevents seasonal or longer-termed dryness which characterizes most of the inland hypersaline ecosystems (Bayly 1972). Thus, animals without resistance to desiccation survive and reproduce both in the Gavish Sabkha and in the Solar Lake. Even though they are exlcuded by high salt concentrations from the deeper water bodies, they find sufficient favorable microenvironments in the littoral areas which are protected against desiccation by the continuous seawater supply. The size of the organism is probably of importance. The majority of species in the Gavish Sabkha and the Solar Lake belong to the meiofauna or to small macrofaunal forms. With about 2 cm lengths, the horsefly larvae represent one of the largest forms. They occur, however, only in small numbers.

The seawater influx apparently contributes to the formation of stable salinity zones along a gradient with increasing values. In the Gavish Sabkha this gradient is horizontal. In the deeper Solar Lake with submersed seawater seepage inlets there is also a vertical salinity profile. The diversity of the fauna follows this gradient in opposite direction, as demonstrated in Figs. 15.6 and 15.7.

In summer, air temperatures often raise above 45° C. Water temperatures were, however, considerably lower. Besides, they did not follow the rapid decrease of the air temperature during the night. It is reasonable to assume that the continuous water influx from the sea stabilizes the temperature regime in the shallow water.

The separation from the sea is a second parameter of primary importance. The coastal bars are barriers preventing the migration of estuarine predators, such as fish, shrimps and other forms into the Gavish Sabkha and the Solar Lake. Also planktonic larval swarms of sessile epibenthic invertebrates are prevented by the coastal bars. This is of importance, because in marine littoral areas, epibenthic sessile invertebrates like calcareous worms, corals, bryozoans and porifers are assumed to compete in space with benthic laminated microbial mats (Pratt 1982). These mats are, however, the main food source and the protective substrates for the fauna of the Gavish Sabkha and the Solar Lake. Here the hypothesis of Dahl (Dahl 1967, Por 1978a, 1980) may be mentioned, according to which organisms adapted to life in extreme environments often pay with a weakness against competition. This phenomenon encompasses both, the fauna and the microflora.

The coastal bars prevent the open supply of seawater with tidal mixing, wave action and sedimentation or erosion processes as well.

A third important environmental feature is the hypersalinity of most of the seawater bodies. The fauna, found in both environments, consists of terrestrial and aquatic species adapted to life in hypersaline environments. Several general characteristics of hypersaline biota are maintained in both environments.:

1. Most of the *genera* and some of the *species* are cosmopolitic. The terrestrial beetles, arachnids and isopods belong to genera which characterize the shorelines of saline aquatic ecosystems nearly worldwide (Bro Larsen 1936, Heydemann 1967, Lengerken 1929, Moore 1978, Schulz 1936). Also such aquatic genera as *Artemia, Ephydra* and *Ochthebius* were reported from other parts of the world (Hase 1926, Jaquin 1956, Kristensen 1971, Swain 1955), with the exception of Australian hypersaline lakes (Bayly 1967, 1972, K. G. McKenzie 1981, K. F. Walker 1973).
2. Most of the major *taxa* are represented by one or two species only with the exception of the insects. This reduction is obvious in both systems being described and indicates the extreme environmental conditions.
3. Based upon this reduction, the *food chains* are simplified. Both systems are grazer dominated. The number of secondary consumers decreases with increasing salinity. Since in Unit II a number of predators is still present (mainly dytiscids), Unit III demonstrates an almost all-grazer community with the exception of few coleopteran larvae (Por 1980).

On the other hand, the preponderance of coleopterans seems to be a peculiar property of the environments described here. Insects are usually represented in hypersaline environments by a wealth of dipterans and hemipterans. Most of the coleopterans living in the Gavish Sabkha and the Solar Lake are grazers. Probably, the wealth of microbial mats supports the high number of coleopterans (Por 1975b).

Though the fauna of the Gavish Sabkha and the Solar Lake is grazer dominated, no massive grazing on the microbial mats was observed, with the exception of the destructive influence of *Pirenella conica* (Gerdes and Krumbein 1984).

The findings on the Gavish Sabkha and the Solar Lake are similar to observations which were made along the Sinai coast during the last 15 years (Por 1969, 1972, 1975b, 1978a, 1980; see also Chap. 6). The stability of seawater seepage evidently contributes to stable environmental conditions and to a climax-like biota in these systems. This is also expressed by the high percentage of species which were found in both environments (Fig. 15.11). This high percentage applies to all of the three ecological units which were defined according to the distributional patterns of the fauna.

The two sections in Fig. 15.11 dealing with species which were found only in the Gavish Sabkha or in the Solar Lake, emphasize their individual differences. In the Gavish Sabkha, the terrestrial fauna and the metahaline aquatic fauna are much richer and diversified than those of the corresponding zones in the Solar Lake. The permanence of groundwater humidity and the stable seawater seepage streams with lower salt concentrations allow the establishment of a stable and permanent fauna there. These conditions also allow a few species from the littoral area of the gulf to live in the Gavish Sabkha carried into this environment with occasional floods. In the Solar Lake, fluctuation of groundwater levels between summer and winter and the temporal state of the seasonal seawater seepage force the fauna to migrate constantly in search of more favorable conditions, thus limiting the establishment of stable populations.

In the Solar Lake, the deeper water of the lagoon allows the existence of richer and diversified fauna which is not found in the lagoon of the Gavish Sabkha. The aquatic fauna of the Solar Lake includes swimming elements which are limited by the shallow waters of the Gavish Sabkha.

The individual differences between both environments are pointed out especially by two species: *Pirenella conica* (Gavish Sabkha) and *Artemia* sp. (Solar Lake). In the Gavish Sabkha, *P. conica* has developed a stable and dense population (see Fig. 11.11, Chap. 11). The short time in which appropriate habitats emerge in the Solar Lake does no allow the formation of stable populations of *P. conica* in this environment. Analyses of cores have shown, however, that the Solar Lake had a *Pirenella* phase some 2,500–3,000 years ago (Krumbein and Cohen 1974, Por 1975b). In that former time, the Solar Lake was a shallow lagoon. The deep basin was caused by a tectonic event about 1,900 years ago.

The second species emphasizing the differences between the two environments is the brine shrimp *Artemia* sp. This species maintains a permanent population in the Solar Lake throughout the year.

Salinity levels in the epilimnion during stratification and in the entire water body during holomixis are within the salinity range in which *Artemia* can fulfill its life-cycle. The absence of *Artemia* from the Gavish Sabkha is a result of the shallowness in the areas fit for their existence (seawater seepage zone) and the inability of *Artemia* resting eggs to hatch osmotically in the high salinity of the other areas of this water body (Dimentman, unpublished).

In conclusion, we may say that the Gavish Sabkha and the Solar Lake are two variations of anchialine hypersaline water bodies (see Chap. 5) located in an arid region and that this common denominator is significantly more important than the individual differences between the Gavish Sabkha and the Solar Lake. Despite the various differences in limnological, microbiological and zoological features both are interesting examples of one and the same: An intermediate ecosystem between the sea and the continent, which is characterized (1) by its evolutionary antiquity (2) by evaporative restrictional stress and (3) by re-invasion from both major environments, marine and terrestrial.

16. Trace Metal Concentrations in Sediments from the Gavish Sabkha

W. Berry Lyons and Henri E. Gaudette

The role played by early diagenetic processes in the chemical transformation, mobilization and speciation of trace metals in nearshore sediments is poorly understood. In addition, little information, either qualitative or quantitative, is available regarding trace metal concentration and accumulation in hypersaline environments. The analysis of hypersaline sediments for trace metals is important due to the fact that recently both Renfro (1974) and J. H. Davis (1977) have suggested that diagenetic interactions in supratidal sabkha environments may be influential in the formation of "Mississippi Valley" and "Copper-Shale" (i.e., Kupferschiefer) type stratiform base metal deposits.

A stratigraphic and regional association between evaporite deposits and stratiform base-metal deposits has long been recognized (Davidson 1965, Roedder 1968, Sasaki and Krouse 1969, Beales and Hardy 1977, Lange and Murray 1977, van Eden 1978). Theide and Cameron (1978) feel that even though causal relationships are unclear, evaporite deposits may have served as sources of metal-rich brines leading to the deposition of economically minable sedimentary metal deposits. G. E. Smith (1976) has recently described stratiform Cu-rich deposits in Texas, Oklahoma and Kansas that may have been formed by an evaporative process. He differs in the idea that blue-green algae (Cyanobacteria) actually served as the loci for metal deposition.

In recent years, many theories have been proposed relating evaporite (especially sabkha) formation and diagenesis to metal deposition. Although coastal evaporites have been studied for the past 20 years, only recently have trace metal data been available permitting us to evaluate the possible role of trace metal deposition in these unique environments (Gibbons 1978, Long and Angino 1982).

This chapter presents trace metal data from sediments of different composition from the Gavish Sabkha. It was our objective to determine: (1) what the concentrations of various trace metals are in these sediments and (2) whether the cyanobacterial mats concentrate metals above the ambient sedimentary background values? It should be noted that the samples from the Gavish Sabkha analyzed by our group were not collected by us but instead dried sediment aliquots were given to us by E. Gavish (in January 1981).

16.1 Analytical Methods

The samples were weighed, placed into precleaned polyethylene bottles and leached for about 18 h with 10% nitric acid at 60 °C. Previous investigators (Carmondy et al. 1973; Thomson et al. 1975) have shown that nitric acid leaching extracts as much as 80%–85% of the total sedimentary metal in nearshore clastic sediments as compared to HF/HNO_3 treatment.

The transition metal iron (Fe) was determined colorimetrically after suitable dilution with MilliQ[T.M.] water (Stookey 1970). The trace metals cadmium (Cd), chromium (Cr), copper (Cu), lead (Pb) and zinc (Zn) were analyzed via flame atomic absorption spectrophotometry after appropriate dilution with MilliQ[T.M.] water. Procedural blanks were also prepared and analyzed for all the metals mentioned above. All the metal concentrations with the exception of Cd and Pb were determined by the standard curve technique; Cd and Pb were determined by standard additions. The precision of the metal analyses determined by the analysis of sabkha samples was: Fe$\pm$8%, Cd$\pm$7%, Cu$\pm$6%, Cr$\pm$5%, Pb$\pm$10%, and Zn$\pm$4%.

16.2 Results and Discussion

Descriptions of the sediment samples and their trace metal concentrations are shown in Table 16.1. Trace metal analysis of the top 2 cm of living cyanobacterial mat from Solar Lake, a hypersaline lake in the Sinai, is also presented for com-

Table 16.1. Concentrations of Fe, Cd, Cr, Cu, Pb and Zn in sediments from Gavish Sabkha and top 1–2 cm of Solar Lake cyanobacterial mat

Sediment type (general description)	Fe mg/g	Cd µg/g	Cr µg/g	Cu µg/g	Pb µg/g	Zn µg/g
Gypsum-rich sediments	1.64	–	8.2	3.4	<1	11
Gypsum-rich sediments	1.50	–	7.2	2.9	<1	7
Mat/gypsum[a]	1.56	–	11.7	5.9	2	24
Gypsum/carbonate	1.11	–	7.3	3.4	4	6
Carbonate/silicate sand	1.69	<0.25	4.5	2.5	8	26
Halite-rich sediments	0.52	<0.25	6.4	3.5	8	4
Gypsum-rich sediments	0.43	–	7.0	3,7	5	4
Gypsum-rich sediments	0.41	–	7.0	3.5	3	5
Carbonate/gypsum	1.59	–	4.5	3.5	3	10
Carbonate/mat[a]	1.49	–	11.4	9.4	2	46
Carbonate/mat[a]	1.07	–	10.8	8.9	5	40
Mat[a]	1.47	–	13.4	10.0	4	44
Mat[a]	1.51	–	5.2	3.8	4	15
Carbonate/gypsum	1.57	–	5.3	3.5	5	14
Mean mat (n=5)	1.42	–	10.5	7.6	3	34
Mean non-mat (n=9)	1.16	–	6.4	3.3	4	10
Solar Lake mat[b]	7.64	0.56	8.0	15.3	5	39
Mean Solar Lake sediments[c]	8.00	<0.36	10	9.4	3.8	27

[a] Sediments containing abundant mat detritus
[b] Top 1–2 cm of living cyanobacterial mat
[c] From Gaudette and Lyons (1984), n=36 with the exception of Cd where n=12

Table 16.2. Average concentrations of metals in various sedimentary rock types and nearshore marine sediments (μg/g dry wt)

Metal	Carbonate[a]	Shale[a]	Sandstone[a]	Nearshore marine[b]
Fe	3,800	47,200	9,800	12,300
Cd	0.035	0.3	–	0.39
Cr	11	90	35	–
Cu	4	45	–	55
Pb	9	20	7	36
Zn	20	95	16	117

[a] Mason (1966)
[b] Hunt (1979), surficial sediments, Long Island Sound, U.S.A. (n = 63)

parison. The geochemistry and hydrology of the Solar Lake has recently been compared and contrasted with the Gavish Sabkha in Gavish (1980).

The metal concentration from Gavish Sabkha sediments are lower than other nearshore marine environments and sedimentary rock types (Table 16.2). This suggests that contemporary sabkha sediments are not preconcentrating metals above normal marine sediments. These data support recent findings in other sabkhas and evaporitic pans throughout the world (Gibbons 1978, Long et al. 1983).

It does appear, however, that the samples containing microbial mat debris have higher mean concentrations of Cr, Cu and Zn than the other samples. For example, the mean concentrations of Cr, Cu and Zn for the old mat samples are 10.5 μg/g, 7.6 μg/g and 34 μg/g respectively, compared to those of 6.4 μg/g, 3.3 μg/g and 10 μg/g for the other sediment types. Only two non mat sediment samples from the Gavish Sabkha were analyzed for Cd (Table 16.1). These sediments yielded results below our detection limit of 0.25 μg/g.

Unlike the Gavish Sabkha sediments, the majority of the Solar Lake sediments are produced primarily by the growth and decay of cyanobacterial mats (Gavish 1980). The top 10–20 cm of sediment from the marginal portions of the Solar Lake are black and rich in organic matter, while the next 50–70 cm are composed primarily of alternating laminations of $CaCO_3$ (produced in situ via organic matter decomposition) and mat detritus (Krumbein et al. 1977). G. E. Jones et al. (1978) have shown that Cyanobacteria can concentrate metals as much as or to a greater extent than other types of bacteria, algae and fungi. The living Solar Lake mat, the mat detritus rich Solar Lake sediments, as well as the mat detritus rich Gavish Sabkha sediments, all have higher mean concentrations of Cr, Cu and Zn than the Gavish Sabkha samples, which contain no mat material (Table 16.1). This may suggest that the Cyanobacterial play a role in sequestering metals in these hypersaline environments. The Fe values are much higher in the Solar Lake sediments compared to Gavish Sabkha, while the Pb values are relatively constant. This may suggest that Pb is not deposited preferentially in the organic matter but rather is associated with an inorganic phase (Gaudette and Lyons 1984) or that Pb is lost from the sediments via the refluxing of brine through the sediments (Long and Angino 1982).

16.3 Conclusions

This work suggests that cyanobacterial mats may play a role in the concentration of trace metals in marine sabkha environments. As this was a preliminary study, much more work needs to be undertaken to better establish the role of Cyanobacteria in trace metal geochemistry and the fate of these metals during the diagenesis of evaporite sediments.

17. Biogeochemistry of Gavish Sabkha Sediments

I. Studies on Neutral Reducing Sugars and Lipid Moieties by Gas Chromatography-Mass Spectrometry

JAN W. DE LEEUW, J. S. SINNINGHE DAMSTÉ, J. KLOK, P. A. SCHENCK, and JAAP J. BOON

The Gavish Sabkha environment is of a startling complexity and the result of many variables which go to extremes (see Chaps. 9 and 11).

As a consequence of all these variable conditions some niches are dominated by Eukaryotes, e.g., algae, protozoa, insects, others by prokaryotes, e.g., cyanobacteria and purple sulfur phototrophs and by archaebacteria, mainly halophiles (see Chaps. 11–15). As biogeochemists, we are interested in the molecular architecture of the organisms involved and how their chemical composition is related to environmental conditions. The few chemical studies on the organisms and microorganisms grown in pure culture indicate that several different biosynthetic pathways are followed and quite peculiar compounds are produced undr special conditions.

Such compounds in turn may act as species markers in the field and – perhaps – as molecular indicators of fossil environmental conditions. In vertical section the sediments are highly variable and reflect the facies distribution presently seen in the Sabkha depression. Mud layers, gypsum deposits and occasionally laminated microbial mats of the type so well developed in the Solar Lake (Krumbein et al. 1977) are observed. These fossil microbial mats are rarely more than about 1 cm thick. In several cores taken at various places they may sum up, however, to 20%–30% of the total sediment of the uppermost 80 cm. Part of the work reported in this chapter is intented as a survey of molecular markers in the top mat of the sediments permanently covered with water (Zone 9 of Chap. 9) where neutral sugars and extractable hydrocarbons, fatty acids, fatty alcohols and sterols were investigated. We also report here a total lipid profile of a gypsum-rich sediment which contains some mud and remains of microbial mat layers. This sample comes from 50 cm depth of a core taken at the outer rim of Zone 9. The results are compared to those from a similar profile of the top mat and suggestions are made for further studies.

17.1 Sampling and Sample Description

The Sabkha sediments used for analysis were collected during field trips in November 1980 and April 1981 after the first and second sheetfloods. Samples were taken by surface scooping with clean steel spoons or by coring with perspex pipes at sites indicated on the sketch map shown in Fig. 17.1.

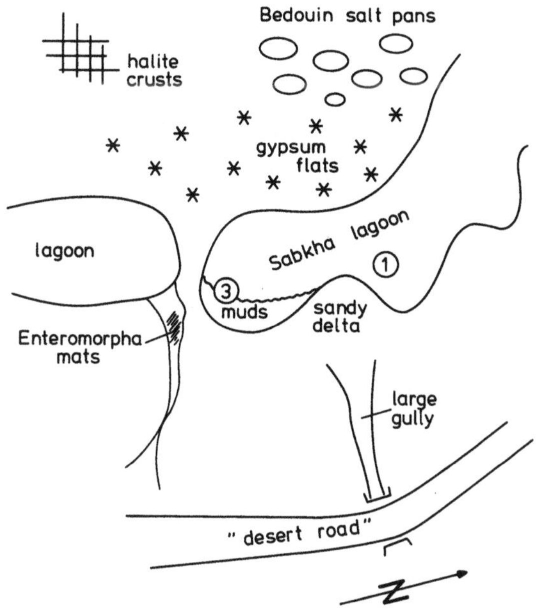

Fig. 17.1. Sketch map of the Sabkha area

The sample used for a survey of neutral sugars was collected at Site 1 in November 1980 in the zone permanently covered with water. The microbial mats previously described by Krumbein et al. (1979 a) were covered by a yellow brown mud layer of about 0.5 cm thickness deposited during the sheetflood in winter 1979–1980.

The sedimentary material used for the sugar analysis has a slimy semi-viscous consistency and represented the decomposed remains of the top mat before the first sheetflood.

The samples used for our survey of lipid moieties were collected in April 1981 at Site 3 at the outer rim of the zone permanently covered with water. A core was taken with a perspex pipe of 10 cm diameter and 70 cm length. After full insertion the core pipe was dug out, carefully extruded and sectioned on the spot. The core profile is shown in Fig. 17.2. The upper 21 cm were black on sectioning but the sedimentology could be studied after oxidation from prolonged exposure to air. The upper 12 cm consisted of red and green laminated microbial mats overlying coarse sand with abundant rock fragments, *Pirenella* shell remains, and an occasional layer of gypsum crystals. Below this section the core showed alternating layers of grey mud and sandy gypsum-rich strata usually slightly red. The upper centimeters consisted of a greenish-brown slime-rich layer which evidently represents the remnants of the section described by Krumbein et al. (1979 a). The mud overlying this layer in the zone permanently covered with water were partially removed here by splashing of waves close to the shore of the lagoon of the Gavish Sabkha. The samples used for lipid analysis were a section of 3B (31–41 mm depth) and section 3F (50–55 cm) as shown in Fig. 17.2. The rationale behind this choice was to evaluate and compare the lipid composition from a laminated mat

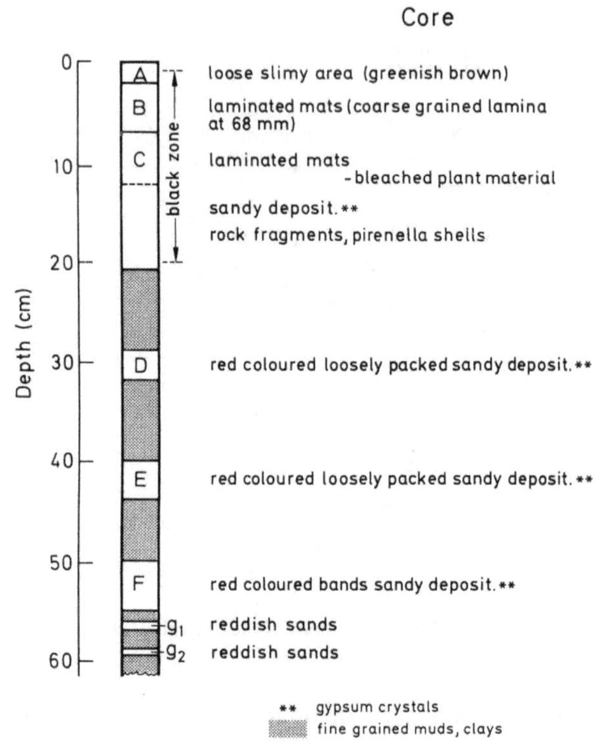

Fig. 17.2. Core description of Site 3

zone (3B) and a greyish sandy deposit with reddish-brown fine grained gypsum bands and some *Pirenella* (3F). The latter depositional environment represents a zone which has not been covered permanently with water and which has undergone influences from the shores, the microbial mats and evaporation events (transient zone from Zone 7 to 8, Chap. 11).

17.2 Analysis of Neutral Sugars

17.2.1 Hydrolysis and Derivatization

Freeze-dried mat material was hydrolyzed in 0.5 N H_2SO_4 in a sealed glass ampoule under nitrogen at 105 °C for 18 h. The hydrolysate was neutralized with $BaCO_3$ and the resultant $BaSO_4$ was removed by centrifugation. The supernatant was adjusted to pH = 9 with NH_3 and $NaBH_4$ added. After 2 h at room temperature the solution was acidified with glacial acetic acid and dried in vacuo. Boric acid was removed by repeated addition of methanol and evaporation to dryness. Acetylation was performed on the dry sample in a solution of acetic acid anhydride-pyridine (1:1) for 2 h at 100 °C. The excess acetylation agent was removed by dessication over P_2O_5 and KOH. The residue was taken up in CH_2Cl_2, washed with water and dried with anhydrous Na_2SO_4.

17.2.2 Gas Chromatography (GC) and Gas Chromatography-Mass Spectrometry (GC-MS)

The alditol acetates were analyzed by capillary GC, as described by Klok et al. (1981). Identification was based on GC retention data and on comparison of the mass spectra obtained by capillary GC-MS with those published by Jansson et al. (1976) and with mass spectra of synthesized standard compounds (Klok et al. 1982).

17.3 Neutral Reducing Sugars from a Decomposing Top Mat in the Gavish Sabkha

The distribution of the neutral sugars is shown in Fig. 17.3. The major components are separated to base line by the glass capillary column used. Apart from the major sugars, a number of methylated sugars could also be identified using capillary GC-MS (Tables 17.1 and 17.2).

Glucose is the major hexose sugar, xylose the major pentose. The absolute abundance of glucose is much lower than the values found by Boon et al. (1983) in the top mat section from Solar Lake. Such lower values are probably caused by the burial of the top mat under a mud layer and the consequent increase in

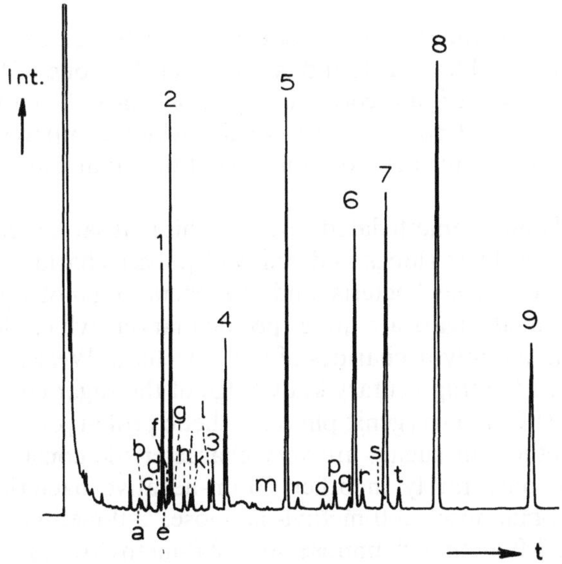

Fig. 17.3. GC -trace of the neutral reducing and methylated sugars from the decomposed top mat

Table 17.1. Neutral reducing sugars in the decomposed top mat in the Gavish Sabkha

Peak number	Identity	Abundance (‰ of dry weight)
1	Rhamnose	0.9
2	Fucose	1.6
3	Ribose	0.3
4	Arabinose	1.0
5	Xylose	2.4
6	Mannose	2.1
7	Galactose	3.0
8	Glucose	4.8
9	Internal standard	–

Table 17.2. Methylated sugars identified in the decomposed top mat from the Gavish Sabkha

a	2-O-methyl-deoxyhexose	l	3-O-methyl-xylose
b	2-O-methyl-rhamnose	m	6-O-methyl-glucose
c	2-O-methyl-fucose	n	2-O-methyl-mannose
d	3-O-methyl-rhamnose	o	2-O-methyl-galactose
e	4-O-methyl-fucose	p	2-O-methyl-glucose
f	3-O-methyl-fucose	q	3-O-methyl-mannose
g	2-O-methyl-ribose	r	3-O-methyl-glucose
h	2-O-methyl-arabinose	s	3-O-methyl-galactose or 4-O-methyl-galactose
i	4-O-methyl-arabinose	t	4-O-methyl-glucose
k	3-O-methyl-arabinose		

chemoorganotrophic degradation processes. The distribution of the nonmethylated neutral sugars resembles that found in the anaerobic zone of the Solar Lake mat sequence. Again xylose is a predominant pentose sugar. The total amount of hydrolyzable sugars is 16.1 ‰ of the dry weight, which compares well with the value found for the anaerobic degradation zone of the Solar Lake mats (Boon et al. 1983).

The methylated and nonmethylated sugars in this mat sample can be expected to derive from extracellular slimes and cell-wall polysaccharides of Cyanobacteria, photosynthetic purple bacteria and the chemoorganotrophic microbial community. Methylated sugars are quite specific markers which derive from the O-specific chains in lipopolysaccharides of these bacteria. Weckesser and Drews (1979) have reviewed the fragmentary knowledge of the sugar composition of lipopolysaccharides (LPS) in oxygenic photosynthetic prokaryotes.

The O-methyl ethers of sugars are very characteristic constituents of these bacteria but occur only rarely in other Gram-negative bacteria. 3-0-methyl-xylose, 2-0-methyl-mannose, 3-0-methyl-mannose, 2-0-methyl-galactose, 2-0-methyl-rhamnose. 3-0-methyl-rhamnose and 2-0-methyl-fucose mentioned in Table 17.2 occur in photosynthetic prokaryotes, but no methylated glucoses have yet been found in these organisms. All the nonmethylated sugars mentioned in Table 17.1 have been described as constituents in the LPS of prokaryotic phototrophs, but never all together. Arabinose is a rare constituent of LPS. Our Sabkha sample represents a complex mixture of prokaryotic remains and a great variety of sugars and methylated sugars is therefore expected.

Little is known about the composition of extracellular slime. The extracellular slimes of *Aphanocapsa halophytica* were found to contain glucose, mannose, galactose and fucose, but additional components may be present because packed column gas chromatography with poor resolution was used for the separation of the alditol acetates (Jones and Yopp 1979). The extracellular slimes are copious in the Sabkha microbial mats and may be an important source of sugar components. More knowledge about the composition of such slimes produced in pure culture is needed.

A third possible source of methylated sugars could be "waste" sugars produced by decomposition of the microbial mat material. This would require the presence of microorganisms which methylate sugars, but otherwise leave them untouched. The feasibility of this scenario remains to be demonstrated. The question as to whether "waste" sugar production is required for the purpose of osmotic regulation, light transfer and/or toxicity protection is discussed in some detail in Chapter 11.

17.4 Lipid Analysis

17.4.1 Extraction

Sample 3F (water content 40%) was freeze-dried before extraction. In a Soxhlet extraction apparatus 85 g of dry sediment was extracted twice with 175 m methanol/toluene (3:1, v/v) for 6 h. The combined extracts were evaporated to a smaller volume, using a rotatory evaporator, diluted in 50 ml CH_2Cl_2 and washed with distilled water to remove inorganic salts. The CH_2Cl_2 solution was dried with anhydrous Na_2SO_4 and concentrated to a volume of 10 ml. A red-colored extract was obtained which represented 1.1 ‰ of dry sediment weight.

Sample 3B (31–41 mm) was wet-extracted (water content 59%). About 38 g of wet microbial mat material was homogenized in a Waring Blender with 140 ml distilled water. The resulting brown slurry was centrifuged briefly and the supernatant was transferred to a separating funnel. The precipitate was extracted successively with 100 ml CH_3OH, 100 ml CH_3OH/CH_2Cl_2 (2:1, v/v), 100 ml CH_3OH/CH_2Cl_2 (1:5, v/v) and 100 ml CH_2Cl_2, using ultrasonication and centrifugation. All supernatants were transferred to the separating funnel. The ternary mixture of water, methanol and dichloromethane was separated into methanol/water (further extracted with 100 ml CH_2Cl_2) and dichloromethane layers. The combined CH_2Cl_2 solutions were washed with distilled water, dried with anhydrous Na_2SO_4 and concentrated to 10 ml. A red extract was obtained which represented 4.0 ‰ of the dry sediment weight.

17.4.2 Saponification

A volume equivalent to 80% of the extract was evaporated to dryness under a stream of nitrogen and refluxed with 4 ml 1N KOH in 96% CH_3OH for 1 h. After boiling, the basic solution was acidified to pH = 3.5 and extracted with CH_2Cl_2. The total lipid extract was washed with distilled water, dried with Na_2SO_4 and reduced to a volume of 10 ml.

17.4.3 Total Lipid Profiles

Of the extracts obtained after saponification 1 ml was used for the determination of the total lipid profiles. The aliquots were reduced to dryness under a stream of nitrogen and immediately treated with diazomethane in diethyl ether to methylate free carboxyl groups. Free hydroxyl groups were silylated by reaction with bis(trimethyl)silylacetamide in ethyl acetate at 60° C for 15 min. Gas chromatography was performed on a Carlo Erba 4160 instrument, as described below.

17.4.4 Separation of Lipid Classes

Of the extracts obained after saponification 2 ml were separated into hydrocarbons, ketons, fatty acids, alcohols and sterols by thin layer chromatography on silica gel, using the elution system described by Skipski et al. (1965). Lipid-rich zones were detected by inspection under UV light after spraying with a 0.05% Rhodamine 6 G solution in ethanol. The bands were scraped off the TLC plate and extracted with ethyl acetate. Fatty acids were methylated with CH_2N_2 in diethyl ether. Alcohols and sterols were silylated with bis(trimethyl)silylacetamide in ethylacetate at 60° C for 15 min.

17.4.5 Gas Chromatography and Gas Chromatography-Mass Spectrometry

Gas chromatography was carried out on a Carlo Erba 4160 instrument equipped with a flame ionization detector and an on-column injection system (Grob 1978, Grob and Grob 1978). A glass capillary column (20 m, i.d. 0.32 mm) coated with SE-52 was used with helium as carrier gas. Samples were injected at 50° C and after solvent elution, the oven temperature was programed from 125 °C to 330° C at 4° C/min. Gas chromatography-mass spectrometry was carried out with a Varian 3700 gas chromatograph connected to a MAT 44 quadrupole mass spectrometer. A glass capillary column (25 m, i.d. 0.23 mm) coated with SE-52 was used with helium as carrier gas. The gas chromatograph was temperature-programed from 125° C to 300° C at a rate of 4° C/min. The capillary column was connected to the mass spectrometer by an open atmospheric split. Electron impact mass spectra were obtained at 80 eV by cyclic scanning with a total cycle time of 2 s. Data acquisition was performed with the MAT 44 microprocessor. Further data handling was carried out with a PDP 11/45 computer.

17.5 Lipids from a Decomposing Top Mat in the Gavish Sabkha

17.5.1 Hydrocarbons

The capillary GC-trace of the hydrocarbon fraction is shown in Fig. 17.4. The peak numbers correspond to the compound numbers mentioned in Table 17.3. Identifications were based on both relative GC retention times and mass spectral data of standard compounds when available. Mass spectra of monomethyl midchain-branched hydrocarbons are characterized by enhanced intensities of alkyl and alkene fragmentations due to cleavage at the tertiary carbon atom (Gelpi et al. 1970). Figure 17.5 shows the mass spectrum of 8-methylpentadecane (Table 17.3, compound 8). The molecular ion at m/z 240 has a low intensity and the enhanced intensities of fragmentations at m/z 126, 127 and at m/z 140 and 141 are indicative of the presence of a methyl group at position 8 in the alkyl chain. The identifications of dimethyl-branched hydrocarbons (Table 17.3, compounds 18, 27 and 28) are tentative. Figure 17.6 shows the mass spectrum of one of the major hydrocarbon components (Table 17.3, compound 18). It has a fragmentation pattern similar to that observed for long-chain dimethyl-branched hydrocarbons with "isoprenoid spacing" as reported to occur in insects (L. L. Jackson and Blomquist 1976 and references cited therein). The enhanced intensities of m/z 98 and m/z 197 indicate the presence of a methyl group at position 6. The slightly enhanced intensity of the mass peak at m/z 169 and the virtual absence of a peak at m/z 183 can be explained by the presence of a second methyl group at position 10. The relative GC retention time observed for this compound with 19 carbon atoms ($M^+ = 268$) is appropriate for a compound with two midchain methyl substituents (L. L. Jackson and Blomquist 1976). Based on the mass spectrum, an isoprenoidtype C_{19}-compound (2,6,10-trimethylhexadecane) might be an alternative structure. However, the mass spectrum of this compound is well known and differs considerably from our spectrum (Haug and Curry 1974). The major hy-

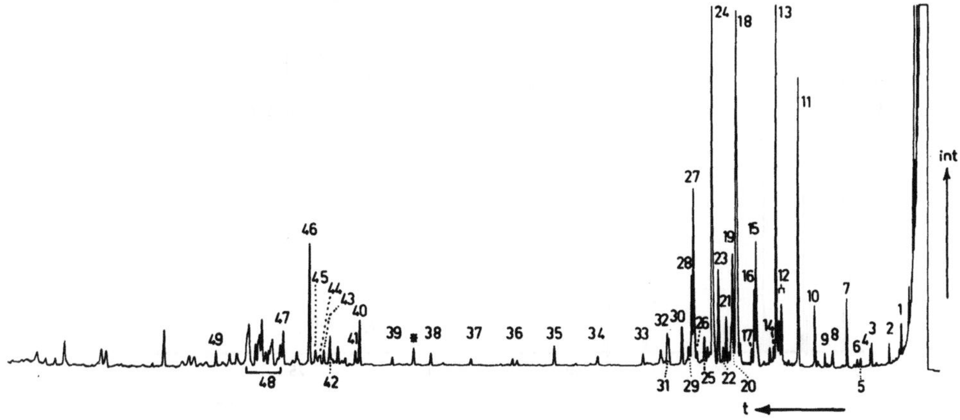

Fig. 17.4. GC-trace of the hydrocarbons from the decomposed top mat

Table 17.3. Hydrocarbon fraction of top mat

1. n-Dodecane	25. Phytadiene isomer
2. n-Tridecane	26. Phytadiene isomer
3. 2,6,10-trimethyldodecatetraene	27. 7,10-dimethyloctadecane
4. n-Tetradecane	28. 6,10-dimethyloctadecane
5. Pentadecane (branched)	29. n-Nonadecane
6. Pentadecane (branched)	30. Phytadiene isomer
7. n-Pentadecane	Phthalate ester
8. 8-methylpentadecane	31. 7-methyleicosane
9. 2-methylpentadecane (iso)	32. 6-methyleicosane
10. n-Hexadecane	33. n-Eicosane
11. 8-methylhexadecane	34. n-Heneicosane
12. n-Heptadecenes	35. n-Docosane
13. n-Heptadecane	36. n-Tricosane
14. 2,6,10,14-tetramethylpentadecane (pristane)	37. n-Tetracosane
15. 6-methylheptadecane	38. n-Pentacosane
16. 5-methylheptadecane	39. n-Hexacosane
17. 4-methylheptadecane	40. Alkene: hydrogenated squalene isomer?
18. 6,10-dimethylheptadecane	41. n-Heptacosane
19. n-Octadecane	42. Tetrahydrosqualene
20. 2,6,10,14-tetramethylhexadecane (phytane)	43. Hydrogenated squalene isomer
21. Phytene isomer?	44. Dihydrosqualene, n-octacosane
22. Phytene isomer	45. Dihydrosqualene isomer
23. Phytene isomer	46. Squalene
24. 8-methylnonadecane	47. n-Nonacosane
9-methylnonadecane	48. Mixture of sterenes and steradienes
7-methylnonadecane	49. n-Hentriacontane
6-methylnonadecane	

* phtalate (corresponding to the peaks labelled with an asterisk)

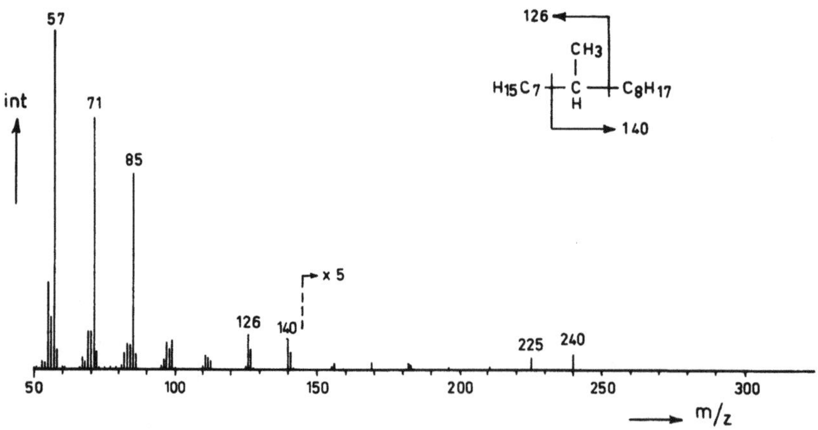

Fig. 17.5. Mass spectrum of 8-methylpentadecane (compound 8)

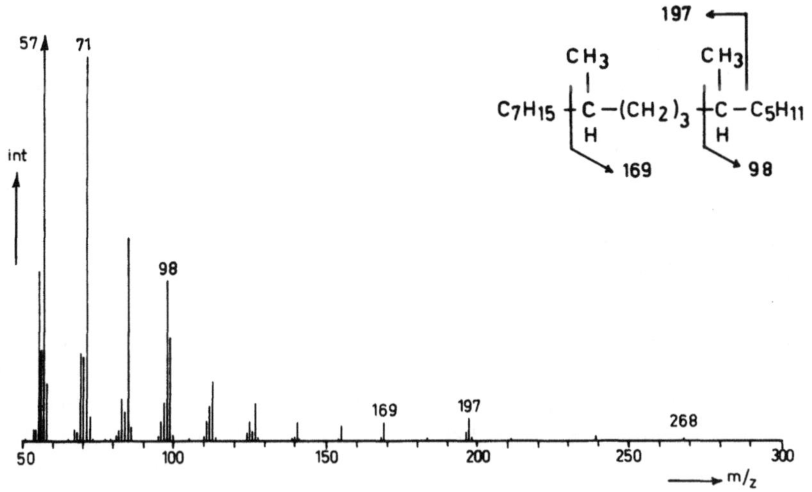

Fig. 17.6. Mass spectrum of compound 18 (6,10-dimethylheptadecane)

drocarbons occur in the relatively small range between 16 and 20 carbon atoms. The proportion of midchain mono- and dimethyl-branched hydrocarbons is surprisingly high. Monomethyl-branched hydrocarbons with less than 20 carbon atoms, such as 4-, 6-, 7- and 8-methyl hydrocarbons have previously been found in cyanobacteria and cyanobacterial mats (Gelpi et al. 1970, Han et al. 1968, Philp et al. 1978, Cardoso et al. 1976, 1978).

It seems reasonable therefore to assign the monomethyl midchain-branched hydrocarbons (5, 6, 8, 11, 15, 16, 17, 24, 31 and 32 in Table 17.3) to an origin from cyanobacteria. An origin from other bacteria, such as purple sulfur bacteria, however, cannot be excluded in this sediment sample. The tentatively identified dimethyl-branched hydrocarbons with chain lengths of C_{19} and C_{20} observed in this sediment (compounds 18, 7 and 28 in Table 17.3) have not been detected in organisms. Long-chain dimethyl-branched hydrocarbons with carbon chain lengths of C_{25} to C_{55} have been found in insect waxes, as mentioned above. An origin for these hydrocarbons from the staphylinid beetle *Bledius* abundant in Zone 4 (see Chaps. 11 and 15) is less likely because the chain-length distribution of the dimethyl-branched alkanes encountered differs from that found in insects. Moreover, a contribution of organic material from *Bledius* to the sediment in Zone 9 would require a considerable facies shift seen in the sedimentology. Therefore, we think that the relatively high amounts of these components in the hydrocarbon fraction point to a bacterial origin. Specific cyanobacteria or other photosynthetic bacteria may biosynthesize these branched alkanes only in hypersaline environments. Experiments with cultures of these bacteria grown under different conditions may help establish their origin. Mono- and polyunsaturated normal and isoprenoid alkenes are only present in small amounts. 2,6,10-Trimethyldodecatetraene (Table 17.3, compound 3) and the isomeric phytadienes (Table 17.3, compounds 25, 26 and 30) are thought to be dehydration products of the corresponding alcohols farnesol and phytol (see alcohol fraction). The other isoprenoid alkenes, the phytene isomers (compounds 21–23) have been encountered before in cyanobacterial mats (Philp et al. 1978, Boon et al. 1983). They may have a cyanobacterial origin but an origin from archaebacteria – in the Gavish Sabkha especially *Halobacterium* sp. – is another possibility (see Chap. 13). Evidence for archaebacterial inputs to the mat is provided by the presence of phytenes and of squalene and di- and tetrahydrosqualenes (compounds 42–46). These partially hydrogenated squalenes occur in several archaebacteria, including *Halobacterium* (e.g., Kramer et al. 1972, Tornabene 1978). In absence of C_{25}-isoprenoid alkanes makes a contribution from methanogenic archaebacteria less likely (Tornabene et al. 1979). The suite of sterenes and steradienes (compounds 48) are diagenetic products of the corresponding sterols (see sterol fraction; see also Boon et al. 1983). It is noteworthy that the n-alkanes are only minor components. The long-chain n-alkanes exhibit odd over even predominance and probably indicate a minor input from higher plant waxes (P. A. Schenck and de Leeuw 1982). n-Alkenes with a relatively short chain length (e.g., compounds 12) are minor components and might originate from cyanobacteria (e.g., Gelpi et al. 1970).

17.5.2 Fatty Acids

The capillary gas chromatogram of the fatty acid fraction is shown in Fig. 17.7, with the peak numbers referring to the identifications given in Table 17.4. It should be noted that the fatty acids were analyzed as their methylesters. In contrast to the relative abundance of midchain-branched hydrocarbons only minor amounts of midchain methylbranched fatty acids are present (components 1, 12, 21). Iso- and anteiso-fatty acids, especially the C_{15} and C_{17} components, are relatively abundant and so is the C_{19} cyclopropyl fatty acid (component 23). These branched and cyclopropyl fatty acids reflect the presence of bacteria – either dead or alive – in this sediment (Philp et al. 1978, Cranwell 1973). Major fatty acid components are the straight-chain C_{16} and C_{18} saturated and unsaturated fatty acids (components 10, 11, 18, 19, 20). These fatty acids are common in virtually all organisms. However, the relatively high intensity of the mono-unsaturated C_{18}

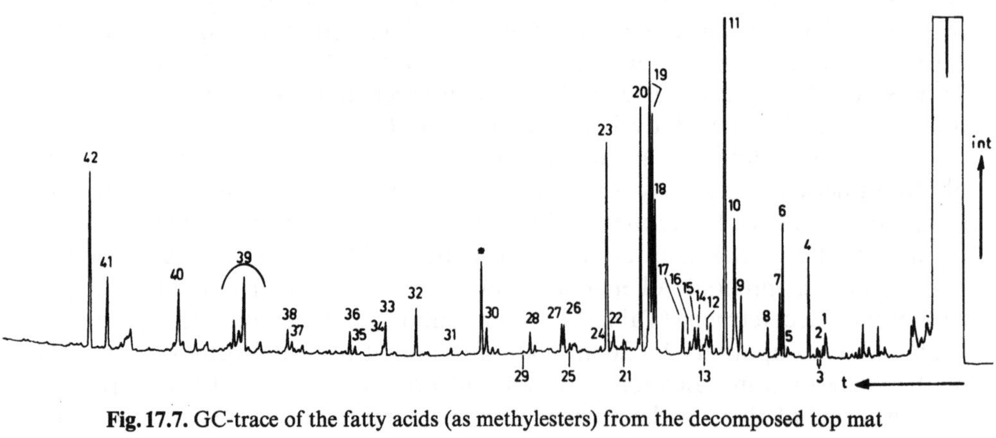

Fig. 17.7. GC-trace of the fatty acids (as methylesters) from the decomposed top mat

Table 17.4. Fatty acid fraction of top mat

1. 10-methyltridecanoic acid	21. 10-methyloctadecanoic acid?
2. *iso*-Tetradecanoic acid	22. *iso*-Nonadecanoic acid
3. n-Tetradecenoic acid	23. Cyclopropylnonadecanoic acid ($\Delta C_{19:0}$)
4. n-Tetradecanoic acid	24. n-Nonadecanoic acid
5. Pentadecanoic acid (branched) 4,8,12-trimethyltridecanoic acid (isoprenoid $C_{16:0}$)	25. n-Eicosenoic acid 26. n-Eicosanoic acid 27. Diterpenoic fatty acid? (m/z 271, 189, 191, 95)
6. *iso*-Pentadecanoic acid	28. Unknown
7. *anteiso*-Pentadecanoic acid	29. n-Heneicosanoic acid
8. n-Pentadecanoic acid	30. n-Docosanoic acid
9. *iso*-Hexadecanoic acid	31. n-Tricosanoic acid
10. n-Hexadecenoic acid	32. n-Tetracosanoic acid
11. n-Hexadecanoic acid	33. Squalene
12. 10-methylhexadecanoic acid	34. n-Pentacosanoic acid
13. Heptadecanoic acid (branched)	35. n-Hexacosenoic acid
14. *iso*-Heptadecanoic acid	36. n-Hexacosanoic acid
15. *anteiso*-Heptadecanoic acid	37. n-Octacosenoic acid
16. n-Heptadecenoic acid	38. n-Octacosanoic acid
17. n-Heptadecanoic acid	39. Free sterols [a]
18. n-Octadecadienoic acid	40. 22-hydroxy-homohopanol
19. n-Octadecenoic acids	41. Δ^7-bishomohopenoic acid
20. n-Octadecanoic acid	42. 17β(H), 21β(H)-bishomohopanoic acid

* phtalate (corresponding to the peaks labelled with an asterisk)
[a] The presence of minor amounts of free sterols in this fraction is due to an incomplete TLC-separation

fatty acids (components 19) and the absence of C_{20} and C_{22} polyunsaturated fatty acids suggest a major contribution from bacteria. (Philp et al. 1978). It is noteworthy that the longer-chain fatty acids (C_{20}–C_{28}) maximize at C_{24}. We have observed this phenomenon before in Solar Lake cyanobacterial mats and also in other lipid classes (see hydrocarbon and fatty alcohol fractions). An origin for these fatty acids from higher plants seems unlikely although such a possibility cannot be ruled out completely (Boon et al. 1983). Two hopanoid fatty acids (compounds 41 und 42) are major components in the high molecular-weight region. These fatty acids originate from bacteria and are probably metabolic products of the C_{35} hopane tetrol (Boon et al. 1983, Ourisson et al. 1979). Component 40 is tentatively identified as 22-hydroxy-homohopanol. This tertiary alcohol occurs in the fatty acid fraction due to its unique retention behavior during thin layer chromatography (Boon et al. 1983). The compound is abundant in Solar Lake cyanobacterial mats and is thought to reflect a contribution from some highly specific bacterium.

17.5.3 Sterols

That part of the capillary gas chromatogram where sterols elute is shown in Fig. 17.8. The peak numbers correspond with the identifications in Table 17.5. The sterols were analyzed as trimethylsilyloxyethers. Most of the components are identified by comparison of their mass spectra with spectra of standards and by

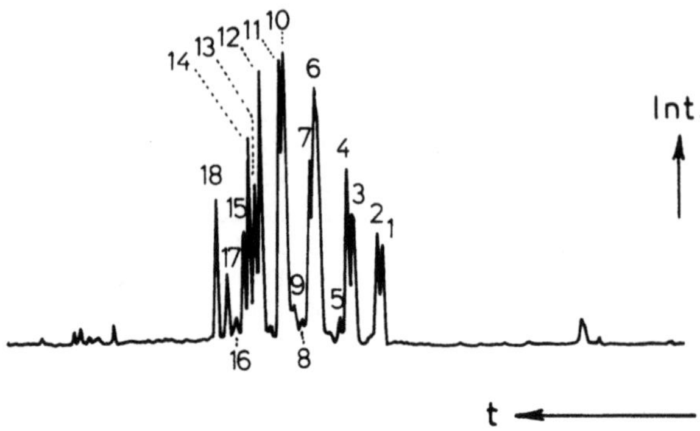

Fig. 17.8. Part of GC-trace representing the sterols (as trimethylsilylethers) from the decomposed top mat

Table 17.5. Sterol fraction of top mat

1. Cholest-5-en-3β-ol	10. 5α(H)-24-methylcholest-7,24(28)-dien-3β-ol
2. 5α(H)-cholestan-3β-ol	11. 5α(H)-24-methylcholest-7-en-3β-ol
3. 24-methylcholest-5,22-dien-3β-ol	12. 24-ethylcholest-5-en-3β-ol
4. 5α(H)-24-methylcholest-22-en-3β-ol	13. 5α(H)-24-ethylcholestan-3β-ol
5. Unknown	14. 24-ethylcholest-5,24(28)-dien-3β-ol
6. 24-methylcholest-5-en-3β-ol	15. 5α(H)-24-ethylcholest-24(28)-en-3β-ol
7. 5α(H)-24-methylcholestan-3β-ol	16. Unknown
8. 24-ethylcholest-5,22-dien-3β-ol	17. 5α(H)-24-ethylcholest-7-en-3β-ol
9. 24-ethylcholest-22-en-3β-ol	18. 5α(H)-24-ethylcholest-7,24(28)-dien-3β-ol

comparison of relative retention times (e.g., Wardroper 1979, Brooks et al. 1968). The identification of some sterol components found in this sediment warrants further discussion. The major components (compounds 10 and 11) and also the last eluting components (compounds 17 and 18) are thought to be Δ^7 and $\Delta^{7,24(28)}$-C_{28} and C_{29} sterols. The mass spectrum of the silylated 5α(H)-24-methylcholest-7,24(28)-dien-3β-ol (compound 10) is characterized by m/z 470 (M^+, 3%), 455 (M^+-15; 4%), 386 (M^+-84; 16%), 380 (M^+-90; 4%), 343 (M^+-127; 40%), 296 (M^+-84-90; 5%), 255 (12%), 253 (20%), 229 (7%), 227 (12%), 215 (11%), 213 (18%) and 55 (100%). The abundant loss of 84 a.m.u. from the molecular ion points to the presence fo a 24-methylene group (Wyllie and Djerassi 1968). The relatively intense mass peaks at m/z 213, 215, 227, 229, 255, and 253 are indicative of the presence of a Δ^7-double bond. Component 11, 5α(H)-24-methyl-7-en-3β-ol has a mass spectrum with $M^{+\cdot}$ = 472 (16%) and m/z 55 (100%) in which m/z 213 (18%), m/z 229 (14%) and m/z 255 (24%) are also indicative for a Δ^7-double bond (Brooks et al. 1968). The elution order observed for these Δ^7-components is in accordance with that given by Brassell (1980).

The Δ^7-5α(H)-sterols with saturated side chains have been observed more often in cyanobacterial mat sediments (Boon et al. 1983). The relatively abundant Δ^7-24-methylene component (compound 10) to our knowledge has not been reported in sediments or organisms before. The Δ^7-24-ethylene component (compound 18) has been reported before (Wright 1981, Cranwell 1982) in sediments and *Dunaliella*. The occurrence of these components may in this case be characteristic of certain bacteria living under extremely saline conditions. The other Δ^5-sterols and 5α(H)-stanols with and without Δ^{22}-double bonds listed in Table 17.5 have also been encountered in other cyanobacterial mat ecosystems (Cardoso et al. 1978, Boon et al. 1983) and in cyanobacteria (Nishimura and Koyama 1977). To our knowledge the biosynthesis of steroids in anoxygenic phototrophs has not been shown. We therefore assume a cyanobacterial origin for these sterols in this sediment. Another phenomenon observed is the relatively high intensity of the 5α(H)-stanols as compared with the corresponding Δ^5-sterols (components 1,2; 3,4; 6,7; 12, 13). The 5α(H)-stanols reflect either a rather specific input of certain organisms or a rapid microbiological conversion of Δ^5-sterols to 5α(H)-stanols. (Boon et al. 1983, Gaskell and Eglinton 1975). An important contribution of organic matter derived from higher organisms seems unlikely due to the relatively low abundance of cholesterol (compound 1). Further research into the sterol composition of pure cultures of cyanobacteria grown under hypersaline conditions is necessary before firm conclusions about the origin of the sterols encountered here and about the pathways of sterol biosynthesis in cyanobacteria in general can be made.

17.5.4 Alcohols

The capillary gas chromatogram of the alcohol fraction is shown in Fig. 17.9. The peak numbers correspond with the numbers in Table 17.6. The alcohols were analyzed as trimethylsilyloxyethers. Some of the major components present are well known as alcohol moieties of chlorophylls. Phytol, the most abundant component (compound 16) probably originates from a variety of chlorophylls such as chlorophyll *a* and *b* and also from bacteriochlorophylls present in anoxygenic photosynthetic bacteria. Another major component (compound 17) is identified as a phytadienol. Although no attempt was made in this study to determine double bond positions we think that this compound is probably $\Delta^{2,10}$-phytadienol, recently described by Steiner et al. (1981) as the alcohol moiety of bacteriochlorophyll *b* in the halophilic photosynthetic bacterium *Ectothiorhodospira halochloris*.

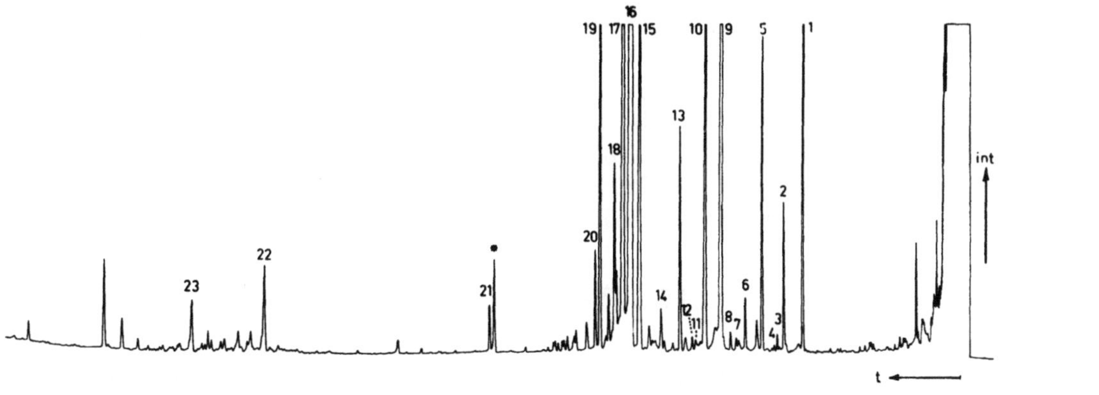

Fig. 17.9. GC-trace of the alcohols (as trimethylsilylethers) from the decomposed top mat

Table 17.6. Alcohol fraction of top mat

1. n-Tetradecanol	14. Phytadiene
2. 2,6,10-trimethyldodecatrienol (farnesol)	15. n-Octadecanol
3. *iso*-Pentadecanol	16. 2,6,10,14-tetramethylhexadecenol (phytol)
4. *anteiso*-Pentadecanol	17. 2,6,10,14-tetramethylhexadecadienol (phytadienol)
5. n-Pentadecanol	18. Diterpenoid alcohol
6. Hexadecanol (branched)	19. 2,6,10,14-tetramethylhexadecatetraenol (geranylgeraniol)
7. *iso*-Hexadecanol	20. Heneicosanol (branched)
8. n-Hexadecenol	21. n-Docosanol
9. n-Hexadecanol	22. Triterpanol (C_{30})
10. Heptadecanol (branched)	23. Triterpanol (C_{30})
11. *iso*-Heptadecanol	
12. *anteiso*-Heptadecanol	
13. n-Heptadecanol	

* phtalate (corresponding to the peaks labelled with an asterisk)

The authors suggest that the occurrence of $\Delta^{2,10}$-phytadienol in this bacterium is an adaption to its extremely saline environment. The relatively high abundance of this isoprenoid alcohol in the extremely saline sedimentary environment of the Gavish Sabkha might support the adaptation hypothesis. Geranylgeraniol (compound 19) is another major alcohol occurring in this alcohol mixture. This alcohol is also biosynthesized by a number of photosynthetic bacteria and incorporated in bacteriochlorophyll *a* and *b* as the alcohol moiety (e.g., I. I. Katz et al. 1972, Foster and Berchtold 1972). Farnesol (compound 2), although much less abundant, may reflect the presence of bacteriochlorophylls *c, d* and *e* (Gloe et al. 1975). The high abundance of the alcoholic components 16, 17, 19 and 2 strongly indicates a major contribution of cyanobacteria and oxygenic photosynthetic bacteria to this sediment. The contribution of bacterial organic matter is also revealed by the presence of iso- and anteiso-C_{15} and -C_{17} alcohols (compounds 3, 4, 7, 11, 12) and by midchain methyl-branched alcohol components (compounds 6, 10 and 20). The exact position of the methyl group cannot be established from the mass spectra because those of straight chain and monomethyl-branched silylated alcohols are virtually identical. Two triterpenoid alcohols (compounds 22, 23) were present in relatively low concentrations. The mass spectral data for these compounds did not allow a more detailed assignment. At this stage we cannot therefore distinguish between higher plant or bacterial origin of these triterpenoids. The saturated straight-chain alcohols (compounds 1, 5, 9, 13, 15, 21) are nonspecific, although the absence of longer straight-chain saturated alcohols excludes an origin from higher plants (Tulloch 1976). They may originate in part from chlorophylls, since Caple et al. (1977) report the occurrence of nonterpenoid alcohol moieties in bacteriochlorophylls.

17.6 Total Lipid Profiles

The gas chromatograms of the total lipid extracts of the decomposed top mat and of the core sample are shown in Figs. 17.10 and 17.11 respectively. The peak

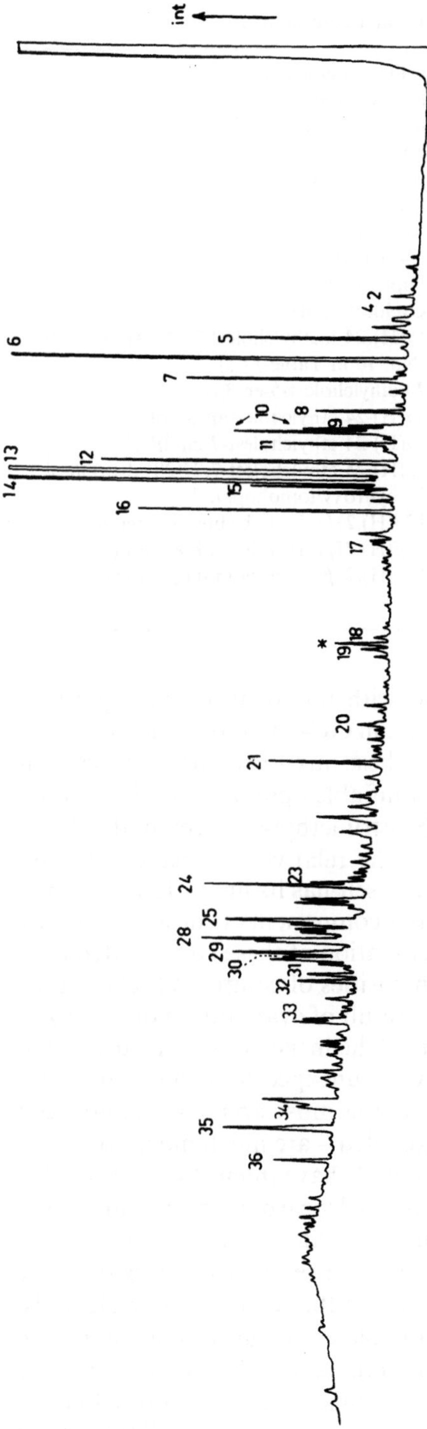

Fig. 17.10. GC-trace of the total lipids (after derivatization) from the decomposed top mat

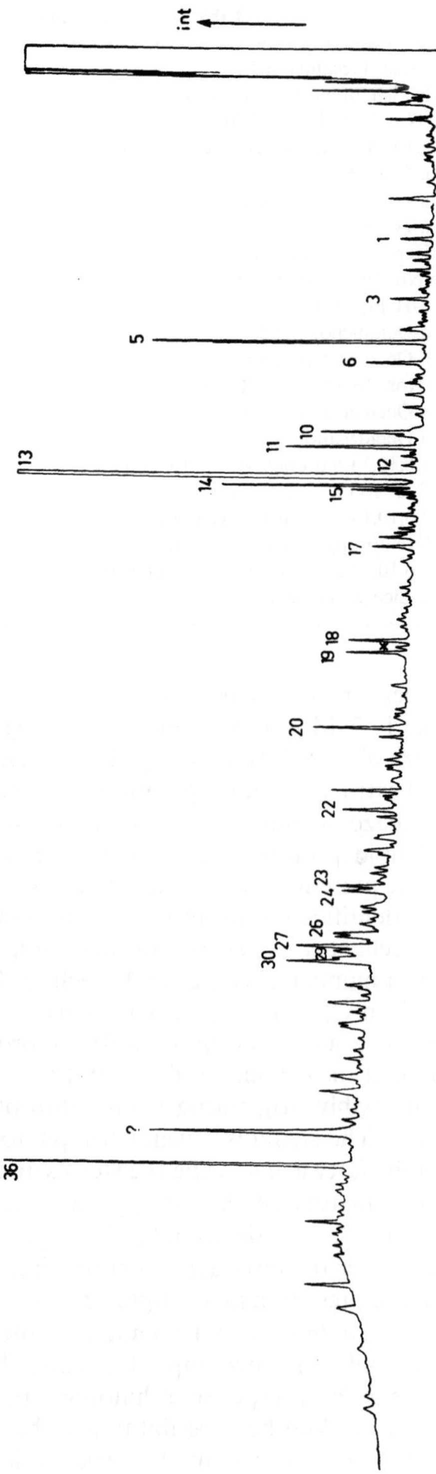

Fig. 17.11. GC-trace of the total lipids (after derivatization) from the core sample

Table 17.7. Total lipids of top mat and core sample

1. n-Tetradecanoic acid
2. Midchain methylnonadecanes (No. 24 in Table 17.3)
3. 6,10,14-trimethylpentadecan-2-one
4. n-Pentadecanol
5. n-Hexadecanoic acid
6. n-Hexadecanol
7. Heptadecanol (branched) (No. 10 in Table 17.6)
8. n-Heptadecanol
9. n-Octadienoic acid
10. n-Octadecenoic acids (Nos. 18 and 19 in Table 17.4)
11. n-Octadecanoic acid
12. n-Octadecanol
13. 2,6,10,14-tetramethylhexadecenol (No. 16 in Table 17.6)
14. 2,6,10,14-tetramethylhexadecadienol
15. Cyclopropylnonadecanoic acid
16. 2,6,10,14-tetramethylhexadecatetraenol
17. n-Eicosanoic acid
18. n-Docosanoic acid
19. n-Docosanol
20. n-Tetracosanoic acid
21. Squalene
22. Δ^2-24-methyl-cholestene
23. Cholest-5-en-3β-ol
24. 5α(H)-cholestan-3β-ol
25. 24-methylcholest-5-en-3β-ol
26. Diploptene
27. Cycloartanone
28. 5α(H)-24-methylcholest-7,24(28)-dien-3β-ol (No. 10 in Table 17.5)
29. 24-ethylcholest-5-en-3β-ol
30. 5α(H)-24-ethylcholestan-3β-ol
31. 5α(H)-24-ethylcholest-7-en-3β-ol
32. 5α(H)-24-ethylcholest-7,24(28)-dien-3β-ol
33. 22-hydroxyhomohopanol
34. 17β(H),21β(H)-Δ^7-bishomohopenoic acid
35. 17β(H),21β(H)-bishomohopanol (?)
36. 17β(H),21β(H)-bishomohopanoic acid

numbers in these chromatograms correspond with the identifications given in Table 17.7. Most compounds were recognized from GC-retention times and mass spectra obtained by capillary GC-MS analyis. These mixtures represent the combined extract of both saponifiable and nonsaponifiable lipids after derivatization with diazomethane and the silylating agent bis-(trimethyl)silylacetamide. These total lipid profiles give a good impression of the relative abundance of compounds present in the extract. This approach in itself has its limitations if a complete identification of all major and minor lipid components is desired. Despite the excellent capillary gas chromatographic separation of these complicated mixtures, a number of compounds coelute. The intensities of compounds eluting before the n-$C_{16:0}$ fatty acid may be reduced as a result of evaporation of more volatile compounds during the work-up procedure. The inspection of the total lipid profile of the decomposed top mat reveals several unexpected phenomena. Phytol, probably originating from chlorophylls, is the most abundant compound present. The alcohols – even when phytol is excluded – are much more abundant than the fatty acids, while the sterols and hopanoids have intensities comparable to the intensity of the n-$C_{16:0}$ fatty acid. The hydrocarbons represent only a minor fraction of the total lipids. In particular, the predominance of the fatty alcohols over the fatty acids is surprising. Normally, in recent sediments the fatty acids are present in much higher concentrations than the corresponding alcohols. This is also the case in the living and fossil cyanobacterial mats of the Solar Lake (Boon and de Leeuw, unpubl. results). The predominance of alcohols over fatty acids may be a response of halotolerant microorganisms to the highly saline environment. Another possibility may be contributions from specific lipids of the anoxygenic phototrophic bacteria, which are abundant in the decomposed top mat. That both the isoprenoid an nonisoprenoid alcohols represent the alcohol moieties of abundantly present bacteriochlorophylls cannot be excluded (Caple et al. 1977). The relatively high concentrations of sterols is intriguing since bacteria are not well known as sterol producers, but an origin from halotolerant cyanobacteria and/or anoxygenic phototrophs is not unlikely, as mentioned earlier.

The total lipid profile of the core sample is different from that of the top mat sample. Although phytol is again the most abundant compound, the other isoprenoid alcohols are much less abundant and the fatty acids predominate over the corresponding alcohols. The absence of $C_{18:2}$ fatty acid should be noted. Both the long-chain fatty acids and alcohols show distribution patterns with a second maximum at C_{24}. No Δ^7-sterols are observed but cycloartanone and bishomohopanic acid are relatively abundant. Several Δ^2-sterenes are present, but the intensities of other hydrocarbons are negligible. The profile is similar to profiles obtained for fossilized cyanobacterial mats in the Solar Lake (Boon and de Leeuw unpubl. results). The large differences – both qualitative and quantitative – observed in the total lipid profiles of the top mat and the core sample mainly reflect the two different subenvironments in the Gavish Sabkha: the decomposed top mat represents Zone 9 and the core sample represents a mixture of Zones 6–8, the outer shore of the lagoon. The absence of Δ^7-sterols, the relatively lower abundances of isoprenoid and other alcohols, and the absence of $C_{18}{}^{:2}$ fatty acid in the core sample probably indicate a much lower contribution of organic matter derived from anoxygenic phototrophic bacteria, relative to cyanobacteria. Other less important differences observed between the two total lipid profiles are due to the diagenetic changes of the organic material in the core sample; e.g., the Δ^2-sterenes are produced by microbiological transformations of sterols and stanols (Boon et al. 1983; P. A. Schenck and de Leeuw 1982).

18. Biogeochemistry of Gavish Sabkha Sediments

II. Pyrolysis Mass Spectrometry of the Laminated Microbial Mat in the Permanently Water-Covered Zone Before and After the Desert Sheetflood of 1979

JAAP J. BOON, JAN W. DE LEEUW, and WOLFGANG E. KRUMBEIN

Prior to the sheetfloods from winters 1979 and 1980, the microbial mats in the permanently water-covered zone in the Gavish Sabkha depression had developed over a long period of time a multilayered complex of different microbial communities. The upper 20 mm of this ecosystem were found to be active in oxygenic and anoxygenic photosynthesis by Krumbein et al. (1979 a). The whole ecosystem was embedded in a translucent viscous slime inhabited mainly by cyanobacteria and purple sulfur bacteria. Both filamentous and coccoid cyanobacteria were present, but the coccoid strains were predominant in the upper layers of the ecosystem.

It was our interest to perform an organic microanalysis of individual layers in the ecosystem using pyrolysis mass spectrometry, and to correlate overall organic chemical characteristics with the microscopic and inorganic data obtained earlier (see also Chaps. 7 and 9).

The usefulness of pyrolysis methods for characterization of organic matter is now clearly established (Meuzelaar et al. 1983, Irwin 1979). Information is obtained from various types of macromolecules, lipids and special cytoplasmic components, such as compatible solutes. Boon et al. (1983) have recently summarized the pyrolysis chemistry of the microbial mats from the Solar Lake (Sinai). Other studies of interest are suspended matter in lakes (van der Meent et al. 1982), sediments of a marine tidal flat ecosystem (Boon and Haverkamp 1979), cyanobacteria in continuous cultures (Boon et al. 1984), bacterial cells walls (Haverkamp et al. 1980, Boon et al. 1983) and soil organic matter (Saiz-Jimenez et al. 1979). Multivariate data analysis methods allow the chemical evaluation of relatively large numbers of samples. Correlation patterns of mass peaks in pyrolysis mass spectra relate to the molecular ion distribution of pyrolysis products from the thermally dissociated macromolecules and their variations in the samples analyzed (Windig et al. 1982 a).

In this chapter we report the pyrolysis mass spectrometric analysis of nine layers in the 20-mm-thick microbial mat sequence present in the permanently water-covered zone of the Gavish Sabkha prior to 1979. The samples were taken from intact mats, transferred to the laboratory in Oldenburg and kept there in dim light prior to our analysis. The intention to sample the same material under field conditions was abruptly cut off by the thunderstorm in December 1979. A field trip in November 1980 revealed that the beautiful sequence had perished and was decomposed partially by anaerobic microorganisms in the dark below the mud. The analyzed sequence from the top mat prior to 1979 is so unique that publication as an example and as pilot study is justified. The result of the decomposition under the mud layer was studied as well, and the pyrolysis mass spectrum is discussed.

18.1 Material and Methods

18.1.1 Sample Description

A depth profile, general characteristics, the predominant microorganisms and the zonation are given by Krumbein et al. (1979 a). A description of the microbial ecosystem (Krumbein et al. 1979 a, Fig. 2) is given in Fig. 18.1 (see also Figs. 11.19 and 11.23 in Chap. 11). The pyrolysis mass spectrometric sample codes (PyMS sample) will be used throughout in this report. The detailed description of the various microorganisms in each layer has been given by Krumbein et al. (1979 a) and was checked for each of the samples by light microscopy. All samples were taken from one mat section, but sample B was chosen from a site in the section relatively rich in *Thiocapsa halophytica*. The mats were transferred to the Oldenburg laboratory after collection in March 1979 by one of us (W. E. Krumbein). They were kept in dim light at a temperature of 28 °C in translucent glass vessels. The samples for pyrolysis analysis were taken from these stored mats by stainless steel spatula, packed in polyethylene vials and cooled in liquid nitrogen. Samples of the decomposed top mat were taken in November 1980.

DEPTH MM	CHARACTERISTICS	DOMINANT MICROORGANISM	ZONE Nr	PYMS SAMPLE
0	WHITE CRUST	NONE VISIBLE		A_1
2	YELLOW SLIME		1	
4	BROWNISH SLIME			A_2
6		SYNECHOCOCCUS		
8	BROWNISH SLIME			
10			2	A_3
12	PINK SLIME	THIOCAPSA		B
14	BROWNISH SLIME	PLEUROCAPSA	3	C_1
16				C_2
	GREEN MAT	LPP — B	4	D
18	PURPLE MAT	PURPLE BACTERIA	5	E
20	GREEN MAT	LPP — B	6	F

Fig. 18.1. The multilayered microbial ecosystem of the Gavish Sabkha lagoon before the sheetflood of December 1979. The caption *zone number* refers to the zonation of Krumbein et al. (1979, Fig. 2). The pyrolysis mass spectrometry samples were coded according to the numbers in the last column (PyMS sample)

These samples were taken with a perspex coring device. The core was protected from light in aluminum foil, brought over to Holland and subsampled within 36 h after sampling. The subsamples were cooled in liquid nitrogen.

18.1.2 Preparation for Pyrolysis Analysis

Samples were freeze-dried and homogenized in glass mortars. About 1 mg of homogenized sample was suspended in methanol immediately before subsampling and application of aliquots to ferromagnetic sample wires used in Curie-point pyrolysis mass spectrometry.

18.1.3 Measurement of the Pyrolysis Mass Spectra

The pyrolysis mass spectrometer used was described by Meuzelaar et al. (1977). The pyrolysis products are generated in vacuum (10^{-6} Torr) by inductive heating of the ferromagnetic sample wire in a high frequency field. The Curie-point of the sample wires was 510 °C. This temperature was reached in the pyrolysis unit 0.1 s after initial generation of the high-frequency field. The pyrolysis products reached the ion source via a heated buffer volume and were subsequently ionized at 14 eV to suppress further electron impact fragmentation. Ions are separated by a quadrupole mass filter (multiple scan procedure) and detected with an electron multiplier. The mass range of the instrument was set at m/z 16 to m/z 160. Samples were analyzed in triplicate. All spectra were normalized for total ion current intensity in order to correct for variations in sample size and minor instrument variations. The reproducibility of the relative peak intensities was within 10%.

18.1.4 Processing of Mass Spectral Data

Numerical comparison of the various spectra was carried out by a combined factor analysis/discriminant analysis procedure which is described in more detail elsewhere (Windig et al. 1982b). In short, the principle of this method is that discriminant functions are extracted from the data set on the basis of mass peak intensities that show a correlated behavior with respect to between-group differences. (A group is a subset of triplicate spectra obtained for a particular sample.) A discriminant function can be considered as a kind of subspectrum that may describe a chemical component present in varying relative amounts in the samples. Such a "discriminant spectrum" contains positive and negative peak intensities due to the fact that the pyrolysis mass spectra used as input data were normalized for total intensity. The first discriminant function (D_1) describes as much as possible of the between-group variance in the data set; D_2 as much as possible of the remaining variance, etc. The intensity of a discriminant function in a particular spectrum is expressed by its "score."

18.2 The Pyrolysis Mass Spectrometry of the Preflood Multilayered Microbial Ecosystem

18.2.1 Discriminant Analysis

The end result of the factor analysis/discriminant analysis procedure is an all-groups scatter plot based upon the first and second discriminant function generated from the pyrolysis mass spectral data. Significant differences between the samples analyzed are maximized in this procedure.

Figure 18.2 shows the scatter plot of the microbial mat layers, each represented by its group centroid and a boundary which describes inner variance. The discriminant functions describe 76% and 13% of the between-group variance respectively. Four areas are discriminated, as shown in Fig. 18.2. Sample B – the *Thiocapsa* layer, and sample E – the purple bacteria mat, are clearly separated from a cluster of filamentous cyanobacteria (samples D and F) and a cluster of coccoid cyanobacteria (samples A_1, A_2, A_3, C_1, and C_2). Some degree of separation between the various cyanobacterial samples is obtained in the third dimension. The third significant discriminant function describes 7% of the between-group variance.

The first conclusion to be drawn from these data is the apparent possibility to discriminate the various major taxonomic groups in the multilayered ecosys-

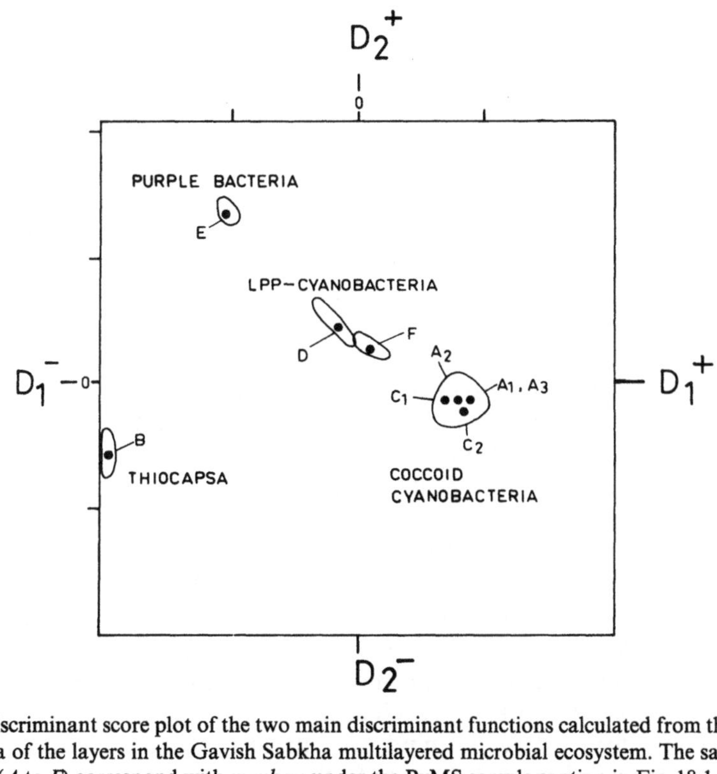

Fig. 18.2. Discriminant score plot of the two main discriminant functions calculated from the pyrolysis mass spectra of the layers in the Gavish Sabkha multilayered microbial ecosystem. The sample codes (*A* to *F*) correspond with *numbers* under the PyMS sample caption in Fig. 18.1

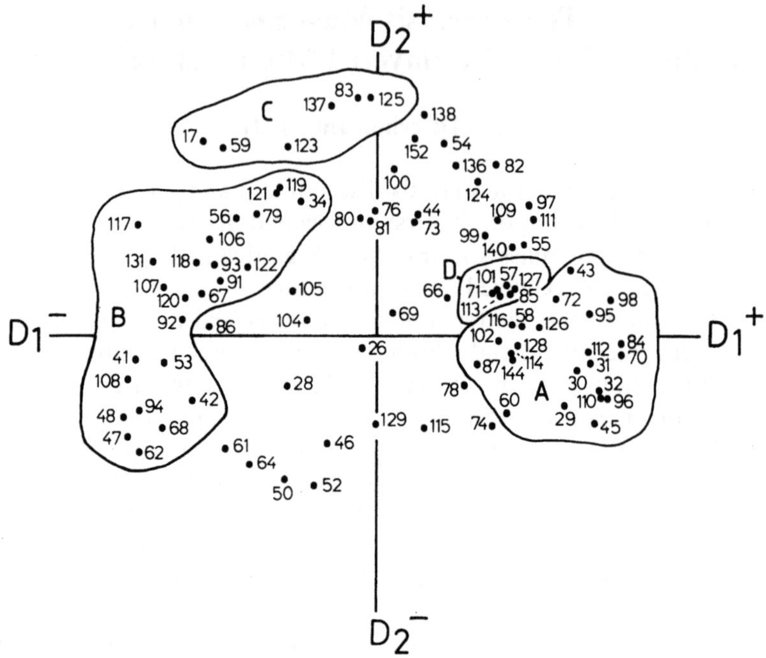

Fig. 18.3. Mass-peak vector plot based on the loadings of the two main discriminant functions D_1 and D_2. The areas outline bundles of mass-peak vectors translated into chemical components (see text)

tem. The chemical background for this discrimination is based of course on the inherent macromolecular composition of each layer and the derived pyrolysis mass spectral data. The discriminant functions describe the variations in chemical composition by sets of correlated mass peaks expressed in mass spectra.

Figure 18.3 is a two-dimensional mass-peak vector plot based on the first and second discriminant function. Each mass value is mapped using the discriminant loadings of these functions. The direction of the mass-peak vectors can be found by drawing imaginary lines through the origin and each of the mass values.

Sets of correlated mass peaks in this plot form vector bundles which in our experience are translatable into chemical compounds and macromolecular components in the samples. Bundles of mass-peak vectors are delineated in Fig. 18.3 as areas of chemical information.

Thus area A delineates mass peaks representing pyrolysis products of polysaccharides. Area B outlines mass values indicative of proteins. General characteristic mass peaks for proteins in pyrolysis mass spectral data are m/z 34, 48, 67, 81, 83, 92, 94, 97, 108, 117, 131 (compare Meuzelaar et al. 1982). Area C outlines m/z 59, 123, 137, 83, 125, a group of masses which is characteristic of N-acetyl amino sugar polymers. A relatively sharp cluster of masses indicative of alkanes is outlined in area D. This cluster contains mostly (low-voltage) electron impact fragmentation mass values of alkanes, such as m/z 57, 71, 85, 99, 13, 127.

With these considerations in mind, we can now evaluate the distribution of the sample clusters in Fig. 18.2. Cluster E, the purple bacteria, is separated from the other clusters by the relative abundance of proteins and N-acetyl amino sugars and the relatively low abundance of polysaccharides.

Cluster B, the *Thiocapsa* layer, is relatively rich in sulfur-containing proteins with unusually high methionine contents. The cluster of the coccoid cyanobacteria is characterized by its high abundance of polysaccharides. Alkanes are components of coccoid and filamentous cyanobacteria.

The LPP-cyanobacteria are both protein and carbohydrate rich with only small indications of N-acetyl amino sugars. The cyanobacteria from layer D are poorer in carbohydrates than those from layer F.

18.2.2 Mass Spectral Fingerprints

Discriminant analysis discriminates the samples by maximizing their significant differences but a general impression is obtained by visual inspection of the complete mass spectral fingerprints. Figures 18.4–18.7 present pyrolysis mass spectra of representatives of each of the four discriminated groups. Microscopic presentations of samples of the four groups are given in Figs. 18.9–18.12. Figure 18.4 is the spectrum of the *Thiocapsa* layer, which is microscopically pure *Thiocapsa* embedded in slime. The spectrum shows an unusually intense m/z 48, which represents methanethiol. This pyrolysis product is known from methionine-rich proteins (Posthumus et al. 1974) and from polymethionine (unpubl. observations). Another intense sulfur-containing pyrolysis product is dimethylsulfide with m/z 62 ($M^{+\cdot}$) and 47 ($M^+ - CH_3$). Whether this dimethylsulfide points to the presence of dimethylpropiothetin (Andreae 1980), as was found in the Solar Lake top mat studies (Boon 1984), or represents a methylation product of the abundant methanethiol during pyrolysis cannot be decided at this point. Obviously organically bonded sulfur plays an important role in the metabolism of the microorganisms in this layer.

Several pyrolysis products of proteins are evident. The series of phenolic compounds with m/z 94, 108, and 122 points to tyrosine-rich proteins, which the microorganisms may have developed as a protection against short-wavelength light.

The mass peaks at m/z 86 ($M^{+\cdot}$) and 68 ($M^+ - H_2O$) are interpreted as markers for β, γ-butenoic acid, a pyrolysis product of poly-β-hydroxy butyric acid (Morikawa and Marchessault 1981). The spectrum of the other purple bacteria layer (Layer E) is shown in Fig. 18.5. This mass spectrum is distinctly different from the *Thiocapsa* spectrum, as was already evident from the discriminant analysis. General protein pyrolysis mass spectrometric characteristics, e.g., m/z 34, 48, 67, 81, 92, 94, 97, 108, 117, and 131 are relatively abundant. A relatively abundant N-acetyl amino sugar polymer component in this sample is evident from m/z 59, 97, 109, 111, 125, 137, and 139 (compare Boon et al. 1981).

The mass peaks at m/z 86 ($M^{+\cdot}$), 68 ($M^+ - H_2O$), 100 ($M^{+\cdot}$) and 82 ($M^+ - H_2O$) point to the presence of a poly-β-hydroxy-alkanoate rich in C_4 and C_5 β-hydroxy acids (Morikawa and Marchessault 1981, Wallen and Rohwedder 1974). Similar pyrolysis products were seen in pyrolysis mass spectra of living top mat from the Solar Lake, and the corresponding pyrolysis products were identi-

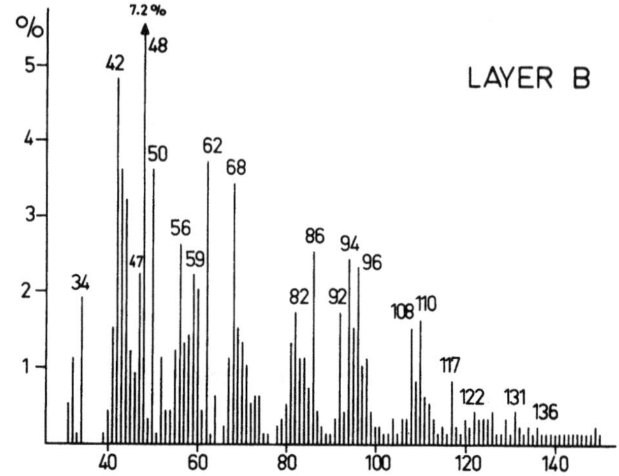

Fig. 18.4

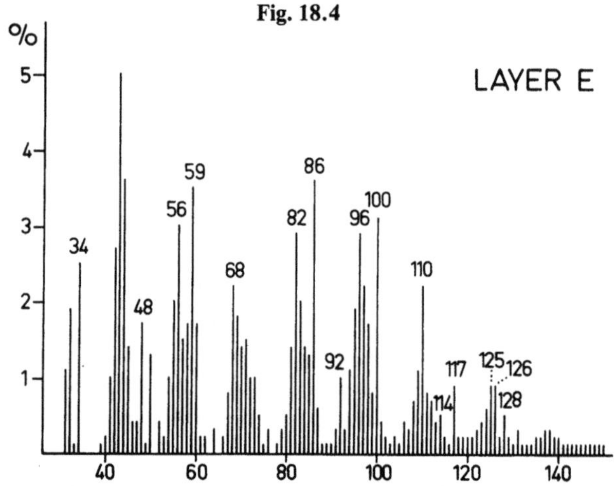

Fig. 18.5

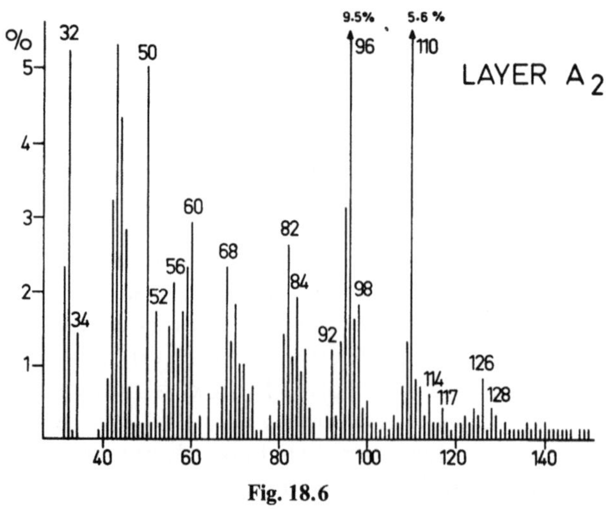

Fig. 18.6

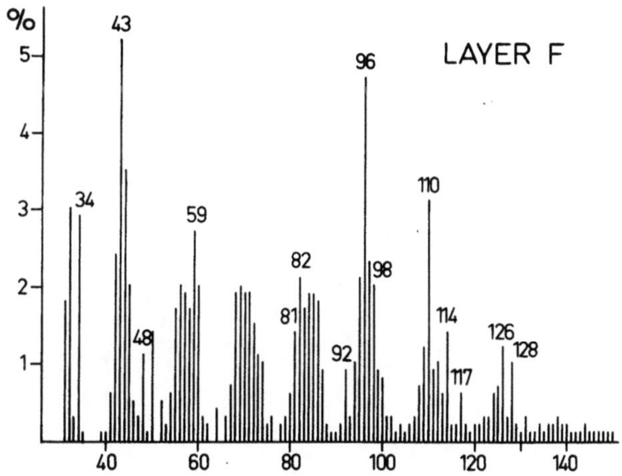

Fig. 18.7

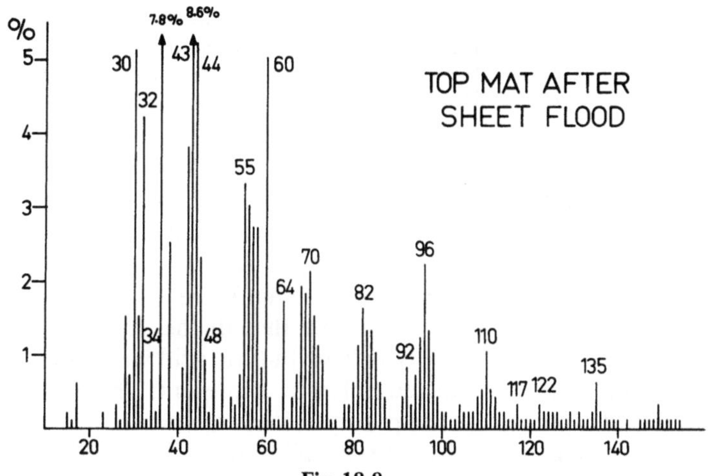

Fig. 18.8

Fig. 18.4. Pyrolysis mass spectrum of the *Thiocapsa halophytica* (tentative name) rich layer (layer B) in the multilayered ecosystem

Fig. 18.5. Pyrolysis mass spectrum of the purple bacterial layer (layer E) in the multilayered ecosystem

Fig. 18.6. Pyrolysis mass spectrum of the *Synechococcus* layer (layer A_2) in the multilayered ecosystem

Fig. 18.7. Pyrolysis mass spectrum of filamentous LPP-B group cyanobacteria layer (layer F) in the multilayered ecosystem

Fig. 18.8. Pyrolysis mass spectrum of the partially decomposed multilayered ecosystem under 5 mm of mud for 11 months

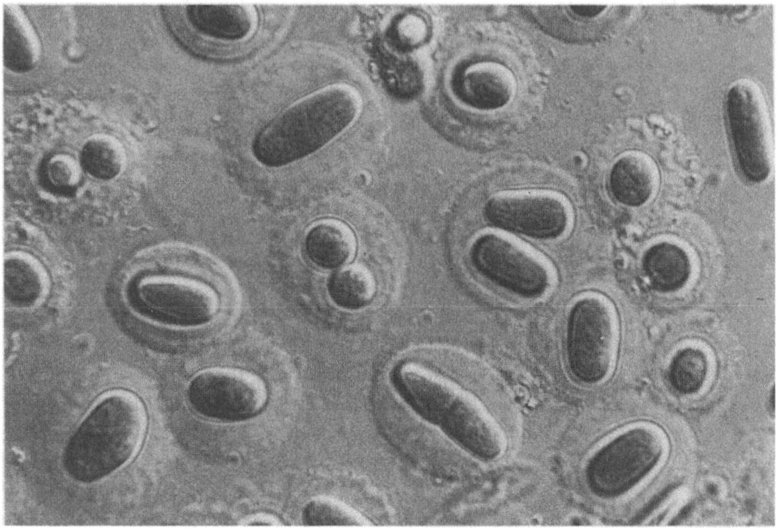

Fig. 18.9. Main cyanobacterium of the uppermost layers of the mat described in Fig. 18.1.: *"Synechococcus"* or *"Gloeothece"*. In the field samples this strain is embedded in large amounts of slime material. A distinction between individual cell clusters can rarely be made. In culture the strains we isolated resemble more or less *Gloeothece* strains described before; cell length between 4 and 7 μm, cell width between 3 and 4 μm

fied by pyrolysis gas chromatography-mass spectrometry (Boon 1984). No evidence was found for elemental sulfur, which is easily detected by this method.

It is possible that the droplets observed in the *Thiocapsa*-like organisms in this layer (Krumbein et al. 1979a; description of layer 5) are in fact the polyhydroxyalkanoates (see Fig. 18.12). It is remarkable that in the *Thiocapsa halophytica* layer (layer B) only evidence for polyhydroxybutyric acid was found, whereas in the purple bacteria of layer E a mixed polyester is present. These differences and the other differences apparent in the pyrolysis mass spectra indicate that layers B and E must have been inhabited by entirely different species of anoxygenic photosynthetic purple bacteria.

Figures 18.6 and 18.7 show the spectra of the layers rich in coccoid and filamentous cyanobacteria respectively (Fig. 18.10). Figure 18.6, which is a typical spectrum of the coccoid cluster (see Fig. 18.2) corresponds to layer A_2, which is described by Krumbein et al. (1979a, layer 2). This layer is practically only, albeit sparsely, inhabited by species of Synechococcus embedded in extracellular slimes (see Fig. 18.9). Some types of LPP-group cyanobacteria are admixed in the lower part of the *Synechococcus* zone (layer A_3). No microorganisms were seen in the upper 2 mm.

The white crust is presumably gypsum. The abundance of extracellular slime in these layers appears to be reflected in the intense (above 5%) mass peaks at m/z 96 and 110. The chemical identity of these mass peaks has been determined by pyrolysis gas chromatography as furfuraldehyde and methylfurfuraldehyde. These pyrolysis products are often generated from polysaccharides. Their high intensity and the much lower intensity of the other carbohydrate pyrolysis prod-

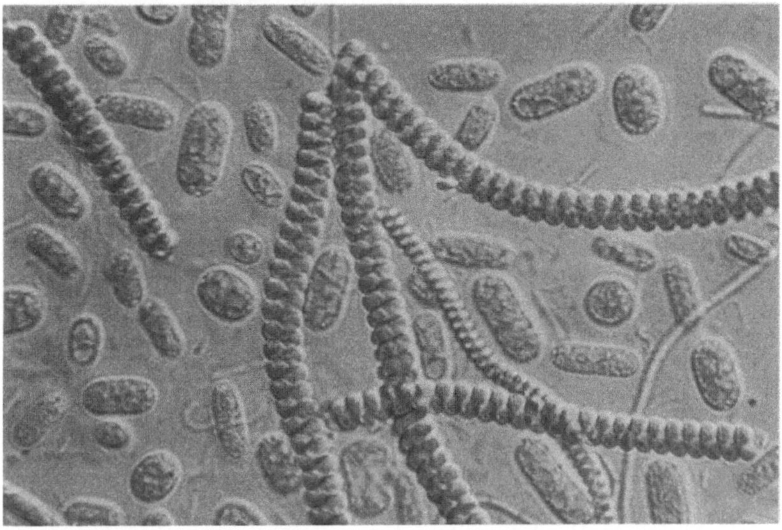

Fig. 18.10. "Synechococcus" and two types of Spirulina from the lower layers of Zone 1 in Fig. 18.1. The *Synechococcus* may turn in culture into the *"Gloeothece"* type shown in Fig. 18.9. Ehrlich and Dor (Chap. 14) describe the coccoid cyanobacterium as *Aphanothece halophytica,* the larger *Spirulina* as *Spirulina subsalsa,* the smaller one as *S. labyrinthiformis*

ucts, such as m/z 58, 60, 72, 74, 82, 84, 98, 102, 114, 126, and 128, point to a specific chemical component.

The pyrolysis mass spectrum of alginic acid (Meuzelaar et al. 1982) is characterized by a very intense m/z 96. The co-occurrence of m/z 96 and 110 in the coccoid spectra suggests a rather specific macromolecule with acidic groups, related to alginates but different with respect to chemical components, which generate methylfurfural upon pyrolysis. This difference in behavior may be a matter of cross-linking. The mass peaks at m/z 114, 126, and 128 are quite specific for various types of neutral pentose, hexose, and deoxyhexose polysaccharides (van der Kaaden et al. 1983, Meuzelaar et al. 1982, Schulten et al. 1981, Ohnishi et al. 1977). Several protein-derived pyrolysis products are apparent.

It is remarkable that chloromethane (m/z 50 and 52) is relatively intense in all the coccoid spectra. The presence of this pyrolysis product is interpreted as a transmethylation reaction product. These peaks correlate with m/z 36 and 38 from HCl, which were deleted in the spectra shown. The evolution of chlorine species is caused by some reaction between the organic matter and chloride salts in the sample during the thermal decomposition. It points to readily available hydrogen and/or methyl radicals in the organic substrate.

Figure 18.7 is a typical spectrum of the cluster of filamentous cyanobacteria (see Fig. 18.11). The pyrolysis mass spectrum shown, is from sample F corresponding to layer 6 of Krumbein's description. The spectrum represents a mixture of pyrolysis products of carbohydrates (note the relatively intense m/z 114, 126, and 128), proteins and N-acetyl amino sugar polymers. The samples from layer F are richer in carbohydrates than the samples from layer D, which show higher-

Fig. 18.11. Two types of LPP (Lyngbya-Plectonema-Phormidium) group B occurring above and below the *Thiocapsa* layer. Ehrlich and Dor (Chap. 12) define them in classical taxonomic terms as *Microcoleus chthonoplastes* (larger diameter) and as an *Oscillatoria* species (smaller diameter, less pigmented). In culture both types are not forming the large bundles characteristic for *Microcoleus* and are thus difficult to interpret in classical taxonomic terms. Therefore we define both of them as LPP-B forms

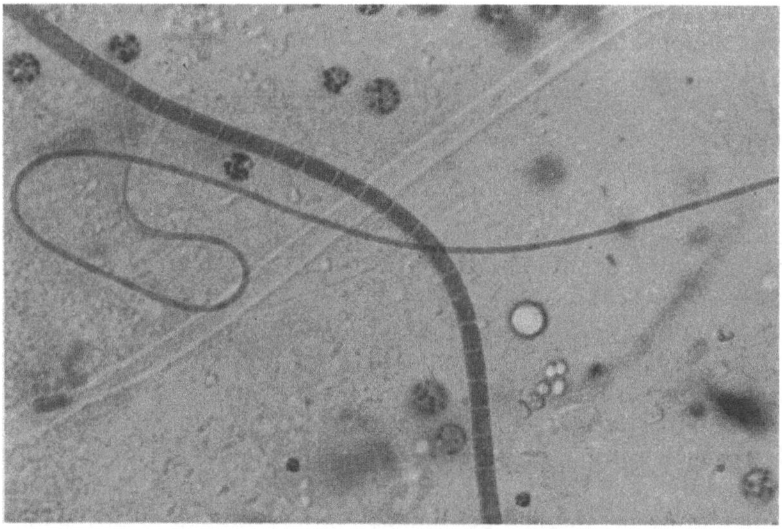

Fig. 18.12. Two different LPP-forms and *Thiocapsa halophytica* from Zones 4, 5, and 6 of Fig. 18.1. The larger LPP form corresponds to *Microcoleus*, the smaller one to *Schizothrix* (see Chap. 14). The large spherical cells of *Thiocapsa* are granulated by sulfur globules and/or poly-betahydroxybutyric acid. The small barely visible granular bodies are decaying *Thiocapsa* cells and some small Chromatium. An empty sheath of *Microcoleus* is crossing from *left to right*. To the right of *Microcoleus* and below an air bubble a small colony of *Pleurocapsa* is visible, which apparently just emerges from beocytes. *Microcoleus* has 3 µm diameter, *Thiocapsa* between 4 and 6 µm diameter. In extreme cases it reaches 7–8 µm. It lives aerobically and anaerobically in the Gavish Sabkha

intensity peaks from proteins and N-acetyl amino sugar polymers. This can be deduced from the results shown in Fig. 18.2. The spectra in layer D resemble the spectra from light-limited LPP cyanobacteria (Boon et al. 1984) except for the high-intensity m/z 96 and 110, but these peaks represent the slimy matrix of the whole ecosystem.

18.3 The Pyrolysis Mass Spectrometry of the Postflood Partially Decomposed Top Mat

A mud layer of about 5 mm was present on the top mat of the permanently water-covered zone of the Sabkha lagoon at the time of sampling. Macroscopic and microscopic inspection of core samples revealed that the viscous extracellular slime of the *Synechococcus* layer in the preflood top mat had completely collapsed. Remains of *Synechococcus* were visible however and even showed red fluorescence of chlorophyll pigments when inspected with a epifluorescence microscope. Although red- and green-colored layers were visible in the upper centimeter of the mat section below the mud layer, the microscopic contents of these layers were brownish filamentous cyanobacterial remains and yellowish "shadows" of purple bacteria.

The pyrolysis mass spectrum of the mud layer (not shown) demonstrated an abundance of elemental sulfur, but an absence of organic remains. The presence of elemental sulfur can be explained as an oxidation product of H_2S, which has diffused upward from actively sulfate-reducing zones below the mud layer.

Figure 18.8 shows the pyrolysis mass spectrum of the upper half-centimeter of the decomposing top mat below the mud. Visual comparison of this spectrum with the spectra shown above demonstrates that significant changes in organic matter composition have occurred. We do not see the abundant m/z 96 and 110, nor significant peaks at m/z 86, 100, 114, 126, and 128, which points to the degradation of extracellular slimes, neutral polysaccharides, and polyhydroxesters.

The mass peaks at m/z 32, 55, 68, 82, 96, and 110 can be interpreted as characteristics for a degraded kind of carbohydrate similar to the material observed in the oldest Solar Lake mats. The mass peaks at m/z 92, 94, 108, and 117 point to remains of proteins.

The intense mass peak at m/z 60 could be acetic acid trapped as salt during sulfate reduction. No elemental sulfur is present. The low m/z 34 (H_2S) points to the absence of polysulfides. The relatively abundant m/z 64 points to pyrite, which releases S_2 during pyrolysis. The intense peak at m/z 30 points to nitrates, as has been demonstrated in pyrolysis mass spectra of sewage (Meuzelaar et al. 1982, Fig. 47).

The results demonstrate the dramatic effect of sheetfloods on the organic matter fossilization process in the mats. Most characteristics of intact mats are replaced and the remainders presumably consist of poorly degradable matter. In our experience the spectrum shown in Fig. 16.8 is the general fingerprint for decomposed cyanobacterial mats. Similar patterns, except for the m/z 30 und 64, have been observed in pyrolysis mass spectra of potential stromatolites from Mururoa (Pacific), Baffin Bay (Texas) and subsurface Solar Lake (northern Sinai).

18.4 Conclusion

The microbial ecosystem found in the permanently water-covered zone of the Gavish Sabkha consists of a number of layers with a distinctly different chemical composition, which could be visualized by pyrolysis mass spectrometry.

19. Carbon Isotope Geochemistry and ^{14}C Ages of Microbial Mats from the Gavish Sabkha and the Solar Lake

MANFRED SCHIDLOWSKI, UDO MATZIGKEIT, W. G. MOOK, and WOLFGANG E. KRUMBEIN

Photosynthetic organisms incorporate carbon with a marked though variable bias in favor of the light carbon isotope (^{12}C) which ultimately derives from a kinetic fractionation effect imposed on the principal assimilatory pathways (see Vogel 1980, O'Leary 1981). As a result, average organic matter is enriched by about 20–30‰ in ^{12}C as compared to oceanic bicarbonate and carbonate, the quantitatively most abundant inorganic carbon species in our environment (Schidlowski 1982, 1983).

While there exists a wealth of carbon isotope data on higher plants, isotope investigations of microbial (algal and bacterial) communities are as yet scant and, apart from culture experiments, confined to a small number of natural occurrences (see Behrens and Frishman 1971, Smith and Epstein 1971, Calder and Parker 1973, Seckbach and Kaplan 1973, Barghoorn et al. 1977, Peters et al. 1981, Estep 1982). This is in marked contrast to possible implications of such work for several important biogeochemical questions. Apart from obtaining relevant information on the fractionations intrinsic to the assimilatory pathways of individual microbial mat-builders, knowledge of the bulk isotopic composition of organic matter from potential stromatolite-forming habitats might allow an assessment of the contribution of the Earth's microbenthos to the global carbon budget. In view of the fact that benthic prokaryotic ecosystems have left their vestiges in the geological record as far back as 3.5×10^9 years ago (Walter 1983), the consequences of such work may be indeed far-reaching.

The scarcity of work hitherto performed on extant microbial mat communities is, in part, due to a corresponding scarcity in the present environment of habitats conducive to an exuberant proliferation of benthic bacterial and algal ecosystems. In contemporary potential habitats the ubiquity of a grazing fauna virtually precludes any large-scale development of microbial carpets except in niches that are hostile to most higher forms of life, in particular hyperthermal and hypersaline environments (cf. Bauld 1981a, b). Accordingly, most of the data hitherto available for microbial communities come from such exceptional habitats.

This chapter summarizes carbon isotope investigations carried out on microbial mats and derivative organic matter from the Gavish Sabkha (Fig. 19.1), a near-coastal brine pool of limited extension with salinities up to 300‰ and more, which is situated near the southern tip of the Sinai Peninsula (Chap. 9) and from the Solar Lake (Sinai) situated approximately 20 km south of the city of Elat. The region is extremely arid with a rainfall of about 10 mm and an evaporation rate of 4.0 m per year. Influx of seawater into the Gavish Sabkha is maintained by see-

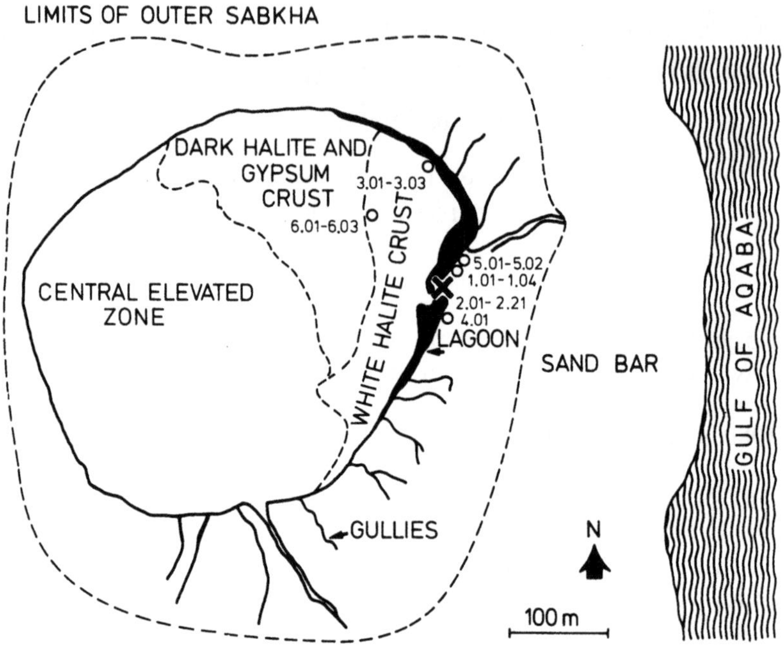

Fig. 19.1. Sketch map of Gavish Sabkha showing localities of samples listed in Table 19.1 (site of large core with samples 2.01 to 2.21 is indicated by *cross)*

page through a slightly elevated (2 m) coastal bar some 400 m wide. The situation of the Solar Lake is similar. The coastal bar, however, is only 100 m wide.

19.1 Sources of Organic Carbon in the Gavish Sabkha and the Solar Lake

In sabkha systems, the high evaporation rates typically cause an upward movement of nutrient-enriched pore waters which may give rise to a marked eutrophication of these environments and, consequently, to impressive rates of primary production (up to 8–12 g organic carbon per m^2 and day; cf. Cohen et al. 1977a, b, Krumbein and Cohen 1977, Krumbein et al. 1979a, see also Chap. 11). It is, therefore, by no means surprising that the biological community of the Gavish Sabkha constitutes a flourishing microbial ecosystem dominated by a sedentary (benthic) microflora well adapted to both hypersaline conditions and heliothermal stress and occasionally interspersed with rare elements of a grazing fauna. The primary producers of this system are, for the most part, photosynthetic prokaryotes, though minor eukaryotic elements like diatoms and green algae have also been reported. As a whole, the environment qualifies as a potentially stromatolite-forming biotope. The rare manifestations of grazing eukaryotic macrobenthos are virtually confined to a few genera of halotolerant gastropods and insects which have successfully invaded zones of moderate salinity (see Chap. 15).

As detailed by Krumbein et al. (1979a) and Gerdes and Krumbein (1983), cyanobacteria are the principal prokaryotic mat-builders, comprising a formidable array of both coccoid (e.g., *Synechococcus, Synechocystis, Gloeocapsa, Gloeothece*) and filamentous genera (*Lynbgya, Phormidium, Anabaena, Microcoleus, Oscillatoria, Schizothrix, Spirulina*). Minor elements of the prokaryotic community are *Chloroflexus* and related flexibacteria as well as halobacteria (see also Chaps. 12 and 13). Occasionally, eukaryotic photosynthesizers, such as diatoms (*Navicula, Amphora*), make their appearance in the upper parts of individual mats, while a marginal belt of lower salinity (50–70‰) at the eastern fringe of the Gavish Sabkha (Zone 5 of Chap. 11) harbors an abundant flora of *Enteromorpha* sp., a eukaryotic green alga often found in estuarine environments. The moderate salinities of this zone have also admitted to this particular habitat grazing gastropods (e.g., *Pirenella* sp.) whose fecal pellets often form conspicuous accumulations on the sediment floor.

In typical multi-layered mats, the predominantly cyanobacterial top stratum is underlain by layers of anoxygenic photosynthetic prokaryotes, notably purple (Chromatiaceae) and green (Chlorobiaceae) sulfur bacteria. Further below follow the domains of chemoautotrophic microorganisms, and finally of sulfate reducers (dominated by *Desulfovibrio* sp.). Thus, we have a relatively clear picture of the individual microorganisms which may contribute to the fractionation of carbon isotopes in the Gavish Sabkha.

Since the microbial inventories of the samples selected for carbon isotope analysis have not been investigated in detail, the measurements listed in Tables 19.1 and 19.2 broadly integrate over the isotopic composition of the microbial community as a whole. This holds particularly for the protokerogenous material of buried (subfossil) mats encountered in the drill cores included in Tables 19.1 and 19.2. In first approximation, the carbon isotope values can be expected to reflect the $^{13}C/^{12}C$ ratios of the prevailing cyanobacterial community with subordinate additions of other prokaryotic and (minor) eukaryotic microbial elements. An analysis of *Pirenella* sp. (Table 19.1, No. 4.01) shows that the isotopic composition of the primary producers has been imparted to the members of the heterotrophic grazing community with but minor change.

19.2 Experimental

After recovery the samples were thoroughly dried in the field in carefully cleaned PVC containers. This worked well for most samples except some extremely salty ones which failed to lend themselves to complete dehydration even in a vacuum desiccator containing silica gel. To keep secondary alterations of the organics at a minimum, care was taken to process the material immediately after return to the laboratory base (6 days after sampling). Storage of the biogenic materials in organic solvents was avoided because of the risk of irreversible contamination.

Table 19.1. Isotope values and selected petrographic background parameters of 34 sediment samples from the Gavish Sabkha. Nos. 2.01–2.21 and 3.01–3.03 are from two cores 2.45 m and 0.45 m deep, respectively, the rest representing surface/near surface samples from various parts of the Sabkha (cf. locality map, Fig. 19.1). Numbered zones mentioned in sample descriptions are those from Chapter 11

No.	Original field No.	Description of sample	Depth below surface (cm)	Water-soluble salt fraction (% dry mass)	C_{org} in salt-free dry mass (%)[a]	$\delta^{13}C_{org}$ (‰, PDB)	$\delta^{13}C_{carb}$ (‰, PDB)	$\delta^{18}O_{carb}$ (‰, SMOW)	$\Delta\delta_c$[b] (‰)
1.01	34	Orange-colored layer of surface mat	0–2	81.5	5.66	− 9.0	+2.9	+32.6	11.9
1.02	35	Green layer below 1.01	0–2	80.7	8.43	− 8.1	+2.0	+33.9	10.1
1.03	36	Purple to brownish horizon below 1.02	0–2	0.0	6.00	− 8.3[e]	+3.1	+34.0	11.4
1.04	37	Lower parts of mat underneath 1.03	0–2	0.0	2.95	− 9.8	+4.7	+34.0	14.5
2.01	3	Layer of gray silt	25	57.8	0.49	− 8.4	+1.3	+33.3	9.7
2.02	4	Layer of yellowish-gray silt	30	72.6	0.26	− 9.1	+3.6	+34.2	12.7
2.03	5	Gray clay horizon	50	36.1	<0.05	n.d.	−0.7	+27.4	–
2.04	6	Evaporite horizon mixed with greenish clay	60	91.7	0.15	n.d.[c]	−0.7	+29.9	–
2.05	7	Evaporite horizon	70	76.4	<0.05	n.d.	+1.3	+34.7	–
2.06	8	Fossil mat	80	43.5	0.81	− 9.2	−0.1	+32.5	9.1
2.07	9	Carbonate-rich horizon	85	65.5	0.30	−11.7[d]	+0.1	+34.2	11.8
2.08	10	Layer dominated by pink-colored gypsum	180	72.1	<0.05	n.d.	+2.9	+34.6	–
2.09	11	Gypsum layer similar to 2.08	186	64.0	0.28	− 8.5	+1.5	+34.3	10.0
2.10	12	Laminated greenish mat with carbonate nodules	213	68.2	1.52	− 8.5	+1.5	+34.6	10.0
2.11	13	Fossil mat composed of alternating red and white microlayers	215	51.6	0.14	n.d.[f]	−1.5	+34.5	–
2.12	14	Laminated greenish mat	220	63.6	0.36	− 7.8	−0.2	+34.8	7.6
2.13	15	Evaporite horizon	228	65.0	0.93	− 9.0	+0.8	+33.2	9.8

2.14	16	Laminated green mat with carbonate	234	78.3	4.30	− 9.3	+1.3	+33.6	10.6
2.15	17/21	Gypsum-rich layer	239	38.7	0.43	−11.7[e]	+0.6	+34.7	12.3
2.16	17/22	Green laminated mat	240	67.2	7.07	− 9.6	+0.6	+34.0	10.2
2.17	17/23	Clay horizon	240	64.2	5.58	− 9.6	−0.2	+30.7	9.4
2.18	17/24	Green laminated mat	241	69.2	3.04	− 8.1	−1.0	+33.7	7.1
2.19	17/25	Mat with carbonate granules	242	48.6	1.71	−10.9[e]	–	–	–
2.20	17/27	Carbonate-bearing clay horizon	243	8.1	0.14	−17.0[d]	−0.8	+36.3	16.2
2.21	17/26	Laminated green mat with carbonate granules	245	33.6	2.14	− 8.6	+1.9	+35.3	10.5
3.01	29	Green crust below surface of brine pool (zone 10)	~0	99.0	<0.05	n.d.	−0.8	+25.7	–
3.02	32	Clay horizon below 3.01 (deposited by 1979 flood)	39	45.6	0.15	−16.8[e]	−0.8	+26.1	16.0
3.03	33	Pigmented portion of older mat buried below 3.02	42	32.8	<0.05	n.d.	−0.5	+26.4	–
4.01	38	Gastropod shells (*Pirenella* sp.) from zone 5	~0	0.0	3.63	− 6.2	+3.3	+32.4	9.5
5.01	41	Nodular surface layer from zone 7	~0	2.8	4.38	n.d.[c]	+1.0	+25.8	–
5.02	42	Orange-colored salt crust with oncoids from zone 7	~0	76.1	1.60	− 6.1	–	–	–
6.01	44	Orange-colored salt crust from Bedouin pool (zone 11)	~0	47.2	1.44	−11.3	0.0	+28.2	11.3
6.02	45	Orange-colored salt crust	~0	51.6	1.09	−10.4[e]	−3.5	+30.2	6.9
6.03	46	Green surface mat from zone 11	~0	63.2	0.39	−14.1[e]	−0.6	+26.9	13.5

[a] Salt-free dry mass still includes evaporite minerals of low solubility (sulfate). C_{org}-content was determined coulometrically on carbonate-free samples by an automated analyzer (Coulomat 701)

[b] $\Delta\delta_c = \delta^{13}C_{carb} - \delta^{13}C_{org}$

[c] Determination failed due to blasting of reaction tubes by sudden release of crystal water from gypsum

[d] Values reliable only within ±1.5‰

[e] Samples processed with 10% H_3PO_4 at 60 °C (results identical within ±1‰ with those obtained in standard procedure conducted at room temperature)

[f] Sample released major amounts of H_2S and SO_2, the ensuing formation of (probably) CS^+ gravely interfering with the mass 44 signal of CO_2

Due to the limits imposed by our currently operated instrumentation, samples with C_{org} <0.05% usually failed to yield quantities of CO_2 adequate for a reliable isotope analysis

Table 19.2. Isotopic composition of organic carbon and carbonate of surface mats and their subfossil equivalents from a 1-m-deep core from Solar Lake (samples marked by *asterisk* analyzed in Mainz, others in Groningen). $\Delta\delta_c = \delta^{13}C_{carb} - \delta^{13}C_{org}$

No.	Original field No.	Description of sample	Depth below surface (cm)	C_{org} (%)	$\delta^{13}C_{org}$ (‰, PDB)	$\delta^{13}C_{carb}$ (‰, PDB)	$\delta^{18}O_{carb}$ (‰, SMOW)	$\Delta\delta_c$ (‰)
1*	48	Surface mat	0	4.6	−5.4	+3.3	+34.6	8.7
2	C1	Surface mat	0 – 3.4	14.9	−5.0	+1.6	n.d.	6.6
3*	48	Lower part of surface mat	3.0	4.6	−5.5	+3.2	+34.9	8.7
4*	48	Lowermost part of surface mat	5.0	4.6	−4.4	+3.3	+35.1	7.7
5	C2	Lowermost part of surface mat	3.4– 7.5	16.1	−4.6	+2.2	n.d.	6.8
6	C3	Buried mat with flood detritus	12.8– 16.4	10.7	−3.5	+4.1	n.d.	7.6
7	C4	Fossil laminated mat	21.7– 26.0	9.6	−5.1	+1.3	n.d.	6.4
8*	51	Fossil mat with borings of *Bledius* sp.	40.0	5.4	−5.5	+3.9	+36.2	9.4
9	C5	Fossil mat with borings of *Bledius* sp.	38.0– 42.0	18.9	−5.1	+4.5	n.d.	9.6
10*	50	Fossil mat overlying clay horizon	50.0	5.4	−8.4	+4.0	+35.3	12.4
11	C6	Fossil mat below clay horizon	54.0– 58.0	20.6	−5.8	+5.0	n.d.	10.8
12	C7	Fossil laminated mat with carbonate	62.0– 66.0	15.6	−6.3	+3.7	n.d.	10.0
13	C8	Fossil mat (rich in ooids and carbonate) above gypsiferous ooid sand	66.0– 72.0	4.2	−6.2	+6.3	n.d.	12.5
14*	49	Dark clay with ooids above ooid-bearing gypsum layer	90.0–100.0	1.3	−4.9	+3.2	+34.8	8.1

For work on the *carbonate fraction,* the primary chloride constituents were removed by washing the samples with distilled water. The residue (which still contained evaporite minerals of low solubility, such as sulfate) was collected on carbonate-free glass-fiber filters (type MN 85/90 BF from Macherey-Nagel). After drying at room temperature, these residues were finely powdered in an agate mortar and treated with anhydrous (100%) phosphoric acid at 35 °C as described by McCrea (1950) and Craig (1953). Omission of the distilled water rinse usually resulted in the release of consiberable quantities of hydrogen chloride along with CO_2.

To separate the *organic carbon fraction* for isotope analysis the samples were, in a first attempt, treated with concentrated hydrochloric acid for about 2 h at 160 °C in a sand bath to obtain quantitative removal of the carbonates. However, this approach failed and had to be abandoned as visible changes of the organic substances could be observed during the procedure (with colors turning from green to black and the concomitant release of a tar-like smell). Therefore, attempts were made with varying concentrations of phosphoric acid at different temperatures and reaction times. An adequate procedure proved to be treatment with 10% H_3PO_4 at room temperature for about 5 h. The bulk of the C_{org} samples listed in Tables 19.1 and 19.2 were processed in this way; only in few cases, which are specifically identified (cf. Table 19.1, footnote e), premature exhaustion of the sample prevented a repetition of the procedure under these standardized conditions. After filtration and drying, the carbonate-free organic residue was combusted in sealed quartz tubes along with an excess of copper oxide at 900 °C for 5 min. The water released during the combustion procedure was removed with a vacuum cold trap at −120 °C.

Carbon isotope ratios of both carbonate and organic carbon were determined for the CO_2 obtained from the described procedures. Measurements were carried out with a modified Varian-MAT CH5 mass spectrometer possessing a double inlet system and two Faraday collectors which permitted successive determinations of the mass ratios 44/45 and 44/46. Results are reported as $\delta^{13}C$ values relative to the conventional PDB (Peedee Belemnite) standard, with

$$\delta^{13}C = \left[\frac{(^{13}C/^{12}C)_{sa} - (^{13}C/^{12}C)_{st}}{(^{13}C/^{12}C)_{st}}\right] \times 1000\ [‰], \tag{1}$$

where sa = sample and st = standard. All values are corrected for ^{17}O (Craig 1957). The mass spectrometric correction factors applied were determined after Deines (1970). Standard deviations of the $\delta^{13}C$ values listed in Tables 19.1 and 19.2 are ±0.1‰ for both organic and carbonate carbon.

Oxygen isotopic compositions of the carbonate samples were also determined and are reported as $\delta^{18}O$ values vs SMOW (Standard Mean Ocean Water), with

$$\delta^{18}O = \left[\frac{(^{18}O/^{16}O)_{sa} - (^{18}O/^{16}O)_{st}}{(^{18}O/^{16}O)_{st}}\right] \times 1000\ [‰] \tag{2}$$

(again, sa = sample and st = standard). Standard deviations of the measurements usually lie in the range of ±0.3‰.

Table 19.3. ^{14}C ages of samples from the Solar Lake as reported by different authors (Friedman et al. 1973a, Krumbein and Cohen 1974, Krumbein et al. 1977, and this paper). The ^{14}C pollution of the upper samples has been probably caused by biological productivity measurements. The other ^{14}C ages reported here agree with previous results made on smaller samples. We have now obtained a better precision for our laminae counts and full evidence for an earthquake close to the beginning of the Christian era, which caused the subsiding of the central part of the Solar Lake. This also caused an interruption of microbial mat formation in the deeper parts of the pond and created heliothermal monomictic conditions, yielding gypsum precipitation in the central parts of the pond and continuation of mat formation only in the marginal shallow water regions (see Krumbein and Cohen 1974)

Sample No.	Depth (cm)	This paper				Friedman et al. (1973a), Geyh in Krumbein and Cohen (1974)	
		Lab. No.	$^{14}C_{age}$ of C_{org}	Lab. No.	$^{14}C_{age}$ of C_{carb}	$^{14}C_{age}$ of C_{org}	$^{14}C_{age}$ of C_{carb}
SL C1	0 – 3.4	GrN-11134	4,618.5 ± 5.8%	GrN-11090	1,650 ± 6%	718 ± 387 BP	–
SL C2	3.4 – 7.5	GrN-11135	97.3 ± 0.6%	GrN-11115	118.4 ± 1.3%	–	–
SL C3	1.28– 1.64	GrN-11142	2,270 ± 80 BP	GrN-11116	2,030 ± 140 BP	–	–
SL C4	21.7 –26.0	GrN-11136	1,760 ± 80 BP	GrN-11091	950 ± 140 BP	–	–
SL C5	32.0 –42.0	GrN-11143	2,340 ± 80 BP	GrN-11117	1,340 ± 80 BP	–	–
Geyh 3	40.0 –50.0	–	–	–	–	1,918 ± 370 BP	–
SL C6	54.0 –58.0	GrN-11144	2,410 ± 70 BP	GrN-11118	2,230 ± 80 BP	–	–
SL C7	62.0 –66.0	GrN-11137	2,480 ± 80 BP	GrN-11092	2,300 ± 90 BP	2,465 ± 155 BP	–
SL C8	66.0 –72.0	GrN-11145	3,210 ± 100 BP	GrN-11119	3,040 ± 70 BP	3,400 ± 230 BP	3,430 ± 90 BP
Beachrock underneath microbial mats		–	–	–	–	4,644 ± 555 BP	3,378 ± 172 BP
Uppermost mats of central part underneath 100 cm gypsum (Core 1)		–	–	–	–	1,910 ± 555 BP	–
Lowermost mats of central part of SL underneath gypsum		–	–	–	–	3,400 ± 180 BP	–
Uppermost mat of central part of SL underneath 60 cm gypsum		–	–	–	–	1,940 ± 130 BP	–
Texel sea-dike seagrass reference sample		GrN-11022	700 ± 70 BP	–	–	–	–

For ^{14}C analysis about 22–25 g of sediment was treated with 1% hydrochloric acid. The CO_2 evolved was purified before ^{14}C measurement. The remaining residue was washed, dried and combusted in pure oxygen. Also this resulting CO_2 was purified. ^{14}C counting was performed in proportional CO_2 counters. The resulting conventional ^{14}C ages, based on the 5568 year half-life, are presented in Table 19.3. Results obtained for the upper samples are reported as ^{14}C activities rather than ages. The measured activities are compared with the oxalic acid standard activity and reported in percentages. Hundred percent ^{14}C refers to the year 1950.

19.3 Results

The results of isotope analyses carried out on both organic carbon and coexisting carbonates from the Gavish Sabkha are summarized in Table 19.1. Apart from this work, a small core drawn from Solar Lake near Elat, Israel, has been investigated for comparison (Table 19.2). Solar Lake harbors abundant microbial mat communities closely related to those found in the Gavish Sabkha (Krumbein and Cohen 1977, Krumbein et al. 1977, Cohen et al. 1980).

It is apparent from the accumulated data that organic carbon from the Gavish Sabkha is extraordinarily heavy as compared to average organic matter, with $\delta^{13}C_{org}$ values ranging from -17.0‰ to -6.1‰ vs PDB (Table 19.1). Eliminating the gastropod sample (No. 4.01), 25 analyses of recent and subrecent (buried) mats and sedimentary organic matter have yielded a mean of -10.0 ± 2.6‰. With our limited background information on the specific microbial source material of individual samples, we have as yet no clue as to an explanation of the small number of more negative readings obtained (with $\delta^{13}C_{org}$ between -14.0‰ and -17.0‰; cf. Table 19.1, Nos. 2.20, 3.02 and 6.03) which markedly deviate from the bulk of the values.

Analyses performed on six samples from Solar Lake at the MPI Mainz and eight samples at Groningen (Table 19.2) have yielded $\delta^{13}C_{org}$ values that are even more positive than those obtained for the Gavish Sabkha, their average of -5.4 ± 1.1‰ ostensibly conveying the impression of an *inverse* fractionation since these organic materials are isotopically *heavier* than atmospheric CO_2 as the ultimate feeder substrate ($\delta^{13}C_{CO2} = -7.0$‰). Microbial mats and protokerogens from this locality surely figure among the heaviest biogenic matter as yet encountered in the Earth's biosphere.

Most of the samples investigated were found to contain sufficient carbonate to enable an isotopic analysis of carbonate carbon to be carried out along with that of organic carbon. $\delta^{13}C_{carb}$ values obtained for the Gavish Sabkha (Table 19.1) span the range of -3.5 to $+4.7$‰ with a mean of $+0.6 \pm 1.7$‰ vs PDB ($n = 31$; again, the gastropod shells of sample No. 4.01 were excluded from the calculation). As indicated by this mean, the bulk of the carbonate values largely coincide with the normal range of marine carbonates (about $\pm 0.5 \pm 2.5$‰; cf. Schidlowski et al. 1975, 1983, Veizer et al. 1980),the faint slant in positive direction stemming from the presence in the sample population of some moderately heavy carbonates. It is, therefore, plausible to assume that the carbonate content of the Sabkha basically derives from seawater seepage, since marine ingressions across the coastal sand bar may be largely excluded even during major floods (see Chap. 9). The small number of isotopically heavy samples ($\delta^{13}C_{carb}$ between +2.0‰ and +4.7‰) were, in all probability, precipitated during intermittent stagnant conditions, since carbonates with such characteristics are typical of closed and semi-closed depositional environments including hypersaline ones (cf. Schwarcz 1969, Schidlowski et al. 1976, Botz and Müller 1981). It should be noted that the carbonates from the Solar Lake core investigated (Table 19.2) are consistently enriched in ^{13}C ($\delta^{13}C_{carb} = +3.5 \pm 1.3$‰ for total carbonate) which average comes very close to that derived from previously reported data (Aharon et al. 1977).

As part of the work on carbonates, the oxygen isotope values of the samples were measured as a matter of routine. It is apparent from the $\delta^{18}O_{carb}$ readings listed in Tables 19.1 and 19.2 and the corresponding means of +32.1 ±3.3‰ vs SMOW (Gavish Sabkha) and +35.2 ±0.6‰ vs SMOW (Solar Lake) that the oxygen isotopic composition of carbonates from these environments is shifted by several per mill in positive direction as compared to marine carbonates of recent and subrecent (Quaternary) age which usually range from +26‰ to +32‰, with an average close to +29‰ (cf. Keith and Weber 1964, Garlick 1969). On the other hand, the $\delta^{18}O_{carb}$ values found coincide, for the most part, with data reported by Degens and Epstein (1964) for carbonates from the Bahama evaporite ponds and the Coorong Lagoon, South Australia (mainly between +31‰ and +36‰; recalculated from original PDB-related values). It seems noteworthy that, with one exception, six readings of $\delta^{18}O_{carb} \leqq +30$‰ listed in Table 19.1 are correlated with $\delta^{13}C_{carb} \leqq 0$‰, this decidedly corroborating the marine pedigree of the isotopically lighter carbonates deposited in the Gavish Sabkha.

From eight Solar Lake samples ^{14}C ages were obtained on the organic as well as the carbonate fractions. The results are given in Table 19.3. For the upper two levels the results are reported as ^{14}C activities (in%). The second level shows an almost recent ^{14}C content, which is difficult to translate into a meaningful ^{14}C age. The extremely high ^{14}C activity of the upper level is probably caused by local biological primary productivity determinations. Between 1971 and 1982 tracer experiments were continuously carried out in the mats and in the water column, many of them as part of advanced microbial ecology courses of the Hebrew University. Possibly, the influence of this type of contamination extended to the second level. The originally obtained age for the surface level of about 700 years (Krumbein and Cohen 1974) fits well with the age given below in Table 19.3 for recent marine sea grasses. The other data present a consistent picture of the age of evolution of microbial mats and correspond very well with former determinations made by Geyh in 1971 (Krumbein and Cohen 1974) for a sampling site about 15 m to the north. Because those datings were performed on much smaller samples, the reported standard deviations were larger than those given in our measurements.

The Solar Lake potential stromatolites thus started to form 3200–3400 years ago and represent the best profile of organic material so far known in nonpolluted areas throughout the past 3000 years (except for surface pollution by radioactive substances). Organic carbon content and aging of organic carbon thus can be studied in these sediments as well as in the Gavish Sabkha sediments over the past 3500 to 5000 years in organic-rich profiles of completely different composition as compared to peat swamps of the northern hemisphere. Also, sulfur isotope studies will be of utmost interest with respect to atmospheric sulfur pollution studies.

19.4 Discussion

19.4.1 Organic Carbon

The most conspicuous geochemical feature of the Gavish Sabkha ecosystem is the occurrence of extremely heavy organic carbon ($\delta^{13}C_{org} = -10.0 \pm 2.6‰$). These findings contrast markedly with the isotopic composition of average biogenic matter (both living and fossil) whose $\delta^{13}C_{org}$ values usually scatter between −20‰ and −30‰ with a mean of perhaps −26‰ to −27‰ (cf. Schidlowski 1982, 1983, Hayes et al. 1983). With $\delta^{13}C_{carb}$ averaging close to zero per mill, the mean isotopic fractionation between organic and carbonate carbon, $\Delta\delta_c$ (cf. Table 19.1, footnote b), is also between 26‰ and 27‰ on a global scale, while 23 coexisting C_{org}-C_{carb} pairs from the Gavish Sabkha listed in Table 19.1 have yielded $\Delta\delta_c = 11.0 \pm 2.5‰$. To account for this obvious anomaly, a brief excursion into the fundamentals of biological carbon isotope fractionation seems to be called for.

The magnitude of the fractionation effect inherent in the common photosynthetic pathways is, for the most part, a composite of the fractionations that occur at two cardinal steps in the assimilation chain, namely, (1) the diffusion of external CO_2 to the carboxylation sites within the photosynthetic tissue, and (2) the first enzymatic carbon-fixing reaction by which CO_2 is incorporated into the carboxyl (COOH) group of an organic acid (for recent overviews see Vogel 1980, O'Leary 1981, Schidlowski 1983). Since the assimilation rates commonly exceed several times those of potential dissimilatory (reverse) reactions (such as decarboxylation during photorespiration), the gross isotopic composition of photosynthetic organisms will be principally determined during these two steps which are graphically represented in Fig. 19.2.

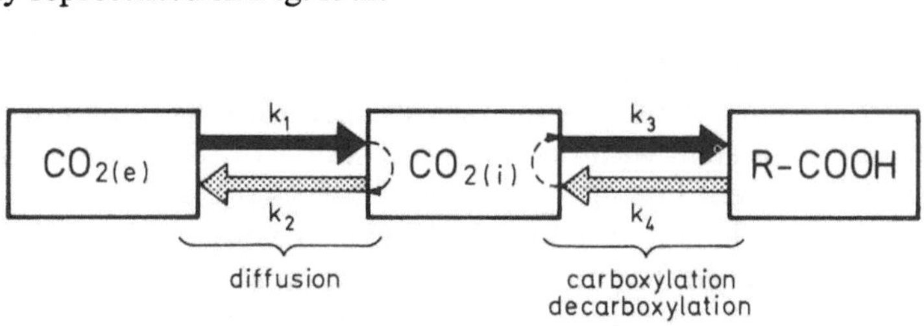

Fig. 19.2. The principal isotope-discriminating steps in the biochemical pathway of autothrophic carbon fixation (*black:* assimilatory reactions; *stippled*: dissimilatory and other reverse reactions). $CO_{2(e)}$ and $CO_{2(i)}$ stand for external and internal CO_2, respectively, k_1–k_4 are the rate constants assigned to individual reactions, and R-COOH represents the product of the initial CO_2-fixing carboxylation. In sum, these reactions lead to a marked enrichment of light carbon (^{12}C) in the cell substance relative to the parent CO_2 substrate. (From Schidlowski 1983)

The initial *diffusion step* entails a fractionation which, in first approximation, may be equated with the value for CO_2 diffusion in air (−4.4‰). Isotope discriminations inherent in the subsequent *enzymatic step* are substantially larger, notably in the case of the ribulose-1,5-bisphosphate (RuBP) carboxylase reaction which channels most of the carbon transfer from the inorganic to the organic world. Here, fractionations mainly span the range from −20‰ to −40‰, the variations observed depending on a number of external variables, such as metal cofactor availability, pH, salinity and temperature (O'Leary 1981, Winkler et al. 1982). With the far majority of photosynthetic biota relying exclusively on this pathway, the large fractionations of the RuBP carboxylase reaction (which feeds CO_2 directly into the Calvin cycle as a C3 compound, phosphoglycerate) have come to dominate the isotopic composition of the biosphere as a whole and, subsequently, the huge reservoir of fossil organic carbon stored in the Earth's sedimentary shell (Schidlowski et al. 1983).

A quantitatively less important enzymatic carboxylation reaction that fixes carbon via phosphoenolpyruvate (PEP) carboxylase as a C4 compound (oxaloacetate) is associated with a minor isotope effect of −2‰ to −3‰ relative to bicarbonate ion (HCO_3^-) that serves as the "active" inorganic carbon species in this process (O'Leary 1981, 1982). PEP carboxylase is the key carboxylating agent in the C4 (Hatch-Slack) pathway of photosynthesis and in plants utilizing Crassulacean acid metabolism. The intrinsic fractionations of some ferredoxin-linked carboxylation reactions (notably those fixing CO_2 as α-ketoglutarate and acetyl coenzyme A in the reductive carboxylic acid cycle of green sulfur bacteria) are as yet unknown.

When compared to the isotopic compositions of some major divisions of higher plants, algae and autotrophic bacteria (Fig. 19.3), the bulk of organic matter from the sabkha environments investigated largely coincides with the uppermost (positive) ranges displayed by C4 plants and, somewhat less pronounced, of selected groups of photosynthetic bacteria (including cyanobacteria). However, a fair number of the Gavish Sabkha values and almost all Solar Lake data markedly exceed the above ranges in positive direction. Since there is sufficient evidence that neither prokaryotic nor eukaryotic members of the two microbial ecosystems practice carbon fixation via the PEP carboxylase reaction (whose poor isotope discriminating properties in conjunction with the utilization of HCO_3^- account for the small overall fractionations in the C4 pathway), the observed coincidence with C4 plants is necessarily a fortuitous one. On the other hand, the partial encroachment of the measurements on the ranges of photosynthetic bacteria is consistent with the microbial source of the organic carbon.

Based on our current knowledge of the pathways of autotrophic carbon fixation entertained by the microbial world (Fuchs and Stupperich 1981), we may state with fair confidence that all eukaryotic and most prokaryotic members of sabkha-type microbial ecosystems sequester carbon via the RuBP carboxylase reaction and the Calvin cycle. This holds particularly for cyanobacteria and most anoxygenic photosynthetic bacteria which are the principal primary producers in such and related biotopes. There seem to be only two exceptions, namely (1) the green photosynthetic sulfur bacteria (Chlorobiaceae) operating the reductive carboxylic acid cycle, a form of assimilatory metabolism unique to photosynthetic

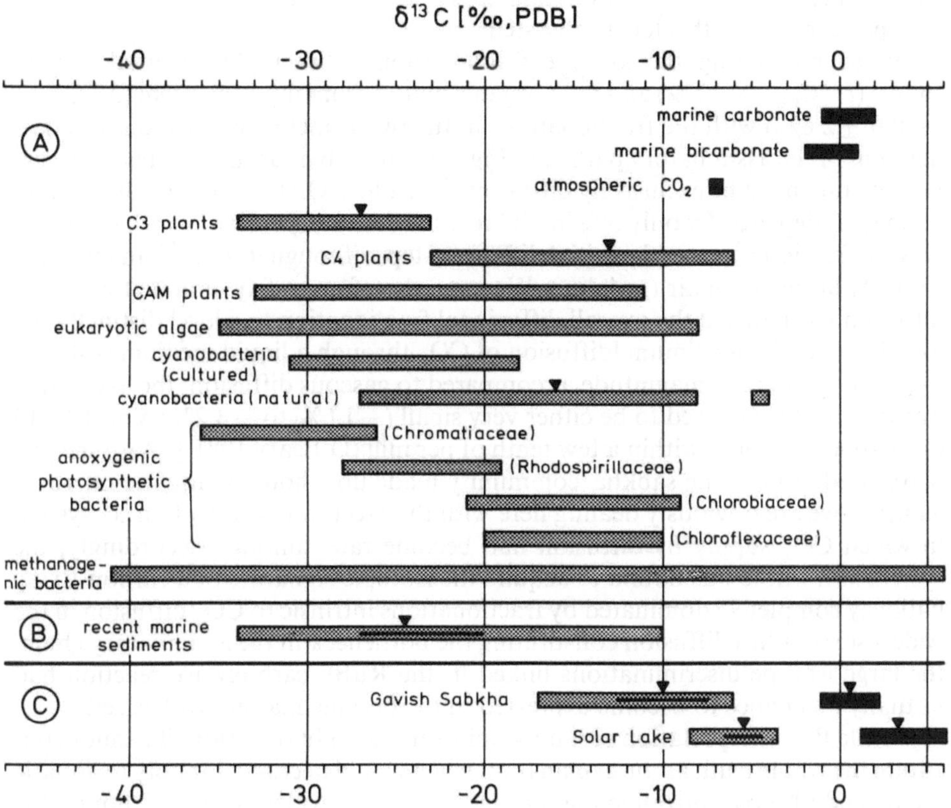

Fig. 19.3 A–C. Isotopic composition of major groups of higher plants and autotrophic microorganisms (**A**) compared to organic matter from recent marine sediments (**B**) and from the two hypersaline environments dealt with in this investigation (**C**) [**A** and **B** are adaptions from Schidlowski (1983) and Deines (1980), respectively; in **B**, > 90% of some 1600 data points represented by the graph lie on *inner black line*]. It is evident that the bulk of $\delta^{13}C_{org}$ values from the Gavish Sabkha and from Solar Lake (**C**) coincide with, or exceed, the positive extremes of the distribution spectra furnished by extant autotrophic organisms (means indicated by *triangles*, standard deviations by *black lines* inserted in total spreads). *Black bars on right side* represent isotope spreads of coexisting carbonates. While the carbonate values from the Gavish Sabkha largely coincide with normal marine $\delta^{13}C_{carb}$ distributions, those from Solar Lake are distinctly heavier

prokaryotes (Buchanan 1979), and (2) the chemoautotrophic methanogens believed to entertain a noncyclic (sequential) pathway based on some of the carbon-fixing reactions also utilized by green photosynthetic bacteria (Thauer and Fuchs 1979).

As can be inferred from the microbial inventory of the Gavish Sabkha given above (Sect. 19.2), C3 (Calvin cycle) photosynthesis prevails among the primary producers of this ecosystem. With this established, the crucial question arising is: *Why are the sizeable isotope fractionations of the RuBP carboxylase reaction not expressed in the average isotopic composition of organic matter from this particular environment?* There seems to be little doubt that these fractionations must have been suppressed by a set of external factors that consequently came to control the isotope economy of this local ecosystem.

When comparing the average fractionation between the external carbon source ($\delta^{13}C_{CO2} = -7‰$) and bulk organic matter from this ecosystem ($\delta^{13}C_{org} = -10.0 \pm 2.6‰$) with the fractionations in the two principal isotope-discriminating steps of the assimilatory pathway (Fig. 19.2) it is obvious that, in this case, carbon fixation must be exclusively diffusion-controlled. On the average, the sabkha biomass is depleted by only 3‰ in ^{13}C relative to atmospheric CO_2 which comes close to fractionations in the initial diffusion step, although it is less than the value for CO_2 diffusion in air (−4.4‰). We may, therefore, safely assume that *liquid* diffusion has reduced the overall diffusional fractionation to a level distinctly below the potential maximum [diffusion of CO_2 through a liquid medium is slower by about 5 orders of magnitude as compared to gaseous diffusion; the associated isotope effect is believed to be either very small (−1.6‰ to −3.2‰; Vogel 1980) or to approach unity within a few tenth of per mill (O'Leary 1981)]. With the primary producers of the sabkha community made up wholly of aquatic microorganisms, we are obviously dealing here with the exceptional case of an ecosystem in which CO_2 supply by diffusion had become rate-limiting. Accordingly, the system as a whole was bound to acquire the isotopic signature of a carbon-fixing pathway completely dominated by fractionations intrinsic to CO_2 diffusion in the widest sense. With diffusion constituting the bottleneck in the assimilatory chain, the large isotope discriminations linked to the RuBP carboxylase reaction had virtually no chance to become expressed in the organic matter synthesized.

While this interpretation certainly points in the right direction, it cannot constitute the whole truth as the explanation given would necessarily apply to aquatic plants and photoautotrophic microorganisms in the widest sense. Although the $\delta^{13}C_{org}$ spreads entertained by algae and bacteria from aquatic habitats are commonly skewed toward positive values and clearly overlap respective ranges from the Gavish Sabkha, their means are clearly tethered to the negative portions of the distribution spectrum (cf. Fig. 19.3). Hence, the factors generally responsible for the synthesis of heavy organics in the diffusion-limited pathway of aquatic photoautotrophs must have been considerably accentuated in the sabkha environment investigated.

There seems to be little doubt that the main factor responsible for such accentuation is the marked depletion of carbon dioxide in hypersaline biotopes. While slow diffusion of CO_2 in water generally sets the stage for an assimilatory scenario in which CO_2 supply may become rate-limiting, dramatically decreasing CO_2 solubilities in heliothermal brines are apt to force carbon definitely into the role of a limiting nutrient (a part normally played by phosphorus or nitrogen). As is illustrated by the solubility functions depicted in Fig. 19.4, concentrations of CO_2 rapidly decrease in aqueous solutions with increasing temperatures and salinities. With average temperatures in Sinai coastal pools fairly high during most of the year, the higher isotherms depicted in Fig. 19.4 probably give the best reflection of actual CO_2 solubilities in such environments.

It follows from the relationships summarized in these diagrams that carbon dioxide will become a rare commodity in hypersaline pools. The low stationary CO_2 concentrations in such biotopes are, moreover, excessively drained as a re-

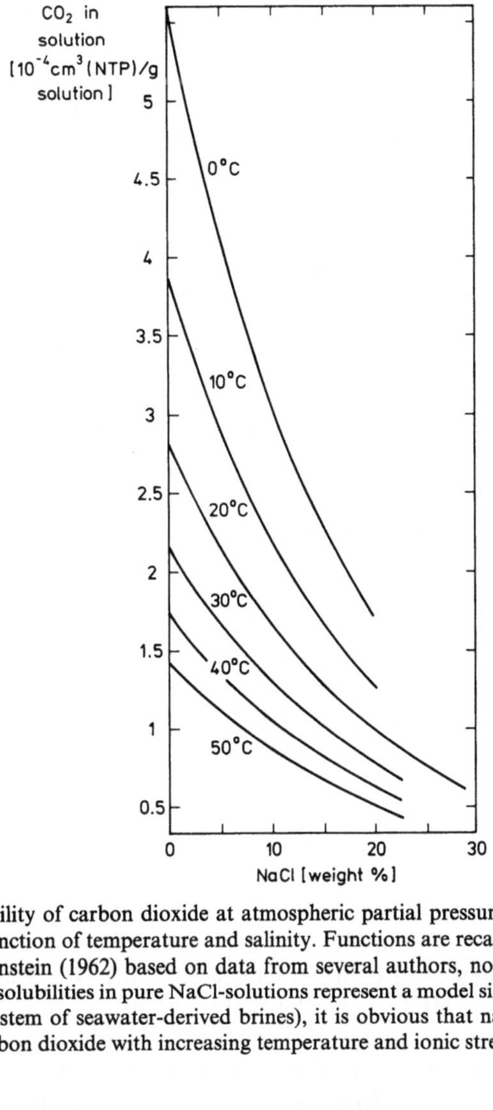

Fig. 19.4. Solubility of carbon dioxide at atmospheric partial pressure (0.003 atm) in aqueous NaCl-solution as a function of temperature and salinity. Functions are recalculated from a graphic synopsis in Landolt-Börnstein (1962) based on data from several authors, notably Harned and Davis (1934). Although CO_2 solubilities in pure NaCl-solutions represent a model situation (as compared to the multicomponent system of seawater-derived brines), it is obvious that natural waters will be depleted in carbon dioxide with increasing temperature and ionic strength of the solution

sult of high productivities typical of sabkha ecosystems, replenishment by the overlying atmospheric reservoir possibly lagging behind equilibrium due to slow diffusion of CO_2 in concentrated salt solutions (Ratcliff and Holdcroft 1963). (Such lags should be particularly pronounced during microbial blooms that impose heavy strains on the local CO_2 reservoir.) Under conditions of an extreme shortage of carbon dioxide, excessive isotope discrimination would become an intolerable luxury for the biological community as a whole; hence, more and more heavy carbon is likely to enter the assimilatory pathway as the ambient inorganic carbon reservoir approaches exhaustion. In markedly eutrophicated environments such as sabkhas the trend to utilize heavy carbon will be further enhanced by high rates of productivity which under these conditions could not be sustained safe by a largely indiscriminate acceptance by primary producers of all inorganic carbon suitable for metabolization (in general, well-nourished plants display more positive $\delta^{13}C$ values than those grown in nutrient-deficient environments, see O'Leary 1981).

Though less pronounced in magnitude, a distinct correlation between decreasing CO_2 availability and decreasing isotope fractionation has also been recorded from culture experiments with populations of marine diatoms (Degens et al. 1968), which relationships was subsequently invoked to account for the wide range of isotopic compositions observed in oceanic phytoplankton (−9‰ to −31‰; Deuser et al. 1968). In these experiments, maximum fractionations of −18‰ to −19‰ vs dissolved CO_2 ($\delta^{13}C_{CO2}(aq) \simeq -8‰$) were obtained under conditions of excess availability of CO_2 and/or HCO_3^-, i.e., when the inorganic carbon reservoir was virtually unaffected by carbon fixation rates in photosynthesis. Conversely, with inorganic carbon exhausted to levels approaching the turnover rates of CO_2 in the biological cycle, fractionations were reduced to an apparent minimum of about −6‰ vs CO_2 (or −14‰ vs HCO_3^-, respectively, as equilibrium fractionations between carbon dioxide and bicarbonate amount to some 7‰ to 10‰ in the range 0° to 30 °C, cf. Mook et al. 1974). As is evidenced by the substantially smaller discriminations shown by our data (Tables 19.1 and 19.2), this apparent minimum fractionation between cell carbon and substrate does not constitute an absolute barrier, as suspected by Deuser et al. (1968), being clearly amenable to further reduction in extremely carbon-deficient habitats. Further demonstrations of the CO_2 concentration effect on carbon isotope fractionations by aquatic bacteria and algae in both artificial (culture) and natural environments have been given by Calder and Parker (1973), Pardue et al. (1976), Vogel (1980), Estep (1982) and Mizutani and Wada (1982).

It seems appropriate to point out that neither salinity nor temperature or pH *as such* seem to be responsible for the observed reduction of biological carbon isotope fractionations [1] in sabkha-type biotopes, but that these factors primarily combine to bring down carbon dioxide concentrations in these environments to substantially lower levels (high salinities and temperatures will directly reduce CO_2 solubilities, while changes in pH affect CO_2 availability via the equilibrium $CO_2+H_2O \rightleftharpoons H_2CO_3 \rightleftharpoons H^+ + HCO_3^-$). As the carbon supply of microbial photoautotrophs in such hypersaline habitats is consequently diffusion-limited, the large fractionations of the RuBP carboxylase reaction of the Calvin cycle should necessarily remain cryptic (as they do in the "compartmented" biochemistry of the C4 pathway; hence the coincidence of our values with those of C4 plants, cf. Fig. 19.3). On the other hand, it has been demonstrated that, with increasing CO_2 pressures, the enzymatic effect is bound to emerge because the rate-limiting step in the assimilatory pathway shifts from diffusion to carboxylation (Fig. 19.5). It is worth noting that Estep et al. (1978 a, b) have reported values between −28‰ and −39‰ for carbon isotope fractionations obtained during in vitro experi-

1 Parenthetically, mention should be made of in vitro experiments with RuBP carboxylase that have revealed a distinct correlation between pH and the magnitude of carbon isotope fractionation in this enzymatic reaction, with discriminations increasing from −18‰ to −35‰ over the pH-range 7.5 to 8.5 (Winkler et al. 1982). However, the pH relevant in this case is primarily that of the *physiological,* not the *external,* environment

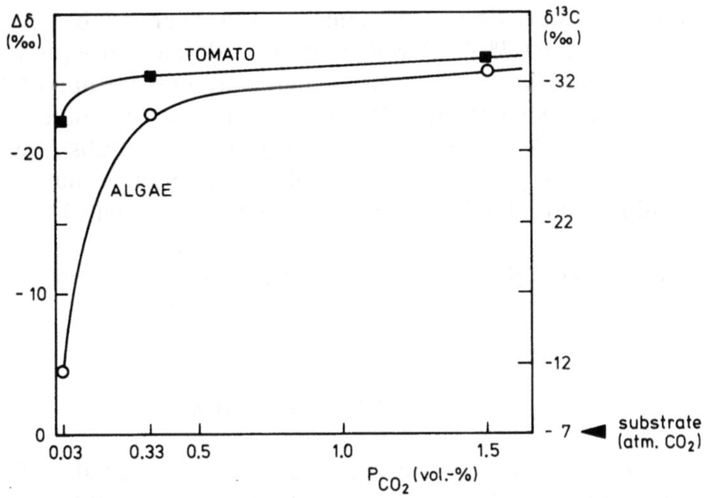

Fig. 19.5. Carbon isotope fractionation in higher plants (tomato) and aquatic algae (*Chlorella* sp.), both utilizing the C3 pathway, as a function of partial pressure of carbon dioxide in ambient air (adapted from Vogel 1980). At atmospheric P_{CO2} (0.03%), fractionation in algae is completely diffusion-controlled (difference $\Delta\delta$ between substrate and cell material $\lesssim$4‰), while the $\Delta\delta$ of tomato plants reflects the formidable fractionation of the RuBP carboxylase reaction. Increasing CO_2 pressures lead to but small increases of $\Delta\delta$ in the enzymatically controlled pathway of higher plants, while exercising a profound impact on the isotopic composition of algae as the rate-limiting step is shifted from diffusion to carboxylation (with a concomitant expression of the large fractionations typical of the enzymatic reaction)

ments with RuBP carboxylase isolated from selected eukaryotic (diatoms) and prokaryotic microorganisms (the latter including *Agmenellum* sp., a coccoid cyanobacterium).

An additional effect responsible for a retarded diffusion of CO_2 into the photosynthetic tissue of the microbial components of benthic mats was most probably exercised by the protective layer of polysaccharide slimes that usually cover sedentary assemblages of prokaryotes (see also Chap. 11). It is a common observation that mat-forming microbiota (notably cyanobacteria) are dispersed in their extracellular slime secretions rather than being in direct contact with the surrounding aqueous environment in the way planktonic forms do (even here, formation of slimy aggregates has been widely noted during plankton blooms, e.g., of *Macrocystis* sp.). It is well known that the diffusivity of gases in such slimes (as generally in polysaccharide solutions) is most drastically reduced, which is also the reason for the observed retention of volatile metabolites – in the form of conspicuous gas bubbles – in microbial mat systems (allowing, inter alia, the coexistence of thermodynamically incompatible gas species such as oxygen and hydrogen sulfide in separate vacuoles; see Krumbein et al. 1979 a).

The $\delta^{13}C_{org}$ mean of -5.4 ± 1.1‰ obtained for the Solar Lake samples (Table 19.2) is still more disconcerting than the respective average for the Gavish Sabkha in that it seems to indicate an *inverse fractionation* of the stable carbon isotopes (with a ^{13}C enrichment in organic matter) if we assume $\delta^{13}C_{CO2} = -7$‰ as value for the carbon source. This apparent anomaly is, however, not necessarily real since coexisting carbonates also appear to be isotopically heavy ($\delta^{13}C_{carb}$ = +3.5±1.3‰). Dissolved carbon dioxide in equilibrium with such carbonates should be about 7‰ to 8‰ lighter at temperatures of 30 °C (cf. Mook et al. 1974), the coexisting CO_2(aq) thus having approximately ranged between −3.5‰ and −4.5‰. With this composition of the substrate, discrimination against ^{13}C is clearly evident also in Solar Lake organic matter as it was in the Gavish Sabkha samples, but the magnitude of the biological isotope effect is further reduced ($\Delta\delta_c = \delta^{13}C_{carb} - \delta^{13}C_{org}$ is 8.9±2.0‰ for Solar Lake, while it had been 11.0±2.5‰ in the case of the Gavish Sabkha and amounts to 25–27‰ for average terrestrial carbon).

19.4.2 Carbonates

As is indicated by the observed spread of $\delta^{13}C_{carb}$ values from the Gavish Sabkha (−3.5‰ to +4.7‰), and notably their mean of 0.6±1.7‰, the carbonates from this environment (Table 19.1) are basically marine by common standards, an interpretation consistent with the seawater seepage model for the origin of this coastal pool. Moderately positive excursions may be best explained as reflecting stagnant interludes in the subrecent history of the brine pool (see heavier values interspersed in core samples No. 2.01 to 2.21) or, in the case of surficial samples, stagnant conditions in local compartments of present surface waters. With moderately heavy carbonates well presented also among the surface samples (see values with sampling depth of 0 cm in Table 19.1), slightly stagnant conditions seem to prevail at least in portions of the present pool. The somewhat aberrant value of −3.5‰ yielded by No. 6.02 is probably due to an incorporation of biogenic CO_2 into the bicarbonate precursor of the present carbonate sample.

In contrast, the consistently heavy values yielded by the Solar Lake carbonates ($\delta^{13}C_{carb}$ = +3.5±1.3‰, see Table 19.2) would suggest that the influx of seawater into this environment was severely impeded, the lake consequently developing the "Lomagundi-type" isotopic characteristics of secluded basins (see Schidlowski et al. 1976). If, due to retarded seawater seepage, the exchange of carbon between a coastal pool and the huge reservoir of marine bicarbonate is substantially reduced, communication with the global exchange reservoir can only be maintained over the gas phase, which is largely insufficient to ensure isotopic equilibrium with the bulk of inorganic carbon of the global environment. With the influx of atmospheric carbon dioxide into such basins heavily curbed as a result of reduced CO_2 solubilities in hypersaline waters, and ^{12}C overrepresented in the outgoing sedimentary carbon flux due to preferential removal with organic carbon, the isotopic composition of the local bicarbonate reservoir will be pushed toward heavy values which are subsequently monitored by the $\delta^{13}C_{carb}$ values of precipitated carbonates. The salient information conveyed by isotopically heavy carbonates is, therefore, that of a limited decoupling of a local carbon pool from the global exchange reservoir (which latter is largely identical with that of marine bicarbonate as this surpasses the mass of atmosphere CO_2 by almost two orders of magnitude).

A restricted basin model for the Solar Lake is further substantiated by the $\delta^{18}O_{carb}$ mean of +35.2+0.6‰ (SMOW) yielded by the carbonates. While this mean is about 6‰ more positive than that of average marine carbonates of Recent and Quaternary age, it coincides with the oxygen isotope spreads reported for other evaporite ponds and lagoons from different localities (see, for instance, Degens and Epstein 1964). In contrast, the $\delta^{18}O_{carb}$ average of $+32.1 \pm 3.3$‰ (SMOW) found for the Gavish Sabkha samples still falls into the upper range of modern carbonates, attesting to a marine affiliation of this facies (with a minor slant toward semi-secluded conditions) that was also evidenced by the bulk of the $\delta^{13}C_{carb}$ values from this environment.

19.5 Implications for the $\delta^{13}C_{org}$ Record of Ancient Stromatolites

It is obvious from the synopsis of terrestrial carbon isotope data given in Fig. 19.3 that the small fractionations maintained in the local carbon cycle of sabkha environments make virtually no impact on the isotope economy of the global carbon budget, notably the $^{13}C/^{12}C$ ratios of organic matter in recent marine sediments. This is fully consistent with the fact (largely self-evident) that hypersaline environments represent miniscule side stages in the contemporary carbon cycle, and that the contributions rendered by their locally exuberant microbial communities are negligible on a global scale. While this statement certainly holds for the present world, we may safely assume that ecological scenarios were profoundly different over most of the geological past.

There is abundant evidence that microbial ecosystems have, in fact, dominated the biosphere during the first 3 billion years of recorded Earth history, with the benthic variant well established since about 3.5 Ga ago, as testified by an impressive stromatolite record (see Awramik 1982, Walter 1983 for recent overviews). It seems reasonable conjecture to envision for such primitive (largely prokaryotic) global ecosystem a state of plenitude, i.e., a situation in which bacteria and algae had proliferated in existing aquatic habitats to ultimate limits set by the availability of environmental resources, notably phosphorus and nitrogen as critical nutrients and light as limiting factor. Planktonic and benthic microorganisms were, therefore, the sole sources of fossil organic matter stored in Precambrian sediments in about the same proportion (0.5–0.6%) as in geologically younger rocks (Schidlowski 1982).

This being the case, it had been noted since long (e.g., Behrens and Frishman 1971, Schopf et al. 1971, Calder and Parker 1973, Eichmann and Schidlowski 1975) that organic carbon from Precambrian stromatolites differed markedly in its isotopic composition from that of modern analogs. While $\delta^{13}C_{org}$ values of ancient stromatolites fall into the range of average organic carbon (e.g., -28.5 ± 3.6‰ for 22 samples reported by Eichmann and Schidlowski 1975), the isotopic composition of extant and subrecent microbial mat communities was shown to be mostly slanted toward heavy values, the extremes being exemplified by the "superheavy" organic matter from the Gavish Sabkha and Solar Lake reported in this chapter. With the insight gained in the course of the present investigation, there may be some simple explanation for this discrepancy.

It is generally accepted that, with the emergence of metazoan heterotrophs some 0.7 Ga ago (cf. Cloud 1976) and the subsequent adaptive radiation of grazing and burrowing animals, the "classical" microbial ecosystem of the Precambrian era suffered a marked decline. As a result of the grazing stress imposed since then on microbial communities in all "normal" aquatic (notably marine) habitats, bacterial and algal mat-builders failed to form stromatolite-type carpets except in niches hostile to potential predators, the prototype of such sanctuaries being hypersaline basins. Accordingly, the bulk of potentially stromatolite-forming environments in the present world are characterized by an extreme degree of salinity. As has been fully set out above, reduced solubility of CO_2 in hypersaline waters (Fig. 19.4) keeps this commodity in check in habitats otherwise abounding in nutrients, thus giving rise to an exceptional situation in which the carbon supply and decreased diffusivity become rate-limiting for the productivity of a local ecosystem. Also, the photosynthetic community of sedimentary environments is considerably condensed. In contrast to plankton blooms, the productivity of benthic microbial mats is concentrated in a very narrow light-limited zone in which productivities occur that either match or outcompete the productivity of the water column above. Furthermore, benthic mats create a distinct borderline between the overlying water body and the interstitial water. This borderline is further accentuated by the fact that the extracellular polysaccharides create a diffusivity barrier of considerable magnitude.

There is, accordingly, little doubt that the large-scale suppression of the isotope-discriminating potential of the common assimilatory pathways prevailing under these conditions ultimately derives from the depletion of carbon dioxide to levels approaching the turnover rates in the biological cycle of such habitats. Since, on the other hand, the same microbial communities had been free to spread throughout "normal" marine environments during the Precambrian (rather than being locked up in hypersaline sanctuaries), the CO_2 concentration effect was irrelevant to the carbon-fixing activities of these ancient forms. Hence, it should not be surprising that organic matter from Precambrian stromatolites displays magnitudes of fractionation that are commonly associated with photosynthetic carbon fixation and clearly dominated by the large enzymatic effect of the RuBP carboxylase reaction of the C3 pathway. As pointed out above, C3 (or "Calvin cycle") photosynthesis is operated by the bulk of microbial primary producers of these ecosystems, a noteworthy exception being green photosynthetic sulfur bacteria (Chlorobiaceae).

It should be stated explicitly that the *ultimate* reason for the observed reduction of biological carbon isotope fractionations in these habitats is the exhaustion of the local CO_2 reservoir, with high salinity values merely being the agent inducing such condition. In other microbial habitats, low CO_2 pressures may be due to different causes. For instance, Estep (1982) was able to show that thermophilic microorganisms growing in hot springs of the Yellowstone National Park, U.S.A., were isotopically heavy when CO_2 concentrations in effluent dissolved gases were low, whereas the same populations yielded lighter carbon with P_{CO2}

in the emanations rising to higher levels. Incidentally, the CO_2 concentration effect (in the form of a postulated CO_2-enriched Precambrian atmosphere) has also been invoked by Mizutani and Wada (1982) in efforts to explain the larger fractionations in the early stromatolite record as compared to the bulk of "modern" stromatolitic environments.

The amount of outgassing of the crust during the Precambrian has been discussed at various occasions. Veizer et al. (1982) have stressed the importance of mantle buffering of the world ocean during the Precambrian without giving data on carbon dioxide. It can be expected, however, from carbon dioxide concentration data in soil atmospheres and in recent sedimentary decay environments, that even under salt-walter conditions the carbon dioxide concentration can be increased by a factor of 10. If we assume that several possibilities exist for higher carbon dioxide levels in the Precambrian (based, e.g., on more extensive outgassing, less buffering by the biosphere, and different biological cycles in the absence of eukaryotic life), we may expect that the Precambrian stromatolitic environment was exposed to higher concentrations of carbon dioxide and thus less limited in terms of diffusivities of this gas into benthic microbial mats. This might provide an additional explanation for the occurrence of lighter biogenic matter in Precambrian stromatolites.

We suggest that subsequent investigations of contemporary stromatolite-forming habitats test our proposal that modern benthic microbial ecosystems are generally confined to CO_2-deficient environments (mostly of the hypersaline type), or whether additional factors are needed to account for the difference in isotopic composition between fossil stromatolites and the bulk of their modern analogs.

These could be lower productivities and/or higher carbon dioxide concentrations in the Precambrian, nonphotosynthetic origin of organic carbon compounds in Precambrian stromatolites or, less likely, different carbon-fixing and fractionating pathways of the ancient (perhaps still anoxygenic) photosynthetic communities.

Note Added in Proof

Another series of extremely heavy carbon isotope values from Solar Lake mats ($\delta^{13}C_{org}$ between -5.3 and -8.6‰) has been most recently reported and discussed by Z. Aizenshtat, G. Lipiner and Y. Cohen [in Y. Cohen et al., *Microbial Mats: Stromatolites* (MBL Lectures in Biology *3*), New York: A. R. Liss, Inc., 1984, pp. 281–312]. For a brief general discourse on the problem of isotopically "superheavy" organic carbon from hypersaline microbial mats see also the article by M. Schidlowski, U. Matzigkeit und W. E. Krumbein in *Naturwissenschaften 71,* pp. 303–308 (1984).

D. Applied Aspects and Paleoecology

20. Introduction

Wolfgang E. Krumbein

The study of the biosphere and specific biogeochemical systems delivers keys not only to the past but also to the future. In a slight change of wording we may say that bygoing biospheres are smoothly shifting and altering the physiognomy of Earth's surface.

The impact of biological systems on the geology and geochemistry of the Earth crust is tremendous. Examples like (1) the "flying mountain" of mineral material in the form of a locust swarm passing the Red Sea in 1888, which equaled the total amount of tin and copper mined in the whole nineteenth century (i.e. 4.4×10^7 tons) (Carruthers 1890), (2) the hot brines welling up in the "Atlantis Deep" of the Red Sea or in the "Hot Vents" of the Pacific creating numerous very specific biota and ecosystems which are basically driven by chemical energy and pile up huge amounts of heavy minerals, and (3) coral reefs that form huge "Goldwasher Sieves" in the framework of which organics and inorganic compounds are trapped, enriched and selectively deposited are very impressive. Only now do we begin to understand the snowball effect of the avalanches of "Marine Snow" sinking and sliding down the slopes of the continents. Each individual particle of marine snow turned out to be a heavy metal processing "factory" with its workers, catalytic surfaces, bioreactors and so forth.

Krumbein (1982) has used the picture of a huge "butter churn" operating in nearshore estuarine and coastal environments. This picture, however, has been brought forward for shallow-water hypersaline systems and their biological mineral resources production potential in geological and economic relation several times before. The context was made for hypersaline stromatolitic systems (Muir 1979) and in economic geology (Renfro 1974). Dexter-Dyer et al. (1984) have connected even such precious metals as gold and uranium with the trapping potential of hypersaline sabkha-type microbial mats. The examples in most cases were stemming from huge Precambrian ore deposits and their relation to microbial mats of hypersaline origin. They have been partially verified, however, by laboratory experiments with recent potential stromatolites from the Gavish Sabkha and Mexican sabkha systems.

Thus in Chapter 21 the paleoecological importance of the study of recent sabkhas and their comparison to ancient systems is stressed, while in Chapter 22 it is tried to give a summary of the applied aspects so far identified and an outlook to future analyses of sabkha systems in two directions.

Firstly, an attempt is made to circumscribe the potential for ore deposit recognition and functional understanding of sedimentary ore and oil deposits on the background of sabkha systems and, secondly, it is tried to give a prognosis of possible biotechnological and bioengineering aspects of sabkha-type microbiology. These are based on the fact that these microbiota live in strongly supersaturated cation and anion solutions and selectively trap some of them, while others are passing through the system completely unaffected. A situation well known to us from biotechnology and hydrometallurgy.

21. A Paleobiological Perspective on Sabkhas

ANDREW H. KNOLL

At first glance, coastal sabkhas appear to be singularly esoteric environments, hardly meriting the painstaking geochemical and ecological study they have received in recent years. Sabkhas are forbiddingly arid, extremely limited in geographical extent, and notable biologically largely for the absence of most organisms that figure prominently in twentieth centuary bestiaries. The microbial mat communities characteristic of sabkhas are easily disrupted by minor environmental fluctuations, leaving the casual observer with the impression that these ecosystems are fragile and geologically evanescent. It is in fact the geological record that gives lie to this impression, documenting the persistence of arid coastal environments and their biotas in both ecological and evolutionary time and demonstrating the critical importance of an understanding of modern sabkha communities for the interpretation of the Earth's earliest records of life.

Coastal evaporites have long been recognized in the Phanerozoic sedimentary record. Historical geology textbooks commonly discuss the Zechstein Sea, the Paradox Basin, and the anhydrites of the Carboniferous Limestone (see Schreiber 1978, A. C. Kendall 1979 for surveys of Phanerozoic evaporites). The same texts, on the other hand, make scant mention of Precambrian arid zone deposits. As recently as 1964, F. Lotze could summarize an authoritative discussion of the question with the statement that "evaporites of proved Precambrian age are not yet known."

The recent blossoming of Precambrian stratigraphy has changed this viewpoint dramatically. Table 21.1 lists Precambrian sabkha successions ranging in age from those found in the Warrawoona Group of Western Australia, at 3,500 Ma the oldest negligibly metamorphosed supracrustal sequence known (Barley et al. 1979, Buick et al. 1981), to the Precambrian/Cambrian boundary evaporites of the Salt Range, Pakistan, and elsewhere. Many of the formation names listed are familiar ones to paleontologists because they are associated with silicified microfossil assemblages that have been of primary importance in the elucidation of life's early history.

On the present day Earth, microbial mats and potential stromatolites are largely confined to sabkhas and a few other restricted coastal environments, hot springs, and geochemically anomalous or otherwise environmentally hostile lakes. This areal restriction is not a consequence of the physiological requirements of mat-building microorganisms; most, if not all, of the cyanobacterial species found in sabkhas are quite capable of growth in normal seawater. Rather, the distribution of microbial mats is correlated with the metabolic capabilities of metazoans. Animals graze on mat populations, disrupt mat formation by bur-

Table 21.1. Selected Precambrian coastal evaporites. Criteria for inclusion are evaporite minerals (or pseudomorphs of evaporites) *and* sedimentological evidence of peritidal depositional environments

Geological unit	Age (Ma)	Location	Reference
Desu Fm.	550–600	Iran	Huckreide et al. (1962)
Punjab Saline Fm.	550–600	Pakistan	Schindewolf and Seilacher (1955)
Hongchunping Fm.	550–650	China	Awramik (pers. comm.)
Noab Fm.	550–650	Namibia	Hartnady (1978)
Narssârssuk Fm.	700	Greenland	Strother et al. (1983)
Bhandar Fm.	700–900	India	Singh (1980)
Society Cliffs Fm.	700–1,100	Baffin Island, Canada	G. D. Jackson and Iannelli (1981)
Shaler Group	700–1,200	Canada	Young (1981)
Burra Group	800	Australia	von der Borch (1980)
Callanna Beds equivalents	800–1,000	Australia	Plumb et al. (1981)
Bitter Springs Fm.	850	Australia	Stewart (1979)
Nosib Fm.	900–1,000	Namibia	Tankard et al. (1982)
Belt Supergroup	1,100–1,400	USA	Ross (1963)
Dismal Lakes Group (Unit 13)	1,200	Canada	Kerans et al. (1981)
Hornby Bay Group (Unit 9)	1,200–1,600	Canada	Kerans et al. (1981)
Wumishan and Yanzhuang Fms.	1,200–1,400	China	Awramik (pers. comm.)
Gaoyuzhuang Fm.	1,400–1,500	China	Zhang (1981)
Balbirini Fm.	1,400–1,575	Australia	Oehler (1978)
Batten Subgroup	1,400–1,575	Australia	Plumb et al. (1981)
Umbolooga Subgroup	1,400–1,575	Australia	Muir (1979)
Karns Fm.	1,400–1,575	Australia	Plumb et al. (1981)
Frere Fm.	1,590–1,710	Australia	Goode (1981)
Upper Mt. Isa Group	1,680–1,750	Australia	Neudert and Russell (1981)
Quilalar Fm.	1,680–1,750	Australia	Plumb et al. (1981)
McNamara Group	1,680–1,750	Australia	Plumb et al. (1981)
Flicking Group	1,680–1,750	Australia	Plumb et al. (1981)
Malbon Group	1,720–1,800	Australia	Plumb et al. (1981)
Umkondo Group	1,800	Zimbabwe	Button (1977)
McLeary Fm.	1,900	Canada	Ricketts and Donaldson (1981)
Kasegalik Fm.	1,900	Canada	Ricketts and Donaldson (1981)
Kona Fm.	1,900–2,100	Canada	Larue (1981)
Batchelor Group	2,000–2,200	Australia	Crick and Muir (1980)
Malmani and Campbellton Subgroups	2,100–2,300	South Africa	Tankard et al. (1982)
Manjeri Fm.	2,700	Zimbabwe	Martin et al. (1980)
Onverwacht Group	3,400–3,500	South Africa	Lowe (1982)
Warrawoona Group	3,400–3,500	Australia	Barley et al. (1979)

rowing, and generally outcompete cyanobacteria for available space on the shallow sea floor. Thus, mats flourish in environments from which potentially disruptive metazoans are excluded, often because of high salinity. Multicellular animals are not necessarily excluded in toto from areas in which potential stromatolites are accreting. Sea snakes ply the water of Shark Bay, Australia, and nematodes and other meiofauna are common in modern mats. It is the *disruptive* fauna whose distribution is antisympathetic to that of microbial mats.

Logically, one might expect that prior to the appearance of multicellular (tissue grade) organisms, microbial mats were much more widely distributed than they are today, and the Precambrian sedimentary record demonstrates that this was indeed the case. Stromatolites record the former presence of microbial mat-building communities across a broad spectrum of supratidal to subtidal shelf environments. The initial decline in stromatolite abundance and diversity coincides in the fossil record with the Ediacarian radiation of soft-bodied metazoans (Awramik 1971, Cloud and Glaessner 1982), and significant further decline correlates with the Ordovician radiation of well-skeletonized sessile suspension feeders, such as articulate brachiopods, bryozoa, crinoids, and corals (Pratt 1982). Even today, the first organisms to colonize a patch of reef destroyed by a tropical storm are blue-greens. Later, they are displaced by overgrowing populations of epifaünal invertebrates (Krumbein, pers. comm.), providing a successional recapitulation of a stratigraphic replacement pattern.

The widespread distribution of stromatolites in Precambrian shallow marine environments has prompted biologists to consider modern sabkha ecosystems mainly as general analogs to ancient benthic systems; however, they are more than that. The most valuable records of ancient microbial mats are those in which the actual mat microbiotas, or parts of them, are preserved as permineralized microfossil associations. These are but a small subset of all Precambrian stromatolites, and they consist in large part of flat laminated or mamillate structures from restricted coastal environments. Thus, modern sabkhas are not only the best available analogs of Precambrian stromatolitic ecosystems in general; they are precise environmental homologs of some of the best-known and paleontologically most significant early microfossil assemblages. It is appropriate to ask why sabkhas have provided favorable environments for microfossil preservation, while many other shelf settings have not.

21.1 Sabkhas and Microfossil Preservation

Following the death of an organism, most of its constituent organic components are rapidly decomposed to the CO_2, H_2O, and inorganic nutrients from which they were originally synthesized. The cycling of biologically important elements is the most fundamental process occurring in ecosystems, and without it, life on Earth could not persist. However, the vast stores of organic matter contained in sedimentary rocks demonstrate that the carbon cycle is leaky and that, under appropriate conditions, some organic matter can escape complete decomposition. Microfossil preservation depends on the early termination of organic degradation.

Microbial decomposition can proceed under either aerobic or anaerobic conditions, but cellular preservation is likely only when oxygen is excluded from the ambient diagenetic environment. Complete anaerobic degradation requires several specialized decomposers operating in succession (Golubic and Barghoorn 1977), and some biological compounds are difficult or impossible to break down in the absence of aerobic respiration. These two considerations suggest that microfossil preservation will be enhanced under environmental conditions that exclude one or more members of the anaerobic decomposer chain, and that the remains which escape complete degradation will be biased in favor of organisms that produce degradationally recalcitrant structures.

The brief statements of the preceding paragraph go a long way toward explaining the biological nature and paleoenvironmental distribution of Precambrian stromatolitic microfossils. Abundant organic matter produced by photoautotrophs in the surface layers of microbial mats serves to deplete oxygen within a few mm of the mat surface. Thus, as a potential stromatolite accretes, cells and extracellular material pass quickly into the anaerobic zone. Hypersaline groundwater circulating through accumulating microbial sediments may exclude one or more members of the anaerobic decomposer chain, preventing complete degradation. Clearly, restricted coastal environments possess two characteristics that together favor preservation. This stands in strong contrast to many open shelf areas where well oxygenated pore waters prevent cell preservation. (For an extended discussion of cyanobacterial mat preservation, see Bauld 1981 a).

Cyanobacteria that live in arid coastal zone mats often secrete tough extracellular sheaths and envelopes. The relative resistance of these structures to autolysis and microbial destruction further enhances the probability of microfossil preservation in sabkhas and hypersaline embayments. It also determines, in part, the nature of preserved assemblages. Sheaths are the predominant structurally preserved remains in Precambrian stromatolitic microbiotas. The actual cellular material originally encompassed by the sheaths is almost invariably shrunken or absent (Figs. 21.6 D, E, 21.7, 21.8, and 21.9 B, C, G, I), although in exceptional instances cellular preservation can be excellent (e.g., the oscillatorian trichomes in Schopf 1968, Fig. 21.9 A). Similarly, eukaryotic and putatively eukaryotic unicells found in mat assemblages are preserved as empty vesicles or as walls containing a small cytoplasmic residue. This differential preservation of Precambrian microorganisms agrees well with observations made on modern mats. A rich diversity of bacterial and other species may be present in surface or near-surface mat layers, but most of these species are no longer represented in more deeply buried mat horizons. What one does find is an enrichment of cyanobacterial envelopes, along with scattered protist cysts and some remains of animals (e.g. *Artemia* mouth parts) and algae (e.g., fragments of diatom frustules).

In an attempt to understand the post-mortem information loss attendant to the fossilization of Atlantic Coastal Plain invertebrate communities, Lawrence (1968) developed an interesting means of comparing his Oligocene oyster beds with modern communities dominated by taxonomically related oysters. The modern community is far more diverse than the fossil assemblage; however when Lawrence subjected a species list of the present community to "theoretical fossilization," he found that the various taphonomic biases accompanying fossilization selectively removed groups of taxa, leaving at the end of the procedure a much-shortened list of species likely to be recorded in the fossil record. This list compares closely with the list of fossils actually found in the Oligocene deposit, suggesting that early Tertiary oyster communities were similar in composition to those still inhabiting coastal environments. A comparable analysis of living and fossil microbial mat communities leads to similar conclusions. On the basis of information on microbial fossilization gleaned from studies of modern mats, one can predict that preservation in arid coastal zones should be good, that blue-greens should dominate fossil assemblages, that many of the bacterial species present in modern mat associations should be missing from the fossil record, and that sheaths should predominate over well-preserved cellular remains. This is exactly what one finds in Precambrian deposits, and it prompts an inference much like that made by Lawrence. In Proterozoic mats, at least, the biological constitution of microbial mat communities was probably quite similar to that of modern communities in comparable environments.

One cannot conclude a discussion of microbial fossilization without making some mention of silicification. Microorganisms do not require petrifaction for preservation; cyanobacterial sheaths and the robust reproductive cysts of planktonic algae (another taphonomic enrichment) are widespread in Upper Proterozoic siltstones and shales (Timofeev 1969, Vidal 1976). Poorly preserved filamentous microfossils can also be observed occasionally in Precambrian stromatolitic carbonates (Horodyski 1975); however, it is only in carbonaceous cherts that the detailed preservation of sabkha microbial mat assemblages has been found to occur. The silica in iron formations such as the Gunflint may be primary, but the siliceous lenses, pods, and discontinuous beds found in Precambrian carbonate successions are, in most if not all cases, diagenetic replacements of originally calcareous sediments. This can be demonstrated in the field where it can often be seen that cherty areas cut across bedding planes. The diagenetic origin of the silica can also be confirmed petrographically.

As in the petrifaction of wood, silicification of microbes does not involve the replacement of organic material by SiO_2. Rather, the partially decomposed microorganisms are filled in and surrounded by a silica matrix. Because it is relatively difficult to recrystallize, compress, or shear, chert preserves delicate organic structures in three-dimensional detail. Carbonates are much more easily recrystallized and far more permeable. Hence, it is small wonder that while cherty areas of formations like the Bitter Springs and Draken often contain abundant microfossils, associated carbonates are barren.

The extent to which the process of silicification introduces a further bias into the Precambrian fossil record is not clear. If silica is to be considered as an evaporitic mineral that might readily have precipitated from the coastal fringes of a pre-diatom and radiolarian ocean saturated with respect to amorphous silica, then an environmental basis in favor of sabkhas is expected. Unfortunately, field studies of Proterozoic stratigraphic successions provide little support for this view, at least in its simplest form. Chert nodules and lenses are found in a variety of carbonates, including those from shelves and basinal environments. The common denominator in these occurrences seems to be the presence of abundant organic matter. A chemical hypothesis advanced by Leo and Barghoorn (1976) to explain the silicification of wood may also explain the close relationship between chert and organic matter in Proterozoic carbonates. Leo and Barghoorn proposed that functional groups, particularly hydroxyl groups, in partially degraded wood form hydrogen bonds with dissolved monosilicic or polysilicic acid in ambient groundwater. As silicic acid molecules build up, they begin to polymerize, with the expulsion of water. In this way, both the exquisite preservation of some petrified woods and the intimate relationship of silica and organic matter in petrifications are elegantly explained. Similar features of stromatolitic microbiotas may best be explained by invoking analogous geochemical processes in Precambrian microbial mat sediments.

Much remains to be learned about the formation of chert nodules in rocks of all ages. With respect to the question of bias, however, perhaps it is fair to conclude that, in a general sense, early diagenetic silicification reenforces the biases introduced by post-mortem organic decomposition.

21.2 The Fossil Record

21.2.1 Eoentophysalis, a Prominent and Persistent Mat Builder

Microbial mat accretion is often associated with the activity of filamentous cyanobacteria, but in certain environments coccoidal blue-greens are also capable of building and stabilizing mats. Of particular importance in living mats are species of the genus *Entophysalis*. *Entophysalis* populations consist of small, globose unicells that divide in three planes to produce hemispherical cell aggregates (Golubic and Hofmann 1976; Fig. 21.3 E). Each cell secretes an extracellular gelatinous envelope, and following division new envelopes are formed within the confines of older ones, producing a characteristic multilamellate structure. The yellow-brown pigment scytonemine colors the envelopes, particularly on the upper surface of the colony. A distinctive brown mammillate surface characterizes *Entophysalis* mats (Fig. 21.3 A); sediments are trapped in small depressions between adjacent mammillae and are stabilized by the gelatinous mat surface. The mat accretes by overgrowth of colonies and, following episodes of rapid sediment influx, by recolonization of the surface by individual cells (Golubic 1976a). On the present day Earth *Entophysalis* mats occur in the lower intertidal zone and in coastal ponds of arid coasts bordering hyypersaline embayments, for example in Shark Bay, Western Australia, and in Abu Dhabi along the Persian (Arabian) Gulf (Golubic 1973, 1976a, b). Because of their robust extracellular envelopes and present environmental distribution, entophysalid blue-greens seem to be excellent candidates for preservation in ancient coastal mat systems.

To avoid circularity in the paleoecological interpretation of fossil entophysalids, it is necessary to recognize ancient sabkha sequences on the basis of geological criteria. As discussed elsewhere in this volume, a variety of mineralogical, stratigraphic, and sedimentary structural features distinguish arid-zone coastal environments (A. C. Kendall 1979). An illustrative example is provided by the latest Proterozoic (ca. 700 Ma) Narssârssuk Formation, northwestern Greenland (Strother et al. 1983).

The most distinctive feature of the Narssârssuk Formation is its cyclical nature (Fig. 21.1). Each cycle begins with tabular bedded limestones, often stromatolitic, which grade upward into cryptalgal laminated dolomites (Fig. 21.4 A) that become increasingly organic rich and gypsiferous as one moves stratigraphically upward. Gypsiferous carbonates are, in turn, overlain by silty red beds that in

Fig. 21.1. Exposed cliff section of the upper Narssârssuk Formation, Saunders Ø, Greenland. Light units are predominantly carbonates; dark are mainly red sandstones and siltstones. Height of cliff is approximately 300 m

many cases are capped by limonitic, highly weathered erosional surfaces. A typical unit cycle is diagramed in Fig. 21.2 (after Strother 1980). This sequence is similar in its salient features to Holocene and earlier Phanerozoic successions deposited during the progradation of sabkhas across somewhat restricted coastal embayments (e.g. P. E. Schenk 1967, C. G. Kendall and Skipwith 1968, Wood and Wolfe 1969, Gill 1977, A. C. Kendall 1979). Sedimentary structures supporting a comparable interpretation for the Narssârssuk Formation include (in addition to stromatolites) "augen" structures, "chicken-wire" textures formed by the growth of anhydrite nodules, desiccation cracks, and occasional intraformational breccias. Other features found in the middle part of the formation include domal carbonate precipitate structures (Fig. 21.4 B) and textures often associated with tufa deposition, especially along the shores of playa lakes (Smoot 1978).

The Narssârssuk carbonate/red bed cycle is repeated dozens of times in the formation; an exact count is impossible because of incomplete exposure. Carbonates are most extensive in the middle member of the Narssârssuk Formation, and it is in this unit, which was deposited on a broad, quiet, shallow subtidal to supratidal, hypersaline carbonate flat, that microfossils are preserved. Mammillate stromatolites found in this sequence are closely comparable to those accreting at present in the lower intertidal zone bordering the Persian Gulf in Abu Dhabi (Fig. 21.3 A, B). The similarities between the ancient and modern physical settings and stromatolite morphologies are complemented by the strong morphological similarity of the fossil and living cyanobacteria found within the stromatolites.

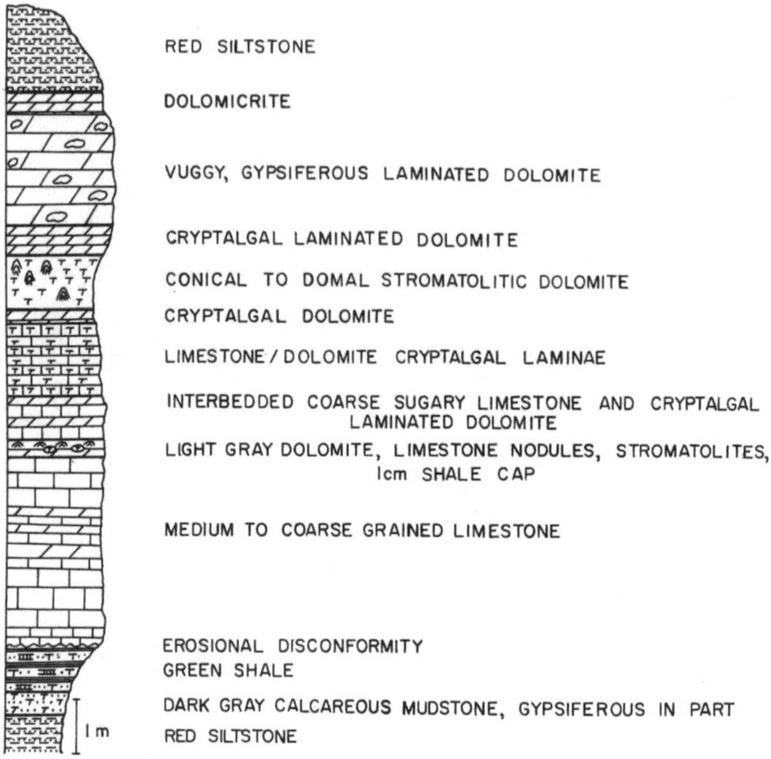

Fig. 21.2. Stratigraphic section of a single sedimentary cycle lower Narssârssuk Formation exposed south of Thule Air Force Base, nortwestern Greenland. (After Strother 1980)

Silicified portions of the Narssârssuk stromatolites contain large, gregarious populations of microfossils assignable to the genus *Eoentophysalis* Hofmann (1976; Fig. 21.3 D). As its name implies, *Eoentophysalis* is distinguishable from modern *Entophysalis* only by its age. Preservation is variable within the stromatolites, and most fossils are preserved as envelopes with little or no cellular (cytoplasmic) contents remaining.

While *Eoentophysalis* is the predominant microfossil in this assemblage, it is not the only organism preserved. Local patches of other simple spheroids occur, as do horizons of filamentous sheaths interpreted as the extracellular sheaths of oscillatorian cyanobacteria that proliferated in local microenvironments within the mammillate mat area. Large colonial microfossils comparable to species of the extant genus *Microcystis* are also common locally. These appear to be allochthonous elements and may have lived in evanescent ponds associated with the *Eoentophysalis* mats. Recognizable purple and green photosynthetic bacteria or other prokaryotes that might be expected to live in subsurface horizons of microbial mats have not been detected. Like the cellular contents of the entophysalid blue-greens, their absence is most likely the result of differential post-mortem destruction.

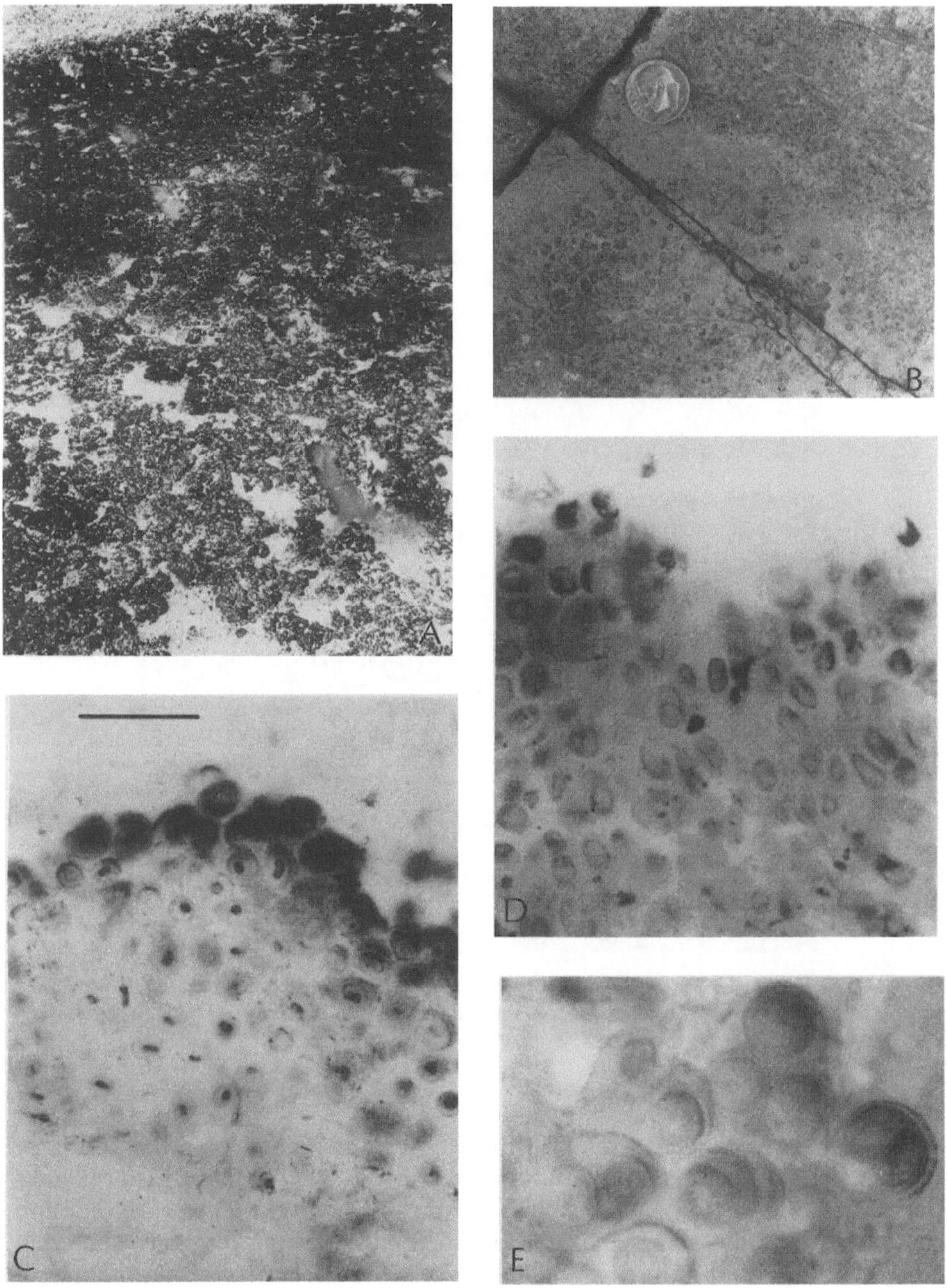

Fig. 21.3. A Mammillate *Entophysalis* mats in intertidal zone, Abu Dhabi. Note footprints for scale. **B** Surface of *Eoentophysalis*-bearing mammillate stromatolites, Narssârssuk Formation. **C** *Eoentophysalis belcherensis* Hofmann from the 1,900-Ma-old Belcher Supergroup, Canada. Note shrunken cell contents within rounded envelopes. **D** *Eoentophysalis* cf. *belcherensis* from the Narsârssuk Formation. **E** *Eoentophysalis major* Ercegovic from Shark Bay, Western Australia. Bar in **C** = 40 μm for **C**, and **D**, and 12.5 μm for **E**. (**A, E** courtesy of S. Golubic; **C** courtesy of H. Hofmann)

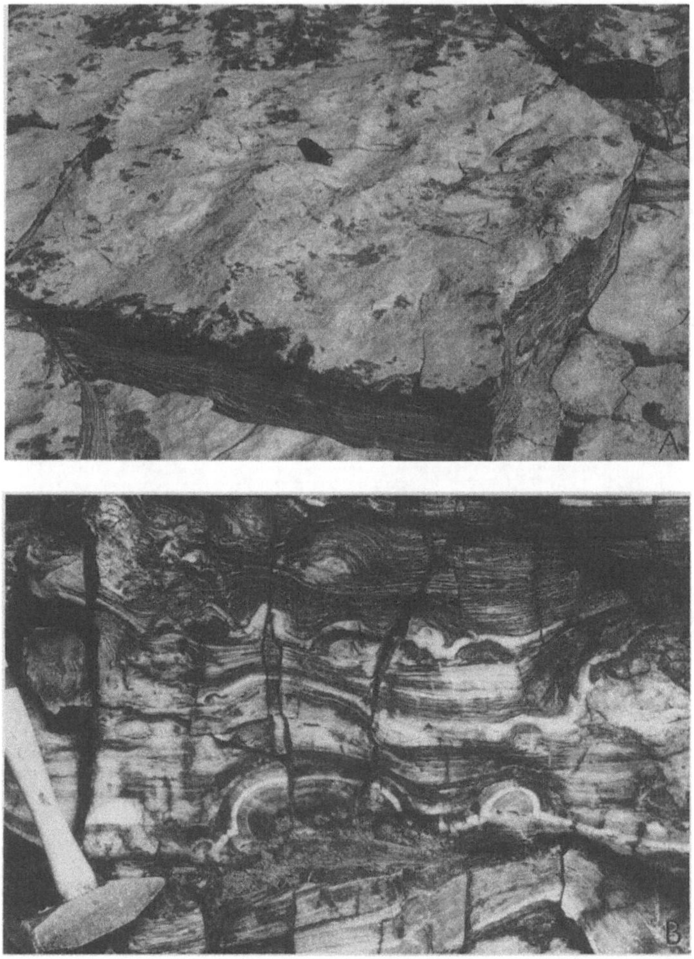

Fig. 21.4. A Flat, laminated stromatolites within the Narssârssuk Formation. Pocket knife provides a scale. **B** Dominal precipitate structures in high intertidal or supratidal dolomites from the Narssârssuk Formation

In short, the observed record of microfossils in mammillate stromatolites of the 700 Ma Narssârssuk Formation is pretty much what would be expected on the basis of the biology and known processes of early diagenetic degradation characteristic of comparable modern environments. This suggests that microbial communities inhabiting arid intertidal environments of the late Precambrian were similar to the biotas living in the same environments today. It is worth noting here that the entophysalid assemblage is not the only association of microfossils recovered from the Narssârssuk Formation. Three additional discrete microbial associations representing other subtidal to supratidal environments have also been reported, demonstrating the microbial heterogeneity within an ancient coastal zone (Strother et al. 1983).

The close comparison between the Narssârssuk biota and its environmentally equivalent modern counterpart is striking, but it is by no means unique. In fact, entophysalid-dominated assemblages in arid coastal environments constitute a major theme in Proterozoic paleobiology. *Eoentophysalis* populations were described by Knoll and Golubic (1979) from the approximately 850-Ma-old Bitter Springs Formation, Australia. The sedimentary environmental setting of the samples available to these authors is not known in detail, but parts of the formation are known to have been deposited in a barred basin evaporative setting (Stewart 1979), and low diversity prokaryotic microfossil assemblages preserved in these units include entophysalid blue-greens (Oehler et al. 1979). It may well be that all preserved Bitter Springs assemblages (most of which do not contain *Eoentophysalis*) derived from a restricted coastal environment.

Entophysalid populations are well preserved in several Middle Proterozoic formations, including the 1,200-Ma-old Dismal Lakes Group, arctic Canada (Horodyski and Donaldson 1980), the 1,400–1,500 Ma-old Gaoyuzhuang Formation of China (Zhang 1981), the approximately 1,500-Ma-old Balbirini Dolomite of Australia (Oehler 1978), and the slightly older Amelia Dolomite, also of Australia (Muir 1976). The arid intertidal to supratidal settings of these fossiliferous horizons are beyond dispute. Particularly in the Australian formations, evaporites and casts of evaporites are abundant (Oehler 1978, Muir 1979). Oehler, in fact, reported the presence of pseudomorphs after gypsum (or anhydrite) within the same laminae as microfossils.

The oldest known entophysalid populations are those found in the 1,900-Ma-old Kasegalik and McLeary formations of the Belcher Supergroup, Belcher Islands, Canada (Hofmann 1976). Although nearly 2 billion years old, these microfossils are beautifully preserved and are morphologically indistinguishable from modern *Entophysalis* populations (Golubic and Hofmann 1976; Fig. 21.3 C). Hofmann (1976) provided a thorough sedimentological description of the fossiliferous Belcher beds, and it is apparent that they, like the much younger Narssârssuk cycles, reflect the repeated progradation of sabkha sediments over subtidal carbonates.

There is, then, a remarkable continuity in the fossil record of entophysalid mat builders, spanning nearly half of the Earth's history. Taphonomic considerations suggest that there should be a preservational bias in favor of these thickly enveloped cyanobacteria from hypersaline coastal habitats, and it is clear that such a bias exists; however, this does not diminish the fact that 2,000 Ma of evolution has not produced a superior competitor in this particular environment. It is unlikely that fossil and modern entophysalid blue-greens were genetically identical. Random and not-so-random processes continually change the genomes of all species. Paleoecology does allow the inference, however, that whatever genetic changes have occurred in the past 2 billion years, they have been of relatively minor consequence to the clone, and that Early Proterozoic and modern entophysalids might be expected to differ mainly in the details of their physiologies. One could hardly make a stronger case for the crucial importance of research on modern sakbhas for an understanding of aspects of the early fossil record.

21.2.2 Mats Built by Filamentous Cyanobacteria: Heterogeneous Associations of Ancient Microorganisms

In a transect across a modern mat-bearing coastline, one encounters several distinct microbial communities, each thriving within a limited environmental subset of the available peritidal range. For example, at Laguna Guerrero Negro, Mexico, Javor, and Castenholz (1981) have observed mats dominated by *Microcoleus chthonoplastes* (with an underlying purple layer of photosynthetic bacteria), *Lyngbya aestuarii,* and *Calothrix crustacea* in the lower, middle, and upper intertidal zones, respectively. In Abu Dhabi, four distinct microbial communities build mats in shallow subtidal to uppermost intertidal environments (Golubic 1976a, Kinsman and Park 1976) and in Shark Bay, Western Australia, as many as six or seven mat communities are present (Golubic 1976a). The general zonation of these communities parallel to the coastline suggests that wetting frequency associated with tidal cycles is a primary determinant of microbial distribution. Ponds and channels within the intertidal zone increase the complexity of community distribution, as do disturbance and chance (Park 1977).

Modern microbial mat communities can be differentiated on the basis of the morphologies of the mats they build. By extrapolation, it might be considered that stromatolites in Precambrian carbonates provide clues to the diversity and distribution of ancient microbial mat communities. Certainly, a strong environmental influence on stromatolite morphology must be acknowledged (e.g., Horodyski 1977b); however, it appears that stromatolite fabric and microstructure (sensu Monty 1976) bear the strong imprint of constituent mat-building populations (Gebelein 1974, Monty 1976). Thus, stromatolite distribution can be helpful in assessing the paleoecological distribution of community types in Proterozoic arid coastal environments and across much wider stretches of ancient platforms and continental margins (e.g., Bertrand-Sarfati 1976).

However useful fabrics and microstructures may be, they remain inferential proxies for microbial communities that have long since disappeared. Biological comparisons between modern sabkhas and their Proterozoic counterparts ultimately require that paleoecological analyses of actual microfossil assemblages be undertaken. Fortunately, such investigations are possible because silicified stromatolites preserve the spatial relationships within and between microfossil populations. By censusing microfossils in individual horizons within laminate structures, it is possible to determine what organisms lived together in recurrent associations, and what the importance of each preserved species may have been to be community (Knoll 1981). In this approach recurrent microfossil associations are recognized as the preserved remnants of microbial communities. One must also recognize the effects of post-mortem degradation in differentially removing some species (thereby augmenting the apparent importance of others) and, as well, in altering the morphologies of different individuals within a single species so as to make them appear quite distinct. These problems are best surmounted by working populations. This often allows one to recognize and deal with problems of with partial degradation and cell division cycles (e.g., Knoll and Golubic 1979).

To illustrate this approach to microbial paleoecology, I will briefly summarize the results of recent work on the Upper Riphean (800–700 Ma) Draken Conglo-

Fig. 21.5. Fossiliferous "flake" conglomerate of the Draken Conglomerate, Spitsbergen. Black areas are silicified rip-up clasts of microbial mats and, subordinately, nonstromatolitic muds. These are set in a sandy to gritty matrix of clastic carbonates. This conglomerate provides a sampling of various microenvironments within the general lagoon habitat in which Draken deposition occurred

merate, Spitsbergen (Knoll 1982). Draken mats accreted in a broad, protected carbonate platform environment. Periodic storms ripped up thin mats and redeposited them locally, forming what Wilson (1961) has called a "flake conglomerate." Individual mat shards, a few millimeter thick and up to 5 or 6 cm long, sit in a matrix of sand-sized micritic clasts. The mat shards are preferentially silicified (Fig. 21.5), and they provide convenient samples for paleoecological analysis. By analogy with modern mats, densely interwoven filament populations are interpreted as the remains of principal mat-building microorganisms. (This recognizes the fact that organisms important to mat accretion may not be preserved; see, for example Horodyski et al. 1977). In the Draken Conglomerate, each mat shard contains a single mat-building population, but different clasts contain different principal builders. In all, three principal mat-building filament types have been recognized. All are sheaths of probable oscillatorian cyanobacteria and are differentiable on the basis of size (Figs. 21.6 and 21.7). One mat type contains a secondary or auxiliary builder, a large oscillatorian or scytonematacean blue-green characterized by multiple, often copious sheaths (Fig. 21.6 E).

Two of the three mat types contain additional microbenthic populations that occur in recurrent and specific association with a given builder population (Figs. 21.6 and 21.9 E, F). These are interpreted as mat dwellers or "guests" that lived in the mats but did not contribute significantly to their construction (Golubic 1976a). Other elements are distributed randomly within and between mat shards and occur in nonstromatolitic mud clasts as well (Figs. 21.6 B, 21.9 D, H). These are considered to be allochthonous microfossils, most probably representing a lagoonal plankton or periphyton community.

In all, 28 taxa have beed reported from the Draken Conglomerate. This diversity seems high in light of the number of preservable species in modern mats; how-

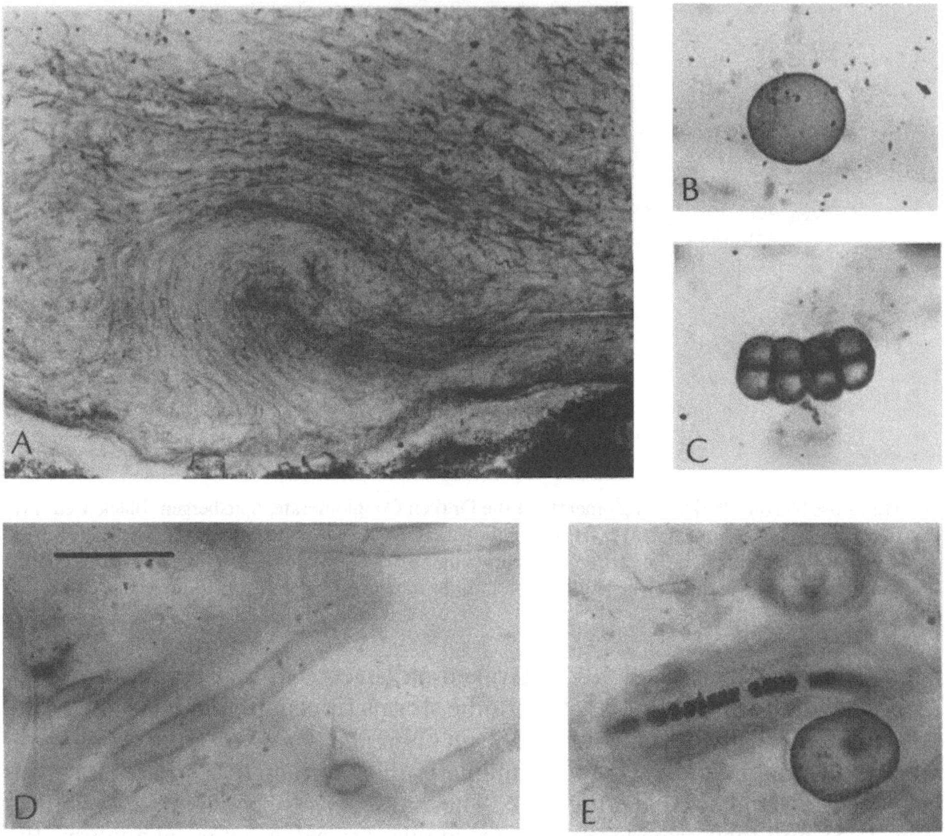

Fig. 21.6 A–E. Microfossils from the Draken Conglomerate. **A** Low-magnification photograph showing the matted fabric of interwoven filamentous microfossil populations within a single mat shard. At higher magnification, each dark wisp can be resolved as a sheath of the oscillatorian form species *Siphonophycus inornatus* Zhang. **B** *Mycococcoides* sp., a probable allochthonous element. **C** *Sphaerophycus wilsonii* Knoll, a mat-dwelling cyanobacterium. **D** *Siphonophycus inornatus* Zhang, a higher-magnification photograph of the same microfossil taxon illustrated in **A**. **E** *Salome svalbardensis* Knoll, a multilamellate filamentous cyanobacterium that is an auxiliary mat builder in certain Draken associations. Note the partially degraded nature of the trichome and the large spheroidal microfossil nearby. Bar in **D** = 125 μm for **A**, 20 μm in **B** and **D**, 12.5 μm for **C**, and 40 μm for **E**

ever, the bulk diversity figure for the Draken assemblage is misleading in that it represents three discrete microbial mat associations containing one to six in situ species, a nonstromatolitic, mud-dwelling microbenthos association of comparable diversity, and a probable planktonic component that accounts for some 40% of the entire observed species richness. That is to say, the diversity, structure, and taxonomic make-up of the Late Riphean Draken mat biotas are comparable to what one sees in recent mats growing in similar environments. As occurs today, the Draken environment was subdivided into domains dominated different microbial communities. For a more complete discussion of this formation, see Knoll (1982).

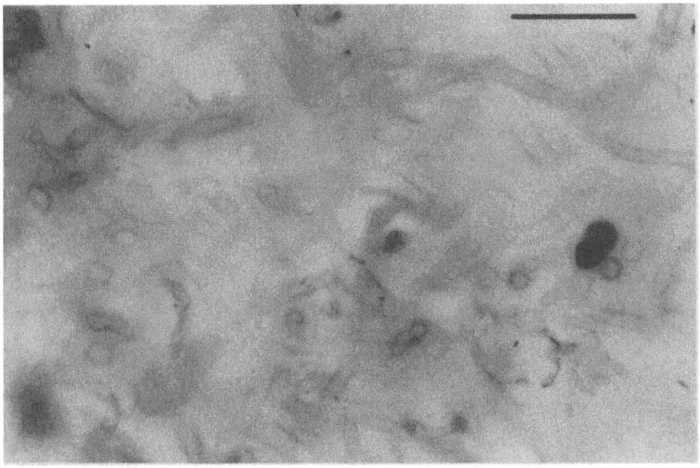

Fig. 21.7. Tightly interwoven sheaths of the mat-building cyanobacterium *Eomycetopsis robusta* Schopf emend. Knoll and Golubic, from the Draken Conglomerate. Bar = 20 µm

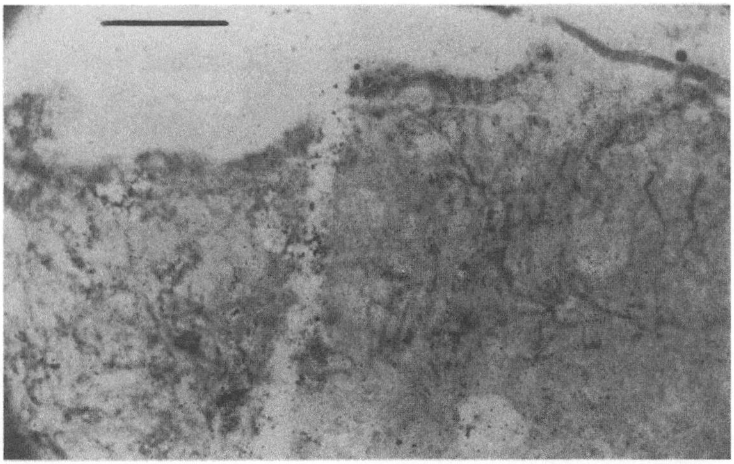

Fig. 21.8. Populations of the mat-building cyanobacterium *Eomycetopsis robusta* Schopf emend. Knoll and Golubic from the Bitter Springs Formation, Australia. Note the alternately vertical and horizontal orientation of individual sheaths. Bar = 100 µm

Other Proterozoic microbiotas can be broken down into recurrent associations in a similar fashion, and with similar results. The well-known Upper Riphean (ca. 850 Ma) Bitter Springs Formation (Schopf 1968, Schopf and Blacic 1971; Figs. 21.8 and 21.9 A, B), for example, can be viewed as a collection of low-diversity mat associations with an admixture of planktonic elements. Samples from the Ross River locality of the formation contain three distinct mat associations plus plankton (Knoll 1981), while other localities have yielded at least five additional benthic associations (Schopf and Blacic 1971, Knoll and Golubic 1979,

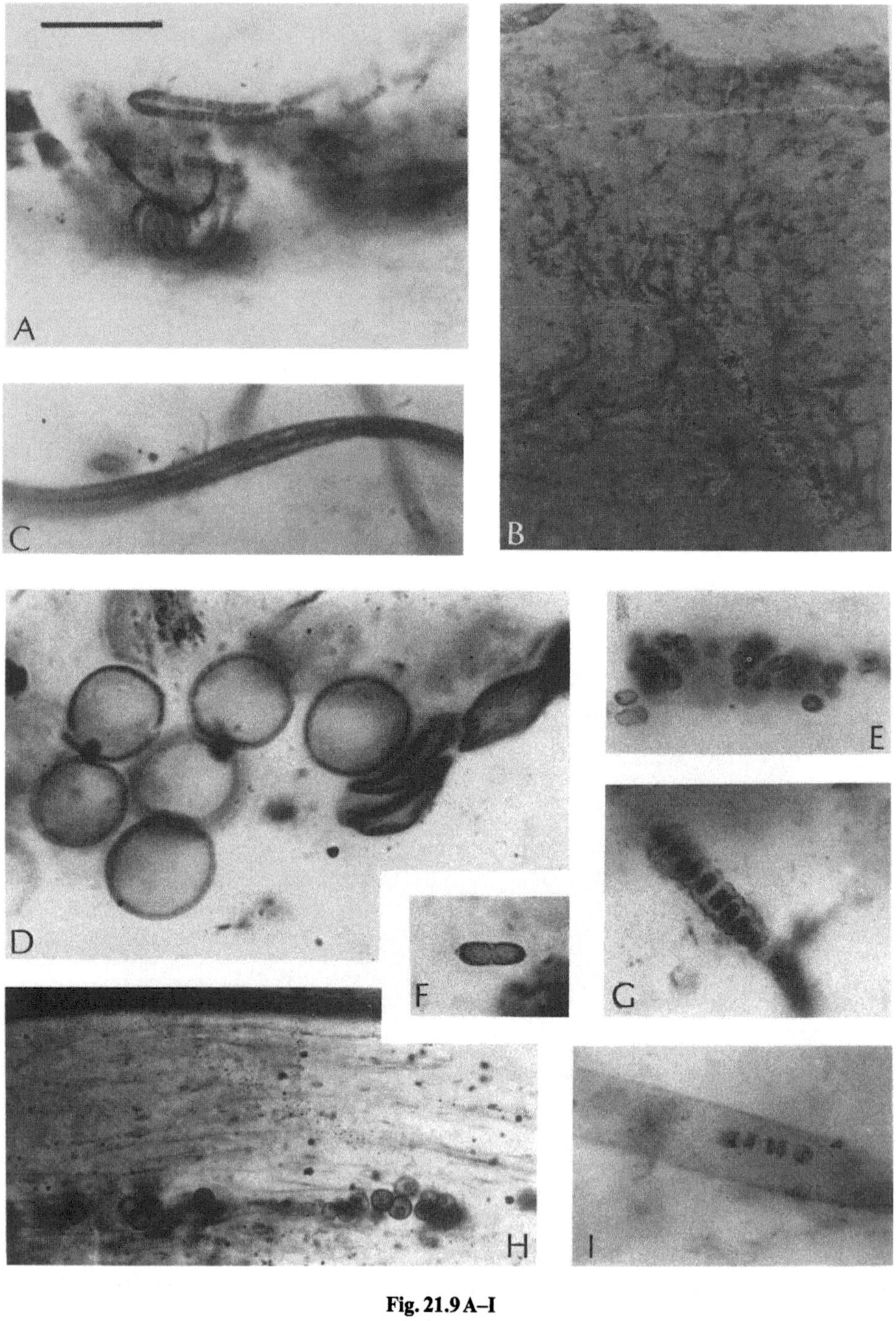

Fig. 21.9 A–I

Oehler et al. 1979, Knoll 1983a). The two dominant mat builders preserved in Bitter Springs cherts are morphologically indistinguishable from two of the major elements of Draken mat associations. This may mean that, like *Phormidium hendersonii* today (Golubic and Focke 1978), some late Precambrian filamentous mat builders had a cosmopolitan distribution. On the other hand, species based on extracellular sheats are form taxa, and it may be that the morphologies of biologically distinct blue-greens have converged as a consequence of early diagenetic changes.

Ecological heterogeneity whithin a single Proterozoic stromatolitic microbiota is the rule rather than the exception. Certainly, it is true for the *Eoentophysalis*-bearing formations mentioned in the previous section. In addition to coccoidal mat builders, those formations all contain filamentous mat associations representing different microenvironments within the arid coastal habitat. The important point is that through careful paleoecological analysis some measure of the microbial heterogeneity characteristic of modern peritidal environments can be recognized even within a sedimentologically homogeneous package of Precambrian microbially laminated rocks.

21.2.3 The Antiquity of the Coastal Sabkha Environment

It has been argued that the habitats in which phylogenetically ancient or primitive organisms now thrive represent the types of sites in which they first evolved or, more broadly, the general conditions of the Earth's surface at the time that these groups appeared. Prokaryotic communities today exist in environments that are very hot, cold, alkaline, acidic, anaerobic, and/or saline. Of these, only O_2-free conditions are widely accepted as likely when prokaryotic phototrophs diversified. In a discussion of angiosperm origins, G. L. Stebbins (1974) cogently argued that the southeast Asian and Pacific rain forests in which primitive dicotyledonous trees persist were not the cradles of early flowering plant evolution, but merely serve as their museum. His metaphor is apt, and it applies equally well to the extreme habitats in which prokaryotes remain conspicuous. There is no a priori reason to believe that prokaryotes (and, hence, life on Earth) necessarily originated in such environments and, thus, little basis for the supposition that the appearance of coastal sabkhas was linked in any significant way to the origin of cyanobacteria or other bacterial groups. Sabkha environments, however, are likely sites for the preservation of early microorganisms, and thus, the

Fig. 21.9. A Cellularly preserved oscillatorian cyanobacteria from the Bitter Springs Formation. **B** Enlarged view of the population illustrated in Fig. 21.8 (Bitter Springs Formation). **C** Two entwined sheaths of *Eomycetopsis robusta* from the Draken Conglomerate **D** *Myxococcoides cantabrigiensis* Knoll, allochthonous elements in the Draken mats. **D, E** *Eosynechococcus brevis* Knoll and *Eosynechococcus medius* Hofmann, respectively; both mat-dwelling microorganisms in the Draken conglomerate **G, I** Partially degraded cyanobacterial filament illustrating the shrinkage of trichome cells prior to their complete disintegration. Specimens come from the 800-Ma-old Hunnberg Formation, Svalbard. **H** Low-magnification photograph of a Draken mat, illustrating the mat texture and a population of spheroidal unicells in one horizon. Bar in **A** = 35 μm for **A,** 30 μm for **B,** 20 μm for **C, G,** and **I,** 12.5 μm for **D,** 15 μm for **E** and **F,** and 55 μm for **H**

paleontological investigation of early arid coastal sequences may well constrain ideas on the timing of the initial biological diversification on the Earth.

From the preceding micropaleontological discussion it is clear that coastal sabkha successions are found throughout the Proterozoic geological column. It is of interest to ask about comparable sequences in the Archean record, because prior to late Archean times the areal extent of continental crust on the Earth's surface was limited (Windley 1977, Knoll 1979). Restricted coastal accumulations, probably deposited on the flanks of subaerially exposed volcanoes or along the margins of protocontinental blocks and uplifted basaltic terrains, have been described from several Archean greenstone belts.

Carbonates within the approximately 2,700-Ma-old Belingwe Greenstone belt, Zimbabwe, (Martin et al. 1980) contain a surprisingly diverse array of stromatolites. A number of sedimentary features, including intraformational edge like breccias, ripple marks, and desiccation cracks collectively suggest that deposition occurred in a tidal flat to lagoonal setting. Large sheaves of bladed crystals associated with the stromatolites further suggest mild evaporative conditions. Recrystallization and possible replacement make it difficult to determine whether the crystals were originally aragonite or gypsum.

Significantly older coastal sabkha deposits are found in the Warrawoona Group near North Pole, Western Australia (Barley et al. 1979, Groves et al. 1981). Zircons within this sequence have been dated at $3{,}452 \pm 12$ Ma (Pidgeon 1978), and Sm-Nd isotope analyses yield ages of $3{,}250 \pm 60$ Ma (Cooper et al. 1982). This makes the Warrawoona sequence not only the oldest Sabkha-bearing sequence known on Earth, but also (along with the Swaziland Supergroup of southern Africa) the oldest negligibly metamorphosed sedimentary sequence of any type known at present. Therefore, sabkha deposits can be traced back virtually to be beginning of the observable stratigraphic record. (The 3,800-Ma-old Isua supracrustals of southwestern Greenland do contain older sedimentary rocks, but these have been metamorphosed to amphibolite grade and are not easily probed for vestiges of sedimentary structures and mineralogical evidence indicative of arid coastal sedimentation.) In the Warrawoona sequence, barite pseudomorphs after gypsum occur both as radiating sheaves of crystals within sandstones and shales and as bedded units. This suggests evaporite accumulation in a local coastal embayment 3,500 Ma ago.

The Warrawoona sequence is of unusual interest not only because of its physical evidence of evaporite deposition, which is important in efforts to understand the early development of ocean chemistry and geochemical cycles, but also because it contains evidence of early biological activity. Carbon isotope ratios in Warrawoona organic matter suggest the existence of a biologically controlled carbon cycle fueled by photosynthesis at the time these rocks were deposited (Hayes 1983). What type of photosynthesis was involved is less clear. Stromatolites of probable microbial mat origin are also present (Lowe 1980, M. R. Walter et al. 1980; see also the critical appraisal by Buick et al. 1981), and both coccoidal and filamentous microfossils have been reported (Awramik et al. 1983). Additional silicified evaporites of comparable age have also been reported from the Onverwacht Group of South Africa, and those also contain biological fingerprints (Lowe 1982). From the meager microfossil record available at present it is not clear whether or not early Archean microbial mat biotas were taxonomically similar to those of the Proterozoic and Recent; however, the combined weight of evidence does seem to indicate that arid coastal zones contained complex prokaryotic communities of some type very early in Earth history. In summary, one runs out of records to examine before one runs out of evidence for sabkhas or prokaryotic microorganisms. Traces of the first sabkhas and (perhaps) their microbiotas must lie in the as yet undiscovered (and perhaps undiscoverable) sedimentary record of the earliest Archean Earth.

21.3 Conclusions

The foregoing discussions indicate that biological, geochemical, and sedimentological studies of recent sabkhas are important not only because they illuminate a most fascinating modern environment, but also because they provide the scientific framework for the interpretation of some of the earliest records of life on Earth. It is a heady thought to realize upon looking at the Gavish Sabkha that one is seeing the Belcher or Bitter Springs biota come to life. Remarkable as that is, one must remember that modern sabkhas provide an extremely limited sampling of modern biology, and that the same may be true of Precambrian examples. Rare examples of microfossil preservation in open shelf mats and stromatolites (e.g., Licari and Cloud 1972, Schopf and Sovietov 1976, Knoll 1984) do suggest that, in general terms, the stromatolites that are widespread in Proterozoic carbonates were biologically similar to those of more restricted environments and, hence, to modern mat communities[1]. However, planktonic microbiotas are quite another matter. The small, morphologically simple phytoplankters associated with restricted coastal mat biotas of late Precambrian age do not even begin to suggest the diversity of planktonic microfossils that abound in contemporaneous shelf deposits. In Upper Proterozoic shales and siltstones, large, robust-walled and, often, ornamented cell remains are common, and in formations 800–650 Ma-old, protozoan and metaphyte fossils occur as well (Vidal 1976, Knoll 1983b, Vidal and Knoll 1983). This does not diminish the paleobiological importance of stromatolitic microbiotas. It only serves to point out that they are only one part of the available Precambrian fossil record.

The close relationship between modern and fossil microbial mat biotas provides numerous opportunities for synergistic research collaborations between paleontologists and microbiologists, sedimentologists and organic geochemists. It is clear from various contributions in this volume and elsewhere in the literature that such programs are thriving and bearing fruit, and that the next decade of research will result in a significantly improved knowledge of both modern and ancient microbial mat ecosystems.

1 The one exception to this appears to be the Gunflint microbiota and other assemblages from stromatolites found in iron formations. The Gunflint biota is qualitatively different from the other fossil and living microbial assemblages discussed in this chapter, and the environment in which Gunflint organisms lived may have no close modern counterpart. The interpretation of this historically and paleontologically significant microbiota remains a subject for debate; thus, it is prudent to exclude iron formation microfossils from the general class of interpretation advocated here

22. Applied and Economic Aspects of Sabkha Systems – Genesis of Salt, Ore and Hydrocarbon Deposits and Biotechnology

Wolfgang E. Krumbein

The Gavish Sabkha contains a volume of water not larger than the swimming pool of an average provincial city, and its immediate surroundings do not exceed the play- and sports grounds on the outskirts of the latter. Still, a lot of interest has focused and will focus in the future on the Gavish Sabkha and other similar systems around the world. This is caused by the following reasons:

1. The huge salt reservoirs of the Permian (Zechstein) in Europe have very striking similarities with sabkhas. Furthermore, many of these salt systems are related to petroleum-producing strata and traps.

2. Recently much interest has focused on marine salt production plants mainly in the Caribbean and the Pacific. Biological factors, and among them almost exclusively those produced and controlled by prokaryotic phototrophic microorganisms, have been found to be most important in the production of large crystal-sized clear and pure salts with as little iron admixture as possible.

3. Practicians of applied geology have realized that many if not all of the huge or and phosphate deposits of the Precambrian and later periods of Earth history, when initially sedimentary, are at least partially interconnected with environmental conditions typical for present-day sabkhas.

4. Petroleum geologists and organic geochemists have concluded that many, if not most, of the hydrocarbon and gas-generating systems in the past are related to sabkha-type evaporative ocean margins.

5. Furthermore, the capacity of many microorganisms to develop very individual osmotic pressure regulation systems, e.g., glycerol, cation pumping, light energy driven membrane potentials (see Chap. 13), has attracted interest of biotechnologists for applications in bioengineering. Several other products and processes of halophilic microorganisms are also of applied interest.

6. Also several aspects of salt-water agriculture, as an alternative using marine microorganism or salt-walter plants, and the study of soil biology under saline conditions (*Mykorrhyza, Rhizobium,* etc.) have encountered growing interest. This interest will most probably increase in the near future.

7. Finally, salt-water microbiology has important aspects in the management of the growing number of desalination plants in semi-arid and arid countries. This concluding chapter thus tries to explain to some extent why the ecology and biogeochemistry of sabkhas is an important field of research and why biochemical and biotechnological aspects have to be involved into these studies.

22.1 Salt Deposits

Salt is one of the first, if not the first, type of fossil mineral deposits which were "industrially" exploited by mankind. Actually, many animals have adapted their behavior pattern long before man existed to the use and exploitation of fossil salt deposits. They were and are resting at "licking places" on their annual migrations in many countries. Salt, gold, and amber, together with volcanic glasses (obsidian) and chert, were probably the first fossil deposits exploited. Man has always looked for such places and have traded in these goods. Salt is essential for life, while the other minerals mentioned so far are far less important.Sabkha-type salt deposits of the Pliocene, Tertiary and earlier periods were and are mined in many places in Africa and Arabia and have been traded on camel backs or carried by man since the very old times. Most of these salt deposits are of shallow sabkha environments and exhibit the typical structures of sabkhas.

The main minerals occurring in the well-known European fossil salt deposits are halite (NaCl), sylvine (KCl), carnallite ($KCl \cdot MGCl_2 \cdot 6\,H_2O$), polyhalite ($K_2SO_4 \cdot MgSO_4 \cdot 2\,CaSO_4 \cdot 2\,H_2O$), kainite ($KCl \cdot MgSO_4 \cdot 3\,H_2O$), several borates and, naturally, gypsum, the water free anhydrite, and the carbonate minerals calcite, aragonite, and dolomite.

These minerals do occur in the Permian, Triassic, and Tertiary salt mines of Middle Europe in varying amounts, mixtures and successions. Generally, salt mineralogists and geologists explain these ratios and sequences by purely physical-chemical factors and talk about evaporation sequences or cycles corresponding to the complete or almost complete evaporation of isolated shallow bays and basins. The occurrence of sulfur, organic material, and sulfides, as well as the astonishing lack of gypsum and anhydride, as compared to the relatively large amounts of carbonates, indicate considerable biogeochemical activities during the depositional processes.

The change of habits and agricultural fertilizing have led to the fact that presently less and less rock salt is mined while the salt-mining industry concentrates on the salts of the highest evaporation series, i.e., the valuable potassium chlorides occurring in the Permian of Germany and in the Oligocene of the Rhine Graben. These correspond to conditions as they prevail presently in the environments of the Natrun oasis depression in Egypt and at the Dead Sea and some coast-marginal sabkhas, as the Gavish Sabkha. The Gavish Sabkha, however, is not yet developed enough to ensure the precipitation of the bitter salts of potassium and magnesium. There is too much reflux of the heaviest brines into the sea, because the microbial mats have not yet sufficiently sealed the underground (see Chap. 9).

On the other hand, living organic matter in the form of planktonic and benthic microbiota serves as well as a salt trap. It is well established, that dolomite is a microbially induced product of early diagenesis, although it does not belong to the bacterially precipitated minerals. Krumbein (1979a) and G.M. Friedman (Chap. 10) have demonstrated that magnesium is bound to the organic material embedded in freshly deposited bacteriogenic carbonates and later reacts with the bicarbonates released from the anaerobic decay processes within the sediments or with Mg-calcite directly under a different physico-chemical and chemical interstitial water regime (see also Krumbein 1979a). Furthermore, potassium and phosphorus are selective-enriched and trapped from the brines flowing through sabkha systems, thus influencing the chemistry and mineralogy of evaporative salt deposits (Czoering 1869, Buschmann 1906, 1909).

Finally, it may be stated that not only the amount and chemical composition of natural salt deposits mined in many places of the world (e.g., the of the relative abundance of valuable potassium-rich layers) but also the applied technology of modern marine salinas depend largely on geomicrobiology and the maintenance of equilibrated biogeochemical conditions.

22.2 Salt Production Plants (Salt Gardens, Salinas)

As pointed out already in Chapter 2, marine salt gardens have been run by mankind for at least 2,500 years. Already in mediaeval times people knew about biological problems in the management of salinas. The technical tools and technical processes, including the relevant terms, were adapted in such a way as to maintain a certain biological equilibrium at the bottom of the artificial sabkha pools (production and maintenance of benthic microbial mats) as well as in the water column (supression of too intensive blooms of *Dunaliella* and of phototrophic purple bacteria or Cyanobacterial, e.g., *Spirulina platensis*). Also the Bedouins of Sinai follow these rules in a very limited and not highly developed way of salt garden management of Dahab and the Gavish Sabkha. The Arabs have additionally run several salt gardens at the ancient Sesostris Suez Canal site and in the Bardowil Lagoon on the Mediterranean coast (see Chaps. 2 and 11).

The relationship between physical, geochemical, and biological factors which influence the occurrence and maintenance of marine evaporitic sediments has been mentioned and summarized by I. S. Davis (1978) and Schneider (1979), while some of the major geological and geochemical textbooks on salts and salt deposits take a purely chemical and geological approach.

On the other hand, the precipitation pattern and behavior of salts from supersaturated brines do not correspond to laboratory simulations (Hermann et al. 1973, Schneider and Hermann 1980, and others). Thorstenson (1969, 1971) stated that already on the inorganic side 27 individual cations and anions are to be considered for precipitation models. At least 19 out of these ions are controlled by biological processes, leading to considerable changes in the precipitation of salts of the evaporation sequence. Berner (1971a) and Mitterer (1971) mention microbially induced changes of pH, redoxpotential, saponification effects and other organic interactions with the precipitation pattern of supersaturated evaporative minerals. Surface phenomena may be enhanced or biased by macromolecule monolayers on precipitating mineral surfaces, and microbial slimes (see Chap. 11) may lead to different patterns of diffusion. Diffusion rates of ions may be changed by a factor of up to 10,000 by biological processes in sabkhas (Krumbein et al. 1979a). Sulfate reduction by *Desulfovibrio* sp. and several other genera of sulfate-

and sulfur-reducing bacteria may reach tremendous dimension in highly productive hypersaline systems, which in turn influence considerably the geochemical balance and thereby salt-deposit structure and dimensions.

One has to keep in mind that in the formation process of carbonate, gypsum, and salt deposits not only carbonate and bicarbonate play an important role, but also the reduction of enormous amounts of sulfate under anaerobic conditions. Sulfate serves as terminal electron acceptor for a variety of bacteria. Thus, the correlation is not between biomass and the evaporite masses accumulating, but between bioenergetics and the evaporites. Biological equations, as they relate to the deposition and dissolution of minerals of the evaporite sequence, have been presented recently by Krumbein and Swart (1983). The mass flow, which can be reached by the interference of biological and evaporative processes in sabkha systems, has been computed by Garrels et al. (1975). In order to exemplify the enormous importance of biological processes in the formation of evaporite deposits it may be justified to use an example from a completely different evaporite environment, namely the Great Salt Lake in Utah.

Garrels et al. (1975) report that three major sources sulfur exist at Salt Lake City, i.e., (1) petroleum refinery and automobile exhaust, (2) emissions of a large copper-smelting unit, and (3) bacteriogenic sulfur released via respiratory sulfate reduction from the evaporative environment of the Great Salt Lake with its plankton and benthic microbial mats. From the changes in sulfur isotope composition of the Salt Lake City atmosphere during a strike in the copper smelter they calculated that the two civilization sources together produced less than 10% of the total content of reduced sulfur in the atmosphere.

Ivanov and Freney (1983) have compiled a more recent inventory of the biological cycle of sulfur as it influences the geochemical cycles of salt deposits through geological time.

Not only the amount and species of salts in sabkha-type salt deposits but also the quality of the salt is biologically controlled, as has been pointed out before (Chap. 9). Schneider (1979), Golubic (1980) and Javor (1979) have studied the individual basins of salt gardens in several different parts of the world concerning the different genera and species of phototrophic bacteria dominating the benthic and planktonic flora. Some of these data have been discussed also in M. R. Walter (1976). Schneider (1979) gives a fairly detailed model of the behavior of iron under such conditions, clearly indicating that the biological system underneath the aerobic zone of oxygenic cyanobacteria is responsible for the color of the final salt and its purity. The presence of halobacteria and square bacteria in the brine reduces the formation of surface salt crusts which otherwise would have to be destroyed by the workers in the salinas.

Furthermore, the experienced workers at the salinas of Rovinj (Yugoslavia) always systematically stir up the reducing mud from the channels between the basins. After the accumulation of piles of salt on the separating dams between the basins the reducing water flows down through the salt and takes away the mobilized iron. It has been observed that the salt produced by unexperienced young hands, who did not follow this procedure, was reddish and of minor quality.

22.3 Heavy Metal and Phosphate Deposits

Already as early as 1530, Paracelsus, a German physician and scientist (Philippus Theophilus Bombastus von Hohenheim), has pointed to the importance of microorganisms, which he called living slimes and threads, in the precipitation of minerals and as indicators of mineral deposits (see Krumbein 1984a). Schneiderhöhn (1955), with reference to some of his earlier works (Schneiderhöhn 1926), clearly states that the important and famous silver and copper deposits of Rammelsberg, Mansfeld and Eisleben were created via bacterial activity in a generally reducing and evaporative environment. The mining of copper and silver in these areas is as old as 700–1,000 years and there are still some active mines in these sedimentary biogenic metal sulfide deposits. The situation he describes sounds very similar to present-day conditions along the shores of the Gulf of Elat. The weathering conditions during the "Old Red" or "Rotliegend" geological periods, names indicating the red colors so attractive for everybody traveling in present-day desert environments (see. Chap. 2), aided by microbial weathering processes (Krumbein 1969, Krumbein and Jens 1981) apparently initiated the biotransfer of large amounts of metals into the evaporating warm seas of these geological periods. According to Schneiderhöhn's analysis of the "Kupferschiefer", shallow side lagoons and basins were constantly evaporating and piling up large amounts of organic material produced by benthic and planktonic microorganisms, as indicated by large marginal bryozoan reef complexes, which on the Russian Platform also produced oil shales (see Sect. 22.3). Hereby the metals dissolved in sporadic rain floods were coming down the mountain slopes in particulate form, or as desert storm dusts were transferred into the reduced dissolved species, absorbed to microbial mats or marine flocculates (marine snow) or accumulated in constantly upward-moving anaerobic sediment layers. Within these environments heavy metal sulfides were precipitated around bacterial colonies, diagenetically transformed and remobilized, until the typical flasery, or disseminated ore deposits were finally stabilized.

Schneiderhöhn (1955) summarized: *bitumen-rich ores are finely disseminated. They contain all sulfidic and carbonatic minerals of copper, zinc, lead, nickel, vanadium, and silver, occasionally even gold. A major part, naturally, is presented as pyritic iron ore. The whole ore is powdered with heaps of tiny spots of sizes between 1 and 40 µm. I assume they are metallized bacteria of the sulfur-reducing kind, alternating with those which form droplets of elementary sulfur. In summary, we may say that the combined activities of organic and inorganic reactions in these basins "filter" the trace elements out of the terrestrial weathering residues and enrich them into these sedimentary metal deposits of the biological cycle of sulfur.*

A few hundred million years later, the coastlines were differently arranged, but the biosphere was as active as ever. Bird's-eye structures, stromatolitic, and oncoidal carbonaceous parts in the Triassic intertidal sandstone flats, which today are exposed in the Red Cliff of Helgoland (North Sea), contain organic reaminders and silver, copper, vanadium, and uranium. These accumulations were undoubtedly produced by sabkha systems existing in the intertidal flats of that period. They have been mined and smelted already in the Chalcolithic period. Muir

(1979) describes similar conditions for huge ore bodies in Australia, and Renfro (1974) has made a survey of the system for several other sites. He calls the process of formation of these evaporite-associated ore bodies a sabkha process without further elaborating the biological processes involved. Dexter-Dyer et al. (1984) have estimated that an enrichment factor of only 50 was needed to accumulate the huge uranium and gold deposits of South Africa under similar conditions prevailing in the Precambrian (Amiel et al. 1972).

Many more examples could be given of salt flat or sabkha related ore deposits, which are increasingly well-documented trapping and binding results of biological systems. It suffices to say with Vernadsky's words pronounced as early as in April 1935 at the Congress on the genesis of iron, manganese, and luminium ores: *There is hardly a single metal for whose production in large concentrations life does not play a substantial role*. Nadson (1903), a Russian geomicrobiologist, has done and published several series of microbiological experiments on the enrichment and precipitation of such ores via microbial activity in saline and hypersaline systems.

Only very recently it has been made clear, that also several of the major phosphate deposits, and especially those that are highly enriched in phosphate are related to the activity of microbial mats. Several of them show signs of very shallow, partially dry, hypersaline conditions. The Negev phosphorites, far example, and some of the phosphorites of Morocco have recently been shown to be precipitated on or outside the cell walls of filamentous and coccoid cyanobacteria in a microbial mat environment (Soudry and Champetier 1983). Krajewski (1981) describes phosphate stromatolites for the Cretaceous of Poland and recently we received samples of phosphatic stromatolites from the Precambrian of Nepal. In laboratory experiments (Gerdes and Krumbein, unpublished results) we were able to show that especially in the so-called stromatolite subfacies of the littoral microbial mats when going through cycles from freshwater influences through hypersaline during tidal and seasonal fluctuations, cyanobacterial mats precipitate large amounts of phosphates. Laboratory cultures convincingly proved these findings.

The huge ore bodies of the Sudbury district in Canada, the gold and uranium deposits in South Africa, Australian ore bodies, and many in the United States and the Soviet Union today are increasingly well documented examples of ore concentrations produced by the biodynamics of sabkha systems. In many of the cases the salt has been remobilized and the environment can only be deduced from mummies and pseudomorphoses, i.e., replacements of salt minerals by other minerals under preservation of the previous mineral morphology. A fascinating but very hypothetical question is the origin of the so-called banded iron formations. Some people say that they have been generated by a more intensive sun irradiation in the Precambrium by photochemical reactions. Others say that reduced iron may have served as electron donor for photosynthetic pathways subsequent to hydrogen sulfide und before hydrogen oxide (water). In this case the question remains, however, what kind of hydrogen donor would have been used by these hypothetical "ferrigenic" (derived from anoxygenic and oxygenic) iron photosynthesizing bacteria. Undoubtedly many banded iron formations are evaporite-related, show traces of biological activity, and can be located at the borderline between reducing and oxidizing conditions at the surface of the Earth. In this case reducing and oxidizing conditions must not necessarily involve the presence of large amounts of free oxygen as the situation is today.

General biological aspects and special cases of biological accumulation of metals and salts are documented and discussed in Trudinger and Swaine (1979). Many shales which contain hydrocarbons on the one hand, and large amounts of mineable metal minerals on the other hand, are attributed to increasingly hypersaline conditions and evaporative environments. Further studies are, however, necessary in many cases to conclusively establish this mutual relationship.

In conclusion, it may be said that a large percentage of the sedimentary ore bodies of many heavy metals is related to the actively trapping and binding activity of microbiota of coastal sabkhas.

22.4 Hydrocarbons

The connection of hydrocarbons (volatile and nonvolatile) to sabkha environments is a rather difficult task. Firstly, most of the hydrocarbon deposits exploited or inventored have migrated from the source-rock into migration-tight and porous bed-rocks or reservoirs. The high mobility thus often hinders a direct correlation to the original productive environment. Secondly, there is still some debate on what are the major source substances of petroleum. In the first half of his century and even after 1945 many petroleum geologists still believed in completely inorganic sources of petroleum. Today we know that all hydrocarbons occurring in the geosphere, be they exploitable (approx. 2% of the total) or not (the remaining 98%), are biogenic. The biological source material is, however, still under discussion. Plankton and benthos of highly productive shallow seas are the favorites presently. It has been pointed out on the other hand that the amount of pollen with all its ethereal compounds, and the oils and waxes driven into the sea from the terrestrial environment, would be largely sufficient to explain the presently known exploitable oil deposits.

It is generally accepted that the transformation of original primary products and organic material embedded in sediments into oil and gas takes place at depths of 2 to 6 km in the Earth's crust. We also know that a minor fraction of less than 1% of the annual biological production, when preserved in sediments, is sufficient to generate the oil deposits known so far. Since not all kinds of organic matter readily transform into hydrocarbons under increased pressure and temperature, it has been even suggested that the very small amounts of chitinous substances derived from marine fungi, crustaceans, and insects would suffice to explain all oil deposits of the world (P. Köll, pers. comm.). It is true that chitin most readily generates a large amount of hydrocarbons of the appropriate chain length when submitted to the pressures and temperatures at 2–6 km depth. It is also conceivable that the cyclic compounds, which make up a much smaller fraction, may be derived from the remainders of planktonic and benthic phototrophic microorganisms. Furthermore, it has been shown that most of the exploitable oil deposits are younger than Precambrian, although a lot of hydrocarbons occur finely dissiminated in Precambrian sediments (McKirdy and Hahn 1982). We have demonstrated in Chapter 15, that a considerable amount of chitin may be preserved in the anaerobic phase of sabkha environments. Thus, this theory cannot be excluded presently. It also could explain the correlation of petroleum source rocks with hypersaline environments, because these guarantee the anaerobic conditions necessary for the preservation of large amounts of the end products of primary and secondary productivity (for references see Schopf 1983, Oehler et al. 1982).

Sabkha environments, oil-producing systems and petroleum reservoir rocks are closely related in many areas of the world on oceanic dynamic margins. The interbedding of sand deposits with evaporative sequences and shales, sealing off the bedrocks against upward migration, are ideally represented in (1) the Persian and Arabian Gulf, (2) the Niger Delta, (3) the Missisippi and (4) the Amazonas Delta, (5) in the Gulf of Suez, (6) the margins of the Permian evaporative system, (7) some of the Jurassic, Cretaceous, and Tertiary ancient deltas and continental margins, and (8) the evaporative facies of Grabens as, e.g., the Rhine Graben. Many of the oil shales not yet exploited have in common that they contain metal sulfide deposits, sulfur, and hydrocarbons of different compositions. Many of them expose as well stromatolithic microbial mat environments, indicating a sabkha-type system prevailing for certain periods.

In contrast to the caustobioliths of continental or terrestrial origin, which may occur at many places on the continental plates, the marine caustobioliths (petroleum, asphalt, and gas) are, whenever their origin can be located, related to marginal conditions and evaporative sinking coastal plains. As for the direct origin of petroleum, it must be stated over and over again that the exploitable sources do make up less than 0.5% of the total organic matter in sedimentary rocks. Therefore it is very difficult to decide which organism group has mainly contributed to the individual petroleum reservoirs. Isotope analysis enabled the definition of some oil shales as terrestrial/continental, while most of the real petroleum deposits and many of the oil shales are definitely related to terrestrial or shallow marine evaporative environments and they often contain stromatolithic rocks in the nearest vicinity of the petroleum-generating rock sequences.

Many of the biomolecules in petroleum resemble those molecules we find in deeper sabkha sediments (see Chap. 17 and 18). Petroleum colloquially is calles "canned solar goods" and the sun shines best and most efficiently in those coastal areas which generate evaporative microbial mat sequences of the sabkha stromatolite type described in Chapters 11 and 21.

Welte (1978) and Krumbein (1978 a) have mentioned the relation of enrichment and selective preservation of organic substances in petroleum and ore deposits. It has been stated that the ratio of phytan and pristan and the dominance of even- over odd-numbered n-alkanes in many cases can be correlated with evaporative environments as the places of petroleum genesis. By the analysis of petroleum deposits and their connection with carbonate or evaporate series it has been established that certain compounds indicate an evaporative biological system as source of the hydrocarbons. Therefore it is certainly important to study the microbiota of evaporative modern sabkha systems in order to better understand the earliest stages of petroleum and ore deposit formation. It is also necessary to study not only the ecological but also the sedimentological frame in order to be able to make prognoses of where to find new oil-bearing deposits.

22.5 Industrially Important Osmoregulators

At about the same time that many of the authors of this book were studying the Gavish Sabkha or materials from it, three experimental technological plants were established at Elat: (1) An experimental Solar Lake with its complex microbiota, which should deliver heat and electrical energy by the establishment of a thermogradient in the near vicinity of two different saline water bodies and a bacteria plate; (2) Researchers of the Weizmann Institute were running several hypersaline pools in order to produce large amounts of *Dunaliella,* harvest them and produce glycerol from them. *Dunaliella* produced glycerol as osmoregulator; (3) Several botanists of the Negev University Ber Sheba have planted a group of desert plants and were irrigating them by seawater irrigation systems enriched with some plant nutrients. The basic principles for the latter two are the capacities of microorganisms and plants to cope with extreme environmental conditions and water stress by osmoregulative functional systems (Avron and Ben-Amotz 1979, Brown 1979). The number of osmoregulators known is steadily increasing and the case of glycerol production in large amounts by *Dunaliella* is only one example out of many. Several of these may be of industrial importance, such as the selective enrichment for some amino acids (e.g., glutamate, aspartite, proline), carbohydrate derivatives, and polyhydric alcohols.

Many anoxygenic and oxygenic photosynthetic bacteria produce betains and several other important compounds. The advantage of producing such compounds from halophilic or halotolerant phototrophic organisms is obvious inasmuch as sun energy is available in unlimited amounts, salt-walter availability in contrast to freshwater has never been a limiting factor, and space is usually available in dry arid coastal regions in contrast to the coastal plains of humid regions. This leads us directly to the question of possible agricultural aspects of saline and hypersaline environments.

22.6 Salt-Water Agriculture

Many southern countries especially in the Third World presently seriously fight with the problems of salted soils. These are caused by intensification of agriculture by fertilizers and large-scale irrigation. Many of these countries produce very specialized crops for the home markets, but to a large extent also for the markets of the industrial nations in the Northern Hemisphere. The problems are caused by increasing storage of evaporite minerals within the soil and at the soil surface by continuous irrigation.

If we consider that almost all the agricultural products cultivated this way are derived from not more than 20 species of wild-type undomesticated plants which have been bred to the needed capacities through centuries, and which are still studied in hundreds of seed- and plant-breeding Research Institutes and thousands of nurseries, it may be allowed to ask the question how many salt-walter-resistant plants could be found which – when bred and cross-bred as intensively as some of the graminaceae or solanaceae presently – would be as useful and nourishing and tasting deliciously as those we use now. On the other hand, it may also be possible to adapt domesticated forms usually thriving and depending on freshwater to higher water salinities (Bloch et al. 1954). The astonishingly high primary productivity of hypersaline environments may be caused by the permanent stress of salinity and may yield even higher growth rates and productivities, when sufficient nutrients are added as fertilizers to hypersaline water.

It is easily possible in many arid coastal areas to direct seawater into large fields, where it could be evaporated in order to concentrate the nutrients in the water. By these techniques fertilizer production could be reduced and the prices of agricultural products could be lowered as long as salinity-resistant plant types were selected and bred accordingly. The study of the biology of sabkhas and hypersaline coastal plains thus can contribute considerable information about alternative agricultural plants, the selective breeding of which would be very valuable in the search for alternatives to the presently increasingly problematic irrigation agriculture in arid countries.

22.7 Biotechnology and Bioengineering

Finally, some of the new methods in biological technical applications shall be mentioned, which are potentially important in their applicability to hypersaline microorganisms.

In Mexiko and Israel mainly, but also in India and some of the Arab countries, numerous companies and experts are already searching for biotechnological applications of microbial growth in saline waters. Large shallow bays and coastal strips, especially depressions behind coastal beach berms in these countries and in Australia could be exploited for the mass cultivation of many cyanobacteria, green algae and other microorganisms.

Recently, D. Giani of the Geomicrobiology Division of the University of Oldenburg has submitted a plan for the mass cultivation of nitrogen-fixing nonheterocystous cyanobacteria on plastic mats, which would float on the water surface and could be harvested by pulling them ashore and stripping the microbial mats from the plastic material. Enough cells would remain to guarantee an inoculum for new growth. This method would enable the production of large amounts of specific proteins for chicken and fish food pellets. The harvesting would be much easier and much less costly than cultivation of plankton in shallow basins, because the separation from the water is costly and difficult by all methods tested so far (centrifugation, filtering, etc.). One of the larger oil companies has already set up several production plants for saline microorganisms in Mexico. They produce mainly ingredients for fish and chicken food. Many of the benthic cyanobacteria of the Gavish Sabkha are excellent candidates for such saline aquaculture systems. Whenever seawater pond aquaculture is tried in arid countries, the salinity increases very fast to values beyond that of seawater. The disadvantage is that halotolerant species have to be selected. The advantage is the resulting higher concentration of nutrients in these evaporating ponds.

The Geomicrobiology Division has also calculated that by these methods large amounts of atmospheric nitrogen could be bound by the well-established capacity of benthic marine and halotolerant cyanobacteria to fix nitrogen under anaerobic to microaerophilic and even aerobic conditions (Stal and Krumbein 1981). The cell mass could be transferred into methane or hydrogen and the remaining cells could be dried easily in the arid climate and used as fertilizer in freshwater agriculture. DNA recombinant techniques would accelerate the progress in selecting and breeding the appropriate strains for such floating-mat or shallow-water planktonic and/or benthic aquaculture systems. E. Holtkamp of the Geomicrobiology Division of the University of Oldenburg is presently involved in studies of the aquaculture food chain project of the Israel Limnological and Oceanographic Research Company. We hope that the data presented in this book and in Cohen et al. (1984) will help to initiate new technologies in all those countries which are faced with the lack of freshwater resources and have ideal conditions for salt-water aquaculture or agriculture.

Much more studies are needed along these lines, considering the fact that hunger and lack of appropriate and simple industries and technologies are especially typical for the countries in which the best developed hypersaline microbial mat systems and sabkhas are located and are presently under study.

Acknowledgements

In a mixed enterprise with biologists, chemists, geoscientists and physicists dealing with extreme environments through geological time it is certainly not astonishing that the number of persons and institutions which have contributed to our work should be considerable.

We wish to start with acknowledgements, we found in an article draft of Eliezer Gavish. To our grief he was not able to transmit his gratefulness, having been taken at the beginning of a life's achievements.

Eli wrote acknowledgements to G.M. Friedman, E. Sass, P.K. Weyl and D.H. Yalon for discussion and advice. He acknowledged especially heartily the late Prof. N. (Choffish) Tadmor. Tadmor, who himself died at a very early age, had shown the place later called the Gavish Sabkha to Eli for the first time. He thus initiated the work on the Gavish Sabkha as D. Friedman of the Elat Aquarium was the initator of work on the Solar Lake. Eli was aided in the field and the laboratory by his wife A. Gavish, and also by M. Zweig, L. Perliss, A. Paul, and N. Tamir.

Eliezer Gavish was supported in his work by the Bat Sheba de Rothchild Fund and by several grants of the Israel Academy of Sciences and the Hebrew University.

Further, many of the individual chapter authors field help and valuable scientific advice by colleagues and friends in Israel, including the whole staff of the H. Steinitz Marine Biology Laboratory Elat, and the late H. Steinitz himself. The Field School (now no longer existing) of the Israeli Friends of Nature at Na'ama-Ophira and all the field guides there were extremely helpful, as well as many Bedouins of the region. Support and help was also received from numerous friends and collaborators of the Israel Oceanographic and Limnological Research Company, namely H. Gordin, G. Kissil, Arie, Robin and many others. Leon (Leonidas) and Avi, and many more of the Elat staff, should be mentioned for their field aid. We would like to acknowledge as well the services of the board of directors of the MBL Elat and particulary of F.D. Por, Z. Reiss and M. Shilo. The list of individual scientists who have contributed in the field and by reading and critically commenting on the manuscripts is long. We wish, however, to express the gratitude of all authors by alphabetically listing the colleagues: Z. Aizenshtat, A. Bein, J. Ben-Asher, R. Bogosh, R.A. Bogomolni, M. Braun, B. Buchbinder, M.R. Buyce, Y. Cohen, C. Druckman, J. Erez, M. Evenari, H. Foner, J. Frank, A. Gavish, E. Gavish, G. Gvirtzman, L. Hardie, M.E. Hines, G.E. Jones, B.B. Jorgensen, D. Kahan, I.R. Kaplan, A. Katz, S. Katz, Y. Kolodny, W.E. Krumbein, R.G. LaFleur, D.T. Long, C. Monty, D. Neev, H.F.

Nelson, Y. Nir, A. Nissenbaum, I. Paperna, C. Peat, F.D. Por, U. Safriel, J.E. Sanders, E. Sass, H.K. Schminke, D. Shearman, M. Shilo, P. Strother, L.X. Zawacki and E. Zohar.

A very special and often underestimated contribution to our work is the aid of specialists in the field of taxonomy. We wish to express our thanks to the following colleagues, who have generously helped to identify many of the microorganisms and macroorganisms we were individually incapable of identifying correctly: C. Besuchet, E.G. Burmeister, Y. Cohen, I. Dor, A. Ehrlich, L. Fishelson, V. Haeseler, J.P. Henry, E. Holtkamp, L. Hottinger, D. Kahan, W.E. Krumbein, J.J. Lee, G.A. Lohse, S. Lorenzen, V. Mahnert, J. Mallwitz, N. Pfennig, F.D. Por, Z. Reiss, R. Rippka, U. Safriel, H. Sturm, H. Trüper, W. Waitzbauer, J. Wiesner.

Many people, among them laboratory technicians, students, field hands, colleagues and even directors and presidents of institutions, have actively helped in performing analyses, cultural work, producing media, processing data, drawing working on electron microscopes and scanning electron microscopes, furnishing photographs and photomicrographs from other work and, last not least, carefully preparing and cross-checking several manuscript drafts and the final typesets. K. Oetken, without knowing a word of English, has typed many English manuscripts and H. Elsken was diligently doing the processing of the reference compilation at the end of the book, using a more or less experimental data bank. Without mentioning the individual universities, all the authors wish to extend their gratitude to the following persons for this tedious but so necessary type of work: V. Arad, B.P. Armstrong, S. Ashkenazi, S.M. Awramik, B. Brand, E. Burkhardt, M.R. Buyce, Y. Cohen, M. Dvorachek, A. Ecker, K. Eylers, R. Flügel, G. Gerdes, S. Golubic, S. Grossberger, C. Hadar, H. Hofmann, E. Holtkamp, M. Hornoff, K. Jens, I.B.B. Jorgensen, B. Katz, R. Knafo, M.F.E.C. Krumbein, C.H.D. Krumbein, A.F.S. Krumbein, G. Kuenen, R. Kutiner, J. Levi, S. Levy, Y. Levy, A. Licht, A. Marziano, U. Matzigkeit, N. McKinney, G.J. Meyburg, A. Miller, S. Moshkovitz, E.H. Nieberg-van Velzen, A. Peer, I. Raether, C. Requadt, A. Rosenfeld, U. Safriel, S. Seufer, L.J. Stal, B. Steinfeld, A. Tom.

Further, our work has been sponsored and aided by many national and international science funding Academies, Institutions and Associations. Substantial scientific grants, travel awards and encouragement have been supplied. We wish to mention especially the following Institutions in Israel: The Israeli Academy of Science, The Hebrew University Jerusalem, The Tel Aviv University, the Israel Geological Survey, The Israel Oceanographic and Limnological Research Company and the Bat-Sheba de Rothschild Fund. The Deutsche Forschungsgemeinschaft with grants Nr. Kr 333/16–17 to W. E. Krumbein, and the current program of Sonderforschungsbereich 73 (Atmospheric Trace Components) have contributed considerable funds, as well as The National Science Foundation of the USA via the following grants: NSF grant EAR 79–11194; DPP 77–06993; DEB 80-04290. NASA has sponsored the work through its Global Biology and Ecology program with grants Nr. NSG-7151; NGL-05-003-003; Further, The Volkswagen Foundation of Germany has sponsored the work by a generous grant given to Y. Cohen and W.E. Krumbein during 4 years. The Minerva foundation has given a travel grant to W. E. Krumbein. Our special thanks are due to ICSU. ICSU has largely motivated and promoted our work via the International Geological Correlation Project 157 on Early Organic Evolution and Mineral and Energy Resources. Within the frame work of this project several national grants have been awarded to the ORGAST and MEGAST groups. We also wish to acknowledge a substantial grant of the "Society of Geomicrobiology and Biogeochemistry" for the color plates which would not have been possible otherwise.

Finally, the Editors and the heirs of Eli Gavish wish to express their gratitude to all contributors, who have generously resigned from any possible honorarium or income from this book in favor of the heirs of Dr. Eliezer Gavish.

References

Adams JE, Rhodes ML (1960) Dolomitization by seepage refluxion. AAPG Bull 44:1912–1920

Aharon P, Kolodny Y, Sass E (1977) Recent hot brine dolomitization in the Solar Lake, Gulf of Elat Isotopic, chemical and mineralogical study. J Geol 85:27–48

Alderman AR (1965) Dolomitic sediments in their environment in the SE of South Australia. Geochim Cosmochim Acta 29:1355–1365

Alderman AR, Skinner HCW (1957) Dolomite sedimentation in the SE of South Australia. Am J Sci 255:561–567

Aller JC (1982) Biosaline research in the United States and Canada. In: San Piertro A (ed) Biosaline research. Plenum, New York, pp 9–18

American Geological Institute (ed) (1972) Glossary of geology. Am Geol Inst, Washington

Amiel AJ, Friedman GM (1971) Continental sabkha in Arava Valley between Dead Sea and Red Sea: significance for origin of evaporites. AAPG Bull 55:581–592

Amiel AJ, Miller DS, Friedman GM (1971) The geochemistry of uranium in aragonitic sediments from a hypersaline lake, Red Sea. Am Geophys Union Trans 52:361

Amiel AJ, Miller DS, Friedman GM (1972) Uranium distribution in carbonate sediments of a hypersaline pool, Gulf of Elat, Red Sea. Israel J Earth Sci 21:187–191

Andreae MO (1980) The production of methylated sulphur compounds by marine phytoplankton. In: Trudinger PA, Walter MR, Ralph BJ (eds) Biochemistry of ancient and modern environments. Aust Acad Sci, Canberra, pp 253–259

Armstrong RE, Walsby AE (1981) The gas vesicle: a rigid membrane enclosing a hollow space. In: Ghosh BK (ed) Organization of prokaryotic cell membranes, vol II. CRC Press, Cleveland, pp 95–129

Ashbel D (1951) Bio-climatic atlas of Israel. Meteorol Dep Jerusalem

Asbhel D (1963) Climatic conditions of Elath. In: "Elath": Israel Explor Soc. 18th Archaeol Convent, Jerusalem, Israel, pp 242–256

Avraham Z Ben, Almagor G, Garfunkel Z (1979) Sediments and structure of the Gulf of Elat (Aqaba) – Northern Red Sea. Sediment Geol 23:239–267

Avraham Z Ben (1983) Structure and evolution of the Gulf of Elat: Israel Geol Soc, 1983 Annual Méeting, p 9

Avron M, Ben-Amotz A (1979) Metabolic adaptation of the alga *Dunaliella* to low water activity. In: Shilo M (ed) Strategies of microbial life in extreme environments. Verlag Chemie, Weinheim, pp 83–92

Awramik SM (1971) Precambrian columnar stromatolite diversity: reflection of metazoan appearance. Science 174:825–827

Awramik SM (1982) The pre-Phanerozoic fossil record. In: Holland HD, Schidlowski M (ed) Mineral deposits and evolution of the biosphere. Springer, Berlin Heidelberg New York, pp 67–81

Awramik SM, Schopf JW, Walter MR (1983) Filamentous fossil bacteria from the Archean of Western Australia. Precambrian Res 20:357–374

Ayyad MA, Ghabbour SJ (1984) Egypt and Sudan. In: Evenari M, Noy-Meir I (eds) Ecosystems of the world, vol XII. Elsevier, Amsterdam, in press

Badr AM, Crossland C (1939) Topography of the Red Sea floor. Publ Mar Biol Stn Ghardaqa (Red Sea) 1:13–20

Ball MM, Shinn EA, Stockmann KW (1967) The geologic effects of Hurricane Donna in South Florida. J Geol 75:583–597

Baltzer F, Conchon O, Freytet P, Purser BH (1982) Un complexe fluviodeltaique sûrsalé et son contexte: originalité du Mehran, SE Iran. Soc Géol Fr Mem 144:27–42
Barghoorn ES, Knoll AH, Dembicki H. Meinschein WG (1977) Variation in stable carbon isotopes in organic matter from the Gunflint iron formation. Geochim Cosmochim Acta 41:425–430
Barley ME, Dunlop JSR, Glover JE, Groves DI (1979) Sedimentary evidence for an Archean shallow-water volcanic-sedimentary facies, eastern Pilbara Block, Western Australia. Earth Planet Sci Lett 43:74–84
Bathurst RGC (1971) Carbonate sediments and diagenesis. Developments in sedimentology, vol XII. Elsevier, Amsterdam
Baudrimont R (1973) Recherches sur les diatomées des eaux continentales de l'Algerie, écologie et paléoécologie. Mem Soc Hist Nat Afr Nord 12:1–263
Bauld J (1981 a) Geological role of cyanobacterial mats in sedimentary environments: a production and preservation of organic matter. BMR J Aust Geol Geophys 6:307–317
Bauld J (1981b) Occurrence of benthic microbial mats in saline lakes. Hydrobiologia 81:87–111
Bayly JAE (1967) The fauna and chemical composition of some athalassic saline waters in New Zealand. New Zealand J Mar Freshwater Res 2:105–117
Bayly JAE (1972) Salinity tolerance and osmotic behavior of animals in athalassic saline and marine hypersaline waters. Annu Rev Ecol Syst 3:233–268
Beales FW, Hardy JL (1977) The problem of recognition of occult evaporites with special reference to SE Missouri. Econ Geol 72:487–490
Beall AO, Fischer AG (1969) Sedimentology. In: Natl Sci Found (ed) Initial reports of the Deep Sea Drilling Project, JOIDES 1, pp 521–593
Begin ZB, Ehrlich A, Nathan Y (1974) Lake Lisan. The Pleistocene precursor of the Dead Sea. Isr Geol Surv Bull:63–1
Behrens EW, Frishman SA (1971) Stable carbon isotopes in blue-green algae mats. J Geol 79:94–100
Ben-Menahem A, Nur A, Vered M (1976) Tectonics, seismicity and structure of the Afro-Eurasian junction – the breaking of an incoherent plate. Phys Earth Planet. Inter 12:1–50
Bentor YK, Bogoch R, Eyal M, Garfunkel Z. Shimron A (1974) Geological map of Sinai, 1:100,000, Jebel Sabbagh sheet
Berner RA (1971a) Bacterial Processes effecting the precipitation of calcium carbonate in sediments. In: Bricker OC (ed) Carbonate cements. Hopkins, Baltimore, pp 247–251
Berner RA (1971 b) Principles of chemical sedimentology. McGraw-Hill, New York
Bernstein R (1964) Salt tolerance of plants. USDA Agr Inf Bull 283:1–23
Bertrand HPI (1972) Larves et nymphes des coleopteres aquatiques du Globe. Paillart, Paris
Bertrand-Sarfati J (1976) An attempt to classify Late Precambrian stromatolite microstructures. In: Walter MR (ed) Stromatolites. Elsevier, Amsterdam
Berytii S (1826) Fragmenta de Cosmogonia et Theologia Phoenicum. Orelis CJ (ed), Hinrichs Leipzig
Blaurock AE, Walsby AE (1976) Crystalline structure of the gas vesicle wall from Anabaena flos-aquae. J Miol Biol 105:183–199
Blaurock AE, Wober W (1976) Structure of the wall of *Halobacterium halobium* gas vesicles. J Mol Biol 106:871–888
Bloch MRD, Kaplan D, Schnerb J (1954) *Juncus maritimus*. a raw material for cellulose. Bull Res Conc Israel:4–192
Bloom AL (1974) Geomorphology of reef complexes. In: Laporte LF (ed) Reefs in time and space selected examples from the recent and ancient. Soc Econ Palentol Mineral Spec Publ 18:1–8
Bogomolni RA, Spudich JL (1982) Identification of a third rhodopsin-like pigment in phototactic *Halobacterium halobium*. Proc Natl Acad Sci USA 79:6250–6254
Bogomolni RA, Baker RA, Lozier RH, Stoeckenius W (1976) Light-driven proton translocations in *Halobacterium halobium*. Biochim Biophys Acta 440:68–88
Bogomolni RA, Baker RA, Lozier RH, Stoeckenius W (1980) Action spectrum and quantum efficiency for proton pumping in *Halobacterium halobium*. Biochemistry 19:2152–2159
Boon JJ, Haverkamp J (1979) Pyrolysis mass spectrometry of benthic marine ecosystem – the influence of *Arenicola marina* on organic matter cycle. Neth J Sea Res 13:457–478
Boon JJ (1984) Tracing the origin of chemical fossils in microbial mats: biogeochemical investigations of Solar Lake cyanobacterial mats using analytical pyrolysis methods. In: Cohen Y, Castenholz RW, Halvorson HO (eds) Microbial mats: stromatolites. Liss, New York, pp 313–343

Boon JJ, Boer WR de, Kruyssen FJ, Wouters JTM (1981 a) Pyrolysis mass spectrometry of whole cells, cell walls and isolated cell wall polymers of *Bacillus subtilis* var *niger* WM. J Gen Microbiol 122:119–127

Boon JJ, Hines H, Burlingame AL, Klok J, Rijpstra WIC, Leeuw JW de, Edmunds KLH, Eglinton G (1983) Organic geochemical studies of Solar Lake laminated cyanobacterial mats. In: Bjoröy M (ed) Advances in organic geochemistry. Wiley, New York, pp 207–227

Boon JJ, Liere L van, Loogman J, Mur LR (in preparation) Pyrolysis mass spectrometry of the cyanobacterium *Oscillatoria agardhi* grown under various limiting conditions in continuous cultures. J Gen Microbiol

Borch CC von der (1965) The distribution and preliminary geochemistry of modern carbonate sediments in the Coorong area, S Australia. Geochim Cosmochim Acta 29:781–799

Borch CC von der (1976) Stratigraphy of stromatolite occurrences in carbonate lakes of the Coorong Lagoon area, South Australia. In: Walter MR (ed) Stromatolites. Elsevier, Amsterdam, pp 413–420

Borch CC von der (1980) Evolution of Late Proterozoic to Early Paleozoic Adelaide Foldbelt, Australia: Comparisons with Post-Permian rifts and passive margins. Tectonophysics 70:115–134

Borowitzka LJ (1981) The microflora-adaptations to life in extremely saline lakes. Hydrobiologia 81:33–46

Botz R, Müller G (1981) Mineralogie, Petrographie, anorganische Geochemie und Isotopen-Geochemie der Karbonatsgesteine des Zechstein 2. Geol Jahrb 47:3–112

Bourdon MC (1925) Aciens Canaux-Anciens sites et ports de Suez. Mem Soc R Geogr Egypte 7:1–156

Bourelly P (1970) Les algues d'eau douce. Initiation à la systématique 3. Baubée, Paris

Bourelly P (1972) Note sur les genres *Pleurocapsa* et *Scopulonema*. In: Desikachary TV (ed) Taxonomy and biology of Blue-green algae. Univ Madras, Banglore

Bowen CC, Jensen TE (1965) Blue-green algae: fine structure of the gas vacuoles. Science 147:1460–1462

Boyko M (1966) Salinity and aridity: New Approaches to old problems. Junk, The Hague

Bramkamp RA, Brown GF, Holm DA, Layne NM jr (1963) Geology of the Wadi As Sirhan Quadrangle, Kingdom of Saudi Arabia. US Geol Surv, Misc Geol Invest Map 1–20A

Brassel SC (1980) The lipids of deep sea sediments: their origin and fate in the Japan Trench. Ph D Thesis Bristol

Braun T (1958) Das Geschichtswerk des Herodotos von Halikarnassos. Insel, Stuttgart

Bro Larsen E (1936) Biologische Studien über die tunnelgrabenden Käfer auf Skallingen. Vidensk Medd Dan Naturhist Foren Khobenhavn 100:1–231

Brock TD (1976) Environmental Microbiology of living stromatolites. In: Walter MR (ed) Developments in sedimentology, vol XX. Elsevier, Amsterdam, pp 141–148

Brock TD (1978) Thermophilic microorganisms and life at high temperatures. Springer, Berlin, Heidelberg, New York

Brock TD (1979) Ecology of saline lakes. In: Shilo M (ed) Strategies of microbial life in extreme environments. Life Sci Res Rep 13, Dahlem Konferenzen. Chemie Verlag, Weinheim, pp 29–47

Brock TD, Petersen S (1976) Some effects of light on the viability of rhodopsin-containing halobacteria. Arch Microbiol 109:199–200

Brockmann C (1950) Die Watt-Diatomeen der schleswig-holsteinischen Westküste. Abh Senckenberg Naturforsch Ges 478:1–26

Brooks CJW, Horning EC, Young JS (1968) Characterization of sterols by Chromography-Mass Sepctrometry of the trimethylsilyl ethers. Lipids 3:391–402

Brown AD (1979) Physiological Problems of Water Stress. In: Shilo M (ed) Strategies of microbial life in extreme environments. Chemie Verlag, Weinheim, pp 65–82

Buchanan BB (1979) Ferredoxin-linked carbon dioxide fixation in photosynthetic bacteria. In: Gibbs M, Latzko E (eds) Encyclopedia of plant physiology, New Ser, vol VI. Springer, Berlin Heidelberg New York, pp 416–424

Bucher WH (1919) On ripples and related sedimentary surface forms and their paleographic interpretation. Am J Sci 47:149–210, 241–269

Buckingham IS (1827) Reisen durch Syrien und Palestina. Neue Bibliothek der wichtigsten Reisebeschreibungen zur Erweiterung der Erd- und Völkerkunde 45. Landes-Industrie-Comptoirs, Weimar

Buick R, Dunlop JSR, Groves DI (1981) Stromatolite recognition in ancient rocks: an appraisal of irregularly laminated structures in an Early Archaean chert-barite unit from North Pole, Western Australia. Alcheringia 5:161–181

Bull WB (1964) Alluvial fans and near-surface subsidence in western Fresno County, California. US Geol Surv Prof Pap 437-A

Burckhardt JL (1824) Johann Ludwig Burckhardt's Reisen in Syrien, Palästina und der Gegend des Berges Sinai. In: Gesenius W (ed) Neue Bibliothek der wichtigsten Reisebeschreibungen zur Erweiterung der Erd- und Völkerkunde 38. Landes-Industrie-Comptoirs, Weimar (also London, 1822)

Buschmann JO Frh von (1906) Das Salz, dessen Vorkommen und Verwertung in sämtlichen Staaten der Erde: Asien, Afrika, Amerika und Australien mit Ozeanien. Engelmann

Buschmann JO Frh von (1909) Das Salz, dessen Vorkommen und Verwertung in sämtlichen Staaten der Erde. Engelmann, Leipzig

Bush P (1973) Some aspects of the diagenetic history of the sabkha in Abu Dabhi. In: Purser BH (ed) The Persian Gulf. Springer, Berlin Heidelberg New York, pp 395–407

Butler GP (1969) Modern evaporite deposition and geochemistry of coexisting brines, the sabkha, Trucial Coast, Arabian Gulf. J Sediment Petrol 39:70–89

Butler GP (1970a) Secondary anhydrite from a sabkha, NW Gulf of California, Mexico. N Ohio Geol Soc 3rd Symp Salt:153–155

Butler GP (1970b) Holocene gypsum and anhydrite of the Abu Dabhi sabkhas Trucial Coast: an alternative explanation of origin. N Ohio Geol Soc 3rd Symp Salt:120–152

Butler GP (1973) Strontium geochemistry of modern and ancient calcium sulphate minerals. In: Purser BH (ed) The Persian Gulf. Springer, Berlin Heidelberg New York, pp 423–452

Butler GP, Harris PMS, Kendall CGStC (1982) Depositional and diagenetic spectra of evaporites – a core workshop. In: Hanford DR, Loucks RG, Davies GR (eds) SEPM Core Workshop 3, Calgary, Canada, pp 33–64

Button A (1977) Stratigraphic history of the Middle Proterozoic Umkondo Basin in the Chipinga area, southeastern Rhodesia. Econ Geol Res Unit Univ Witwatersrand Int Circ, Witwatersrand SA

Byblos P von (1887) Sanchuniathon's Phönizische Geschichte, Lübeck

Calder JA, Parker PL (1973) Geochemical implications of induced changes in ^{13}C fractionation by blue-green algae. Geochim Cosmochim Acta 37:133–140

Caple M, Chow H, Burns RM, Strouse CE (1977) Models of pigment aggregation based on crystallographic investigations: recent results pertinent to the composition of the light-harvesting system of green sulfur bacteria. Chlorophyll BNL-50530:56–63

Cardoso J, Brooks PW, Eglinton G, Philipp RP (1976) Lipids of recently deposited algal mats at Laguna Mormona, Baja California. In: Nriagu J (ed) Environmental biogeochemistry. Ann Arbor Sci Publ. Ann Arbor, pp 149–174

Cardoso J, Watts CD, Maxwell JR, Goodfellow R, Eglinton G, Golubic S (1978) A biogeochemical study of the Abu Dhabi algal mats: a simplified ecosystem. Chem Geol 23:273–291

Carless TC (1837) Memoir on the Gulf of Akabah and the head of the Red Sea. Bombay Geogr Proc 1837:27–59

Carmondy DJ, Pearce JB, Yasso WF (1973) Trace metals in sediments of New York Bight. Mar Pollut Bull 4:132–135

Carruthers GT (1890) Locusts in the Red Sea. Nature 41:153–156

Casadio R, Stoeckenius W (1980) Effect of protein-protein interaction on light adaptation of bacteriorhodopsin. Biochemistry 19:3374–3381

Casadio R, Gutowitz H, Mowery P, Taylor M. Stoeckenius W (1980) Light-dark adaptation of bacteriorhodopsin in triton-treated purple membrane. Biochim Biophys Acta 590:13–23

Castenholz RW (1973) Ecology of blue-green algae in hot springs. In: Carr NG, Whitton BA (eds) The biology of blue-green algae. Blackwell, Oxford, pp 379–414

Caumette P (1978) Participation des bactéries phototrophes sulfooxydantes dans le métabolisme du soufre en milieu lagunaire méditerranéen (Étang du Prevost). Ph D thesis, Montpellier

Chace FA, Manning RB (1972) Two new caridean shrimps, one representing a new family, from marine pools on Ascension Island (Crustacea, Decapoda, Natantia). Smithson Contrib Zool 131:1–18

Chapman VJ (1974) Salt marshes and salt deserts of the world, 2nd edn. Cramer, Lehre

Chapman VJ (1975) The salinity problem in General, its maintenance, and distribution with special reference to natural halophytes. In: Poljakoff-Mayber A, Gale J (eds) Plants in saline environments. Ecol Stud 15. Springer, Berlin Heidelberg New York, pp 1–24

Chapman VJ (1977) Introduction. In: Chapman VJ (ed) Ecosystems of the world, vol I. Elsevier, Amsterdam, pp 1–29
Cholnoky BJ (1968) Die Ökologie der Diatomeen in Binnengewässern. Cramer, Lehre
Clementz H (1979) Des Flavius Josephus jüdische Altertümer. Fourier, Wiesbaden (Nachdruck)
Cloud PE (1976) Beginnings of biospheric evolution and their biogeochemical consequences. Paleobiology 2:351–387
Cloud PE, Glaessner MF (1982) The Ediacaran Period and System: Metazoa inherit the earth. Science 218:783–792
Cockayne L (1958) The vegetation of New Zealand, Engelman, London
Cohen S (1975) Red Sea Diver's guide. Red Sea Diver's Publ, Tel Aviv
Cohen Y (1975) Dynamics of procaryotic photosynthetic communities of the Solar Lake. Unpubl Ph D thesis, Hebrew Univ, Jerusalem
Cohen Y, Krumbein WE, Shilo M (1977 a) Solar Lake (Sinai) 2 Distribution of photosynthetic microorganisms and primary production. Limnol Oceanogr 22:609–620
Cohen Y, Krumbein WE, Shilo M (1977 b) Solar Lake (Sinai) 3 Bacterial distribution and production. Limnol Oceanogr 22:621–634
Cohen Y, Krumbein WE, Goldberg M, Shilo M (1977 c) Solar Lake (Sinai) 1 Physical and chemical limnology. Limnol Oceanogr 22:597–608
Cohen Y, Aizenshtat Z, Stoler A, Jorgensen BB (1980) The microbial geochemistry of Solar Lake, Sinai. In: Ralph JB, Trudinger PA, Walter MR (eds) Biogeochemistry of ancient and modern environments. Springer, Berlin Heidelberg New York, pp 167–172
Cohen Y, Castenholz RW, Halvorson HO ed (1984) Microbial mats: stromatolites. MBL Lectures in Biology 3, Liss New York, pp 498
Collins NC (1977) Ecological studies of terminal lakes, their relevance to problems in limnology and population biology. Proc Int Conf Terminal Lakes, Utah State Press, Salt Lake City
Cooper JA, James PR, Rutland RWR (1982) Isotopic dating and structural relationships of granitoids and greenstones in the East Pilbara, Western Australia. Precambrian Res 18:199–236
Copland BJ (1967) Environmental characteristics of hypersaline lagoons. Contrib Mar Sci Univ Tex 12:207–218
Copland BJ, Jones RS (1966) Community metabolism in some hypersaline waters. Tex J Sci:188–205
Craig H (1953) The geochemistry of stable carbon isotopes. Geochim Cosmochim Acta 3:53–92
Craig H (1957) Isotopic standards for carbon and oxygen and correction factors for mass spectrometric analyses of carbon dioxide. Geochim Cosmochim Acta 12:133–149
Cranwell PA (1973) Branched-chain and cyclopropanoid acids in a recent sediment. Chem Geol 11:307–313
Cranwell PA (1982) Lipids of aquatic sediments and sedimenting particulates. Prog Lipid Res 21:271–308
Crick IH, Muir MD (1980) Evaporite and uranium mineralization in the Pine Creek Geosyncline. Int Uranium Symp Pine Creek Geosyncline, 1979. Proc Int At Energy Agency, pp 531–542
Crossland C (1939) Reports on the Preliminary Expedition for the exploration of the Red Sea in the RRS Mahabith 1934–1935: Narrative and list of stations. Publ Mar Biol Stn Ghardaqa (Red Sea) 1:3–11
Cumberland DD (1720) Sanchoniatho's Phoenician history. Wilkin, London
Cummins KW (1973) Trophic relations of aquatic insects. Annu Rev Entomol 18:183–206
Currie R (1955) The "Red" Sea?: Port of Aden Annual 1954/55
Curtis R, Evans G, Kinsman DJJ, Sherman DJ (1963) Association of dolomite and anhydrite in the Recent sediments of the Persian Gulf. Nature (London) 197:679–680
Czoering V v (1869) Salz aus dem Großen Bittersee im Suezkanal. Verh K K Geol Reichsanst Wien 10:222–223
Dahl E (1967) Ecological salinity boundaries in poikilohaline waters. Oikos 7:1–21
Danin A (1981) The impact of geomorphology and climatic conditions on the vegetation of salt marshes along the Mediterranean coast of Israel and Sinai. Actas III Congr Optima Ann Jard Bot Madrid 37:269–275
Danin A (1983) Desert vegetation of Israel and Sinai. Cana Publ, Jerusalem
Danon A, Stoeckenius W (1974) Photophosphorylation in *Halobacterium halobium*. Proc Natl Acad Sci USA 71:1234–1238

Darwin C (1839) Journal of the researches into the geology and natural history of the various countries visited by HM Ship Beagle. Appleton and Company, New York

Davidson CF (1965) A possible mode of origin of strata-bound copper ores. Econ Geol 60:942–954

Davies GR (1970a) Algal laminated sediments, Gladstone Embayment, Shark Bay, W. Australia. AAPG Mem 13:169–205

Davies GR (1970b) Carbonate bank sedimentation, eastern Shark Bay, Western Australia. In: Logan BW, Davies GR, Read JF, Cebulski DE (eds) Carbonate sedimentation and environments. Shark Bay, Western Australia. AAPG Mem 13, pp 85–167

Davis JH (1977) Genesis of the SE Missouri lead deposits. Econ Geol 72:443–450

Davis JS (1978) Biological communities of a nutrient enriched salina. Aquat Bot 4:23:42

Deacon GER (1952) Preliminary hydrological report: The Manihine Expedition to the Gulf of Aqaba 1948–1949. Br Mus (Nat Hist) Zool Bull 1:150–162

Deffeyes KS, Lucia FJ, Weyl PK (1965) Dolomitization of Recent and Plio-Pleistocene sediments by marine evaporite waters on Bonaire, Netherlands Antilles. In: Pray LC, Murray RC (eds) SEPM Spec Publ 13 Tulsa, pp 71–88

Degens ET, Epstein (1964) Oxygen and carbon isotope ratios in coexisting calcites and dolomites from recent and ancient sediments. Geochim Cosmochim Acta 28:23–44

Degens ET, Guillard RRL, Sackett WM, Hellebust JA (1968) Metabolic fractionation of carbon isotopes in marine plankton – 1 Temperature and respiration experiments. Deep-Sea Res 15:1–9

Deines P (1970) Mass spectrometer correction factors for the determination of small isotopic composition variations of carbon and oxygen. Int J Mass Spectr Ion Phys 4:283–295

Deines P (1980) The isotopic composition of reduced organic carbon. In: Fritz P, Fontes JC (eds) Handbook of environmental isotope geochemistry, vol I. Elsevier, Amsterdam, pp 329–406

Desikachary TV (1959) Cyanophyta. Indian Council Agron, Res Monogr Algae, New Delhi

Deuser WG, Degens ET, Guillard RRL (1968) Carbon isotope relationships between plankton and sea water. Geochim Cosmochim Acta 32:657–660

Dexter-Dyer B, Kretzschmar M, Krumbein WE (1984) Possible microbial pathways in the formation of Precambrian ore deposits. J Geol Soc London 141:251–262

Dimentman C, Spira Y (1982) Predation of Artemia cysts by water tiger larva of the genus *Anacaena* (Coleoptera, Hydrophilidae). Hydrobiologia 97:163–165

Dörjes J, Hertweck G (1975) Recent biocoenoses and ichnocoenoses in shallow-water marine environments. In: Frey RW (ed) The study of trace fossils. Springer, Berlin Heidelberg New York, pp 459–491

Dor I (1967) Algues des sources thermales de Tiberiade. Bull Sea Fish Res Stn 48:3–29

Dor I (1974) Considerations about the composition of benthic algal flora in Lake Kinneret. Hydrobiologia 44:255–264

Dor I (1975) The blue-green algae of the mangrove forest of Sinai. Rapp Comm Int Mer Medit 23:109–110

Dor I (1984) Epiphytic blue-green algae (Cyanobacteria) of the Sinai mangal: Considerations on vertical zonation and morphological adaptations. In: Por FD, Dor I (eds) Hydrobiology of the mangal. Dr. W. Junk Publishers, The Hague, pp 35–54

Dubertret ML (1932) Geologie – Les formes structurales de la Syrie et de la Palestine; leur origine. C R Acad Sci 195:66–68

Ebach J (1979) Weltentstehung und Kulturentwicklung bei Philo von Byblos. Kohlhammer, Stuttgart

Eckstein Y (1970) Physicochemical Limnology and geology of a meromictic pond on the Red Sea shore. Limnol Oceanogr 15:363–372

Eden JG van (1978) Stratiform copper and zinc mineralization in the Cretaceous of Angola. Econ Geol 73:1154–1161

Edward IES, Gadd CJ, Hammond NGL, Sollberger E (1975) History of the middle east and the aegean region. The Cambridge ancient history, vol II. Cambridge Univ Press, Cambridge

Ehrenberg CG (1830) Neue Beobachtungen über blutartige Erscheinungen in Ägypten, Arabien und Sibirien, nebst einer Übersicht und Kritik der früher bekannten. Ann Phys Chem 94:477–514

Ehrenberg CG (1832) Die geographische Verteilung der Infusionsthierchen in Nord-Afrika und West-Asien beobachtet auf Ehrenbergs und Hemprichs Reisen. Abh Physikal Kl Akad Wiss Berlin 1829:1–20

Ehrenberg CG (1834 a) Beiträge zur physiologischen Kenntnis der Corallenthiere im allgemeinen und besonders des rothen Meeres nebst einem Versuch zur physiologischen Systematik derselben. Abh K Akad Wiss Berlin 1831:225–380

Ehrenberg CG (1834b) Über die Natur und Bildung der Corallenbänke des Rothen Meeres. Abh K Akad Wiss Berlin 1832:381–438

Ehrlich A (1975) The diatoms from the surface sediments of the Bardawil Lagoon (N Sinai). Nova Hedwigia Z Kryptogamenkd 53:253–277

Ehrlich A (1978 a) The diatoms of the hyperhaline Solar Lake (NE Sinai). Isr J Bot 27:1–13

Ehrlich A (1978 b) Living and subfossil diatoms of a hyperhaline heliothermic pond on the western shore of the Gulf of Elat (NE Sinai). Pol Arch Hydrobiol 25:131–134

Ehrlich A, Ortal R (1979) The influence of salinity on the benthic diatom communities of the Lower Jordan River. Nova Hedwigia Z Kryptogamenkd 64:325–334

Eichmann R, Schidlowski M (1975) Isotopic fractionation between coexisting organic carbon-carbonate pairs in Precambrian sediments. Geochim Cosmochim Acta 39:585–595

Elenkin AA, Danilov A (1916) Recherches cytologiques sur les cristaux et les grains de secretion dans les cellules de *Symploca muscorum* (Ag). Gom Bull Jard Imp Bot Pierre Grand 16

Emery KO (1964) Sediments of Gulf of Aqaba (Eilat). In: Miller RL (ed) Papers in marine geology. Shepard Commem Vol. Macmillan, New York, pp 257–273

Endo R (1961) Phylogenetic relationships among the calcareous algae. Saitama Univ Sci Rep Ser B Endo Commem:1–48

Engel HSJ (1979) Die Vorfahren Israels in Ägypten. Knecht, Frankfurt

Epstein E, Kingsbury RW, Norlyn JD, Rush DW (1979) Production of food crops and other biomass by seawater culture. In: Hollaender A (ed) The biosaline concept. Plenum, New York, pp 77–99

Epstein SA, Friedman GM (1983) Depositional and diagenetic relationships between Gulf of Elat (Aqaba) and Mesozoic of United States east coast offshore. AAPG Bull 67:953–962

Erez Y (1972) Multivariate analysis of biogenic constituents in the sediments of Ras Burka, Gulf of Elat. M Sc Thesis Jerusalem

Estep MF (1982) Stable isotope composition of algae and bacteria that inhabit hydrothermal environments in Yellowstone National Park. Ann Rep Dir Geophys Lab Carnegie Inst Washington 1981-2:403–410

Estep MF, Tabita FR, Baalen C van (1978 a) Purification of ribulose-15-bisphosphate carboxylase and carbon isotope fractionation by whole cells and carboxylase from *Cylindrotheca* sp (Bacillariophyceae). J Phycol 14:183–188

Estep MF, Tabita FR, Parker PL, Baalen C van (1978 b) Carbon isotope fractionation by ribulose-15-bisphosphate carboxylase from various organisms. Plant Physiol 61:680–687

Eugster HP, Hardie LA (1975) Sedimentation in an ancient playa-like complex, the Wilkins Peak Member of the Green River Formation of Wyoming. Bull Geol Soc Am 86:319–334

Evans G, Schmidt V, Bush P, Nelson H (1969) Stratigraphy and geologic history of the sabkha, Abu Dhabi, Persian Gulf. Sedimentology 12:145–159

Evenari M (1984) Adaptations of plants and animals to desert conditions. In: Evenari M, Noy-Meir I (eds) Desert ecosystems of the world, vol XII. Elsevier, Amsterdam, in press

Evenari M, Gutterman Y (1973) Some notes on *Salvadora persica* L in Sinai and its use as a toothbrush. Flora 162:118–125

Evenari M, Gutterman Y (1976) Observations on the secondary succession of three plant communities in the Negev desert Israel I Artemisetum herbae albae. Etud Biol Veg Hommage Prof P Chouard. Jaques R (ed), CNRS Paris

Evenari M, Yaalon DH, Gutterman Y (1974) Notes on soils with vesicular structure in deserts. Z Geomorph NF 18:162–173

Evenari M, Shanan L, Tadmor N (1982) The Negev: The challenge of a desert, 2nd edn. Harvard Univ Press, Cambridge USA

Fabricius B (1883) Der Periplus des Erythräischen Meeres. Veit, Leipzig

Fairbridge RW (1961) Eustatic changes in sea level. Phys Chem Earth 4:99–185

Fairbridge RW, Teichert C (1948) The low isles of the Great Barrier Reef: a new analysis. Geogr J 3:67–88

Farrar JA (1907) Literary Forgeries London, London

Feinbrunn-Dothan N (1978) Flora Palestina. Isr Acad Sci Hum Jerusalem

Felger RS, Mota-Urbina JC (1982) Halophytes: New sources of nutrition. In: San Pietro A (ed) Biosaline research. Plenum, New York, pp 473–477

Ferrar HT (1914) Note on a mangrove swamp at the mouth of the Gulf of Suez. Cairo Sci J 8:23–24

Fischer H (1979) Osmotic behavior of diatoms in a hypersaline lake in comparison with tidal diatoms. Nova Hedwigia 64:251–264

Fishelson L (1973) Ecological and biological phenomena influencing coral species composition on the reef tables at Eilat (Gulf of Aqaba, Red Sea). Mar Biol 19:183–196

Folk RL, Ward W (1957) Brazos River bar: a study in the significance of grain size parameters. J Sediment Petrol 27:3–26

Forskal P (1775 a) Descriptions animalium, avium, amphibiorum, piscium, insectorum, vermium quae in itinere orientali observavit. In: Niebuhr C (ed) Möller, Kopenhagen, pp 7–15

Forskal P (1775 b) Flora aegyptiaco-arabica Sive descriptiones plantarum, quas per Aegyptumn inferiorem et Arabiam felicem detexit, illustravit Petrus Forskal. In: Niebuhr C (ed). Möller, Kopenhagen, pp 184–200

Forskal P (1776) Icones rerum naturalium, quas in itinere orientali depingi curavit Petrus Forskal, In: Niebuhr C (ed). Möller, Kopenhagen, pp 4–15

Foster CH, Berchtold GA (1972) Esterifying alcohols in the chlorophylls of purple photosynthetic bacteria. A new chlorophyll, bacterio-chlorophyll (99), all-trans-geranylgeranyl bacteriochlorophyllide a. J Am Chem Soc 94:7938–7939

Fox GE, Stackebrandt E, Hespell RB, Gibson J, Maniloff J, Dyer TA, Wolfe RS, Balch WE, Tanner RS, Magrum LJ, Zablen LB, Blakemore R, Gupta R, Bonen L, Lewis BJ, Stahl DA, Luehrsen KR, Chen KN, Woese CR (1980) the phylogeny of prokaryotes. Science 209:457–472

Fraas O (1867) Aus dem Orient. Geologische Beobachtungen. Ebner und Seubert, Suttgart

Frank J. Shimkin B, Dowse H (1981) SPIDER – A modular software system for electron image processing. Ultramicroscopy 6:343–358

Frémy P (1930) Les Mycophycées de l'Afrique Equatoriale française, Arch Bot Mem 3:1–508

Frémy P (1934) Cyanophycées des côtes d'Europe. Mem Soc Sci Nat Math Cherbourg 41:1–234

Frémy P, Nasr AH (1938) Two new cyanophyceae from the Red Sea. Bull Fac Sci:18–31

Freund R (1965) A model for the structural development of Israel and adjacent areas since Upper Cretaceous times. Geol Mag 102:189–205

Freund R, Garfunkel Z (1976) Guidebook to excursion along the Dead Sea rift. Dep Geol, Jerusalem

Freund R, Zak I, Garfunkel Z (1968) Age and rate of sinistral movement along the Dead Sea rift. Nature (London) 220:253–255

Freund R, Garfunkel Z, Zak I, Goldberg M, Weissbrod T, Derin B (1970) The shear along the Dead Sea rift. Philos Trans R Soc London Ser A 267:107–130

Fricke HW (1976) Bericht aus dem Riff. Ein Verhaltensforscher experimentiert im Meer. Piper, München

Friedman GM (1959) Identification of carbonate minerals by staining methods. J Sediment Petrol 29:87–97

Friedman GM (1962) On sorting, sorting coefficients, and the lognormality of the grain-size distribution of sandstones. J Geol 70:737–753

Friedman GM (1965) A fossil shoreline reef in the Gulf of Elat (Aqaba). Isr J Earth Sci 14:86–90

Friedman GM (1968) Geology and geochemistry of reefs, carbonate sediments and waters, Gulf of Aqaba (Elat), Red Sea. J Sediment Petrol 38:895–919

Friedman GM (1969) Depositional environments in carbonate rocks. In: Friedman GM (ed) SEPM Spec Publ 14, p 209

Friedman GM (1972) Significance of Red Sea in problem of evaporites and basinal limestones. AAPG Bull 56:1072–1086

Friedman GM (1975) The making and unmaking of limestones or the downs and ups of porosity. J Sediment Petrol 45:379–398

Friedman GM (1978) Solar Lake: A sea-marginal pond of the Red Sea (Gulf of Aqaba or Elat) in which algal mats generate carbonate particles and laminites. In: Krumbein WE (ed) Environmental biogeochemistry and geomicrobiology. The aquatic environment 1. Ann Arbor Sci, Ann Arbor, pp 227–235

Friedman GM (1980) Review of depositional environments in evaporite deposits and the role of evaporites in hydrocarbon accumulation. Bull Cent Rech Explor-Prod ELF-Aquitaine 4:589–608

Friedman GM (1982) Coexisting terrigenous sea-marginal fans and reefs at the shore of the Gulf of Aqaba. Abstr Int Assoc Sedimentol, 11th Int Congr Sedimentol, Hamilton Ontario

Friedman GM, Brenner IB (1977) Progressive diagenetic elimination of Strontium in Quaternary to Late tertiary coral reefs of Red Sea: Sequence and time scale. Stud Geol AAPG 4:353–355

Friedman GM, Foner HA (1982) pH and Eh changes in sea-marginal algal pools of the Red Sea and their effects on carbonate precipitation. J Sediment Petrol 52:41–46

Friedman GM, Gavish E (1971) Mediterranean and Red Sea (Gulf of Aqaba) beachrocks. In: Bricker OP (ed) Carbonate cements. Johns Hopkin Univ Press, Baltimore, pp 13–16

Friedman GM, Sanders IE (1967) Origin and occurrence in dolostone. In: Chilingar GV, Bissell HJ, Fairbridge RW (eds) Carbonate rocks. Elsevier, Amsterdam, pp 267–348

Friedman GM, Sanders IE (1978) Principles of sedimentology. Wiley, New York

Friedman GM, Amiel AJ, Braun M, Miller DS (1968) Are mud-cracked algal laminated sediments evidence of supratidal origin? New evidence from tidal lakes in the Sinai Peninsula. Geol Soc Am Spec Pap 121:103

Friedman GM, Amiel AJ, Braun M, Miller DS (1973 a) Generation of carbonate particles and laminites in algal mats – example from sea-marginal hypersaline pool, Gulf of Aqaba, Red Sea. AAPG Bull 57:541–557

Friedman GM, Sneh A, Owen RW (1973 b) Generation of laminated gypsum in sea-marginal pool, Red Sea (abs.). AAPG-SEPM Symp, Anaheim, CA. AAPG Bull 57:780

Friedman GM, Amiel AJ, Schneidermann N (1974) Submarine cementation in reefs: example from the Red Sea. J Sediment Petrol 44:816–825

Friedman S (1960) Oceanographic data obtained in the Indian Ocean, Gulf of Aden and the Red Sea during the Vema cruise 14 an Vema cruise 16. Tech Rep Lamont Geol Obs CU-10-60, AT (30-1):1808

Fritsch FE (1945) Structure and reproduction of the algae. Cambridge Univ Press, Cambridge

Fuchs G, Stupperich E (1981) Wege der autotrophen CO_2-Fixierung in Bakterien. Forum Mikrobiol 4:198–201

Fürer FA (1900) Salzbergbau- und Salinenkunde. Vieweg, Braunschweig

Gale J (1975) The combined effect of environmental factors and salinity on plant growth. In: Poljakoff-Mayber A, Gale J (eds) Plants in saline environments. Ecol Stud 15. Springer, Berlin Heidelberg New York, pp 186–199

Gale J, Pojakoff-Mayber A (1970) Interrelation between growth and photosynthesis of salt bush (*Atriplex halimus* L) grown in saline media. Aust J Biol Sci 23:937–945

Gale J, Naaman R, Poljakoff-Mayber A (1970) Growth of *Atriplex halimus* L in sodium chloride salinated culture solutions as affected by the relative humidity of the air. Aust J Biol Sci 23:947–952

Ganor A, Markovitz R, Kessler J, Rosnan B (1973) The climate of Sinai. Isr Meteor Serv (Hebrew) Beit Dagan

Garfunkel Z (1970) The tectonics of the Western Margins of the Southern Arava. Thesis, Jerusalem (Hebrew, English summary)

Garfunkel Z, Bartov J, Eyal Y, Steinitz G (1973) Raham conglomerate – new evidence for Neogene tectonism in the southern part of the Dead Sea rift. Geol Mag 111:55–64

Garlick GD (1969) The stable isotopes of oxygen. In: Wedepohl KH (ed) Handbook of geochemistry, vol II/1. Springer, Berlin Heidelberg New York, pp 8-B-1–8-B-27

Garrels RM, Mackenzie FT, Hunt C (1975) Chemical cycles and the global environment. Kaufman, Los Altos

Garrett P (1970) Phanerozoic stromatolites; non competitive ecologic restriction by grazing and burrowing animals. Science 169:171–173

Gary M, McAfee R Jr, Wolf CL (eds) (1972) Glossary of geology. American Geological Institute, Washington, DC

Gaskell SJ, Eglinton G (1975) Rapid hydrogenation of sterols in a contemporary lacustrine sediment. Nature (Condou) 254:209–211

Gassé F (1975) Levolution des lacs de lAfar Central (Ethiopie et TFAI) du Plio-Pléistocence à lActuel; Réconstruction des paléomilieux laucustres à partir de l'ètude des Diatomees (in French). Doct Sc these Université Paris

Gaudette HE, Lyons WB (1984) Trace metal concentrations in modern marine sabkha sediments. In: Cohen Y, Castenholz RW, Halvorson HO (eds) Microbial Mats: Stromatolites, Liss New York

Gavish E (1974a) Mineralogy and geochemistry of a coastal sabkha near Nabek, Gulf of Elat. In: Gill D (ed) Abstracts of papers presented at the 1972/73 seminar of the Geol Survey of Israel Jerusalem, pp 18–19

Gavish E (1974b) Geochemistry and mineralogy of a recent sabkha along the coast of Sinai, Gulf of Suez. Sedimentology 21:397–414

Gavish E (1975a) Example of a sea-marginal hypersaline pool, Gulf of Aqaba, Red Sea. AAPG Bull 57:541–557

Gavish E (1975b) Recent coastal sabkhas marginal to the gulfs of Suez and Elat, Red Sea. Rapp Comm Int Mer Medit 23:129–130

Gavish E (1975c) Recent and Holocene beachrocks along the coasts of Sinai, gulfs of Elat and Suez. Rapp Comm Int Mer Medit 23:131–132

Gavish E (1980) Recent sabkhas marginal to the Southern coasts of Sinai, Red Sea. In: Nissenbaum A (ed) Hypersaline brines and evaporite environments. Elsevier, Amsterdam, pp 233–251

Gavish E, Friedman GM (1969) Progressive diagenesis in Quarternary to Late Tertiary carbonate sediments: sequence and time scale. J Sediment Petrol 39:980–1006

Gavish E, Braun M, Bein A, Buchbinder B, Friedman GM (1969) Neogene-Recent sediments in the Ras Muhammad area. Isr Geol Soc Proc:41–44

Gavish E, Krumbein WE, Tamir N (1978) Recent clastic (carbonate) sediments and sabkhas marginal to the gulfs of Eilat and Suez. Field excursion guidebook IAS 10th Int Congr, Jerusalem II, pp 309–332

Gebelein CD (1974) Biologic control of stromatolite microstructure: implications for Precambrian time stratigrapy. Am J Sci 274:575–598

Geitler L (1925) Cyanophyceae. In: Pascher A (ed) Süßwasserflora Deutschlands, Österreichs und der Schweiz 12. Fischer, Jena

Geitler L (1932) Cyanophyceae. Kryptogamenflora von Deutschland, Österreich und der Schweiz:14–1. Akademie Verlag, Leipzig

Geitler L (1942) Schizophyceae. In: Engler A, Prantl K (eds) Die natürlichen Pflanzenfamilien. Duncker & Humboldt, Berlin

Gelpi E, Schneider H, Mann J, Oro J (1970) Hydrocarbons of geochemical significance in microscopic algae. Phytochemistry 9:603–612

Gerdes G, Krumbein WE (1984) Animal communities in recent potential stromatolites of hypersaline origin. In: Cohen Y, Castenholz RH, Halvorson HO (eds) Microbial Mats: Stromatolites. Liss, New York, pp 59–83

Gerdes G, Krumbein WE, Reineck HE (1982) Grenzgänger des Lebens. Ökologische Studien an zwei strandnahen Salzseen am Golf von Aqaba (Sinai). Nat Mus 112:309–323

Gibbons MJ (1978) The geochemistry of sabkha and related depostis. Ph D Thesis, Newcastle UK

Gill D (1977) Salina A-1 sabkha cycles in the late Silurian paleogeography of the Michigan basin. J Sediment Petrol 47:979–1017

Girdler RW, Styles P (1974) Two stage Red Sea floor spreading. Nature (London) 247:7–11

Glennie KW (1970) Desert sedimentary environments. Developments in sedimentology, vol. 15. Elsevier, Amsterdam

Glennie KW, Evamy DB (1968) Dikaka: plants and plant root structures associated with aeolian sand. Palaeogeogr Palaeoclimatol Palaeoecol 4:77–87

Gloe A, Pfennig N, Brockman J jr, Trowitzsch W (1975) A new bacteriochlorophyll from brown-colored chlorobiaceae. Arch Microbiol 102:103–109

Glover ED, Sippel RF (1967) Synthesis of magnesium calcites. Geochim Cosmochim Acta 31:603–613

Gochnauer MB, Kushner (1969) Growth and nutrition of extremely halophilic bacteria. Can J Microbiol 15:1157–1165

Goldberg M (1958) The geology of the Tiran Island. M Sc Thesis Jerusalem (English abstract)

Goldberg M (1970) Notes on the Neogene and Pleistocene section and the elevated terrace problem in the Sharm-e-Sheikh region. Israel Geol Soc Proc (Central and Southern Sinai) 1970: 39–40

Goltermann HL (1969) Methods for chemical analysis of fresh waters. IBP handbook 8. Blackwell, Oxford

Golubic S (1973) The relationship between blue-green algae and carbonate deposits. In: Carr NG, Whitton BA (eds) The biology of blue-green algae. Blackwell, Oxford, pp 434–472

Golubic S (1976a) Organisms that build stromatolites. In: Walter MR (ed) Stromatolites. Developments in Sedimentology, vol 20. Elsevier, Amsterdam, pp 113–126

Golubic S (1976b) Taxonomy of extant stromatolite-building cyanophytes. In: Walter MR (ed) Stromatolites. Developments in sedimentology, vol 20. Elsevier, Amsterdam, pp 127–140

Golubic S (1979) Einführung in die Probleme der numerischen Taxonomie bei Cyanophyten. In: Krumbein WE (ed) Cyanobakterien – Bakterien oder Algen? Univ Oldenburg, Oldenburg, pp 15–32

Golubic S (1980) Halophily and halotolerance in cyanophytes. Orig Life 10:169–183

Golubic S, Barghoorn ES (1977) Interpretation of microbial fossils with special reference to the Precambrian. In: Flügel E (ed) Fossil Algae. Springer, Berlin Heidelberg New York, pp 1–14

Golubic S, Focke JW (1978) *Phormidium hendersonii* Howe: Identity and significance of a modern stromatolite building microorganisms. J Sediment Petrol 48:751–764

Golubic S, Hofmann HJ (1976) Comparison of Holocene and mid-Precambrian Entophysalidaceae (Cyanophyta) in stromatolitic algal mats: cell division and degradation. J Paleontol 50:1074–1082

Gomont M (1892) Monographie des Oscillariées. Ann Sci Nat Bot:15–263

Goode ADT (1981) Proterozoic geology of Western Australia. In: Hunter DR (ed) Precambrian of the southern hemisphere. Elsevier, Amsterdam

Goreau TF (1959) The ecology of Jamaican coral reefs, Species composition and zonation. Ecology 40:67–90

Goreau TF (1963) Calcium carbonate deposition by coralline algae and corals in relation to their roles as reef-builders. N Y Acad Sci Annu 109:127–169

Goreau TF, Hartman WD (1963) Boring sponges as controling factors in the formation and maintenance of coral reefs. In: Sognnaes RF (ed) Mechanisms of hard tissue destruction, Publ 75. Am Assoc Adv Sci, Washington, pp 25–54

Greenberg-Fertig I (1966) List of Palestine plants with data on their geographic distribution (Hebrew). Hebrew Univ, Jerusalem

Griffiths JF (1972) Climates of Africa, Elsevier, Amsterdam

Grob K (1978) On-column injection on to glass capillary columns, part 2. J High Res Chromatogr 1:263–267

Grob K, Grob K jr (1978) On-column injection on to glass capillary columns, part 1. J Chromatogr 151:311–320

Groot K de (1973) Geochemistry of tidal flat brines at Umm Said SE Qatar, Persian Gulf. In: Purser BH (ed) The Persian Gulf. Springer, Berlin Heidelberg New York, pp 377–394

Groves DI, Dunlop JSR, Buik R (1981) An early habitat of life. Sci Am 245:64–73

Gross MG (1972) Oceanography. A view of the earth. Prentice-Hall, Englewood Cliffs, NJ

Gupta RK, Sakena SK (1968) Resource survey of *Salvadora oleoides* Decne and *S persica* Linn for non-edible oil in Western Rajasthan. Trop Ecol 9:140–152

Gvirtzman G, Friedman GM (1977) Sequence of progressive diagenesis in coral reefs. AAPG Stud Geol 4:357–380

Gvirtzman G, Buchbinder B, Sneh A, Nir Y, Friedman GM (1977) Morphology of the Red Sea fringing reefs: a result of the erosional pattern of the last-glacial low-stand sea level and the following Holocene recolonization. MBRB (Mem Bur Rech Geol Min) 89:480–491

Hall J (1975) Bathymetric chart of the Straits of Tiran. Isr J Earth Sci 24:69–74

Hall J, Avraham Z Ben (1978) New bathymetric map of the Gulf of Elat (Aqaba). Abstr 10th Int Congr Sediment 1:285

Han J, McCarthy ED, Calvin M (1968) Hydrocarbon constituents of the blue-green algae *Nostoc muscorum, Anacystis nidulans, Phormidium luridium* and *Chlorogloea fritschii*. J Chem Soc C:2785–2791

Handford CR, Louks RG, Davies GR (1982) Depositional and diagenetic spectra of evaporites – a core workshop. SEPM Core Workshop No 3:395

Harned HS, Davis R (1943) The ionization constant of carbonic acid in water and the solubility of carbon dioxide in water and aqueous salt solutions from 0 to 50. J Am Chem Soc 65:2030–2037

Hart CW jr, Manning RM (1981) The cavernicolous caridean shrimps of Bermuda. J Crust Biol 1:441–456

Hartmann R, Oesterhelt D (1977) Bacteriorhodopsin mediated photophosphorylation in *Halobacterium halobium*. Eur J Biochem 77:325–335

Hartnady CJH (1978) The stratigraphy and structure of the Naukluft nappe complex Precambrian. Res Unit Univ Cape Town Annu Rep:14/15-163

Hase A (1926) Zur Kenntnis der Lebensgewohnheiten und der Umwelt des marinen Käfers *Ochthebius quadricollis* MULSANT (Hydrophilidae). Int Rev Gesamten Hydrobiol Hydrogr 16:141–179

Hattersley PW (1983) The distribution of C3 and C4 grasses in Australia in relation to climate. Oecologia (Berl) 57:113–128

Hauck F (1885) Die Meeresalgen Deutschlands und Österreichs. Kryptogamenflora von Deutschland, Österreich und der Schweiz: 2. Akademie Verlag, Leipzig

Haug P, Curry DJ (1974) Isoprenoids in a Costa Rican seep oil. Geochim Cosmochim Acta 38:601–610

Haverkamp J, Meuzelaar HLC, Beuvery EC, Boonekamp P, Tiesjema RH (1980) Characterization of *Neisseria meningitidis* capsular polysacchaarides containing sialic acids by pyrolysis-mass spectrometry. Anal Biochem 104:407–418

Hayes JM (1983) Geochemical evidence bearing on the origin of aerobiosis, a speculative interpretation. In: Schopf JW (ed) The origin and evolution of earth's earliest biosphere: An interdisciplinary study. Princeton Univ Press, Princeton, NJ

Hayes JM, Kaplan IR, Wedeking KW (1983) Precambrian organic geochemistry: Preservation of the record. In: Schopf JW (ed) The Earth's earliest biosphere: Its origin and evolution. Princeton Univ Press, Princeton, NJ

Helgerson SL, Requadt C, Stoeckenius W (1983) *Halobacterium halobium* photophosphorylation: Illumination-dependent increase in adenylate energy charge and phosphorylation potential. Biochemistry 22:5746–5753

Helsinger MH, Friedman GM (1973) Diagenetic modification and cementation in carbonate sediments at intertidal levels. Geol Soc Am Abstr Program 5:177

Hemprich WF, Ehrenberg CG (1828) Reisen in Aegypten, Libyen, Nubien und Dongala. In: Ehrenberg CG (ed) Naturgeschichtliche Reisen durch Nord-Afrika und West-Asien in den Jahren 1820 bis 1825. Mittler, Berlin

Herman Y (1965) Evidence of climatic changes in Red Sea cores. Means of correlation of Quaternary successions. Int Assoc Quat Res (INQUA), 7th Congr Proc 8:325–348

Hermann AG, Knake D, Schneider J, Peters H (1973) Geochemistry of modern sea water and brines from salt pans. Mineral Petrol 40:1–24

Herodotus (450 BC) Das Geschichtswerk des Herodot v Halikarnassos. Insel, Leipzig

Heuglin Th von (1860) Reise in Nordost-Afrika und längs des Rothen Meeres im Jahre 1857. Petermanns Geogr Mitt 1860:325–362

Heydemann B (1967) Die biologische Grenze Land-Meer im Bereich der Salzwiesen. Steiner, Wiesbaden

Hirsch P (1978) Microbial mats in a hypersaline Solar Lake: Types composition and distribution. In: Krumbein WE (ed) Environmental biogeochemistry and geomicrobiology. Ann Arbor Sci, Ann Arbor, pp 189–201

Hof T (1935) Investigations concerning bacterial life in strong brines. Extr Rec Trav Bot Neerl XXXI:92–173

Hof T, Frémy P (1932) On Myxophyceae living in strong brines. Rec Trav Bot Neerl 30:140–162

Hofmann HJ (1976) Precambrian microflora, Belcher Islands, Canada: significance and systematics. J Paleontol 50:1040–1073

Holster WT (1966) Diagenetic polyhalite in recent salt from Baja California. Am Mineral 51:99-109

Holthuis LB (1963) On red coloured shrimps (Decapoda, Caridea) from tropical land-locked saltwater pools. Zool Meded 38:261–279

Holthuis LB (1973) Caridean shrimps found in land-locked saltwater pools of four Indopacific localities (Sinai Peninsula, Funafuti Atoll, Maui and Hawaii Islands) with description of one new genus and four new species. Zool Verh 1:128–148

Horodyski RJ (1975) Stromatolites of lower Missoula Group (Middle Proterozoic), Belt Supergroup, Glacier National Park, Montana. Precambrian Res 2:215–254

Horodyski RJ (1977 a) Lyngbya mats at Laguna Mormona, Mexico: Comparison with protozoic stromatolites. J Sediment Petrol 47:1305–1320

Horodyski RJ (1977 b) Environmental influences on columnar stromatolitic branching patterns: examples from the Middle Proterozoic Belt Supergroup, Glacier National Park, Montana. J Paleontol 51:661–671

Horodyski RJ, Donaldson JA (1980) Microfossils from the Middle Proterozoic Dismal Lakes Group, Arctic Canada. Precambrian Res 11:125–159

Horodyski RJ, Bloeser B, Haar S von der (1977) Laminated algal mats from a coastal lagoon, Laguna Mormona, Baja California, Mexico. J Sediment Petrol 47:680–696

Horowitz A (1967) The geology of Museri Island (Dakhlak Aechipelago), Southern Red Sea. Israel Jour Earth Sci, vol 16, pp 74–83

Houwink AL (1956) Flagella, gas vacuoles and cell wall structure in *Halobacterium halobium;* an electron microscopic study. J Gen Microbiol 15:146–150

Hoye K (1907) Recherches sur la moisissure da Bacalao et quelques autres microorganismes halophiles. Bergens Mus Arbok 1906 12:3–64

Hsü KJ, Schneider J (1973) Progress report on dolomitization: hydrology of Abu Dhabi sabkhas, Arabian Gulf. In: Purser BH (ed) The Persian Gulf. Springer, Berlin Heidelberg New York, pp 409–422

Hsü KJ, Siegenthaler C (1969) Preliminary experiments on the hydrodynamic movements induced by evaporation and their bearing on the dolomite problem. Sedimentology 12:11–25

Huckreide R, Kürsten M, Venzlaff H (1962) Zur Geologie des Gebietes zwischen Kerman und Sagard (Iran). Beih Geol Jahrb 51:1–97

Hülsemann JC (1955) Grossrippeln und Schrägschichtungsgefüge im Nordsee-Watt und in der Molasse. Senckenbergiana Lethaea 26:359–388

Hume WF (1906) The topography and geology of the Peninsula of Sinai (southeastern portion). Cairo, Surv Dep Egypt, Natl Print Dep, Cairo pp 11–280

Humm JH, Wicks RS (1980) Introduction and guide to the marine bluegreen algae. Wiley, New York

Hunt CD (1979) The role of phytoplankton and particulate organic carbon on trace metal deposition in Long Island Sound. Ph D Thesis, Univ New Haven, Connecticut

Hustedt F (1930a) Die Kieselalgen Deutschlands, Österreichs und der Schweiz, p 1 in Rabenhorst's (ed) Kryptogamenflora von Deutschland, Österreich und der Schweiz 7: Akad Verlag, Leipzig

Hustedt F (1930b) Bacillariophyta. In: Pascher A (ed) Die Süßwasserflora Mitteleuropas. Fischer, Jena

Hustedt F (1953) Die Systematik der Diatomeen in ihren Beziehungen zur Geologie und Ökologie nebst einer Revision des Halobien-Systems. Sven Bot Tidskr 47:509–519

Hustedt F (1957) Die Diatomeenflora des Flußsystems der Weser im Gebiet der Hansestadt Bremen. Abh Naturwiss Ver 34:181–440

Hustedt F (1959) Die Kieselalgen Deutschlands, Österreichs und der Schweiz. In: Rabenhorst's L (ed) 7. Akademie Verlag, Leipzig, p 2

Hustedt F (1961–1966) Die Kieselalgen Deutschlands, Österreichs und der Schweiz. p 3 In: Rabenhorst's L (ed) 7. Akademie Verlag, Leipzig, p 3

Ideler L (1816) Über die Längen- und Flächenmaße der Alten. Abh K Akad Wiss Berlin Jg 1812/1813:121–200

Illing LV (1954) Bahaman calcareous sands. AAPG Bull 38:1–95

Illing LV, Taylor JCM (1967) Discussion. London Inst Min Metall Trans Sect B 76:883–884

Illing LV, Wells AJ, Taylor JCM (1965) Penecomtemporary dolomite in the Persian Gulf, Dolomitization and Limestone diagenesis: a symposium. In: Pray LC, Murray RC (eds) SEPM Spec Publ 13:89–111

Imbrie J, Buchanan H (1965) Sedimentary structures in modern carbonate sands of the Bahamas. p 149–172 In: Middleton GV (ed) Primary sedimentary structures and their hydrodynamic interpretation. SEPM Publ 12:149–172

Imhoff JF, Sahl HG, Soliman GSH, Trüper HG (1979) The Wadi Natrun: Chemical Composition and Microbial Mass Developments in Alkaline Brines of Eutrophic Desert Lakes. Geomicrobiol J 1:219–234

Ingvorsen K, Jorgensen BB (1979) Combined measurement of oxygen and sulfide in water samples. Limnol Oceanogr 24:390–393

Irwin WJ (1979) Analytical pyrolysis – an overview I and II. J Anal Appl Pyrol 1:3–27–89–123

Ivanov MV, Freney JR ed (1983) The global biogeochemical sulphur cycle. Wiley Chichester, pp 470

Jackson GD, Iannelli TR (1981) Rift-related cyclic sedimentation in the Neohelikian Borden Basin, Northern Baffin Island. Geol Surv Can Pap 81–10:269–302

Jackson LL, Blomquist GJ (1976) Isert waxes. In: Kolattukudy PE (ed) Chemistry and biochemistry of natural waxes. Elsevier, Amsterdam, pp 207–208

Jansson PE, Kenne L, Liedgren H, Lindberg B, Lönngren J (1976) A practical guide to the methylation analysis of carbohydrates. Chem Commun 8:1–74

Jaquin A (1956) Recherches biologiques sur *Ochthebius quadricollis* MULSANT (Coléopt Hydrophil). Bull Soc Hist Nat Afr Nord 1–2:271–290

Javor BJ (1979) Ecology, physiology, and carbonate chemistry of blue-green algal mats, Laguna Guerrero Negro, Mexico. Ph D thesis, Univ Oregon

Javor B, Requadt C, Stoeckenius W (1982) Box-shaped halophilic bacteria. J Bacteriol 151:1532–1542

Javor BJ, Castenholz RW (1981) Laminated microbial mats, Laguna Guerrero, Negro, Mexico. Geomicrobiol J 2:237–273

Jelgersma S (1971) Sea-level changes during the last 10,000 years. In: Steers JP (ed) Introduction to coastline development. Macmillan, London, pp 25–48

Johansen DA (1940) Plant microtechnique. McGraw-Hill, New York

Johnson JH (1954) An introduction to the study of rock-building algae and algal limestones. Colo Sch Mines Q 49:117

Jones GE, Murray L, Carr NG (1978) Trace element composition of five cyanobacteria. In: Krumbein WE (ed) Environmental biogeochemistry geomicrobiology, vol III. Ann Arbor Sci, Ann Arbor, pp 967–973

Jones HL (1959) The geography of Strabo, Harvard Univ Press, Cambridge

Jones JH, Yopp JH (1979) Cell wall constituents of *Aphanotece halophytica*. J Phycol 15:62–66

Jorgensen BB, Cohen Y (1977) Solar Lake (Sinai) 5 The sulfur cycle of the benthic cyanobacterial mats. Limnol Oceanogr 22:657–666

Jorgensen BB, Revsbech NP, Blackburn TH, Cohen Y (1979) Diurnal cycle of oxygen and sulfide microgradients and microbial photosynthesis in a cyanobacterial mat sediment. Environ Microbiol 38:46–58

Kaaden A van der, Haverkamp J, Boon JJ, Leeuw JW de (1983) Analytical pyrolysis of carbohydrates. I. Chemical interpretation of matrix influence on the pyrolysis-mass spectra of amylose using pyrolysis-gas chromatography-mass spectrometry. J Anal Appl Pyrol 5:199–220

Kalecsinszky A v (1904) Über die ungarischen warmen und heißen Kochsalzseen als natürliche Wärmeakkumulatoren sowie über die Herstellung von warmen Salzseen und Wärmeakkumulatoren. Math Nat Ber Ungarn 19:51–54

Kalkowsky E (1908) Oolith und Stromatolith im norddeutschen Buntsandstein. Z Dtsch Geol Ges 60:68–125

Karsten CJB (1846) Lehrbuch der Salinenkunde. Reimer, Berlin

Kassas M (1957) On the ecology of the Red Sea coastal land. J Ecol 45:187–203

Kassas M (1965) Studies on the ecology of the Red Sea coastal Land II. The district from El Galala El Zibliya to Hurghada. Bull Soc Geogr Egypte 38:155–193

Kassas M (1967) On the ecology of the Red Sea littoral salt marsh. Egypt Ecol Monogr 37:297–316

Kassas M (1981) The national plan for nature conservation. Work Pap Pres Egypt Nat Comm IUCN (Arabic)

Kassas M, Zahran MA (1962) Studies on the ecology of the Red Sea coastal land I The district of Gebel Ataqa and El-Galala El Bahariya. Bull Soc Geog Egypte 35:129–175

Kastner M (1982) When does dolomitization occur and what controls it? Abstr Int Sed Congr, Hamilton, Canada

Katz A, Sass E, Starinsky A, Holland HD (1972) Strontium behaviour in the aragonite-calcite transformation: an experimental study at 40–98 °C. Geochim Cosmochim Acta 36:481–496

Katz JJ, Stain HH, Harknes AL, Studiec MH, Sree WA, Janson TR, Cope BT (1972) Esterifying alcohols in the chlorophylls of purple photosynthetic bacteria. A new chlorophyll, bacteriochlorophyll (99), all-trans-geranylgeranyl bacteriochlorophyllide a. J Am Chem Soc 94:7938–7939

Kazmierczak J, Krumbein WE (1983 a) Identification of calcified coccoid cyanobacteria forming stromatoporoid stromatolites. Lethaia 16:207–213

Kazmierczak J, Krumbein WE (1983 b) The Preservability of Coccoid Cyanobacteria Forming Stromatoporoid Stromatolites. Abstr 3rd North Am Paleontol Convent, Montreal

Keith ML, Weber JM (1964) Carbon and oxygen isotopic composition of selected limestones and fossils. Geochim Cosmochim Acta 28:1787–1816

Kelier P, Henis Y (1970) The effect of sodium chloride on some physiological groups of microorganisms inhabiting a highly saline soil. Isr J Agric Res 20:71–75

Keller C (1983) Die Fauna im Suez-Kanal und die Diffusion der mediterranen und erythräischen Thierwelt. Neue Denkschr Allg Schweiz Ges Gesamten Naturwiss 28:3–38

Keller W (1955) Und die Bibel hat doch recht. Econ, Düsseldorf

Kendall AC (1979) Continental supratidal, sabkha, evaporites. In: Walker RG (ed) Facies models, Geol Assoc Can, Toronto, p 145

Kendall CGStC, Skipwith PA d'E (1968) Recent algal stromatolites of a Persian Gulf lagoon. J Sediment Petrol 38:1040–1058

Kendall CGStC, Skipwith PA d'E (1969) Holocene shallow water carbonate and evaporite sediments of Khor al Bazam, Abu Dhabi, southwest Persian Gulf. AAPG Bull 53:841–869

Kenyon KM (1980) Die Bibel im Licht der Archäologie (originally The bible and recent archeology; British Museum Publ 1978). Patmos, Düsseldorf

Kerans C, Ross GM, Donaldson JA, Geldsetzer HJ (1981) Tectonism and depositional history of the Helikian Hornby Bay and Dismal Lakes groups, District of Mackenzie. Geol Surv Can Pap 81-10:157–182

Kessel M, Cohen Y (1982) Ultrastructure of square bacteria from a brine pool in Southern Sinai. J Bacteriol 150:851–860

Kester DR, Pitkowicz (1977) Natural and anthropogenic changes in the global carbon dioxide system. In: Stumm W (ed) Global chemical cycles and their alterations by man. Dahlem Konf Berlin, pp 99–120

Kinsman DJJ (1964) The recent carbonate sediments near Halat el Bahrani, Trucial Coast, Persian Gulf. Deltaic and shallow marine deposits. In: Developments in sedimentology, vol 17. Elsevier, Amsterdam, pp 185–192

Kinsman DJJ (1966) Gypsum and anhydrite of recent age, Trucial Coast, Persian Gulf. Second Symposium on Salt 1, Cleveland, Ohio. North Ohio Geol Soc 1:302–306

Kinsman DJJ (1969a) Modes of formation, sedimentary associations and diagnostic features of shallow water and supratidal evaporites. AAPG Bull 53:830–840

Kinsman DJJ (1969b) Interpretation of Sr concentrations in carbonates minerals and rocks. J Sediment Petrol 39:486–508

Kinsman DJJ (1970) Early diagenesis of carbonate sediments in a supratidal setting. Princeton Univ Dep Geol, API Res Proj, 3rd Semi-Annu Prog Rep, Princeton, NJ

Kinsman DJJ, Park RK (1976) Algal belt and coastal sabkha evolution, Trucial Coast, Persian Gulf. In: Walter MR (ed) Stromatolites. Elsevier, Amsterdam, pp 421–433

Kinsman DJJ, Patterson RJ (1973) Dolomitization process in sabkhas environments. AAPG Bull 57:788–789

Klausewitz W (1975) Fische aus dem Roten Meer XV *Cabillus anchialinae,* eine neue Meergrundel von der Sinai Halbinsel (Pisces, Gobiidae, Gobiinae). Senckenbergiana Biol 56:203–207

Klebahn H (1895) Gasvakuolen, ein Bestandteil der Zellen der wasserblütenbildenden Phycochromaceen. Flora (Jena) 80:241–282

Klebahn H (1919) Die Schädlinge des Klippfisches. Mitt Inst Allg Bot Hamburg 4:11–68

Klebahn H (1922) Neue Untersuchungen über die Gasvakuolen. Jahrb Wiss Bot 61:535–589

Klebahn H (1925) Weitere Untersuchungen über die Gasvakuolen. Ber Dtsch Bot Ges 43:143–159

Klebahn H (1929) Über die Gasvakuolen der Cyanophyceen. Verh Int Ver Limnol 4:408–414

Klein H, Lapid J, Major E (1970) The dissolved gases in the solar pond south of Elat. Isr J Chem 8:535–546

Klok J, Nieberg-van Velzen EH, Leeuw JW de, Schenck PA (1981) Capillary gas chromatographic separation of monosaccharide as their alditol, acetates. J chromatogr 207:273–275

Klok J, Cox HC, Leeuw JW de, Schenck PA (1982) Analysis of synthetic mixture of partially methylated alditol acetates by capillary gas chromatography, gas chromatography-electrin impact mass spectrometry and gas chromatography-chemical ionization mass spectrometry. J Chromatogr 253:55–64

Knoll AH (1979) Archean photoautotrophy: some alternatives and limits. Orig Life 9:313–327

Knoll AH (1981) Paleoecology of late Precambrian microbial assemblages. In: Niklas KJ (ed) Paleobotany, paleoecology, and evolution. Prager, New York, p 17

Knoll AH (1982) Microorganisms from the late Precambrian Draken Conglomerate, Ny Friesland, Spitsbergen. J Paleontol 56:755–790

Knoll AH (1983 a) Microbiotas of the late Precambrian Hunnberg Formation, Nordaustlandet, Svalbard. J Paleontol (in press)
Knoll AH (1983 b) Ancient microbial ecosystems. In: Krumbein WE (ed) Microbial geochemistry. Blackwell, Oxford, pp 287–315
Knoll AH (1984) Microbiotas of the late Precambrian Hunnberg Formation, Nordaustlandet, Svalbard. J Paleontol 58: 131–162
Knoll AH, Golubic S (1979) Anatomy and taphonomy of a Precambrian algal stromatolite. Precambrian Res 10:115–151
König R, Winkler G (1972/77) Plinius Secundus d Ä Naturkunde; Botanik: Bäume XII/XIII:177–181. Heimeran, München
Konopka A, Brock TD, Walsby AE (1978) Buoyancy regulation by planktonic blue-green algae in Lake Mendota, Wisconsin. Arch Hydrobiol 83:524–537
Kossmann R (1877) Zoologische Ergebnisse einer im Auftrage der königlichen Academie der Wissenschaften zu Berlin ausgeführten Reise in die Küstengebiete des rothen Meeres. Engelmann, Leipzig
Krajewski KP (1981) Phosphate microstromatolites in the high-tatric Albian limestones in the Polish Tatra Mts. Bull Acad Pol Sci 29:175–183
Kramer JKG, Kushwaha SC, Kates M (1972) Structure determination of the squalene, dihydrosqualene and tetrahydrosqualen in *Halobacterium cutirubrum*. Biochim Biophys Acta 270:103–110
Krebs JC (1978) Ecology: The experimental analysis of distribution and abundance. Harper and Row, New York
Kristensen I (1970) Competition of three cyprinodont fish species in the Netherlands Antilles. Stud Fauna Curacao 32:82–101
Kristensen I (1971) Preference of euryhaline species for brackish and supersaline waters. View Milieu 3rd Symp Eur Biol Mar Suppl:22–811
Krumbein WE (1969) Über den Einfluß der Mikroflora auf die exogene Dynamik (Verwitterung und Krustenbildung). Geol Rundsch 58:333–363
Krumbein WE (1975) Biogenic Monohydrocalcite spherules in lake sediments of Lake Kivu (Africa) and the Solar Lake (Sinai). Sedimentology 22:631–634
Krumbein WE (1978 a) Geomikrobiologische Prozesse bei der Anreicherung nutzbarer Minerale und sedimentärer Lagerstätten. Erdoel Kohle Erdgas Petrochem 31:147–151
Krumbein WE (1978 b) Oxygen and H2S, do they coexist in stromatolitic microbial ecosystems? Poster Session Abstr 2nd Int Congr Ecol, Jerusalem
Krumbein WE (1979 a) Photolithotrophic and chemoorganotrophic activity of bacteria and algae as related to beachrock formation and degradation (Gulf of Aqaba, Sinai). Geomicrobiol J 1:139–203
Krumbein WE (1979 b) Calcification by bacteria and algae. In: Trudinger PA, Swaine DJ (eds) Biogeochemical cycling of mineral-forming elements. Elsevier, Amsterdam, pp 47–68
Krumbein WE (1982) Biogeochemistry and geomicrobiology of lagoons and lagoonary environments. Coastal lagoon research, present and future. UNESCO Tech Pap Mar Sci 33:81–109
Krumbein WE (1983 a) Stromatolites – challenge of term through space and time. Precambrian Res 20:493–531
Krumbein WE (1983b) Biogene Lamination – Stromatolith und Biostrom. Festschr 60. Geburtstag E Rutte. Weltenburger Akad, Kelheim, pp 131–141
Krumbein WE (1984 a) Auf den Schultern des Riesen. Vom Zeitgeist in der Erforschung geomikrobiologischer Zusammenhänge. In: Degens ET, Krumbein WE, Prashnowsky AA (eds) Festband Georg Knetsch Mitt Hbg Geol Staatsinst 56, pp 435–460
Krumbein WE (1984 b) Microbial geochemistry. Blackwell, Oxford
Krumbein WE, Cohen Y (1974) Biogene, klastische und evaporitische Sedimentation in einem mesothermen, monomiktischen ufernahen See (Golf von Aqaba). Geol Rundsch 63:1035–1065
Krumbein WE, Cohen Y (1977) Primary Production, mat formation and lithification: Contribution of oxygenic and facultative anoxygenic cyanobacteria. In: Flügel E (ed) Fossil algae. Springer, Berlin Heidelberg New York, pp 37–56
Krumbein WE, Jens K (1981) Biogenic rock varnishes of the Negev Desert (Israel) an ecological study of iron and manganese transformation by cyanobacteria and fungi. Oecol 50:25–38
Krumbein WE, Lange-Giele C (1979) Calcification in a coccoid cyanobacterium associated with the formation of desert stromatolites. Sedimentology 26:593–604

Krumbein WE, Swart PK (1983) The microbial carbon cycle. In: Krumbein WE (ed) Microbial geochemistry. Blackwell, Oxford, pp 5–62

Krumbein WE, Cohen Y, Shilo M (1977) Solar Lake (Sinai) 4 Stromatolitic cyanobacterial mats. Limnol Oceanogr 22:635–656

Krumbein WE, Buchholz H, Franke P, Giani D, Giele C, Wonneberger K (1979 a) O_2 and H_2S coexistence in stromatolites. A model for the origin of mineralogical lamination in stromatolites and banded iron formations. Naturwissenschaften 66:381–389

Krumbein WE, Rippka R, Waterbury JB (1979 b) Schematische bakteriologische Gliederung der Cyanophyten im Vergleich zur phykologischen. In: Krumbein WE (ed) Cyanobakterien – Bakterien oder Algen? Univ Oldenburg, Oldenburg

Kurochkin BI (1960) The cause of reddening of salt water in some salt lakes. Mikrobiologiya 29:422–427

Labourg PJ (1975) Contribution à l'hydrologie des étangs saumâtres de la région d'Arcachon Déscription des phénomènes d'eaux blanches. Bull Soc Limnol Bordeaux 5:1–8

Landin (1976) Methods of sampling aquatic beetles in transitional habitats at water margins. Freshwater Biol 6:81–87

Landolt-Börnstein (1962) Zahlenwerte und Funktionen aus Physik, Chemie, Astronomie, Geophysik und Technik. Gleichgewichte außer Schmelzgleichgewichten; 6th edn, Springer, Berlin Heidelberg New York, II/2:1–175

Lange IM, Murray RC (1977) Evaporite brine reflux as a mechanism for moving deep water brines upward in the formation of Mississippi Valley-type base metal deposits. Econ Geol 72:107–109

Larsen H (1967) Biochemical aspect of extreme halophilism. Adv Microb Physiol I:97–132

Larsen H (1980) Ecology of hypersaline environments. In: Nissenbaum A (ed) Hypersaline brines and evaporitic environments. Elsevier, Amsterdam, pp 23–29

Larsen H, Omang S, Steensland H (1967) On the gas vacuoles of the halobacteria. Arch Mikrobiol 59:197–203

Larue DK (1981) The Chocolay Group, Lake Superior region, USA: Sedimentologic evidence for deposition in basinal and platform settings on an early Proterozoic craton. Bull Geol Soc Am 92:417–435

Laughton AS, Whitmarsh RB, Jones MT (1970) The evolution of the Gulf of Aden. Philos Trans R Soc London Ser A 267:227–266

LaViolette PE, Frontenac TR (1967) Temperature, salinity and density of the world's seas: Arabian Sea, Persian Gulf, and Red Sea, Nav Oceanogr Off, Washington

Lawrence DE (1968) Taphonomy and information losses in fossil communities. Geol Soc Am Bull 79:1315–1330

Leach HB, Chandler HPC (1968) Aquatic coleoptera. In: Usinger RL (ed) Aquatic insects of California. Univ Calif Press, Berkeley, pp 293–371

Lebedev AN (1970) The climate of Africa, Part 1: Air temperatures, precipitation (trans from Russian). In: Lebedev AN (ed). Isr Program Sci Translations, Jerusalem, p 482

Lengerken H (1929) Die Salzkäfer der Nord- und Ostsee mit Berücksichtigung der angrenzenden Meere sowie des Mittelmeeres, des Schwarzen und des Kaspischen Meeres. Z Wiss Zool 135:1–162

Leo RF, Barghoorn ES (1976) Silicification of wood. Bot Mus Leafl 25:1–47

LePichon X, Francheteau J, Bonnin J (1973) Plate tectonics, Developments in geotectonics, Vol VI. Elsevier, Amsterdam

Levitt J (1972) Response of plants to environmental stress. Academic Press, London New York

Lewy Z (1972) Recent and Senonian oncolites from Sinai and southern Israel. Isr J Earth Sci 21:193–199

Licari GR, Cloud PE (1972) Prokaryotic algae associated with Australian Proterozoic stromatolites. Proc Natl Acad Sci USA 69:2500–2504

Liebermann D (1967) Synthesis of Dolomite. Nature (London) 213:241–245

Logan BW, Cebulski DE (1970) Sedimentary environments of Shark Bay, Western Australia. p 1–97 In: Logan BW, Davies GR, Read JF, Cebulski DE (eds) Carbonate sedimentation and environments, Shark Bay, Western Australia. AAPG Mem 13, pp 1–97

Long DT, Angino EE (1982) The mobilization of selected trace metals from shales by aqueous solutions: Effects of temperature and ionic strength. Econ Geol 77:646–652

Long DT, Lyons WB, Gaudette HE (1983) Trace metal accumulations in modern marine sabkhas. In: Cohen Y, Castenholz RW, Halvorson HO (eds) Microbial mats: Stromatolites, Liss, New York, pp 425–434

Lotze F (1964) The distribution of evaporites in space and time. In: Nairn AEM (ed) Problems in paleoclimatology. Interscience, London, p 491
Lowe DR (1980) Stromatolites 3,400-Myr old from the Archean of Western Australia. Nature (London) 284:441–443
Lowe DR (1982) Comparative sedimentology of the principal volcanic sequences of Archaean greenstone belts in South Africa, Western Australia and Canada: implications for crustal evolution. Precambrian Res 17:1–30
Lowry OH, Rosebrough NJ, Farr AL, Randall RJ (1951) Protein measurement with the Folin phenol reagent. J Biol Chem 193:265–275
Loya Y (1972) Community structure and species diversity of hermatypic corals at Eilat, Red Sea. Mar Biol 13:100–123
Loya Y, Slobodkin LB (1971) The coral reefs of Eilat (Gulf of Eilat, Red Sea). Symp Zool Soc London, no 28. Academic Press, London New York, pp 117–139
Lucia FJ (1968) Recent sediments and diagenesis of South Bonaire, Netherlands Antilles. J Sediment Petrol 38:845–858
Luksch J (1898) Berichte der Commission für Ozeanographische Forschungen im Roten Meere (nördliche Hälfte) 1895–1896 VI Physikalische Untersuchungen. Denkschr K Akad Wiss Wien 65:351–422
Luksch J (1901) Expeditions SM Pola in das Rote Meer. Denkschr Akad Wiss XVIII, Phys Untersuch 69:337–398
Mandoli DF, Briggs WR (1982) Optical properties of etiolated plant tissues. Proc Natl Acad Sci USA 79:2902–2906
Markovitz R (1973) Climate in South Sinai, Shlomo district. Nature Protect Soc (Hebrew) Tel Aviv
Martin A, Nisbet EG, Bickle MJ (1980) Archean stromatolites of the Belingwe Greenstone Belt, Zimbabwe (Rhodesia). Precambrian Res 13:337–362
Mason B (1966) Principles of geochemistry, 3rd edn. Wiley, New York
Mazor E (1969) The solar pond south of Elat. Mada 14:247–249 (in Hebrew)
McCrea JM (1950) On the isotope chemistry of carbonates and a paleotemperature scale. J Chem Phys 18:849–857
McKee ED (1959) Storm sediments on a Pacific atoll. J Sediment Petrol 29:354–364
McKenzie JA (1980) Holocene dolomitization of calcium carbonate sediments from the coastal sabkhas of Abu Dhabi UAE: a stable isotope study. J Geol 89:185–198
McKenzie JA, Hsü KJ, Schneider JF (1980) Movement of surface waters under the sabkha, Abu Dhabi UAE, and its relation to evaporative dolomite genesis. In: Zenger DH, Dunhan RJ, Ethington (eds) SEPM Spec Publ 28, pp 11–30
McKenzie KG (1981) Palaebiogeography of some salt lake faunas. Hydrobiologia 81/82:407–418
McKinney TF, Friedman GM (1970) Continental shelf sediments of Long Island, New York. J Sediment Petrol 40:213–248
McKirdy DM, Hahn JH (1982) Composition of Kerogen and Hydrocarbons in Precambrian Rocks. In: Holland HD, Schidlowski M (eds) Mineral deposits and the evolution of the biosphere. Springer, Berlin Heidelberg New York, pp 123–154
Meent D van der, Los A, Leeuw JW de, Schenck PA, Haverkamp J (1983) Size fractionation and analytical pyrolysis of suspended particles from the river Rhine delta. In: Bjoröy M (ed) Advances in organic geochemistry 1981. Wiley, New York, pp 336–349
Meister FE (1968) Untersuchungen über Zusammenhänge zwischen Diatomeenführung und Sedimentaufbau, dargestellt an Seeablagerungen der Lisan Formation bei Jericho Palästina und dem Lempa-Becken-El Salvador (in German) Ph D Thesis, RWH Aachen
Mendelsohn CV, Schopf JW (1982) Proterozoic microfossils from the Sukhaya Tunguska, Shorikha and Yudoma Formations of the Siberian Platform, USSR. J Paleontol 56:42–83
Mergner H (1971) Structure, ecology and zonation of Red Sea reefs (in comparison with South Indian and Jamaican reefs). Symp Zool Soc London 28:141–161
Mergner H, Schuhmacher H (1974) Morphologie, Ökologie und Zonierung von Korallenriffen bei Aqaba (Golf von Aqaba, Rotes Meer). Helgol Wiss Meeresunters 26:238–358
Merton RK (1965) On the shoulders of giants. Macmillan, New York
Meterological Service (1952) Climatological data for the Negev. Meteorol Notes, State of Israel, Ser A 4:12
Meteorological Service (1956) Climatological normals, part 2, winds. Meteorol Notes, State of Israel, Ser A d:15

Meteorological Service (1967) Climatological standard normals of rainfall, 1931–1960. Meteorol Notes, State of Israel, Ser A:21

Meuzelaar HLC, Kistemaker PG, Eshuis W, Boerboom HA (1977) Automated pyrolysis mass spectrometry: application to the differentation of microorganisms. In: Daly NR (ed) Advances in mass spectrometry, vol 7 B. Heyden, London, pp 1452–1456

Meuzelaar HLC, Haverkamp J, Hileman FD (1982) Pyrolysis mass spectrometry of recent and fossil biomaterials. Compendium and altas. Elsevier, Amsterdam

Migahid AM, Shafei AM el, Ahman AA el, Mamouda MA (1959) Ecological observations in western and southern Sinai. Bull Soc Geogr Egypt 32:165–206

Milliman JD (1966) Submarine Lithification of Carbonate sediments. Science 153:994–997

Milliman JD, Emery KO (1968) Sea levels during the past 35,000 years. Science 162:1121–1123

Mitterer RM (1971) Influence of natural organic matter on $CaCO_3$ precipitation. In: Bricker OC (ed) Carbonate cements. Hopkins, Baltimore, pp 252–258

Mizutani H, Wada E (1982) Effect of high atmospheric CO_2 concentration on ^{13}C of algae. Orig Life 12:327–390

Mohamed AF (1940) The Egyptian exploration of the Red Sea. Proc R Soc Lond Ser B 128:305–316

Monty CLV (1976) The origin and development of cryptalgal fabrics. In: Walter MR (ed) Stromatolites. Elsevier, Amsterdam

Mook WG, Bommerson JC, Staverman WH (1974) Carbon isotope fractionation between dissolved bicarbonate and gaseous carbon dioxide. Earth Planet Sci Lett 22:169–176

Moore J (1978) Microhabitats of seashore *Bledius* (Coleoptera, Staphilinidae). Coleopt Bull 32:65–66

Morcos SA (1970) Physical and chemical oceanography of the Red Sea. Oceanogr Mar Biol Annu Rev 8:73–202

Morikawa H, Marchessault RH (1981) Pyrolysis of bacterial polyalkanoates. Can J Chem 59:2306–2313

Morris RC, Dickey PA (1957) Modern evaporite deposition in Peru. AAPG Bul 41:241–274

Movers FC (1841/56) Die Phönizier. Untersuchungen über die Religion und die Gottheiten der Phönizier 1:3. Weber, Bonn

Mover FC (1849) Das phönizische Alterthum. Die Phönizier 2–3. Dümmler, Berlin

Mudie PJ (1974) The potential uses of halophytes. In: Reinold RJ, Queen WH (eds) Ecology of salt marshes. Academic Press, London New York, pp 565–597

Müller G (1962) Zur Geochemie des Strontiums in ozeanen Evaporiten unter besonderer Berücksichtigung der sedimentären Coelestinlagerstätte von Hemmelte-West (Spd-Oldenburg). Geologie Beih 35 11:1–90

Müller O (1899) Baciullariaceen aus den Natronthälern von El-Kab. Ober-Ägypten Hedwigia:38:274

Muir MD (1976) Proterozoic microfossils from the Amelia Dolomite, McArthur Bassin, Northern Territory. Alcheringa 1:143–158

Muir MD (1979) A sabkha model for the deposition of part of the Proterozoic McArthur group of the northern territory, and its implications for mineralisation. BMR J Aust Geol Geophys 4:149–162

Murray RC (1969) Hydrology of South Bonaire, NA – A rock selective dolomitization model. J Sediment Petrol 39:1007–1013

Nadson G (1903) Die Mikroorganismen als geologische Faktoren. Petersburger Arb Komm Erf Min Seen, Slavjansk, Petersburg

Natterer K (1889) Chemische Untersuchungen Exped S M Pola in das Rothe Meer, nördliche Hälfte (1895–1896). Ber Komm Oceanogr Forsch Denkschr K Akad Wiss Wien Math-Naturwiss Cl 55:445–572

Natterer K (1898) Berichte der Commission for oceanographische Forschungen im Roten Meere (nördliche Hälfte) 1895–1896 IX Chemische Untersuchungen. Denkschr K Akad Wiss Wien 65:445–526

Natterer K (1901) Chemische Untersuchungen von Wasser- und Grundproben. Ber Komm Oceanogr Forsch Denkschr K Akad Wiss Wien Math-Naturwiss Cl 59:297–309

Natterer K, Luksch J (1901) Expedition SM Schiff Pola in das Rote Meer 1897–1898 XV Chemische Untersuchung von Wasser- und Grundproben. Denkschr K Akad Wiss Wien 69:297–309

Negev A (1983) Tempel, Kirchen und Zisternen (Engl original ms exists). Calwer, Stuttgart

Nesteroff WD (1955) Les récifs coralliens du Banc Farsan Nord (Mer Rouge). Résultats Scientifiques des Campagnes de la Calypso, I Campagne en Mer Rouge. Masson et Cie, Paris

Neudert MK, Russell RE (1981) Shallow water and hypersaline features from the Middle Proterozoic Mt Isa sequence. Nature (London) 293:284–286
Neumann AC (1966) Red Sea. In: Fairbridge RW (ed) McGraw Hill Yearb Sci Technol. McGraw Hill, New York, pp 344–347
Neumann AC, McGill DA (1962) Circulation in the Red Sea in early summer. Deep-Sea Res 8:223–235
Niebuhr BG (1779) Reise und Beobachtungen durch Egypten, Arabien und andere Gegenden des Morgenlands. Steiner, Bern
Niebuhr BG (1781) Reise und Beobachtungen durch Egypten, Arabien und andere Gegenden des Morgenlands. Steiner, Bern
Niebuhr BG (1816) Über die Geographie Herodots. Abh K Akad Wiss Berlin Jg 1812–1813:209–224
Nieman RH, Poulsen LL (1971) Plant growth suppression on saline media: interaction with light. Bot Gaz 132:14–19
Nir D (1971) Marine terraces of southern Sinai. Geogr Rev LXI:32–50
Nishimura M, Koyama T (1977) The occurrence of stanols in various living organisms and the behaviour of sterols in contemporary sediments. Geochim Cosmochim Acta 41:379–387
Nissenbaum A (1970) Chemical analyses of Dead Sea and Jordan River Water, 1778–1830. Isr J Chem 8:281–287
Nissenbaum A (1977) The physicochemical basis of legends of the Dead Sea. In: Greer DC (ed) Desertic terminal lakes. Utah State Univ, Logan, pp 159–168
Nissenbaum A (1978 a) Dead sea asphalts – Historical aspects. AAPG Bull 62:837–844
Nissenbaum A (1978 b) Legends of a dead sea. Rehovot 8:30–35
Nissenbaum A (1979) Life in a Dead Sea – Fables, allegories, and scientific search. BioScience 29:153–157
Nissenbaum A (1980) Searching for oil in the Holy Land. Rehovot 9:57–61
Nissenbaum A, Presley BJ, Kaplan IR (1972) Early diagenesis in a reducing fjord, Saanich Inlet, British Columbia I. Chemical and isotopic changes in major components of interstitial water. Geochim Cosmochim Acta 36:1007–1027
Norlyn JD, Epstein E (1982) Barley production: Irrigation with seawater on coastal soil. In: San Pietro A (ed). Plenum Press, New York, pp 525–529
Odum HT, Odum EP (1955) Trophic structure and productivity of a windward coral reef community on Eniwetok Atoll. Ecol Monogr 25:291–320
Oehler DZ (1978) Microflora of the middle Proterozoic Balbirini Dolomite, (McArthur Group) of Australia. Alcheringa 2:269–309
Oehler DZ, Oehler JH, Stewart AJ (1979) Algal fossils from a late Precambrian, hypersaline lagoon. Science 205:388–390
Oehler JH, Arneth JD, Eglinton D, Golubic S, Hahn JH, Hayes JM, Hoefs JW, Hollerbach A, Junge CE, Krumbein WE, McKirdy DM, Schidlowski M, Schopf JW (1982) Reduced carbon compounds in sediments. In: Holland HD, Schidlowski M (ed) Mineral deposits and the evolution of the biosphere. Springer, Berlin Heidelberg New York, pp 289–307
Oerstedt AS (1841) Beretning om en exkursionen til Trindelen. Naturhist Tidskr 3:552–569
Oesterhelt D, Stoeckenius W (1971) Rhodopsin-like protein from the purple membrane of *Halobacterium halobium*. Nature (London) 233:149–152
Oesterhelt D, Stoeckenius W (1973) Functions of a new photoreceptor membrane. Proc Natl Acad Sci USA 70:2835–2857
Ohnishi A, Kato K, Takagi E (1977) Pyrolytic formation of 3-hydroxy-2-penteno-1,5-lactone from xylan, xylan-oligo-saccharides and methylxylopyranosides. Carbohydr Res 58:387–395
Oldfather CH (1958) Diodorus of sicily. Harvard Univ Press, Cambridge
O'Leary MH (1981) Carbon isotope fractionation in plants. Phytochemistry 20:553–567
O'Leary MH (1982) Phosphoenolpyruvate carboxylase: an enzymologist's view. Annu Rev Plant Physiol 33:297–315
Oppenheimer CH, Master M (1965) On the solution of quartz and precipitation of dolomite in seawater during photosynthesis and respiration. Z Allg Mikrobiol 5:48–51
Orelli JC (1826) Sanchuniathonis Berytii Fragmenta Cosmogonia et Theologia Phoenicum. Hinrichs, Leipzig
Oren A (1983 a) *Halobacterium sodomense* sp nov, a Dead Sea halobacterium with an extremely high magnesium requirement. Int J Syst Bacteriol 33:381–386

Oren A (1983b) Bacteriorhodopsin-mediated CO_2 photoassimilation in the Dead Sea. Limnol Oceanogr 28:33–41

Oren A, Shilo M (1981) Bacteriorhodopsin in a bloom of halobacteria in the Dead Sea. Arch Microbiol 130:185–187

Oren OH (1962a) A note on the hydrography of the Gulf of Eylath. Sea Fish Res St Div Fish State of Israel Bull 30:3–14

Oren OH (1962b) The Israel south Red Sea expedition. Nature (London) 194:834–1137

Ourisson G, Albrecht P, Rohmer M (1979) The hopanoids. Paleochemistry and biochemistry of group of natural products. Pure Appl Chem 51:709–729

Pardue JW, Scalan RS, Baalen C van, Parker PL (1976) Maximum carbon isotope fractionation in photosynthesis by blue-green algae and a green alga. Geochim Cosmochim Acta 40:309–312

Park RK (1977) The preservation potential of some recent stromatolites. Sedimentology 248:485–506

Parker BC, Simmons GM jr, Love FG, Wharton RA jr, Seaburg KG (1981) Modern stromatolites in antartic dry valley lakes. Bio Science 31:656–661

Parkes K, Walsby AE (1981) Ultrastructure of a gas vacuolate square bacterium. J Gen Microbiol 126:503–506

Parrot A (1954) Cahiers d'Archeologie biblique 1954–1955, 5 vols (also translated into German 1956–1961). Delachaud & Niestle, Neuchatel

Patterson RJ, Kinsman DJJ (1977) Marine and continental groundwater sources in a Persian Gulf sabkha. In: Frost JG, Weiss MP, Saunders JE (eds) AAPG Stud Geol 4:381–390

Patterson RJ, Kinsman DJJ (1981) Hydrologic framework of a sabkha along Arabian Gulf. AAPG Bull 65:1457–1475

Paultre K (1804) Karl Paultre's Franz Offiziers bei der leichten Artillerie, vormals Adjudanten des Obergenerals Kleber in Ägypten. Geographische Nachrichten von Syrien; mit einer Karte von Syrien. Aus dem Französischen. Landes-Industrie-Comptoir, Weimar

Pearcy RW (1983) The light environment and growth of C3 and C4 tree species in the understory of a Hawaiian forest. Oecologia (Berl) 58:19–25

Pearcy RW, Calkin HW (1983) Carbon dioxide exchange of C3 and C4 tree species in the understory of a Hawaiian forest. Oecologia (Berl) 58:26–32

Pearcy RW, Ehleringer JR (1984) Comparative ecophysiology of C3 and C4 plants. Plant Cell Envir 7:1–13

Pennak RW (1953) Freshwater invertebrates of the United States, Ronald Press, New York

Peragallo H, Peragallo M (1897) Diatomées marines de France et des districts voisins. Tempère, Grez-sur-Loling

Perath I (1966) Note on some semi-continental-lacustrine outcrops on the Elat shore. Isr J Earth Sci 15:131–133

Perthuisot JP (1975) La Sabkha el Melah de Zarsis Génèse et evolution d'un bassin salin paralique. Trav Lab Géol, Ecole Norm Sup Paris 9:1–252

Perthuisot JP (1980) Sites et processus de la formation d'evaporites dans la nature actuelle. Bull Cent Rech Explor-Prod ELF Aquitaine 4:207–233

Perthuisot JP, Jauzein A (1978) Le Khor el Aadid, lagune surslee de l'Emirat de Qatar. Rev Geogr Phys Geol Dyn 20:347–358

Peryt TM (1981) Phanerozoic oncoids – an overview. Facies 4:197–214

Peters KE, Rohrback BG, Kaplan IR (1981) Carbon and hydrogen stable isotope variations in kerogen during laboratory-simulated thermal maturation. AAPG Bull 65:501–508

Petter HFM (1931) On bacteria of salted fish. Proc Acad Sci Amsterdam 34:1417–1423

Petter HFM (1932) Over roode en andere bakterien van getzouten visch. Doct Thesis, Univ Utrecht

Pfeifer F, Weidinger G, Goebel W (1981) Genetic variability in *Halobacterium halobium*. J Bacteriol 145:375–381

Philp RP, Brown S, Calvin M, Brassell S, Eglinton G (1978) Hydrocarbon and fatty acid distributions in recently deposited algal mats at Laguna Guerrero, Baja California. In: Krumbein WE (ed) Environmental biogeochemistry and geomicrobiology, vol I. Ann Arbor Sci Publ, Ann Arbor, pp 255–270

Phillips JD, Ross DA (1970) Continuous seismic reflexion profiles in the Red Sea. Philos Trans R Soc London Ser A 267:143–152

Phleger FB (1969) A modern evaporite deposit in Mexico. AAPG Bull 53:824–829

Phleger FB, Ewing GC (1962) Sedimentology and oceanography of coastal lagoons in Baja California Mexico. Geol Soc Am Bull 73:145–182

Pia J (1927) Thallophyta. In: Hirmer M (ed) Handbuch der Palaeobotanik, vol I. Hirmer, Leipzig, pp 1–136
Picard L (1958) History of Mineral Research in Israel. Isr Econ Forum 11:10–38
Picotti M (1930) Richerche di oceanografia chimica esequite delle R Nave Ammiraglio Magnaghi Tabelle generali dell analisi clorometriche e dei data di temperatura, salinita et densita. Ann Idrogr 11:73–115
Pidgeon RT (1978) 3,450-my-old volcanics in the Archaean layered greenstone succession of the Pilbara Blocks, Western Australia. Earth Planet Sci Newslett 37:421–428
Pierre C (1982) Teneurs en isotopes stables et conditions de génèse des évaporites marines: applications a quelques milieux actuels et au Méssinien de la Méditerranée. Ph D thesis, Univ Paris-Orsay
Plumb KA, Derrick GM, Needham RS, Shaw RD (1981) The Proterozoic of Northern Australia. In: Hunter DR (ed) Precambrian of the southern hemisphere. Elsevier, Amsterdam
Poljakoff-Mayber A (1975) Morphological and anatomical changes in plants as a response to salinity stress. In: Poljakoff-Mayber A, Gale J (eds) Plants in saline environments Ecol Stud 15. Springer, Berlin Heidelberg New York, pp 97–117
Poljakoff-Mayber A, Gale J (1975) General discussion. In: Poljakoff-Mayber A, Gale J (eds) Plants in saline environments. Ecol Stud 15. Springer, Berlin Heidelberg New York, pp 193–199
Pomeroy LR, Darley WM, Dunn EL, Gallagher JL, Haines E, Whitney DM (1981) Primary production. In: Pomeroy LR, Wiegert RG (eds) The ecology of a salt marsh. Springer, Berlin Heidelberg New York, pp 55–67
Por FD (1968a) Copepods of some land-locked basins on the islands of Entedebir and Nocra (Dhlak Archipelago, Red Sea). Sea Fish Res Stu Haifa Bull 49:32–50
Por FD (1968b) Solar Lake on the shores of the Red Sea. Nature (London) 218:860–861
Por FD (1969) Limnology of the heliothermal Solar Lake on the coast of Sinai (Gulf of Elat). Verh Int Ver Theor Angew Limnol 17:1031–1034
Por FD (1972) Hydrobiological notes on the high-salinity waters of the Sinai Peninsula. Mar Biol 14:111–119
Por FD (1975a) A typology of the nearshore seepage pools of Sinai. Rapp Comm Int Mer Medit 23:103
Por FD (1975b) The Coleoptera-dominated fauna of the hypersaline Solar Lake (Gulf of Elat, Red Sea). 10th Eur Symp Mar Biol Ostend 2:563–573
Por FD (1978a) Hypersaline ecosystems and food webs as an expression of Dahl's competitive principle. Abstr 2nd Int Congr Ecol, Israel
Por FD (1978b) Lessepsian migration. Springer, Berlin Heidelberg New York
Por FD (1979) The Copepoda of Di Zahav pool (Gulf of Elat, Red Sea). Crustaceana 37:13–30
Por FD (1980) A classification of hypersaline waters based on trophic criteria. Mar Ecol 1:121–132
Por FD, Dor I (1975) Ecology of the metahaline pool of Di Zahav, Gulf of Elat, with notes on the Siphonocladacea and the typology of near-shore marine pools. Mar Biol 29:37–44
Por FD, Tsurnamal M (1973) Ecology of the Ras Muhammad Crack in Sinai. Nature (London) 241:43–44
Por FD, Dor I, Amir A (1977) The mangal of Sinai: Limits of an ecosystem. In: Kinne O, Bulnheim HP (eds) Helg Wiss Meeresunters 30. Int Helgol Symp Ecosyst Res, pp 295–314
Posnjak E (1940) Deposition of calcium sulfate from seawater. Am J Sci 238:559–568
Posthumus MA, Boerboom AJH, Menzelaar HLC (1974) Analysis of biopolymers by Curie-point pyrolysis in direct combination with low voltage electron impact ionization mass spectrometry. In: West AR (ed) Advances in mass spectrometry, vol VI:397–402 Appl Sci Publ UK, pp 397–402
Potts M (1980) Blue-green algae (Cyanophyta) in marine coastal environments of the Sinai peninsula; distribution, zonation, stratification and taxonomic diversity. Phycologia 19:60–73
Potts M, Krumbein WE (1978) Desert stromatolites; genetic control of calcification in calcrete by blue-green algae. Abstr 10th Int Congr Sedimentol, Jerusalem 1978, pp 521–522
Powers RW (1962) Arabian Upper Jurassic carbonate reservoir rocks. In: Ham W (ed) Classification of carbonate rocks, Mem I. AAPG, pp 122–192
Pratt BR (1982) Stromatolite decline – a reconsideration. Geology 10:512–515
Purdy EG (1963) Recent calcium carbonate facies of the Great Bahama Bank. J Geol 71:334–355; 472–497
Purdy EG (1974) Reef configurations: cause and effect. In: LaPorte LF (ed) Reefs in time and space, selected examples from the recent and ancient. SEPM Spec Publ 18:9–76
Purser BH (1973) The Persian Gulf. Springer, Berlin Heidelberg New York

Purser BH (1980) Les paléosabkhas du Miocéne inf. dans le SE de l'Iran. Bull Cent Rech Explor-Prod ELF Aquitaine 4:235–244
Purser BH, Evans G (1973) Regional sedimentation along the Trucial Coast, SE Persian Gulf. In: Purser BH (ed) The Persian Gulf. Springer, Berlin Heidelberg New York, pp 211–231
Purser BH, Seibold E (1973) The principal environmental factors influencing Holocene sedimentation and diagenesis in the Persian Gulf. In: Purser BH (ed) The Persian Gulf. Springer, Berlin Heidelberg New York, pp 1–10
Purser BH, Azzawi Al, Hassini N Al, Baltzer F, Hassan K, Orszag-Sperber F, Plaziat JC, Jacoub S, Younis W (1982) Sedimentation et évolution du complexe deltaique Tigre-Euphrate. Soc Géol Fr Mém 144:207–216
Quennell AM (1959) Tectonics of the Dead Sea rift. Int Geol Congr 20 (Mexico) Assoc Serv Geol Afr:385–405
Rains DW (1979) Salt tolerance of plants: strategies of biological systems. In: Hollaender A (ed) The biosaline concept. Plenum Press, New York, pp 47–67
Ratcliff GA, Holdcroft JG (1963) Diffusivities of gases in aqueous electrolyte solutions. Trans Inst Chem Eng 41:315–319
Raumer K von (1862) Die Sinaitische Halbinsel. Petermanns Geogr Mitt:34–35
Rayss T (1959 a) Considérations sur la flore algale de la peninsule du Sinai Ecologie des algues marines. Colloq Int CNRS 81:167–175
Rayss T (1959 b) Contribution à la connaissance de la flore marine de la Mer Rouge. Bull Sea Fish Res Stn Haifa Isr 23:1–32
Rayss T, Dor I (1963) Nouvelle contribution à la connaissance des algues marine de la mer rouge. Bull Sea Fish Res Stn Haifa Isr 34:11–42
Red Sea and Gulf of Aden Pilot (1967) 11th ed
Reiss Z, Hottinger L (1984) The Gulf of Aqaba. In: Ecological Studies, Vol. 50. Springer, Berlin, Heidelberg New York Tokyo, pp 354
Reiss Z, Luz B, Almogi-Labin A, Halicz E, Winter A, Wolf M (1980) Late Quaternary paleooceanography of the Gulf of Aqaba (Elat), Red Sea. Quat Res 14:294–308
Remane A (1940) Einführung in die zoologische Ökologie der Nord- und Ostsee. In: Grimpe G, Wagler E (eds) Die Tierwelt der Nord- und Ostsee. Becker & Erler, Leipzig
Renfro AR (1974) Genesis of evaporite associated-stratiform metaliferous deposits – a sabkha process. Econ Geol 69:33–45
Revsbech NP, Jorgensen BB (1983) Microelectrode studies of the photosynthesis and O_2, H_2S and pH profiles of a microbial mat. Limnol Oceanogr 28:1062–1074
Ricketts BD, Donaldson JA (1981) Sedimentary history of Belcher Group of Hudson Bay. Geol Surv Can Pap 81–10:235–254
Riedl R, Ozretic B (1969) Hydrobiology of marginal caves Part 1, General problems and introduction. Int Rev Gesamten Hydrobiol 54:661–683
Rippka R, Deruelles J, Waterbury JB, Herdman M, Stanier RY (1979) Generic assignments, strain histories and properties of pure cultures of canobacteria. J Gen Microbiol 111:1–61
Ritter K (1820) Die Vorhalle Europäischer Völkergeschichten vor Herodotus, um den Kaukasus und an den Gestaden des Pontus. Reimer, Berlin
Ritter K (1836) Die Erdkunde von Asien. Reimer, Berlin
Ritter K (1847) Die Erdkunde, quoted from Schick AP (ed) Berlin 1958
Vergleichende Erdkunde der Sinai-Halbinsel, von Palästina und Syrien. Reimer Berlin
Ritter K (1850) Die Erdkunde von Asien, vol VIII. 2. Abt: Die Sinai-Halbinsel, Palästina und Syrien. Erdk Allg Vergl Geogr 15/3
Ritter K (1852) Einleitung zur allgemeinen vergleichenden Geographie, und Abhandlungen zur Begründung einer mehr wissenschaftlichen Behandlung der Erdkunde. Reimer, Berlin
Ritter K (1861) Geschichte der Erdkunde und der Entdeckungen. In: Daniel HA (ed). Reimer, Berlin
Roberts HH, Murray SP (1983) Gulfs of Northern Red Sea: Depositional settings of distinct siliclastic-carbonate interfacies (abstr). AAPG Annu Convent, Dallas, p 153
Robinson G (1837) Travels in Palestine and Syria. Colburn, London
Roedder E (1968) Temperature, salinity, and the origin of the ore-forming fluids at Pine Pt, NW Territories, Canada, from fluid inclusion studies. Econ Geol 63:439–450
Rosenan N (1951) The measurement of evaporative power in Israel. Meteorol Notes 1, Meteorol Serv Isr, Ser A:1–11

Rosenan N (1963) The map of evaporation from water surface in Israel. Atlas of Israel, Tel Aviv, Surv Isr
Rosenan N, Mane U (1970) Climatic regions, radiation, evaporation, wind and sharav. Atlas of Israel, Ministry of Labor, Jerusalem
Ross CP (1963) The Belt Series in Montana. US Geol Surv Prof Paper 346, p 122
Rouse JE, Sherif N (1980) Major evaporitic deposition from groundwaters remobilised salts. Nature (London) 285:470–472
Rühs H (1811) Georg Viscount Valentia's und Heinrich Salt's Reisen nach Indien, Ceylon, dem rothen Meere, Abessinien und Aegypten. In: Sprengel MC, Sprengel TF (eds) fortgesetzt von Ehrmann TF. Bibliothek der neuesten und wichtigsten Reisebeschreibungen zur Erweiterung der Erdkunde 45. Landes-Industrie-Comptoirs, Weimar
Rüppell E (1829) Reisen in Nubien, Kordofan und dem peträischen Arabien, Wilmans, Frankfurt am Main
Rüppell E (1838) Reise in Abessinien (Contains a section on Sinai). Schmerber, Frankfurt in Commission auf Kosten d Verf
Russeger J (1841) Reisen in Europa, Asien und Afrika: Reise in Griechenland, Unteregypten, im nördlichen Syrien und südöstlichen Kleinasien. Schweizerbarth, Stuttgart
Russegger J (1843 a) Reise in Europa, Asien und Afrika. Schweizerbarth, Stuttgart
Russegger J (1843 b) Reise in Egypten. Nubien und Ost-Sudan. Schweizerbarth, Stuttgart
Russegger J (1847) Reisen in Europa, Asien und Afrika, mit besonderer Rücksicht auf die naturwissenschaftlichen Verhältnisse der betreffenden Länder, unternommen in den Jahren 1835 bis 1841. Schweizerbarth. Stuttgart
Russegger J (1848) Reisen in Europa, Asien und Afrika 1835–1841. Reise in der Levante und in Europa. Schweizerbarth, Stuttgart
Saiz-Jimenez C, Haider K, Meuzelaar HLC (1979) Comparisons of soil organic matter and its fractions by pyrolysis mass spectrometry. Geoderma 22:25–37
Sanders JE, Friedman GM (1967) Origin and occurrence of limestones. In: Chillingar GV, Bissell HJ, Fairbridge RW (eds) Carbonate rocks. Elsevier, Amsterdam, pp 165–265
Sasaki A, Krouse HR (1969) Sulfur isotopes and the Pine Point lead-zinc mineralization. Econ Geol 64:718–730
Sass E, Weiler Y, Katz A (1972) Recent sedimentation and oolite formation in the Ras Matarma Lagoon, Gulf of Suez. In: Stanley DJ (ed) The Mediterranean Sea. Dowden, Hutchinson and Ross, Stroudsburg, pp 279–292
Scheer G (1971) Coral reefs and coral genera in the Red Sea and Indian Ocean. Symp Zool Soc London 28:329–367
Schenck PA, Leeuw JW de (1982) Molecular organic geochemistry. In: Hutzinger O (ed) Handbook of environmental chemistry, vol 1B. Springer, Berlin Heidelberg New York, pp 111–129
Schenck PE (1967) The Macumber Formation of the Maritime Provinces, Canada – a Mississipian analogue to recent strandline carbonates of the Persian Gulf. J Sediment Petrol 37:365–376
Schick AP (1958) Tiran: the Straits, the Island, and its terraces. Isr Explor J 8:120–130–189–196
Schidlowski M (1982) Content and isotopic composition of reduced carbon in sediments. In: Holland HD, Schidlowski M (eds) Mineral deposits and the evolution of the biosphere. Springer, Berlin Heidelberg New York, pp 103–122
Schidlowski M (1983) Biologically mediated isotope fractionations: biochemistry, geochemical significance, and preservation in the Earth's oldest sediments. In: Ponnampereuma C (ed) Cosmochemistry and the origin of life. Reidel, Dordrecht, pp 277–322
Schidlowski M, Eichmann R, Junge CE (1975) Precambrian sedimentary carbonates: carbon and oxygen isotope geochemistry and implications for the terrestrial oxygen budget. Precambrian Res 2:1–69
Schidlowski M, Eichmann R, Junge CE (1976) Carbon isotope geochemistry of the Precambrian Lomagundi carbonate province, Rhodesia. Geochim Cosmochim Acta 40:449–455
Schidlowski M, Hayes JM, Kaplan IR (1983) Isotopic inferences of ancient biochemistries: carbon, sulfur, hydrogen and nitrogen. In: Schopf JW (ed) Earth's earliest biosphere: its origin and evolution. Princeton Univ Press, Princeton, pp 149–186
Schindewolf OH, Seilacher A (1955) Beiträge zur Kenntnis des Kambriums in der Salt Range (Pakistan). Abh Akad Wiss Mainz 10:257–446
Schleiden MJ (1858) Die Landenge von Sues. Engelmann, Leipzig

Schleiden MJ (1875) Das Salz. Engelmann, Leipzig
Schmidt A, Schmidt F, Fricke F, Heiden H, Müller O, Hustedt F (1892) Atlas der Diatomaceenkunde. Reisland, Leipzig
Schneider J (1979) Stromatolithische Milieus an Salinen der Nord-Adria (Secovlje, Portoroz, Jugoslawien) In: Krumbein WE (ed) Cyanobacterien – Bakterien oder Algen? 1st Oldenburger Symp Cyanobakterien, 1977, Oldenburg
Schneider J, Herrmann AG (1980) Saltworks – Natural laboratories for microbiological and geochemical investigations during the evaporation of seawater. In: Coogan AH, Hauder L (eds) 5th Symp Salt 2. North Ohio Geol Soc, Ohio, pp 371–381
Schneiderhöhn H (1926) Erzführung und Gefüge des Mansfelder Kupferschiefers. Metall Erz 23:143–146
Schneiderhöhn H (1955) Erzlagerstätten. Fischer, Jena
Schobert B, Lanyi JK (1982) Halorhodopsin is a light driven chloride pump. J Biol Chem 257:10306–10313
Schopf JW (1968) Microflora of the Bitter Springs Formation, late Precambrian, central Australia. J Paleontol 42:651–688
Schopf JW (1983) Earth's earliest biosphere – its origin and evolution. Princeton University Press, Princeton
Schopf JW, Blacic JM (1971) New microorganisms from the Bitter Springs Formation (Late Precambrian) of the north-central Amadeus Basin, Australia. J Paleontol 45:925–960
Schopf JW, Sovietov YK (1976) Microfossils in Conophyton from the Soviet Union and their bearing on Precambrian stratigraphy. Science 193:143–146
Schopf JW, Oehler DZ, Horodyski RJ, Kvenvolden KA (1971) Biogenicity and significance of the oldest known stromatolites. J Paleontol 45:477–485
Schreiber BC (1978) Environments of subaqueous gypsum formation. In: Dean WE, Schreiber BC (eds) Marine evaporites. SEPM, Tulsa, p 43
Schubert GH von (1939) Reise in das Morgenland in den Jahren 1836 und 1837. Palm & Enke, Erlangen
Schulten HR, Bahr U, Görtz W (1981) Pyrolysis field, ionization mass spectrometry of carbohydrates, part B polysaccharides. J Anal Appl Pyrol 3:229–241
Schulz E (1936) Das Farbstreifensandwatt und seine Fauna, eine ökologisch-biozönotische Untersuchung an der Nordsee. Kieler Meeresforsch 1:359–378
Schwarcz HP (1969) The stable isotopes of carbon. In: Wedepohl KH (ed) Handbook of geochemistry, vol II/1. Springer, Berlin Heidelberg New York, pp6-B-1-6-B-15
Seckbach J, Kaplan IR (1973) Growth pattern and 13C/12C isotope fractionation of Cyanidium caldarium and hot spring algal mats. Chem Geol 12:161–169
Seetzen UJ (1808 a) Auszug aus einem Schreiben des Russisch-Kaiserlichen Cammer-Assessors Dr UJ Seetzen. In: Frh von Zach F (ed) Monatliche Correspondenz zur Beförderung der Erd- und Himmels-Kunde 17. Beckersche Buchhandlung, Gotha, pp 132–163
Seetzen UJ (1808b) Beyträge zur Geographie Arabiens. In: Frh von Zach F (ed) Monatliche Correspondenz zur Beförderung der Erd- und Himmels-Kunde 18. Beckersche Buchhandlung, Gotha, pp 374–448
Seetzen UJ (1808 c) Bestimmung der geographischen Lage des Sinai, aus Beobachtungen des Russisch-Kaiserlichen Cammer-Assessors Dr UJ Seetzen. In: Frh von Zach (ed) Monatliche Correspondenz zur Befürderung der Erd- und Himmels-Kunde 14. Beckersche Buchhandlung, Gotha, pp 194–208
Seetzen UJ (1812) Auszug aus einem Schreiben des Russ Kais Kammer-Assessor Dr UJ Setzen. In: Frh von Zach F (ed) Monatliche Correspondenz zur Beförderung der Erd- und Himmels-Kunde 26. Beckersche Buchhandlung, Gotha, pp 380–399
Seetzen UJ (1813) Auszug aus einem Schreiben des Russ Kais Kammer-Asessors Dr UJ Seetzen. In: Frh von Zach F (ed) Monatliche Correspondenz zur Beförderung der Erd- und Himmels-Kunde 27. Beckersche Buchhandlung, Gotha, pp 60–79
Shearman DJ (1963) Recent anhydrite, gypsum, dolomite and halite from the coastal flats of the Arabian shore of the Persian Gulf. Proc Geol Soc London 1607:63
Shearman DJ (1970) Recent halite rock, Baja California, Mexico. Inst Min Metall 79:155–162
Shepard FP (1973) Submarine geology, 3rd edn. Harper and Row, New York

Shinn EA (1963) Spur and groove formation on the Florida reef tract. J Sediment Petrol 33:291–303
Shinn EA (1964) Recent dolomite, Sugerloaf Key, Florida. In: Ginsburg RN (ed) Field Guide Book Geol Soc Am 26. p 33
Shinn EA (1968) Selective dolomitization of recent sedimentary structures. J Sediment Petrol 38:612–616
Shinn EA (1973 a) Carbonate coastal accretion in an area of longshore transport NE Qatar, Persian Gulf. In: Purser BH (ed) The Persian Gulf. Springer, Berlin Heidelberg New York, pp 179–192
Shinn EA (1973 b) Sedimentary accretion along the leewards, SE coast of Qatar Peninsular Persian Gulf. In: Purser BH (ed) The Persian Gulf. Springer, Berlin Heidelberg New York, pp 199–210
Shinn EA, Ginsburg RN, Lloyd RM (1965) Recent supratidal dolomite from Andros Island, Bahamas. In: Murray LC, Pray RC (eds) Dolomitization and limestone genesis, a symposium. Soc Econ Paleontol Mineral Spec Publ No 13 Tulsa Oklahoma, pp 112–123
Shomer-Ilan A, Nissenbaum A, Waisel Y (1981) Photosynthetic pathways and the ecological distribution of the Chenopodiaceae in Israel. Oecologia 48:244–248
Silverman M, Simon M (1974) Flagellar rotation and the mechanism of bacterial motility. Nature (London) 249:73–74
Simon RD (1978) *Halobacterium* strain 5 contains a plasmid which is correlated with the presence of gas vacuoles. Nature (London) 273:314–317
Simon RD (1981) Morphology and protein composition of gas vesicles from wild type and gas vacuole defective strains of *Halobacterium salinarium* strain 5. J Gen Microbiol 125:103–111
Simonsen R (1962) Untersuchungen zur Systematik und Ökologie der Bodendiatomeen der westlichen Ostsee. Int Rev Gesamten Hydrobiol 1:1–144
Singh IB (1980) Precambrian sedimentary sequences of India: their peculiarities and comparison with modern sediments. Precambrian Res 12:41–436
Sket B, Iliffe TM (1980) Cave fauna of Bermuda. Int Rev Gesamten Hydrobiol 65:871–882
Skinner HCW (1963) Precipitation of calcian dolomites and magnesian calcites in the south-east of South Australia. Am J Sci 261:449–472
Skipski VP, Smolowne AF, Sullivan RC, Barclay M (1965) Separation of lipid classes by their thin layer chromatography. Biochim Biophys Acta 106:386–396
Smith BN, Epstein S (1971) Two categories of 13C/12C ratios for higher plants. Plant Physiol 47:380–384
Smith GE (1976) Sabkha and tidal flat facies control of stratiform copper deposits in north Texas. In: Wolf KH (ed) Handbook of strata-bound and stratiform ore deposits 6, pp 407–446
Smoot JP (1978) Origin of the carbonate sediments in the Wilkins Peak Member of the lacustrine Green River Formation (Eocene), Wyoming, USA. Spec Publ Int Assoc Sediment 2:109–127
Sneh A (1978) Sedimentary environments of the northern gulfs of the Red Sea: unpublished Ph D thesis. Troy, NY
Sneh A, Friedman GM (1973) Recent and Scenonian oncolites from Sinai and southern Israel (Lewy, 1972) – Discussion. Isr J Earth Sci 22:59–60
Sneh A, Friedman GM (1980) Spur and groove patterns on the reefs of the northern Gulfs of the Red Sea. J Sediment Petrol 50:981–986
Somers GF (1979) Natural halophytes as a potential resource for new salt-tolerant crops: some progress and prospects. In: Hollaender A (ed) The biosaline concept. Plenum, New York, pp 101–115
Sorokin YI (1970) Interrelations between sulphur and carbon turnover in meromictic lakes. Arch Hydrobiol 66:391–446
Sorokin YI, Kadota H (1972) Techniques for the assessment of microbial production and decomposition in fresh waters. IBP Handb 23
Soudry D, Champetier Y (1983) Microbial processes in the Negev phosphorites (Southern Israel). Sedimentology 30:411–423
Spira Y (1981) Ecology of the Solar Lake with emphasis on its coleopteran fauna. M Sc thesis, Hebrew Univ Jerusalem
Spira Y, Rijn J van (1982) Feeding of *Artemia* sp in the Solar Lake. Abstr Isr J Zool
Spiro B (1971) Early diagenesis of corals. M Sc Thesis, Jerusalem
Sprengel K (1822) Theophrast's Naturgeschichte der Gewächse. Hammerich, Altona
Stal LJ, Krumbein WE (1981) Aerobic nitrogen fixation in pure cultures of a benthic marine *Oscillatoria* (cyanobacteria). FEMS Microbiol Lett 11:295–298

Stal LJ, Krumbein WE, van Gemerden H (1984) Das Farbstreifen-Sandwatt – Ein laminiertes mikrobielles Ökosystem im Wattenmeer. Mitt Emdener Naturforsch Ges 7:1–60
Stanier RY, Kunisawa R, Mandel M, Cohen-Bazire G (1971) Purification and properties of unicellular blue-green algae (Order Chroococcales). Bacteriol Rev 35:171–205
Stebbins GL (1974) Flowering plants, evolution above the species level. Belknap, Cambridge, Ma
Steiner R, Schäfer W, Blos I, Wieschoff H, Scheer H (1981) '2,10-phytadienol as esterifying alcohol of bacteriochorophyll b from *Ectothiorhodospira halochloris*. Z Naturforsch 36c:417–420
Steudner (1861) Die Deutsche Expedition bei den Moses-Quellen im Peträischen Arabien. Petermanns Geogr Mitt:427–429
Steven AC, TenHeggeler B, Muller R, Kistler J, Rosenbusch JP (1977) Ultrastructure of a periodic protein layer in the outer membrane of *E coli*. J Cell Biol 72:292–301
Stewart AJ (1979) A barred-basin marine evaporite in the upper Proterozoic of the Amadeus Basin, central Australia. Sedimentology 26:33–62
Stock JH, Vermeulen JJ (1982) A representive of the mainly abyssal family Pardaliscidae (Crustacea, Amphipoda) in cave waters of the Cicos Islands. Bijdr Dierk 52:3–12
Stoddart DR (1971) Geology and morphology of reefs. In: Stoddart DR (ed) Regional variations in Indian Ocean coral reefs. Academic Press, London New York, pp 3–38
Stoeckenius W (1978) Speculations about the evolution of halobacteria and of chemiosmotic mechanisms. Academic Press, London New York
Stoeckenius W (1981) Walsby's square bacterium: Fine structure of an orthogonal procaryote. J Bacteriol 148:352–360
Stoeckenius W, Bogomolni RA (1982) Bacteriorhodopsin and related pigments of halobacteria. Annu Rev Biochem 52:587–615
Stoeckenius W, Kunau WH (1968) Further characterization of particulate fractions from lysed cell envelopes of *Halobacterium halobium* and isolation of gas vacuole membranes. J Cell Biol 34:355–395
Stoeckenius W, Lozier RH, Bogomolni R (1979) Bacteriorhodopsin and the purple membrane of halobacteria. Biochim Biophys Acta 505:215–278
Stolz JF (1984) The effects of catastrophic inundation (1977–1983), On the composition and ultrastructure of a stratified microbial mat community, Laguna Figuero, Baja California, Mexico. Ph D Thesis, Univ Boston
Stookey LL (1970) Ferrozine – A new spectrophotometric reagent for iron. Anal Chem 42:779–781
Stresemann E (1959) Hemprich und Ehrenberg „Reisen zweier naturforschender Freunde im Orient, geschildert in ihren Briefen aus den Jahren 1819 bis 1826. Abh Dtsch Akad Wiss Berlin 1954:1–178
Strickland JDH, Parsons TR (1968) A practical handbook of seawater analysis. Bull Fish Res Board Can, Ottawa
Strogonov BP (1964) Physiological basis of salt tolerance (English translation). Isr Program Sci Transl, Jerusalem
Strother PK (1980) Microbial Communities from Precambrian Strata. Thesis, Cambridge, USA
Strother PK, Knoll AH, Barghoorn ES (1983) Micro-organisms from the Late Precambrian Narrsârssuk Formation, Northwestern Greenland. Palaeontology 26:1–32
Swain FM (1955) Ostracoda of San Antonio Bay, Texas. J Paleontol 29:561–646
Taeckholm V (1974) Students' flora of Egypt. Cairo Univ, Cairo
Tankard AJ, Jackson MPA, Eriksson KA, Hobday DK, Hunter DR, Minter WEL (1982) Crustal evolution of Southern Africa. Springer, Berlin Heidelberg New York
Taraschewski H, Paperna I (1981) Distribution of the snail *Pirenella conica* in Sinai and Israel and its infection by Heterophyidae and other trematoda. Mar Ecol Progr Ser 5:193–205
Teas HJ (1979) Silviculture in saline water. In: Hollaender A (ed) The biosaline concept. Plenum, New York, pp 117–161
Teas HJ (1982) Saline silviculture. In: San Pietro A (ed) Biosaline research. Plenum, New York, pp 369–381
Tewari A (1967) Coastal vegetation and its utility. In: Krishnamurthy V (ed) Sea, salt and plants. Cent Salt Mar Chem Res Inst Bhdagnavar, pp 334–339
Thauer RK, Fuchs G (1979) Methanogene Bakterien. Naturwissenschaften 66:89–94
Thiede DS, Cameron EN (1978) Concentration of heavy metals in the Elk-Point Evaporite Sequence, Saskatchewan. Econ Geol 73:405–415

Thompson RW (1968) Tidal-flat sedimentation in the Colorado River delta NW Gulf of California. Geol Soc Am Mem 107:1–187

Thomson J, Turekian KK, McCaffrey RJ (1975) The accumulation of metals in and release from sediments of Long Island Sound. In: Cronin LE (ed) Estuarine research, vol I. Academic Preiss, London New York, pp 28–44

Thorson G (1975) Bottom communities (sublittoral and shallow shelf). Treat Mar Ecol Paleoecol I:67–461

Thorstenson DC (1969) Equilibration distribution of small organic molecules in natural waters. Ph D Thesis, Univ Evanston

Thorstenson DC (1971) A chemical model for early diagenesis in Devil's Hole, Harrington Sound, Bermuda. In: Bricker OP (ed) Carbonate cements. Hopkins, Baltimore, pp 285–291

Timofeev BV (1969) Proterozoic Phaeromorphs. Nauka, Leningrad (in Russian)

Tornabene TG (1978) Non aerated cultivation of *Halobacterium cutirubrum* and its effect on cellular squalenes. J Mol Evol 11:253–257

Tornabene TG, Langworthy TA, Holzer G, Oro J (1979) Squalenes, phytanes and other isoprenoids as major neutral lipids of methanogenic and thermomoacidophilic Archaebacteria. J Mol Evol 13:73–83

Trudinger PA, Swaine DJ (1979) Biogeochemical cycling of mineral-forming elements. Elsevier, Amsterdam

Trüper HG, Imhoff JF (1981) The genus *Ectothiorhodospira*. In: Starr MP, Stolp H, Trüper HG, Balows A, Schlegel HG (eds) The prokaryotes. A handbook on habitats, isolation, and identification of bacteria. Springer, Berlin Heidelberg New York, pp 274–278

Tulloch AP (1976) Chemistry of waxes of higher plants. In: Kolattukudy PE (ed) Chemistry and biochemistry of natural waxes. Elsevier, Amsterdam, pp 236–249

Udden JA (1924) Laminated anhydrite in Texas. Geol Soc Am Bull 35:347–354

US Navy Hydrographic Office (1960) Summary of oceanographic conditions in the Indian ocean. Oceanogr Anal Div, Mar Sci Dep, Spec Publ, Washington DC

Veizer J, Compston W, Hoefs J, Nielsen H (1982) Mantle buffering of the early oceans. Naturwissenschaften 69:173–180

Veizer J, Holser WT, Wilgus CK (1980) Correlation of 13C/12C and 34S/32S secular variations. Geochim Cosmochim Acta 44:579–587

Vercelli F (1927) Richerche di oceanografia fisica esequite della R Nave Ammiraglio Magnaghi (1923–24), part 4. La temperatura e la salinita. Ann Idrogr 11:1–66

Vercelli F (1931) Nuove richerche sulli correnti marine nel Mar Rosso. Ann Idrogr 12:1–74

Vidal G (1976) Late Precambrian microfossils from the Visingsö Beds in southern Sweden. Fossils Strata 9:1–57

Vidal G, Knoll AH (1983) Proterozoic plankton. Geol Soc Am Mem 161:265–277

Visher GS (1969) Grain size distributions and depositional processes. J Sediment Petrol 39:1074–1106

Vogel JC (1980) Fractionation of the carbon isotopes during photosynthesis. Sitzungsber Heidelb Akad Wiss Math-Naturwiss Kl Jg 1980 (3) 111–135

Volcani BE (1940) Studies on the microflora of the Dead Sea (in Hebrew). Ph D Thesis, Jerusalem

Wagenfeld F (1836) Sanchuniathon's Urgeschichte der Phönizier in einem Auszuge aus der wieder aufgefundenen Handschrift. Hofbuchhandlung, Hannover

Wagenfeld F (1837 a) Sanchuniathon's Phönizische Geschichte, Rohdensche Buchhandlung, Lübeck

Wagenfeld F (1837 b) Sanchuniathon's Historiarum Phoeniciae. Schünemann, Bremen

Wagenfeld FW (1896) Wagenfeld, Friedrich W. Allg Dtsch Biogr 40, Leipzig

Wagner G, Hartmann R, Oesterhelt D (1978) Potassium uniport and ATP synthesis in *Halobacterium halobium*. Eur J Biochem 89:169–179

Waisel Y (1972) Biology of halophytes. Academic Press, London New York

Waisel Y, Agami M (1979) Halophytes of Israel. Yad Ha'chamisha Press, Kfar Chabad (Hebrew)

Walker JE, Walsby AE (1983) Molecular weight of gas-vesicle protein from the planktonic cyanobacterium *Anabaena flos-aquae* and implications for structure of the gas vesicle. Biochem J 209:809–815

Walker KF (1973) Studies on a saline lake ecosystem. Aust J Mar Freshwater Res 24:21–71

Wallen LL, Rohwedder WK (1974) Poly-β-hydroxyalkanoates from activated sludge. Environ Sci Technol 8:576–579

Walsby AE (1969) The permeability of blue-green algal gas-vacuole membranes to gas. Proc R Soc London Ser B 173:233–255
Walsby AE (1971) The pressure relationships of gas vacuoles. Proc R Soc London Ser B 178:301–326
Walsby AE (1972) Structure and function of gas vacuoles. Bacteriol Rev 36:1–32
Walsby AE (1974) The identification of gas vacuoles and their abundance in the hypolimnetic bacteria of Arco Lake, Minnesota. Microb Ecol 1:51–61
Walsby AE (1976) The buoyancy-providing role of gas vacuoles in an aerobic bacterium. Arch Microbiol 109:135–142
Walsby AE (1980) A square bacterium. Nature (London) 283:69–71
Walsby AE (1981 a) Cyanobacteria: planktonic gas-vacuolate forms. In: Starr M, Stolp H, Trüper H, Balows A, Schlegel HG (eds) The prokaryotes. Springer, Berlin Heidelberg New York
Walsby AE (1981 b) Gas-vacuolate bacteria (apart from cyanobacteria). In: Starr M, Stolp H, Trüper H, Balows A, Schlegel HG (eds) The prokaryotes. Springer, Berlin Heidelberg New York
Walsby AE (1982) Cell-water and cell-solute relations. In: Carr NG, Whitton BA (eds) The biology of cyanobacteria. Blackwell, Oxford, pp 237–262
Walsby AE, Klemer AR (1974) The role of gas vacuoles in the microstratification of a population of Oscillatoria agardhii var isothrix in Deming Lake, Minnesota. Arch Hydrobiol 74:375–392
Walsby AE, Rijn J van, Cohen Y (1983 a) The biology of a new gas-vacuolate cyanobacterium, *Dactylococcopsis salina* sp nov, in Solar Lake. Proc R Soc London Ser B 217:417–447
Walsby AE, Utkilen HC, Johnsen IJ (1983 b) Buoyancy changes of a red coloured *Oscillatoria agardhii* in Lake Gjersjoen, Norway. Arch Hydrobiol (in press)
Walsh GE (1974) Mangroves: A review. In: Reinhold RJ, Queen WH (eds) Ecology of halophytes. Academic Press, New York, pp 51–174
Walter H, Breckle SW (1983) Ökologie der Erde, vol I. Fischer, Stuttgart
Walter MR ed (1976) Stromatolites. Developments in Sedimentology 20, Elsevier Amsterdam, p 790
Walter MR (1983) Archean stromatolites: evidence of the Earth's earliest benthos. In: Schopf JW (ed) Earth's earliest biosphere: Its origin and evolution. Princeton Univ Press, Princeton, pp 187–213
Walter MR, Buick R, Dunlop JSR (1980) Stromatolites 3,400–3,500 Myr old from the North Pole area, Western Australia. Nature (London) 284:443–445
Walther J (1888 a) Die Korallenriffe der Sinaihalbinsel. Abh Math Phys Cl Saechs Ges Wiss 24:9–339
Walther J (1888b) Die Korallenriffe der Sinaihalbinsel. Abh Math Phys Cl Saechs Ges Wiss 24:439–505
Walther J (1891) Die Denudation in der Wüste und ihre geologische Bedeutung. Untersuchungen über die Bildung der Sedimente in den ägyptischen Wüsten. Abh Math Phys Cl Saechs Ges Wiss 27:348–569
Walther J (1893–94) Einleitung in die Geologie als historische Wissenschaft: III Lithogenesis der Gegenwart. Fischer, Jena
Walther J (1912) Das Gesetz der Wüstenbildung in Gegenwart und Vorzeit. Quelle & Meyer, Leipzig
Walther J (1916) Zum Kampf in der Wüste am Sinai und Nil. Quelle & Meyer, Leipzig
Walther J (1924) Das Gesetz der Wüstenbildung in Gegenwart und Vorzeit. Quelle & Meyer, Leipzig
Wardroper AMK (1979) Aspects of the geochemistry of polycyclic isoprenoids. Ph D Thesis, Bristol
Wear RG, Holthuis LB (1977) A new record for the anchialine shrimp L gur uvae (Borradaile, 1899) (Decapoda, Hippolytidae) in the Philippines with notes on its morphology, behaviour and ecology. Zool Meded 51:125–140
Weckesser J, Drews G (1979) Lipopolysaccharides of photosynthetic prokaryotes. Annu Rev Microbiol 33:215–239
Weidinger G, Klotz G, Goebel W (1979) A large plasmid from *Halobacterium halobium* carrying genetic information for gas vacuole formation. Plasmid 2:377–386
Weissbrod T (1969) The Paleozoic of Israel and adjacent countries. Geol Surv Isr Bull 48:32
Wells AJ (1962) Recent dolomite in the Persian Gulf. Nature (London) 194:274–275
Wellsted JR (1840) Travels to the City of the Caliphs, along the Shores of the Persian Gulf and the Mediterranean Including a Voyage to the Coast of Arabia, and a Tour on the Island of Socotra. Colburn, London
Welte DH (1978) Erdöl und Kohle-Fossile Energierohstoffe und Zeugnisse vergangenen Lebens. Erdoel Kohle Erdgas Petrochem 31:139–146
Wenzel M (1978) Geschichte der wissenschaftlichen Expeditionen in den Indischen Ozean. Staatsexamensarbeit Bochum

West IM, Ali YA, Hilmyne (1979) Primary gypsum nodules in a modern sabkha on the Mediterranean coast of Egypt. Geology 7:354–358
Wilbert N, Kahan D (1981) Cilates of Solar Lake on the Red Sea shore. Arch Protistenkd 124:70–95
Wildgruber G, Thomm M, Koenig H, Ober K, Ricchiuto T, Stetter KO (1982) *Methanoplanus limicola*, a plate-shaped methanogen representing a novel family, the Methanoplanaceae. Arch Microbiol 132:31–36
Wilson CB (1961) The Upper Middle Hecla Hoek Rocks of Ny Friesland, Spitsbergen. Geol Mag 98:89–116
Windig W, Hoog GS de, Haverkamp J (1982a) Chemical characterization of yeast and yeast-like fungi by factor analysis of their pyrolysis mass spectra. J Anal Appl Pyrol 3:213–220
Windig W, Haverkamp J, Kistemaker PG (1982b) Interpretation of sets of pyrolysis mass spectra by discriminant analysis and graphical rotation. Anal Chem 55:81–87
Windley BF (1977) The evolving continents. Wiley, London
Winkler FJ, Kexel H, Kranz C, Schmidt HL (1982) Parameters affecting the $^{13}CO_2/^{12}CO_2$ isotope discrimination of the ribulose-15-bisphosphate carboxylase reaction. In: Schmidt HL, Förstel H, Heinzinger K (eds) Stable sotopes Anal Chem Symp Ser 11. Elsevier, Amsterdam, pp 83–89
Winter K (1978) Photosynthetic pathways in plants of coastal and inland habitats of Israel and Sinai. Flora 167:1–34
Winter K (1981) C4 plants of high biomass in arid regions of Asia – occurrence of C4 phytosynthesis in Chenopodiaceae and Polygonaceae from the Middle East and USSR. Oecologia 48:100–106
Winter K, Throughton JH (1978) Photosynthetic pathways in plants of coastal and inland habitats of Israel and Sinai. Oecologia (Berl) 167:1–34
Winter K, Throughton JH, Card KA (1976) 013C values of grass species collected in the northern Sahara desert. Oecologia 25:115–123
Winter K, Wallace BJ, Stocker GC, Roksandic Z (1983) Crassulacean acid metabolism in Australian vascular epiphytes and some related species. Oecologia 57:129–141
Wissowa G (1909) Paulys Real-Encyclopädie der classischen Altertumswissenschaft. Metzlersche Buchhandlung, Stuttgart
Wissowa G (1912) Paulys Real-Encyclopädie der classischen Altertumswissenschaft. In: Kroll W (ed). Metzlersche Buchhandlung, Stuttgart
Wollschläger H, Bartsch E (1971) Karl Mays Orientreise 1899/1900. Dokumentation. Jahrb Karl-May-Ges 1971:165–215
Wood GV, Wolfe MJ (1969) Sabkha cycles in the Arab/Darb formation of the Trucial Coast of Arabia. Sedimentology 12:165–191
Woronichin NN (1926) Zur Biologie der bitter-salzigen Seen in der Umgebung von Pjatigorsk (nördl Kaukasus). In: Thienemann A (ed) Arch Hydrobiol, vol XVI. Schweizerbarth, Stuttgart, pp 628–643
Wright JLC (1981) Minor and trace sterols of *Dunaliella teriolecta*. Phytochemistry 20:2403–2405
Wright SR, Burton HR (1981) The biology of antarctic saline lakes. Hydrobiologia 82:319–338
Wyllie SG, Djerassi C (1968) Mass spectrometry in structural and stereochemical problems CXLVI Mass spectrometric fragmentations typical of sterols with unsaturated side chains. J Org Chem 33:305–313
Yopp JH, Tindall DR, Miller DM (1978) Isolation, purification and evidence for a halophilic nature of the blue-green alga *Aphanothece halophytica* Frémy (Chroococcales). Phycologia 17:171–178
Young GM (1981) The Amundsen Embayment, Northwest Territories; relevance to the Upper Proterozoic evolution of North America. Geol Surv Can Pap 81-10:203–218
Zahran MA (1965) Distribution of mangrove vegetation in UAR (Egypt). Bull Inst Deserte Egypte 15:6–11
Zahran MA (1967) On the ecology of the east coast of the Gulf of Suez I Littoral salt marshes. Bull Inst Deserte Egypte 17:225–251
Zahran MA (1977) Africa. Wet formations on the African Red Sea coast. In: Chapman VJ (ed) Ecosystems of the world, vol I. Elsevier, Amsterdam, pp 215–231
Zak I, Freund R (1966) Recent strike slip movements along the Dead Sea Rift. Isr J Earth Sci 15:33–37
Zalcman D, Por FD (1975) The food weeb of Solar Lake (Sinai coast, Gulf of Elat). Rapp Comm int Mer Medit 23:133–134

Zanardini J (1858) Plantarum in mari rubro hucusque collectarum enumeratio. Mem Reale Istituto Veneto 7:89
Zeitzschel B (1973) The biology of the Indian Ocean. Springer, Berlin Heidelberg New York
Zhang Y (1981) Proterozoic stromatolite microfloras of the Gaoyuzhuang Formation (Early Sinian: Riphean). J Paleontol 55:485–503
Ziegler H, Batanouny KH, Sankhla N, Vyas OP, Stichler W (1981) The photosynthetic pathways types of some desert plants from India, Saudi Arabia, Egypt and Iraq. Oecologia 48:93–99
Zohary M (1962) Plant life of Palestine. Ronald Press, New York
Zohary M (1966) Flora Palestina, part 1. Isr Acad Sci Hum, Jerusalem
Zohary M (1972) Flora Palestina, part 2. Isr Acad Sci Hum, Jerusalem
Zohary M (1976) A new analytical flora of Israel. Am Oved, Tel Aviv (Hebrew)
Zohary M, Orshansky G (1947/49) Structure and ecology of the vegetation in the Dead Sea region of Palestine. Palest J Bot Jerusalem Ser 4:177–206

Taxonomic Index

Subject Index

Z. Reiss, L. Hottinger

The Gulf of Aqaba

Ecological Micropaleontology

1984. 207 figures. VIII, 354 pages
(Ecological Studies, Volume 50)
ISBN 3-540-13486-7

Contents: Introduction. - Synopsis. - The Gulf of Aqaba - a Rift-Shaped Depression. - A Desert-Enclosed Sea. - Shell Producers in the Water Column. - The Sea Bottom - a Mosaic of Substrates. - Benthic Foraminifera: Response to Environment. - 150,000 Years Gulf of Aqaba. - References. - Taxonomic Index. - Subject Index.

This volume is an integrative summary of the results of an interdisciplinary research program on the ecology of planktic and benthic Foraminifera, Pteropoda and Coccolithophorida in the Gulf of Aqaba, an important model area for subtropical, oligotrophic marine environments of oceanic gyre-center character. The composition, distribution and seasonality of the various groups are analyzed against the background of their environmental setting, as well as in terms of life strategies and adaptations. Field observations by diving and by submersible as well as long-range synoptic sampling, combined with laboratory observations on living individuals and with stable oxygen and carbon isotope studies, provide a uniquely comprehensive understanding of the ecological relations of some the major groups of organisms widely used in the reconstruction of ancient marine environments. Based on the study of recent populations, fossil assemblages recovered from several deep-sea cores in the Gulf of Aqaba are interpreted in terms of paleoceanographic changes which took place during the last 150,000 years in the Gulf as a result of global and regional glacial-interglacial climate fluctuations. Richly illustrated, the book will prove of value to micropaleontologists, geologists and oceanographers, as well as to ecologists and evolutionists.

MIX
Papier aus verantwortungsvollen Quellen
Paper from responsible sources
FSC® C105338

If you have any concerns about our products,
you can contact us on
ProductSafety@springernature.com

In case Publisher is established outside the EU,
the EU authorized representative is:
Springer Nature Customer Service Center GmbH
Europaplatz 3, 69115 Heidelberg, Germany

Printed by Libri Plureos GmbH
in Hamburg, Germany